APPLIED SUBSURFACE GEOLOGICAL MAPPING

Daniel J. Tearpock
and
Richard E. Bischke

PRENTICE HALL PTR, Upper Saddle River, New Jersey 07458

Library of Congress Cataloging-in-Publication Data

Tearpock, Daniel J.
    Applied subsurface geological mapping / by Daniel J. Tearpock and
Richard E. Bischke.
        p.   cm.
    Includes bibliographical references and index.
    ISBN 0-13-859315-9
    1. Petroleum--Geology. 2. Geological mapping.   I. Bischke,
Richard E. II. Title.
TN870.5.T38 1990
550'.22'3--dc20   90-35994
                                                                 CIP

Editorial/production supervision
  and interior design: *Jacqueline A. Jeglinski*
Cover design: *Lundgren Graphics*
Manufacturing buyers: *Kelly Behr and Susan Brunke*

 ©1991 by Prentice-Hall PTR
A Simon & Schuster Company
Upper Saddle River, NJ 07458

The publisher offers discounts on this book when ordered
in bulk quantities. For more information, write:
        Special Sales/College Marketing
        Prentice-Hall, Inc.
        College Technical and Reference Division
        Englewood Cliffs, NJ 07632

Printed in the United States of America
20 19 18 17 16 15 14 13 12 11

ISBN 0-13-859315-9

Prentice-Hall International (UK) Limited, *London*
Prentice-Hall of Australia Pty. Limited, *Sydney*
Prentice-Hall Canada Inc., *Toronto*
Prentice-Hall Hispanoamericana, S.A., *Mexico*
Prentice-Hall of India Private Limited, *New Delhi*
Prentice-Hall of Japan, Inc., *Tokyo*
Simon & Schuster Asia Pte. Ltd., *Singapore*
Editora Prentice-Hall do Brasil, Ltda., *Rio de Janeiro*

This book is dedicated to
Emery Steffenhagen,
a great geologist and friend

CONTENTS

CHAPTER 5   INTEGRATION OF GEOPHYSICAL DATA
IN SUBSURFACE MAPPING   94

---

prepared by Clay Harmon

CHAPTER 7   FAULT MAPS   195

Contents

**CHAPTER 9   STRUCTURAL GEOMETRY
    AND BALANCING   400**

Contents

CHAPTER 10   ISOPACH MAPS   488

FOREWORD

The petroleum geologist is vital to the economic security of oil-producing nations. The geologist's ability to effectively explore and find oil and gas in the future will have a profound impact on oil-producing nations. Therefore, there is going to be a vital need in the years to come for well-trained, well-educated petroleum geologists.

Applied Subsurface Geological Mapping is a complete and thorough exposition on the subject of subsurface petroleum geologic methods. It explains and illustrates graphically the principles necessary to successfully search for and develop deposits of oil and gas. It covers a wide range of topics in subsurface mapping and is written and illustrated in an easy to understand manner. The material is designed not only for the student of petroleum geology but is appropriate for the experienced geologist who needs to sharpen his tools.

The authors of the book are uniquely qualified to write a comprehensive work on applied subsurface geological mapping. Dan Tearpock has extensive practical petroleum geological and engineering experience having worked as a development geologist for several companies and as a consultant. In addition, he has organized training courses and taught subsurface mapping to many geologists, geophysicists, and engineers. The co-author, Dr. Richard E. Bischke, has a PhD in Geology from Columbia University. He served as Chief Geophysicist for International Exploration for over twelve years and has most recently served on the Research Staff of Princeton University. Through the expertise and experience of both, they have combined to write a text that will quickly become the "Bible" of subsurface mapping.

Frank W. Harrison, Jr.
Consulting Geologist
Past President AAPG (1981–82)

PREFACE

This textbook begins where many others end. There are many textbooks covering numerous geologic subjects; however, since Margaret S. Bishop's classic textbook (1960), there has not been a complete and detailed book on the subject of **subsurface mapping.**

Subsurface geological maps are the most important and widely used vehicle to explore for and develop hydrocarbon reserves. Geologists, geophysicists, and engineers are expected to understand the many aspects of subsurface mapping and to be capable of preparing accurate subsurface maps. Yet, the subject of subsurface mapping is probably the least taught of all petroleum-related subjects. Many colleges and universities do not teach applied subsurface mapping courses, and over the past decade, many company-sponsored training programs have been curtailed or eliminated.

In the coming era of increased efficiency, we must become more aware of our limitations. This involves questioning our methods and thinking more about our interpretation techniques. We need to consider the tools we have at our disposal to support our interpretations and to generate a better quality product. Inaccurate procedures, unjustified shortcuts, and limited mapping skills will result in a poor product.

The United States, for example, presently imports about 50% of its domestic energy. This is occurring at a time when domestic energy output is falling. This shortfall seems to be the result of the inability to locate giant oil and gas fields in the United States. Therefore, in order to hold the line on these shortfalls, we must better locate and extract the reserves that have been left behind in the thousands of existing oil and gas fields.

It has been estimated that 30% or more of future reserve additions will be found in areas that are maturely developed. These reserves will come from redeveloping older oil and gas fields. This future potential is to be found in proved producing reservoirs, reservoirs with proved reserves behind pipe, various types of accumulations (attic, infill, and untested fault blocks), wildcatting in and around the fields and in deeper stratigraphic sections than the current limit within the area.

The techniques presented in this book, from correctly mapping well data to structural balancing, are ideally suited for the more **detailed second look.** Nevertheless, the methods presented are totally relevant to all exploration and exploitation efforts.

This book is intended to fill a major void that currently exists in the area of applied subsurface geologic mapping. We have organized a wealth of information that exists in the literature, in addition to material that has never been published. This book was prepared as a result of our experience in applied petroleum exploration and exploitation, as well as our extensive experience in teaching subsurface mapping and balancing courses to geologists, geophysicists, and engineers in the energy industry (worldwide), and at the college and university level.

We present a variety of subsurface mapping and cross section techniques applicable in the four major petroleum-related tectonic settings: extensional, compressional, diapiric salt, and wrench tectonics. The detailed techniques presented throughout the book are intended to expand your knowledge and improve your skills in preparing geologic interpretations. The knowledge of the principles and techniques presented, plus a burning desire to explore the unknown, will ensure your success.

The textbook is specifically designed for geologists, geophysicists, and engineers who prepare subsurface geological maps. The book should also be beneficial to supervisors, managers, technical assistants, and other persons who have a requirement for the use, preparation, or evaluation of subsurface maps.

Good luck and good prospecting.

Daniel J. Tearpock
Richard E. Bischke

ACKNOWLEDGMENTS

REVIEWERS

We wish to extend our appreciation to the many persons who provided assistance and encouragement during this project. We are grateful to the following persons for their review of the text either in part or in its entirety. Their corrections, suggestions, and insight proved invaluable during the preparation of the textbook, and their suggestions were liberally incorporated into the text. Margaret S. Bishop, Professor of Geology (University of Houston, retired); William G. Brown, Consultant and Professor of Geology (Baylor University); Wendalin Frantz, Professor of Geology (Bloomsburg University); James Harris, Geological Engineering Consultant; Frank Harrison, Consultant and Past President of AAPG (1981–1982); Ron Hartman, Geophysicist and Past President of International Exploration; Brian Lock, Professor of Geology (University of Southwestern Louisiana); John Mosar, Fulbright-Hayes Fellow (Princeton University); Gary Rapp, Manager—Offshore Exploitation and Development (Amerada Hess Corporation); Mike Sewell, Reservoir Engineer (New Field Exploration Company); John Shaw, Princeton University; Emery Steffenhagen, Petroleum Geologist (retired).

CONTRIBUTORS

We are grateful to the numerous companies, organizations, professional societies, and especially to our colleagues and friends for their contributions, which have improved the quality of this textbook. American Association of Petroleum Geologists; American Journal of Science; Margaret S. Bishop, Professor of Geology (University of Houston, retired); Wayne Boeckelman Sr., Drafting Supervisor (Atwater Consultants, Ltd.); William G. Brown, Consultant and Professor of Geology (Baylor University); Chinese Petroleum Institute; Colorado School of Mines; Denver Geophysical Society; Earth Resources Foundation (University of Sydney); Eastman Christensen; Gyrodata Incorporated; Gardes Directional Drilling; Gulf Coast Association of Geological Societies; Gulf Publishing Company; Clay Harmon, Geophysicist and Venture Partner (NewField Exploration Company); James Harris, Geological Engineering Consultant; Steve Hook, Advanced Geologist (Texaco, USA); IHRDC; Jebco Seismic, Inc., Houston, TX; Brian Lock, Professor of Geology (University of Southwestern Louisiana); Lafayette Geological Society; Don Medwedeff, Senior Research Geologist (Arco Oil and Gas); John Mosar, Fulbright-Hayes Fellow (Princeton University); Van Mount, Senior Research Geologist (Arco Oil and Gas); Muzium Brunei; New Orleans Geological Society; Harvey Pousson, Associate Professor of Mathematics (University of Southwestern Louisiana); Prentice-Hall, Inc.; Princeton University; Rocky Mountain Association of Geologists; Royal Society of London; Sandefer Oil and Gas, Inc.; Ted Snedden, Geologist Basin Systems Research (Texaco E&P Technology Division); John Suppe, Blair Professor of Geology (Princeton University); Swiss Geological Society; Tenneco Oil Company; Texaco USA, Eastern E&P Region; TGS Offshore Geophysical Company; Mike Welborne, Reservoir Engineer and Operations Manager (Gas Transportation Corporation); W. W. Norton and Company; and Hongbin Xiao, Ph.D. candidate (Princeton University).

DRAFTING

A textbook is enhanced by the quality of the figures contained in it. We are indebted to several outstanding draftspeople for their understanding, cooperation, and especially their conscientiousness in the preparation of the hundreds of line drawings contained in this text. Sharon Light (Chevron USA) has superior drafting skills and did an outstanding job in the preparation of numerous figures, and was responsible for the special illustrations. Steve Nelson (independent draftsman) was extremely patient with our many changes; the figures he prepared illustrate his exceptional drafting talent. Thank you both.

SUPPORT PERSONNEL

We are grateful to the many individuals who provided support in the form of typing, word processing, reproduction, data collection, organization, and secretarial assistance: Elsie Bischke, Karen Davis, Doris Good, Susie Melacen, Laurie Rowe, Danielle Tearpock, Nicole Tearpock, Paula Tearpock, and Laura Washispack.

SPECIAL RECOGNITION FROM DANIEL J. TEARPOCK

I would especially like to thank Emery Steffenhagen for his continued support and encouragement during the preparation of this manuscript. He has served as an exceptional reviewer and mentor. I thank my friend and colleague Dick Bischke for agreeing to co-author this book, and for his many contributions throughout the text, especially his mathematical derivation of the equations relating vertical separation to throw. I wish to thank Margaret S. Bishop for her excellent reviews and her contributions to the text from her previously published textbook on subsurface mapping; Clay Harmon for his conscientious effort in the preparation of Chapter 5 (Integration of Geophysical Data in Subsurface Mapping); Jim Harris for his critical reviews, suggestions, extensive help, and preparation of a number of the figures; and Jeff Sandefer and the entire staff of Sandefer Oil and Gas, Inc., for their extensive support during the preparation of the manuscript.

SPECIAL RECOGNITION FROM RICHARD E. BISCHKE

I would like to thank Ron Hartman, who, through his compassion and wisdom, has acted as an excellent reviewer and mentor. Particular thanks go to John Suppe and Hongbin Xiao for all their help and for allowing us to publish critical portions of their unpublished data. Thanks to Don Medwedeff, Steve Hook, Van Mount, and Ted Snedden for allowing us to publish portions of their unpublished work. Lastly, I would like to thank Dan Tearpock and my wife Elsie for their friendship and hard work over the years.

CHAPTER 1

INTRODUCTION TO SUBSURFACE MAPPING

TEXTBOOK OVERVIEW

This textbook begins where many others end. It focuses on subsurface geological mapping techniques and their application to the petroleum industry. These techniques are important and applicable to other fields of study, and geologists, geophysicists, engineers, and students in related fields such as mining, groundwater, or waste disposal should benefit from this text as well.

The objectives of subsurface petroleum geology are to find and develop oil and gas reserves. These objectives are best achieved by the use and integration of all the available data and the correct application of these data. This textbook covers the construction of subsurface maps and cross sections based upon data obtained from seismic sections and well logs. It is concerned with correct mapping techniques and how to use these techniques to generate the most reasonable subsurface interpretation that is consistent with all the data.

Subsurface geological maps are perhaps the most important vehicle used to explore for undiscovered hydrocarbons and to develop proved hydrocarbon reserves. However, the subject of subsurface mapping is probably the least discussed yet most important aspect of petroleum exploration and exploitation. As a field is developed from its initial discovery, a large volume of well log, seismic, and production data are obtained. With these data, the accuracy of the subsurface interpretation is improved through time. The most accurate interpretation for any specific oil or gas field can only be prepared after the field has been extensively drilled and most of the hydrocarbons have been depleted. However, accurate and reliable subsurface interpretations are required throughout the life of the field.

During the life of a producing field, many management decisions are based on the interpretation geologists present on subsurface maps. These decisions involve investment capital to purchase leases, permit and drill wells, and workover or recomplete wells, just to name a few. An exploration or exploitation prospect generator must employ the best and most accurate methods available to find and develop hydrocarbon reserves at the lowest cost per net equivalent barrel. Therefore, when preparing subsurface maps, it is essential to use all the available data, evaluate all possible interpretations, and use the most accurate mapping techniques to arrive at a finished product that is consistent with correct geologic models.

Subsurface geologists and geophysicists have the formidable task of mapping unseen structures that may exist thousands of feet beneath the earth's surface. In order to map these structures accurately, the mapper must have a good understanding of the basic principles of structural geology, stratigraphy, sedimentation, and other related geological fields. The mapper must also be thoroughly familiar with the structural style of the region being worked. Since all subsurface maps are an interpretation based on **limited data**, the geologist, geophysicist, or engineer must use (1) imagination, (2) an understanding of local structures, (3) an ability to visualize in three dimensions in order to evaluate the various possible alternate interpretations and decide on the most reasonable, and (4) correct subsurface mapping techniques. These techniques are essential for preparing accurate and technically correct maps.

There are many textbooks on structural geology, tectonics, stratigraphy, sedimentology, structural styles, petroleum geology, and many other related geologic subjects. Since Bishop's classic work (1960), there has been no single source textbook on applied subsurface mapping techniques, and many new developments have occurred since her text was released. In addition, during the 1970s and 1980s, several developments in the field of structural geology have advanced this science to the extent that structural techniques have enhanced subsurface mapping, particularly in compressional regions. The objective of this book is to present a variety of subsurface mapping and cross-section techniques applicable in various settings, including extensional, compressional, diapiric salt, and wrench tectonics. The detailed mapping techniques illustrated throughout the book are intended to expand your knowledge and improve your skills in preparing geological interpretations using a variety of maps and cross sections.

All energy companies expect positive economic results through their exploration and exploitation efforts. Some companies are more successful than others. Many factors lead to success, including advanced technology, aggressive management, experience, and serendipity. A significant underlying cause of success that is often overlooked or taken for granted, however, is the **quality** of the subsurface geological mapping. The application of the numerous mapping techniques presented in this book should improve the quality of subsurface interpretations. This improved quality should positively impact any company's economic picture. This is accomplished by:

1. Developing the most reasonable subsurface interpretations for the area being studied, even in areas where the data are sparse or absent.
2. Generating more accurate and reliable exploration and exploitation prospects (thereby reducing the associated risk).
3. Correctly integrating geological, geophysical, and engineering data to establish the best development plan for a field discovery.
4. Optimizing hydrocarbon recoveries through accurate volumetric reserve estimates.

5. Planning a more successful development drilling, recompletion, and workover depletion plan.
6. Accurately evaluating and developing any required secondary recovery programs.

THE PHILOSOPHICAL DOCTRINE OF SUBSURFACE MAPPING

Our philosophical doctrine of subsurface mapping is designed to provide geologists, geophysicists, and petroleum/mining engineers with the tools necessary to prepare the most reasonable subsurface interpretations. In our quest for hydrocarbons, we are always searching for excellence. The material contained in this book can serve as a teaching medium, as well as a source of reference for conducting subsurface investigations beginning with the initial stages of exploration, continuing through field development, and ending at enhanced recovery. Our basic subsurface mapping philosophy, which is discussed in detail throughout the book, is summarized as follows.

1. One essential requirement is a good understanding of the basic principles of structural geology, petroleum geology, stratigraphy, and other related disciplines.
2. Correct mapping techniques and methods are essential to prepare accurate and geologically reasonable maps and cross sections.
3. Accurate correlations (well log and seismic) are paramount for reliable geologic interpretations.
4. Fault and structure map integration is a must for accurate construction of completed structure maps in faulted areas.
5. Balanced cross sections are required to prepare a reasonably correct restoration of complexly deformed structures.
6. Multiple horizon mapping is essential to support the integrity of any structural interpretation.
7. Interpretive contouring is the most acceptable method of contouring subsurface structure maps.
8. All of the subsurface data must be used to develop a reasonable and accurate subsurface interpretation.
9. The documentation of all work is an integral part of the work.
10. Sufficient time must be allotted to conduct a detailed subsurface mapping study.

 1. A good understanding of the basic principles of structural geology, petroleum geology, stratigraphy, sedimentology, and other related disciplines is required to conduct detailed subsurface geologic studies. The more geology one knows, the more reasonable the resulting interpretation. In addition to a good understanding of geologic principles, a subsurface mapper must be familiar with the structural style of the region being worked. The accuracy of any interpretation increases when the interpretation is made to fit the known style of the structures found in the area of study.

 2. Correct mapping techniques and methods are required to prepare accurate and geologically reasonable maps and cross sections. Maps and cross sections are the primary vehicles used to organize, interpret, and present available subsurface information. The reliability of any subsurface interpretation presented on these maps and cross sections is directly related to the use of accurate and correct mapping techniques.

Perhaps, too often, we become complacent with our methods, do not question our procedures, and use interpretation techniques simply because they are fashionable. There is no substitute for examining and understanding the methods we use.

3. Accurate correlations (well log and seismic) are paramount for reliable geologic interpretations. All geologic interpretations have their foundation in correlation work, whether it be from well log or geophysical data. Therefore, it is important that extreme care be taken in all correlation work. Any errors that occur in the correlation process may be incorporated into the final geologic interpretation.

4. Fault and structure map integration is a must for accurate construction of completed structure maps in faulted areas. Faults play an important role in the trapping of hydrocarbons. Therefore, a reasonable structural interpretation in most faulted areas is dependent on an accurate understanding of the three-dimensional fault pattern. The final structural concept must be developed from the interpretation and construction of fault surface maps and the integration of these fault maps with structure maps.

5. Balanced cross sections are required to prepare a reasonably correct restoration of complexly deformed structures. Structural interpretations are not cast in stone. If a cross section does not volume or area balance, then the interpretation cannot be correct from a simple geometric point of view. Thus, balancing can direct the interpreter toward the correct interpretation.

Structural balancing provides a better understanding of how structures form, what they presently look like, how and where fluids may have entered the structures, and where hydrocarbons may presently exist. Structural balancing is an integral part of many geologic interpretations.

6. Multiple horizon mapping is essential to support the integrity of any structural interpretation. To provide an accurate and reasonable structural interpretation with the available subsurface data, a geologist must present a picture that is sound and geologically plausible. This cannot be accomplished, especially in complex areas, without providing three-dimensional validity to the interpretation. One of the best methods for providing this validity is to map several structural horizons at various levels. This procedure demonstrates that the interpretation fits at all levels mapped and is geologically reasonable.

7. Interpretive contouring is the most acceptable method of contouring subsurface structures. This method of contouring gives the mapper geologic license to prepare a map to reflect the best interpretation of the area, while honoring the available control points. Maps are prepared to illustrate possible structural patterns that are consistent with those found in the tectonic setting being studied.

8. All subsurface data must be used to develop a reasonable and accurate subsurface interpretation. In mapping the subsurface, ordinarily it is necessary to work with scattered and limited data and attempt to obtain the maximum information from these limited data.

An important thought to keep in mind while compiling and using subsurface data are the physical limits and accuracies of that data. A degree of confidence must be built into any final interpretation, but the end result can only be as good as the data that are input. The causes of inaccuracies of data can range from blindly accepting previously constructed maps to not investigating questionable directional surveys, mud logs, electric logs, core reports, velocity functions, production data, and so on. Some of these problems may be unavoidable due to historical methods of record keeping or the poor quality of data. At times, however, it may be possible to determine whether the data are out of date, incomplete, or subject to error.

Questionable data may represent the only data that are currently available. In other cases, more reliable data are available and should be sought. In any case, the physical limits and possible inaccuracies of the data should be noted when presenting completed work. Always acknowledge questionable findings and possible discrepancies in a final presentation.

9. The documentation of work is an integral part of that work. All subsurface mapping projects will provide better results if the data collected and generated are documented in some format that can easily be referenced, used, or revised. The importance of good, accurate documentation cannot be overemphasized. During the course of a subsurface geologic study, a significant volume of data are collected and generated. These data include information such as sand tops and bases, fault cuts, net sand counts, correction factors, etc. All completed work needs to be supported by all the raw data, whether obtained from commercial services or internally generated.

Many persons may, at some time, need to review the subsurface data, including the person conducting the work, supervisors, managers, other members of an organized study team, or persons inheriting the area of study in the future. There are numerous types of data sheets available for documenting mapping parameters. Everyone should become familiar with the forms in use in their company and use them on a regular basis.

10. Sufficient time must be allowed to conduct an accurate and detailed subsurface geologic and engineering study. Often geologic studies are done in haste with insufficient time to use all the data, evaluate alternate solutions, and apply the correct mapping techniques. Murphy's Law tells us that too often there is not enough time or money to complete a project correctly the first time, but there is always enough time and money to redo it at a later date.

TYPES OF SUBSURFACE MAPS AND CROSS SECTIONS

When conducting any detailed subsurface geologic study, a variety of maps and cross sections may be required. The numerous techniques available to use in the preparation of these maps and sections are discussed in subsequent chapters.

As mentioned earlier, the primary focus of this book is on the maps and cross sections used to find and develop hydrocarbons. However, the techniques are applicable to many other related geologic fields. The following is a list of the types of maps and sections discussed in this book.

SUBSURFACE MAPS. Structure, structural shape, porosity top and base, fault surface, unconformity, salt, net sand, net hydrocarbon, net oil, net gas, interval isopach, isochore, facies, and palinspastic maps.

CROSS SECTIONS. Structure, stratigraphic, problem solving, final illustration, balanced, and correlation sections. Conventional and isometric fence diagrams and three-dimensional models are also presented.

CHAPTER 2

CONTOURING AND CONTOURING TECHNIQUES

INTRODUCTION

A wide variety of subsurface maps are discussed in the following chapters. Each map presents a specific type of subsurface data obtained from one or more sources. The purpose of these maps is to present data in a form that can be understood and used to explore for, develop, or evaluate energy resources such as oil and gas.

It might seem elementary to have a chapter on contouring since most geologists are taught the basics of contouring in several introductory geology courses. However, there are two good reasons for this chapter. First, part of our audience includes members of the geophysical and petroleum engineering disciplines. They may have had little, if any, training in basic contour mapping. Second, because the understanding and correct application of contouring and contouring techniques is of paramount importance in establishing a solid foundation in subsurface mapping, a review of contouring is appropriate.

The majority of subsurface maps use the *contour line* as the vehicle to convey the various types of subsurface data. By definition, a contour line is a line that connects points of equal value. Usually this value is compared to some chosen reference, such as sea level in the case of structure contour maps. In preparing subsurface maps, we are dealing with data beneath the earth's surface which cannot be seen or touched directly. Therefore, the preparation of a geologically reasonable subsurface map requires interpretation skills, imagination, an understanding of three-dimensional geometry, and the use of correct mapping techniques.

Any map that uses the contour line as its vehicle for illustration is called a *contour map*. A contour map illustrates a three-dimensional surface or solid in two-dimensional plan (map) view. Any set of data that can be expressed numerically can be contoured.

The following list shows examples of contourable data and the associated contour map.

Data	Type of Map
Elevation	Structure, Fault, Salt
Thickness of sediments	Interval Isopach (isochore)
Percentage of sand	Percent Sand
Feet of pay	Net Pay Isopach
Pressure	Isobar
Temperature	Isotherm
Lithology	Isolith

If the same set of data points to be contoured is given to several geologists, the individually contoured maps generated would likely be different. Differences in an interpretation may be the result of experience, imagination, and interpretive abilities (three-dimensional thinking). Yet the use of all the available data and an understanding and application of the basic principles and techniques of contouring should be the same. These principles and actual techniques are fundamental to the construction of a mechanically correct map.

In this chapter, the importance of visualizing in three dimensions and the basic rules of contouring are discussed. In addition, various techniques of contouring are illustrated and certain important guidelines identified.

THREE-DIMENSIONAL PERSPECTIVE

In this section, we show how three-dimensional surfaces are represented by contours in map view. A good understanding of three-dimensional geometry is essential to any at-

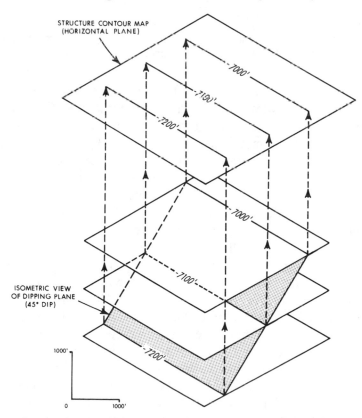

Figure 2-1 Isometric view of dipping plane intersecting three horizontal planes. Modified from Appelbaum. Geological & Engineering mapping of Subsurface: A workshop course by Robert & Appelbaum.

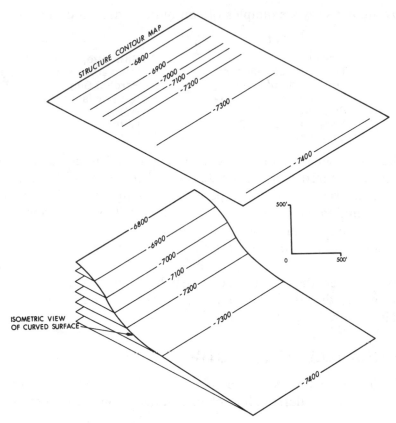

Figure 2-2   Isometric view of a curved surface intersecting a finite number of evenly spaced horizontal planes. Modified from Appelbaum. Geological & Engineering mapping of Subsurface: A workshop course by Robert & Appelbaum

tempt at reconstructing a picture of the subsurface. Some geologists have an innate ability to visualize in three dimensions, while others do not. One of the best ways to develop this ability is to practice perceiving objects in three dimensions.

In its simplest form, we can view a plane dipping in the subsurface with respect to the horizontal. (The reference datum for all the examples shown is sea level.) Figure 2-1 shows an isometric view of a plane dipping at an angle of 45 deg with respect to the horizontal and a projection of that dipping plane upward onto a horizontal surface to form a contour map. This dipping plane intersects an infinite number of horizontal planes; but, for any contour map, only a finite set of evenly spaced horizontal plane intersections can be used to construct the map. (Example: For a subsurface structure contour map, the intersections used may be 50 ft, 100 ft, or even 500 ft apart.) By choosing evenly spaced finite values, we have established the contour interval for the map.

Next, it is important to choose values for the contour lines that are easy to use. For example, if a 100-ft contour interval is chosen, then the contour line values selected to construct the map should be in even increments of 100 ft, such as 7000 ft, 7100 ft, 7200 ft, etc. Any increment of 100 ft could be chosen, such as 7040 ft, 7140 ft, and 7240 ft. This approach, however, makes the map more difficult to construct and harder to read and understand. In Fig. 2-1, a 100-ft contour interval was chosen for the map (the minus sign in front of the depth value indicates the value is below sea level). The intersection of each horizontal plane with the dipping plane results in a line of intersection projected into map view on the contour map above the isometric view. This contour map is a two-dimensional representation of the three-dimensional dipping plane.

Now we shall complicate the picture by introducing a dipping surface that is not a plane but is curved (Fig. 2-2). The curved surface intersects an infinite number of horizontal planes, as did the plane in the first example. Each intersection results in a line of intersection which everywhere has the same value. By projecting these lines into plan view onto a contour map, a three-dimensional surface is represented in two dimensions. If we consider the curved surface as a surface with a changing slope, then the spacing of the contours on the map is representative of the change in slope of the curved surface. In other words, steep slopes are represented by closely spaced contours, and gentle slopes are represented by widely spaced contours (Fig. 2-3). This relationship of contour spacing to change in slope angle assumes that the contour interval for the map is constant.

Finally, look at a three-dimensional subsurface formation that is similar in shape to a topographic elongated anticline and a contour map of that surface (Fig. 2-4). The contour map graphically illustrates the subsurface formation in the same manner that a topographic map depicts the surface of the earth. By using the ability to think in three dimensions, it is possible to look at the contour map and visualize the formation in its true subsurface three-dimensional form.

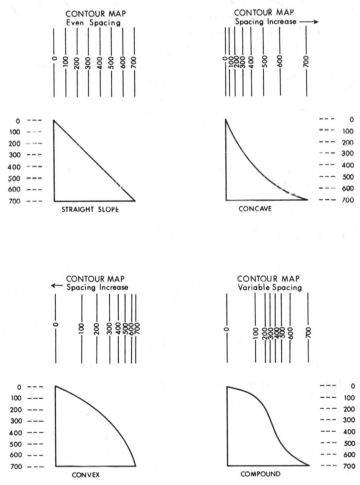

Figure 2-3 The spacing of contour lines is a function of the shape and slope of the surface being contoured.

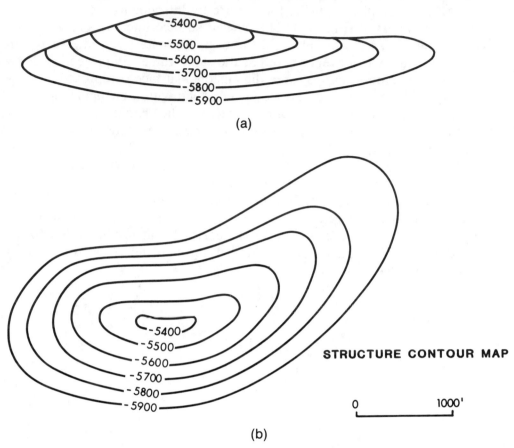

Figure 2-4 A three-dimensional view of a surface structure and the contour map representing the structure.

RULES OF CONTOURING

There are several rules which must be followed in order to draw mechanically correct contour maps. This section lists these rules and discusses a few exceptions.

1. *A contour line cannot cross itself or any other contour except under special circumstances* (see rule 2). Since a contour line connects points of equal value, it cannot cross a line of the same value or lines of different values.

2. *A contour line cannot merge with contours of the same value or different values.* Contour lines may appear to merge or even cross where there is an overhang, overturned fold, or vertical surface (Fig. 2-5). With these exceptions, the key word is **appear**. Consider a vertical cliff that is being mapped. In map view the contours appear to merge, but in three-dimensional space these lines are above each other. For the sake of clarity, contours should be dashed on the underside of an overhang or overturned fold.

3. *A contour line must pass between points whose values are lower and higher than its own value* (Fig. 2-6).

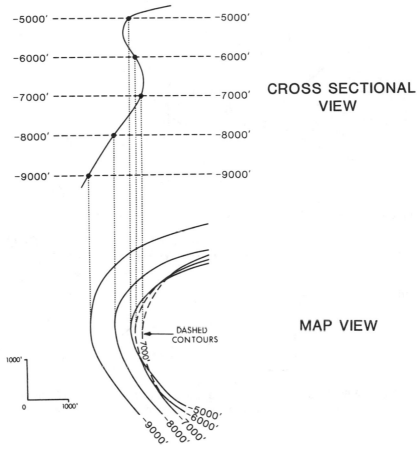

Figure 2-5 To clearly illustrate a three-dimensional overhang or overturned fold, dash the contours on the underside of the structure. (From Tearpock and Harris 1987. Published by permission of Tenneco Oil Company.)

4. *A contour line of a given value is repeated to indicate reversal of slope direction.* Figure 2-6 illustrates the application of this rule across a structural high (anticline) and a structural low (syncline).

5. *A contour line on a continuous surface must close within the mapped area or end at the edge of the map.* Geologists often break this rule by preparing what is commonly referred to as a "postage stamp" map. This is a map that covers a very small area when compared to the areal extent of the structure.

These five contouring rules are simple. If they are followed during mapping, the result will be a map that is mechanically correct. In addition to these rules, there are other guidelines to contouring which make a map easier to construct, read, and understand.

1. All contour maps should have a chosen reference to which the contour values are compared. A structure contour map, as an example, usually uses mean sea level as the chosen reference. Therefore, the elevations on the map can be referenced as being above or below mean sea level. A negative sign in front of a depth value means the elevation is below sea level (i.e., −7000 ft).

2. The contour interval on a map should be constant. The use of a constant contour

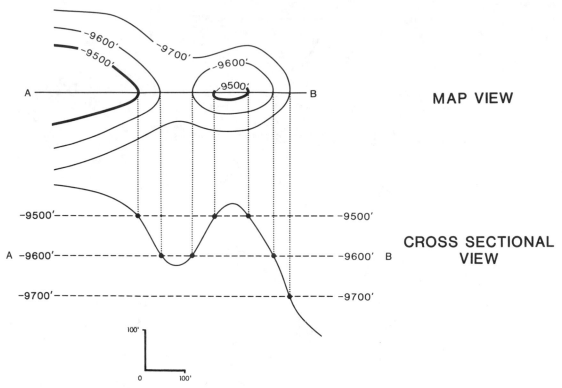

Figure 2-6 A contour line must be repeated to show reversal of slope direction. (From Tearpock and Harris 1987. Published by permission of Tenneco Oil Company.)

interval makes a map easier to read and visualize in three dimensions because the distance between successive contour lines has a direct relationship to the steepness of slope. Remember, steep slopes are represented by closely spaced contours and gentle slopes by widely spaced contours (see Fig. 2-3). If for some reason the contour interval is changed on a map, it should be clearly indicated. This can occur where a mapped area contains both very steep and gentle slopes, such as those seen in areas of saltdome uplift. The choice of a contour interval is an important decision. Several factors must be considered in making such a choice. These factors include the density of data, the practical limits of data accuracy (i.e., directional surveys), the steepness of slope, the scale of the map, and its purpose. If the contour interval chosen is too large, small closures with less relief than the contour interval may be overlooked. If the contour interval is too small, however, the map can become too cluttered and reflect inaccuracies of the basic data.

3. All maps should include a graphic scale (Fig. 2-7). Many people may eventually work with or review a map. A graphic scale provides an exact reference and gives the reviewer an idea of the areal extent of the map and the magnitude of the features shown. Also, it is not uncommon for a map to be reproduced. During this process, the map may be reduced or enlarged. Without a graphic scale, the values shown on the map may become useless.

4. Every fifth contour should be wider than the other contours and it should be labeled with the value of the contour. This fifth contour is referred to as an *index contour*. For example, with a structure contour map using a 100-ft contour interval, it is customary to thicken and label the contours every 500 ft. And at times it may be necessary to label other contours for clarity (Fig. 2-7).

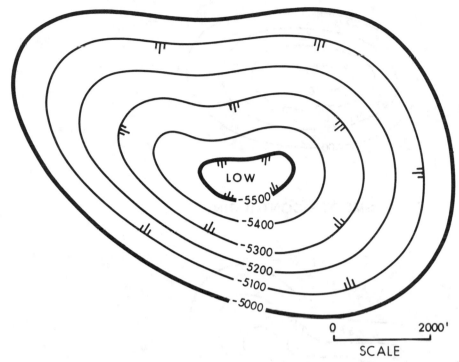

Figure 2-7 Closed depressions are indicated by hachured lines.

5. Hachured lines should be used to indicate closed depressions (Fig. 2-7).
6. Start contouring in areas with the maximum number of control points (Fig. 2-8).
7. Construct the contours in groups of several lines rather than one single contour at a time (Fig. 2-8). This should save time and provide better visualization of the surface being contoured.
8. Initially, choose the simplest contour solution that honors the control points and provides a realistic interpretation.
9. Use a smooth rather than undulating style of contouring unless the data indicate otherwise (Fig. 2-9).
10. Initially, a map should be contoured in pencil with the lines lightly drawn so they can be erased as the map requires revision.

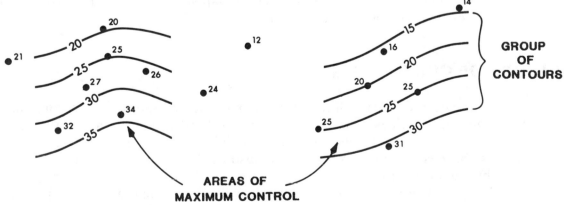

Figure 2-8 Begin contouring in areas of maximum control using groups of contour lines.

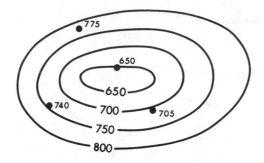

SMOOTH CONTOURING

(a)

UNDULATED CONTOURING

(b)

Figure 2-9 A smooth style of contouring is pre-ferred over an undulating style.

11. If possible, prepare your contour maps on some type of transparent material such as mylar or vellum. Often, several individual maps have to be overlaid one on top of the other (see Chapter 8—Structure Mapping). The use of transparent material makes this type of work easier and faster.

METHODS OF CONTOURING

As mentioned previously, different contoured interpretations can be constructed from the same set of values. The differences in the finished maps may be the result of the geologists' experience levels, interpretive abilities, or other individual factors. This section establishes that the differences can also be the result of the method of contouring used by each geologist.

Unlike topographic data, which are usually obtainable in whatever quantity needed to construct very accurate contour maps, data from the subsurface is scarce. Therefore, any subsurface map is subject to individual interpretation. The amount of data, the areal

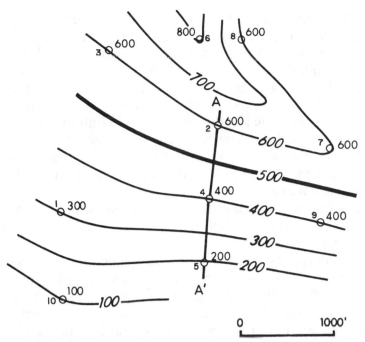

Figure 2-10 Mechanical contouring method. (Modified from Bishop 1960. Published by permission of author.)

extent of that data, and the purpose for which a map is being prepared, may dictate the use of a specific method of contouring. There are four distinct methods of contouring that are commonly used. These methods are (1) mechanical, (2) equal spaced, (3) parallel, and (4) interpretive (see Rettger 1929; Bishop 1960; and Dennison 1968).

1. *Mechanical Contouring.* By using this method of contouring, one may assume that the slope or angle of dip of the surface being contoured is uniform between points of control and that any change occurs at the control points. Figure 2-10 is an example of a mechanically contoured map. With this approach, the spacing of the contours is mathematically (mechanically) proportioned between adjacent control points. We use

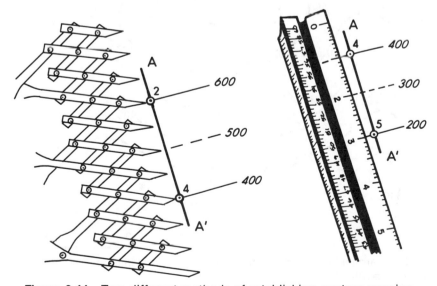

Figure 2-11 Two different methods of establishing contour spacing.

line A-A' in Figs. 2-10 and 2-11 to illustrate this method of contouring. Wells No. 2, 4, and 5 lie on line A-A' with depths of 600 ft, 400 ft, and 200 ft, respectively. The contour interval used for this map is 100 ft. First, we can see that the 600-ft, 400-ft, and 200-ft contour lines pass through Wells No. 2, 4, and 5. Next, we need to determine the position of the 500-ft and 300-ft contour lines. Remembering that this method assumes a uniform slope or dip between control points, we can use ten-point proportional dividers or an engineer's scale to interpolate the location of these two contour lines. The 500-ft contour line lies midway between the 600-ft and 400-ft contour lines. Likewise, the 300-ft contour line is placed midway between the 400-ft and 200-ft contour lines. When this procedure is repeated for all adjacent control points, the result is a mechanically contoured map that is geometrically accurate.

Mechanical contouring allows for little, if any, geologic interpretation. Even though the map is mechanically correct, the result may be a map that is geologically unreasonable, especially in areas of sparse control.

Although mechanical contouring is not recommended for most contour mapping, it does have application in a few areas. When there is a sufficient amount of well control, such as in a densely drilled mature oil or gas field, this method may be acceptable since there is little room for interpretation. Also, this method is commonly employed in litigations and unitization because it minimizes individual bias in the contouring. Finally, this method may be a good first step when beginning work in a new geographic area.

2. *Parallel Contouring.* With this method of contouring, the contour lines are drawn parallel or nearly parallel to each other. This method does not assume uniformity of slope or angle of dip as in the mechanical contouring method. Therefore, the spacing between contours may vary (Fig. 2-12).

As with the previous method, if honored exactly, parallel contouring may yield an

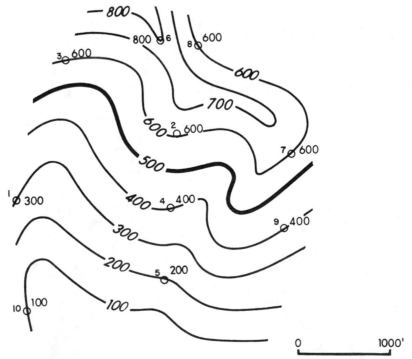

Figure 2-12 Parallel contouring method. (Modified from Bishop 1960. Published by permission of author.)

unrealistic geologic picture. Figure 2-13 shows a map that has been contoured using this method. Notice that the highs appear as bubble shaped structures with the adjoining synclines represented as sharp cusps. This map depicts an unreasonable geologic picture.

Although this method may yield an unrealistic map, it does have several advantages over mechanical contouring. First, the method allows some *geologic license* to draw a more realistic map than one constructed using the mechanical method because there is no assumption of uniform dip. Also, this method is not as conservative as true mechanical contouring. Therefore, it may reveal structures that would not be represented on a mechanically contoured map.

3. *Equal-Spaced Contouring.* This method of contouring assumes uniform slope or angle of dip over an entire area or at least over an individual flank of a structure. Sometimes this method is referred to as a special version of parallel contouring. Equal-spaced contouring is the least conservative of the three methods discussed so far.

To use this method, choose closely spaced wells and determine the slope or angle of dip between them. Usually, the slope or angle of dip chosen for mapping is the steepest found between adjacent control points. Once the dip is established, it is held constant over the entire mapped area. In the example shown in Fig. 2-14, the dip rate between Wells No. 2 and 4 was used to establish the rate of dip for the entire map.

Since the equal-spaced method of contouring is the least conservative, it may result in numerous highs, lows, or undulations that are not based on established points of control, but are the result of maintaining a constant contour interval. The advantage to

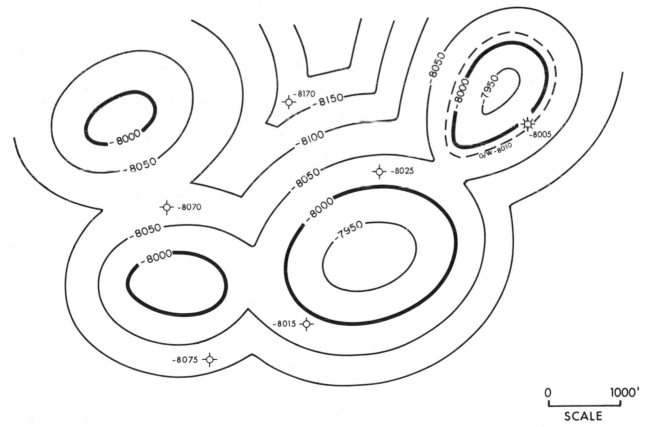

Figure 2-13 An example of an unrealistic structure map constructed using the parallel contouring method.

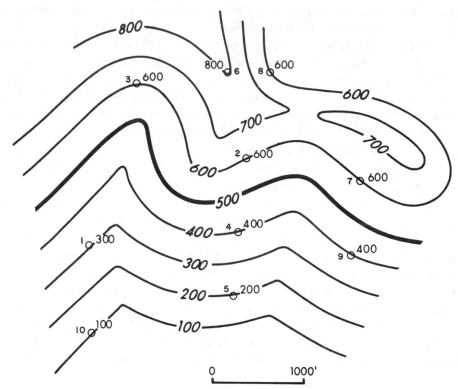

Figure 2-14 Equal spaced contouring method. (Modified from Bishop 1960. Published by permission of author.)

this method, in the early stages of mapping, is that it may indicate a **maximum** number of structural highs and lows expected in the study area. One assumption that must be made in using this method is that the wells used to establish the slope or rate of dip are not on opposite sides of a nose or on opposite flanks of a fold. In Fig. 2-14, Wells No. 6 and 8 are on the northern flank of this structure; therefore, neither well can be used with a well on the southern flank to establish the rate of dip. These two wells can be and were used to establish the rate of dip for contouring of the northern flank of this southeast trending structural nose.

   4. *Interpretive Contouring.* With this method of contouring, the geologist has extreme geologic license to prepare a map to reflect the best interpretation of the area of study, while honoring the available control points (Fig. 2-15). No assumptions, such as constant bed dip or parallelism of contours, are made when using this method. Therefore, the geologist can use experience, imagination, ability to think in three dimensions, and an understanding of the structural and depositional style in the geologic region being worked to develop a realistic interpretation. Interpretive contouring is the most acceptable and the most commonly used method of contouring.

   As mentioned earlier, the specific method chosen for contouring may be dictated by such factors as the number of control points, the areal extent of these points, and the purpose of the map. It is essential to remember that, no matter which method is used in making a subsurface map, *the map is not correct.* No one can really develop a correct interpretation of the subsurface with the same accuracy as that of a topographic map. What is important is to develop the most *reasonable and realistic interpretation* of the subsurface with the available data.

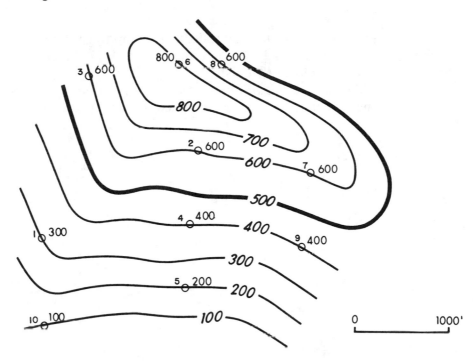

Figure 2-15 Interpretive contouring method. (Modified from Bishop 1960. Published by permission of author.)

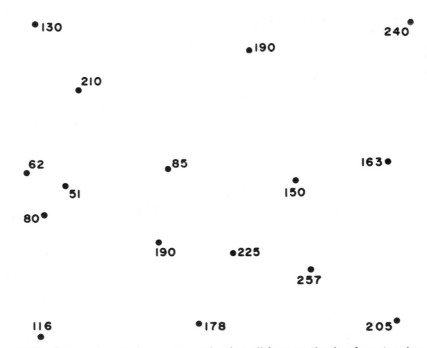

Figure 2-16 Data points to be contoured using all four methods of contouring. (Reproduced from *Analysis of Geologic Structures* by John M. Dennison, by permission of W. W. Norton & Company, Inc. Copyright © 1968 by W. W. Norton & Company, Inc.)

As an exercise in contouring methods, use vellum to contour the data points in Fig. 2-16 using all four methods and, compare the results. Which method results in the most optimistic picture? the most pessimistic?

Special guidelines are used in contouring fault, structure, and isopach maps. Additional guidelines for these maps are discussed in the appropriate chapters.

CHAPTER 3

DIRECTIONALLY DRILLED WELLS AND DIRECTIONAL SURVEYS

INTRODUCTION

A directionally drilled or deviated well is defined as a well drilled at an angle less than 90 deg to the horizontal (Fig. 3-1). Wells are normally deviated intentionally in response to a predetermined plan; however, straight holes often deviate from the vertical due to bit rotation and natural deviation tendencies of subsurface formations.

The technique of controlled directional drilling began in the late 1920s on the US Pacific coast (LeRoy and LeRoy 1977). Through the use of controlled directional drilling, a wellbore is deviated along a preplanned course to intersect a subsurface target horizon(s) at a specific location (Fig. 3-2). Our primary interest in directionally drilled wells centers around their application to subsurface mapping.

APPLICATION OF DIRECTIONALLY DRILLED WELLS

There are a number of reasons to drill a directional well. Some of the more common applications are shown in Fig. 3-3. The most common application is the drilling of offshore wells from a single platform location [Fig. 3-3a]. The use of a single platform from which multiple wells are drilled improves economics and simplifies production facilities.

Onshore, wells are commonly deviated due to inaccessibility to the surface location directly over the subsurface target. Buildings, towns, cities, rivers, and mountains are the kinds of surface obstructions that require the drilling of a deviated well.

One very important safety application of a deviated well is the drilling of a relief well to kill a well that has blown out [Fig. 3-3(f)]. There are other applications of deviated wells, but they are beyond the scope of this textbook and are not discussed.

COMMON TYPES OF DIRECTIONALLY DRILLED WELLS

There are many complex factors that go into the design of a directionally drilled well; however, most deviated wells fall into one of two types. The most common type is a simple ramp well [Fig. 3-1(a)], sometimes called an "L" shaped hole. These wells are drilled vertically to a predetermined depth and then deviated to a certain angle which is usually held constant to total depth (TD) of the well. Many wells are drilled with an "S" shaped design. With an "S" shaped hole, the well begins as a vertical hole and then builds to a predetermined angle, maintains this angle to a designated depth and then the angle is lowered again, often going back to vertical [Fig. 3-1(b)].

General Terminology

The terms used to describe various aspects of a directionally drilled well are defined here and illustrated in Fig. 3-1.

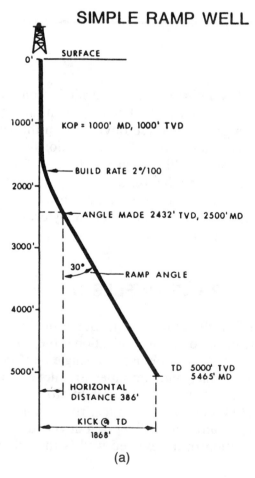

SIMPLE RAMP WELL

(a)

Figure 3-1 Diagrammatic cross section illustration (a) a simple ramp or "L" shaped well, and (b) a more complicated "S" shaped well. (Published by permission of Tenneco Oil Company.)

"S" SHAPE WELL

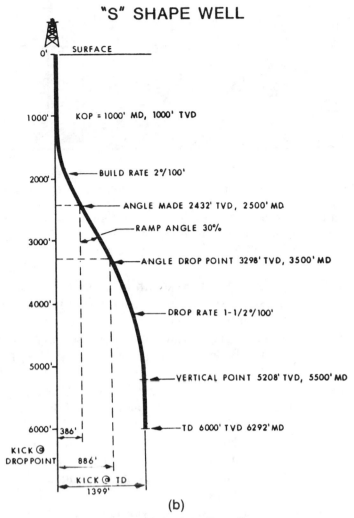

SURFACE

0'

1000' — KOP = 1000' MD, 1000' TVD

2000' — BUILD RATE 2°/100'

ANGLE MADE 2432' TVD, 2500' MD

RAMP ANGLE 30°

3000' — ANGLE DROP POINT 3298' TVD, 3500' MD

4000' — DROP RATE 1-1/2°/100'

5000' — VERTICAL POINT 5208' TVD, 5500' MD

6000' — TD 6000' TVD 6292' MD

386'

KICK @ DROP POINT

886'

KICK @ TD

1399'

(b)

Figure 3-1 (*continued*)

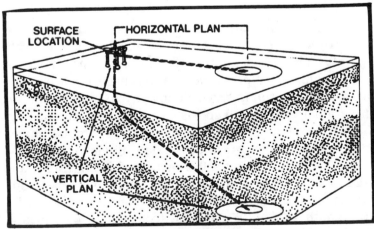

Figure 3-2 Block diagram showing the vertical and horizontal plan views of a directionally drilled well to a predetermined subsurface target. (Published by permission of Eastman Christensen.)

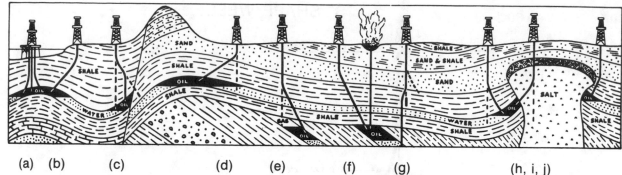

Figure 3-3 Applications of directional drilling. (a) Multiple wells offshore or from artificial islands. (b) Shoreline drilling. (c) Fault control. (d) Inaccessible location. (e) Stratigraphic trap. (f) Relief well control. (g) Straightening hole and side tracking. (h, i, j) Saltdome drilling. (From LeRoy and LeRoy 1977. Published by permission of the Colorado School of Mines.)

KOP = Kick-off Point = Depth of initial deviation from vertical measured as measured depth (MD), true vertical depth (TVD), or subsea true vertical depth (SSTVD).

Build Rate = Build Angle = Rate at which the angle changes during deviation. It is usually expressed in degrees per 100 ft drilled. Example: 2 deg per 100 ft.

Ramp Angle = Hold Angle, Drift Angle, Angle of Deviation = Angle from the vertical that a well maintains from the end of the build through the ramp segment of the well.

BHL = Bottom Hole Location = Horizontal and vertical coordinates to the total depth point usually measured from the surface location.

Drop Rate = Rate at which the ramp angle changes in degrees per 100 ft. Measured in "S" shaped holes.

Vertical Point = The depth where the well is back to vertical measured as MD, TVD, or SSTVD.

DIRECTIONAL WELL PLAN

A variety of data go into the design of a directionally drilled well, including the depth and distance from the surface location to each subsurface target, diameter of the target, kick-off point, build rate, platform location, lease lines, hole size, and total depth of the well. Once preliminary studies indicate the need for a deviated well, most companies rely on a directional drilling service company to prepare the final directional plan.

A directional well design consists of both vertical and horizontal plans. Figure 3-4 shows the horizontal and vertical plans for the Diamond Shamrock Well No. 1 in West Cameron Block 75 offshore Gulf of Mexico. Reviewing Fig. 3-4a, we see that the kick-off point for this well is about 950 ft measured depth (MD), and the build rate averages about 2 deg per 100 ft to a maximum deviation angle of 45°30′ at a measured depth of 4800 ft. Figure 3-4b shows the horizontal plan for the well. The plane of the proposed direction is south 46 deg 25 min west from the surface location with the bottom hole location (BHL) 10,872 ft from the surface location. The well is drilled to a measured depth of 19,484 ft, which is equal to a true vertical depth of 15,695 ft.

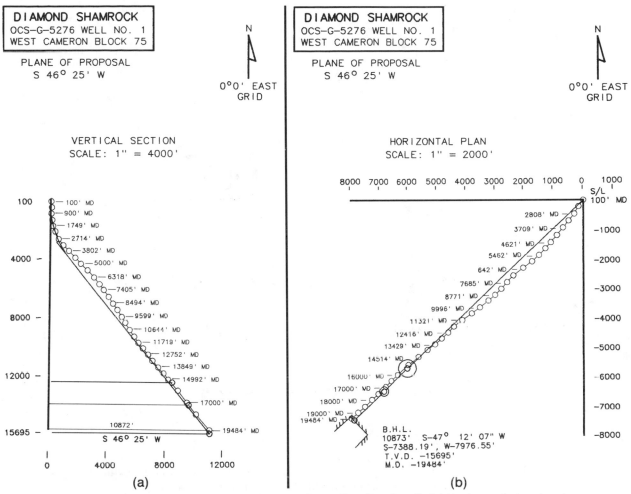

Figure 3-4 (a) Vertical section plan for a directional well. (b) Horizontal plan for the same directional well shown in Fig. 3-4a. (Published by permission of Gardes Directional Drilling.)

Figure 3-5 illustrates the vertical plan for a well that has a maximum deviation angle of 94 deg and a maximum build angle of 14 deg per 100 ft. The wellbore starts out vertical, and at a depth of 1659 ft (TVD) it is horizontal. Such wells are not commonly drilled; however, with technique and equipment advances, high angle wells (60 deg–80 deg) are becoming more common. A well such as the one shown in Fig. 3-5 is referred to as a horizontal wellbore. Lateral (horizontal) drilling is used for various purposes, including (1) to parallel low permeability sand beds to maximize production; (2) to restrict unwanted water flow; and (3) to optimize drilling in coal seam methane operations.

Commonly, deviated wells are drilled with a build rate of 2 deg per 100 ft of hole drilled. Figure 3-6 shows the scaled chart for a 2 deg per 100 ft build rate. Such charts are used to make a quick estimate of well design after structure maps have been made on target horizons. For example, from the chart, a target horizon located at a TVD of 10,000 ft and a horizontal distance of 4000 ft from the platform location requires the drilling of a well with a deviation angle of approximately 23 deg to a measured depth of 10,800 ft. Such charts are available from directional service companies ranging in build rate from 1 deg to 5 deg per 100 ft.

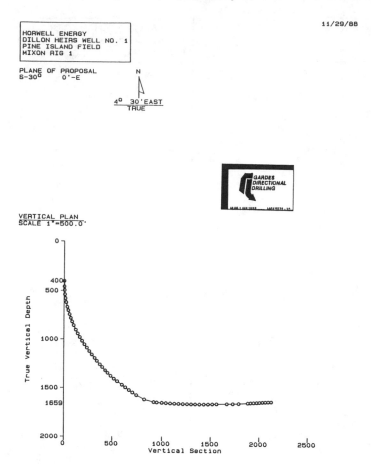

Figure 3-5 Vertical plan for a nearly horizontal well with a maximum deviation angle of 94 deg. (Published by permission of Gardes Directional Drilling.)

DIRECTIONAL TOOLS USED FOR MEASUREMENTS

The tools that are used to survey directionally drilled wells fall into two categories: magnetic and nonmagnetic.

Magnetic Surveys

Magnetic is a generic term used to describe several surveying methods which use a magnetic compass for direction and therefore must be run inside a special nonmagnetic drill collar to negate the effects of the drill pipe. An example of such a survey is the "single shot" magnetic survey. This device records on a heat-resistant film disc the magnetic direction and inclination angle of the wellbore at specific depth intervals. A "multi-shot" survey uses a film strip to record several readings of hole angle and direction at different depth intervals.

Nonmagnetic Surveys

A nonmagnetic survey, sometimes referred to as a drift indicator or Totco, is usually run in vertical or nearly vertical wells where directional information is not required. This tool generally consists of a housing or barrel, a motion indicator, a timer, a punch, and a

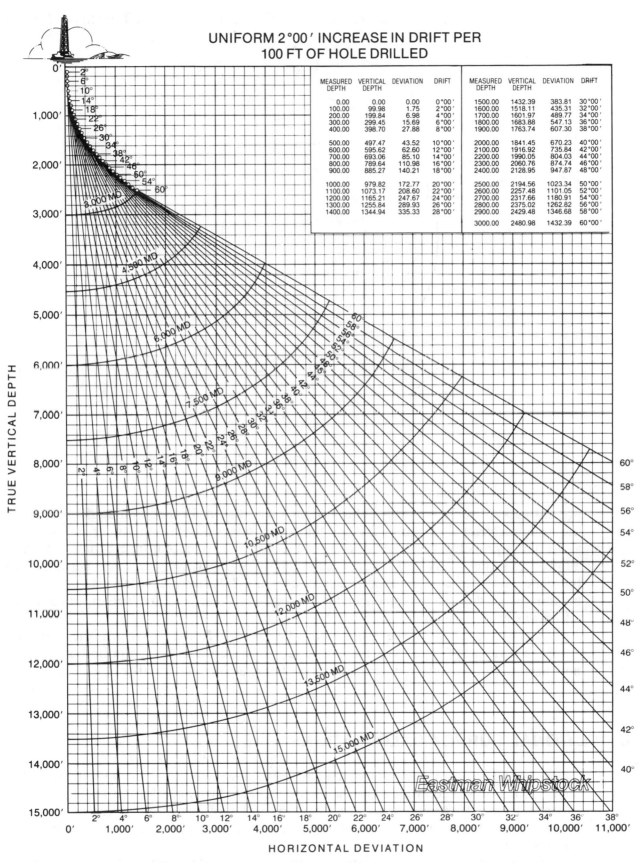

UNIFORM 2°00′ INCREASE IN DRIFT PER
100 FT OF HOLE DRILLED

MEASURED DEPTH	VERTICAL DEPTH	DEVIATION	DRIFT	MEASURED DEPTH	VERTICAL DEPTH	DEVIATION	DRIFT
0.00	0.00	0.00	0°00′	1500.00	1432.39	383.81	30°00′
100.00	99.98	1.75	2°00′	1600.00	1518.11	435.31	32°00′
200.00	199.84	6.98	4°00′	1700.00	1601.97	489.77	34°00′
300.00	299.45	15.69	6°00′	1800.00	1683.88	547.13	36°00′
400.00	398.70	27.88	8°00′	1900.00	1763.74	607.30	38°00′
500.00	497.47	43.52	10°00′	2000.00	1841.45	670.23	40°00′
600.00	595.62	62.60	12°00′	2100.00	1916.92	735.84	42°00′
700.00	693.06	85.10	14°00′	2200.00	1990.05	804.03	44°00′
800.00	789.64	110.98	16°00′	2300.00	2060.76	874.74	46°00′
900.00	885.27	140.21	18°00′	2400.00	2128.95	947.87	48°00′
1000.00	979.82	172.77	20°00′	2500.00	2194.56	1023.34	50°00′
1100.00	1073.17	208.60	22°00′	2600.00	2257.48	1101.05	52°00′
1200.00	1165.21	247.67	24°00′	2700.00	2317.66	1180.91	54°00′
1300.00	1255.84	289.93	26°00′	2800.00	2375.02	1262.82	56°00′
1400.00	1344.94	335.33	28°00′	2900.00	2429.48	1346.68	58°00′
				3000.00	2480.98	1432.39	60°00′

TRUE VERTICAL DEPTH

HORIZONTAL DEVIATION

Figure 3-6 Scaled chart for a build rate of 2 deg per 100 ft of hole drilled.
(Published by permission of Eastman Christensen.)

27

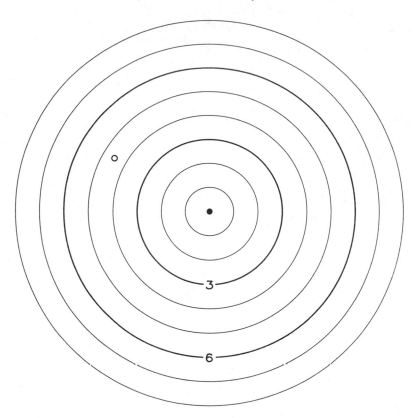

Figure 3-7 Example of a Totco survey. Well deviation angle 4.5 deg.

printed paper disc. The unit is either run on a wire line or dropped on a drill bit. When the motion sensor determines that the tool is no longer moving, the timer is activated, and after a predetermined length of time the punch is released. The punch, which is allowed to swing freely and act as a plumb bob, drops vertically and punches a hole in the paper disc, which is marked in degrees. Figure 3-7 shows a Totco disc scaled to a maximum of 8 deg. The hole punched in the Totco indicates that the well in which it was taken was at an inclination angle of 4.5 deg.

One of the newer surveys is called a Gyroscopic Survey. Because the magnetic compass is replaced by a gyro compass disc, the system can be run in both cased and noncased holes. The gyro system can be set up as a single or multi-shot instrument.

DIRECTIONAL SURVEY CALCULATIONS

There are a number of methods which have been developed to calculate the directional survey for a deviated well. These methods include (1) tangential, (2) trapezoidal, (3) average angle, (4) minimum curvature, and (5) radius of curvature. These methods are used to calculate the three-dimensional location of a directional wellbore anywhere along its entire length.

Three measurements go into the directional survey: (1) measured depth, (2) deviation angle, and (3) direction of deviation. These measurements taken at specific depth intervals are used to calculate the directional survey of a well. Figure 3-8 shows a portion of the

DIRECTIONAL SURVEY

(1)	(2)	(3)	(4)	(5)	(6)	(7)		(8)		(9)
		TRUE				T O T A L				DOG LEG
SUBSEA	MEAS.	VERTICAL	DRIFT	DRIFT	VERTICAL	R E C T A N G U L A R		C L O S U R E S		SEVERITY
DEPTH	DEPTH	DEPTH	ANGLE	DIREC	SECTION	C O O R D I N A T E S		DISTANCE	DIRECT	DEG/100FT
FEET	FEET	FEET	D M	D M	FEET	FEET	FEET	FEET	D M	
5299.88	5700	5330.88	31 30	N 37 0 E	1490.00	1390.73 N	538.16 E	1491.22	N 21 9 E	1.04
5385.14	5800	5416.14	31 30	N 32 0 E	1541.27	1433.77 N	567.75 E	1542.09	N 21 36 E	2.61
5470.52	5900	5501.52	31 15	N 25 0 E	1593.10	1479.50 N	592.57 E	1593.76	N 21 50 E	3.65
5556.01	6000	5587.01	31 15	N 20 0 E	1644.95	1527.41 N	612.42 E	1645.61	N 21 51 E	2.59
5641.95	6100	5672.95	30 15	N 14 0 E	1695.73	1576.28 N	627.36 E	1696.54	N 21 42 E	3.23
5728.23	6200	5759.23	30 30	N 10 0 E	1745.28	1625.73 N	637.87 E	1746.39	N 21 25 E	2.04
5814.50	6300	5845.50	30 15	N 8 0 E	1794.24	1675.67 N	645.78 E	1795.81	N 21 5 E	1.04
5900.88	6400	5931.88	30 15	N 8 0 E	1842.79	1725.56 N	652.79 E	1844.91	N 20 43 E	0.00
5987.59	6500	6018.59	29 30	N 10 0 E	1891.02	1774.76 N	660.59 E	1893.71	N 20 25 E	1.25
6075.26	6600	6106.26	28 0	N 12 0 E	1937.98	1821.97 N	669.76 E	1941.17	N 20 11 E	1.78
6164.16	6700	6195.16	26 30	N 12 0 E	1982.85	1866.75 N	679.28 E	1986.50	N 19 60 E	1.50
6254.23	6800	6285.23	25 0	N 13 0 E	2025.50	1909.17 N	688.68 E	2029.58	N 19 50 E	1.56
6344.86	6900	6375.86	25 0	N 16 0 E	2067.24	1950.08 N	699.26 E	2071.66	N 19 44 E	1.27
6435.40	7000	6466.40	25 15	N 18 0 E	2109.43	1990.68 N	711.68 E	2114.07	N 19 40 E	0.89
6525.75	7100	6556.75	25 30	N 20 0 E	2152.15	2031.20 N	725.63 E	2156.92	N 19 40 E	0.89
6616.10	7200	6647.10	25 15	N 20 0 E	2194.92	2071.47 N	740.29 E	2199.77	N 19 40 E	0.25
6706.73	7300	6737.73	24 45	N 20 0 E	2237.11	2111.18 N	754.74 E	2242.03	N 19 40 E	0.50
6797.82	7400	6828.82	24 0	N 20 0 E	2278.30	2149.96 N	768.86 E	2283.30	N 19 41 E	0.75
6889.17	7500	6920.17	24 0	N 20 0 E	2318.90	2188.18 N	782.77 E	2323.98	N 19 41 E	0.00
6980.53	7600	7011.53	24 0	N 20 0 E	2359.50	2226.40 N	796.68 E	2364.65	N 19 41 E	0.00

Radius of Curvature Calculation

Figure 3-8 Part of the directional survey for a deviated well. (Published by permission of Gardes Directional Drilling.)

directional survey from a deviated well. The tabular printout for this directional survey has nine separate columns of data for each survey point in the well.

Column	Data
1	Subsea depth of wellbore in feet
2	Measured depth of wellbore in feet
3	True vertical depth of wellbore in feet
4	Angle of wellbore deviation
5	Direction of wellbore (true bearing)
6	Distance in feet from the surface location along the proposed directional path
7	True bearing and distance of each survey point from the surface location in rectangular coordinates
8	True bearing and distance from surface location directly to each survey point
9	Maximum change in hole angle in degrees per 100 ft

If the cased or surface section of the hole was surveyed with a nonmagnetic survey such as a Totco, the angle of the cased portion of the hole will be displayed on the survey along with an estimate of the maximum possible deviation of this portion of the hole, as shown in Fig. 3-9. The 146.75 ft highlighted on the survey indicate that if the wellbore drift in the cased portion of the hole was all in the same direction, the well at a depth of 3513 ft could be up to a maximum of 146.75 ft from the surface location. Notice at the end of the survey that it indicates that the bottom of the hole lies within a circle of

DIRECTIONAL SURVEY DATA
BAYOU FER BLANC FIELD, LAFOURCHE REALTY CO. WELL NO. B-3

MEASURED DEPTH	DRIFT ANGLE	TVD DEPTH	COURSE DEVIATION FEET	DIRECTION	COURSE COORDINATES NORTH	SOUTH	EAST	WEST	TOTAL COORDINATES NORTH	SOUTH	EAST	WEST
1000.00	1-45	999.53	30.54	CASING								
2000.00	2-30	1998.58	43.62	CASING								
3000.00	2-45	2997.43	47.98	CASING								
3513.00	2-45	3509.84	24.61	CASING								

MAXIMUM POSSIBLE DEVIATION OF CASED HOLE AT THIS POINT IS 146.75 FEET.

4000.00	2- 0	3996.54	17.00	S 39 W	0.	13.21	0.	10.70	0.	13.21	0.	10.70
4500.00	1- 0	4496.47	8.73	S 36 E	0.	7.06	5.13	0.	0.	20.27	0.	5.57
5000.00	0-30	4996.45	4.36	N 82 E	0.61	0.	4.32	0.	0.	19.66	0.	1.25
5500.00	0-30	5496.43	4.36	S 64 W	0.	1.91	0.	3.92	0.	21.57	0.	5.17
6000.00	0-45	5996.39	6.54	S 26 W	0.	5.88	0.	2.87	0.	27.46	0.	8.04
6500.00	1- 0	6496.31	8.73	S 20 W	0.	8.20	0.	2.98	0.	35.66	0.	11.02
7000.00	1- 0	6996.23	8.73	S 85 W	0.	0.76	0.	8.69	0.	36.42	0.	19.71
7500.00	1-15	7496.11	10.91	S 77 W	0.	2.45	0.	10.63	0.	38.87	0.	30.34
8000.00	2- 0	7995.81	17.45	S 24 W	0.	15.94	0.	7.10	0.	54.81	0.	37.44
8500.00	2- 0	8495.73	8.33	S 30 W	0.	0.	0.	3.00	0.	30.45	0.	45.00
9000.00	1-45	8995.50	15.27	S 25 W	0.	13.84	0.	6.45	0.	64.29	0.	51.45
9500.00	2-15	9495.11	19.63	S 33 W	0.	16.46	0.	10.69	0.	80.75	0.	62.14
10000.00	2-30	9994.64	21.81	S 58 W	0.	11.56	0.	18.50	0.	92.31	0.	80.64
10500.00	1-45	10494.41	15.27	S 75 W	0.	3.95	0.	14.75	0.	96.26	0.	95.39
11000.00	2-30	10993.93	21.81	N 85 W	1.90	0.	0.	21.73	0.	94.36	0.	117.11
11500.00	1-45	11493.70	15.27	S 75 W	0.	3.95	0.	14.75	0.	98.31	0.	131.86
12000.00	2- 0	11993.39	17.45	N 38 W	13.75	0.	0.	10.74	0.	84.56	0.	142.60
12500.00	2- 0	12493.09	17.45	S 40 W	0.	13.37	0.	11.22	0.	97.93	0.	153.82
13000.00	2- 0	12992.78	17.45	S 82 W	0.	2.43	0.	17.28	0.	100.36	0.	171.10
13500.00	1-30	13492.61	13.09	N 64 W	5.74	0.	0.	11.76	0.	94.62	0.	182.86
13700.00	1- 0	13692.58	3.49	N 7 E	3.46	0.	0.43	0.	0.	91.15	0.	182.44

THE BOTTOM OF THE HOLE LIES WITHIN A CIRCLE OF RADIUS 146.75 FEET WITH ITS CENTER
LOCATED 203.94 FEET SOUTH 63 DEGREES 27 MINUTES WEST OF THE SURFACE LOCATION

Figure 3-9 A directional survey from a well in which the surface casing was surveyed with a Totco providing deviation angle but not direction. Notice, at 3513 ft, the maximum possible deviation of the cased hole is 146.75 ft. This assumes that the wellbore deviation in this portion of the hole was in one direction.

radius 146.75 ft with its center located 203.94 ft south 63 deg 27 min. west of the surface location. Such information may be important in fault, structure, and isopach mapping (see Fig. 7-57 in Chapter 7).

DIRECTIONAL SURVEY UNCERTAINTIES

The error introduced by the method of calculation becomes almost academic when the other directional survey uncertainties are considered. Tenneco Oil Company conducted a detailed study of directional survey uncertainties in 1980. An important conclusion from the study indicates that there is a 90% certainty that any directional well will have an error of 35 ft or less TVD and 140 ft or less in departure. This conclusion is drawn irrespective of MD, hole angle, or survey type. This means that wells with hydrocarbon contacts that vary up to 35 ft TVD may well be in the same reservoir with the variations due merely to survey error, rather than such geological events as permeability barriers or faults.

VERTICAL UNCERTAINTY*

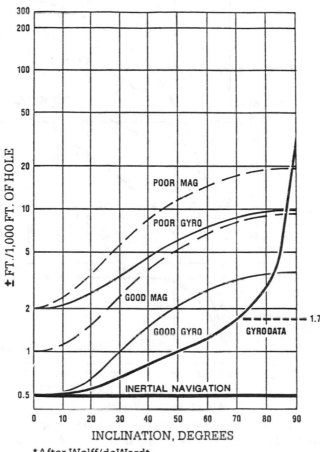

*After Wolff/deWardt

Figure 3-10 Expected vertical uncertainty in a deviated well considering various types of surveys. (Modified from Wolff and de Wardt 1981. Published by permission of the Journal of Petroleum Technology and Gyrodata, Inc.)

The diagrams shown in Figs. 3-10 and 3-11 are modified from Wolff and DeWardt (1981). These diagrams give an indication of expected uncertainty when MD, hole angle, and survey type are taken into consideration.

Survey errors can be corrected in some cases in fields that have hydrocarbon/water contacts. Wells are very often "corrected" to fit the contact. This is usually done by selecting a contact that fits most of the wells and then adjusting the depth of the well(s) that does not fit. An example of how to adjust a well is shown in Fig. 3-12. The water contact in Wells No. 1 and 2 is at a depth of −9738 ft, whereas the contact in Well No. 3 is at a depth of −9748 ft. Since data from two wells are in agreement with a water contact at −9738 ft, Well No. 3 is adjusted upward 10 ft to correct the water level from −9748 ft to −9738 ft. Not only is the water level corrected, but the structural depth of the sand (top and base) is also corrected upward 10 ft. Therefore, the top of the sand at a depth of −9720 ft becomes −9710 ft. An understanding of directional survey errors can at times eliminate the need for a "production fault" to explain discrepancies in water levels.

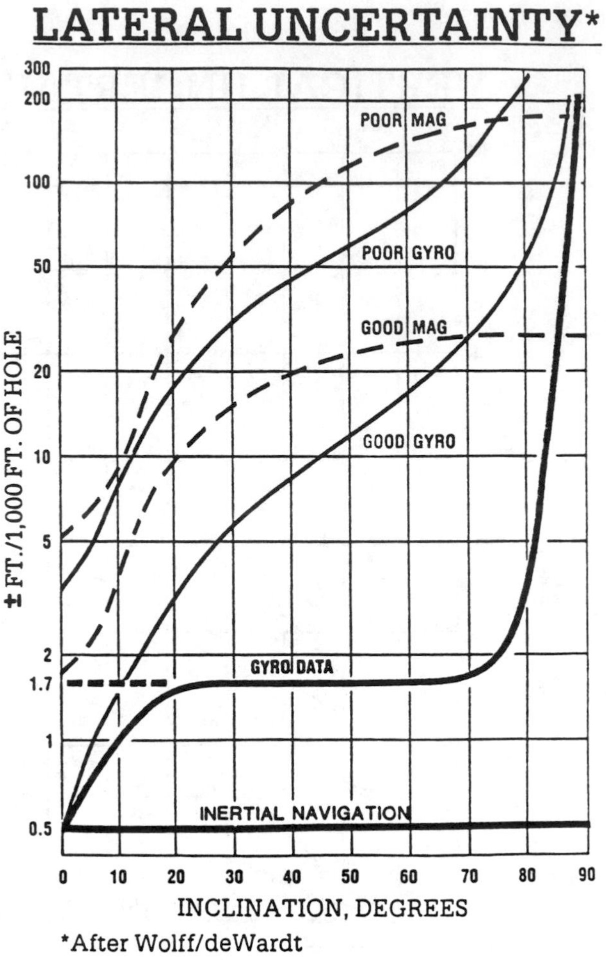

LATERAL UNCERTAINTY*

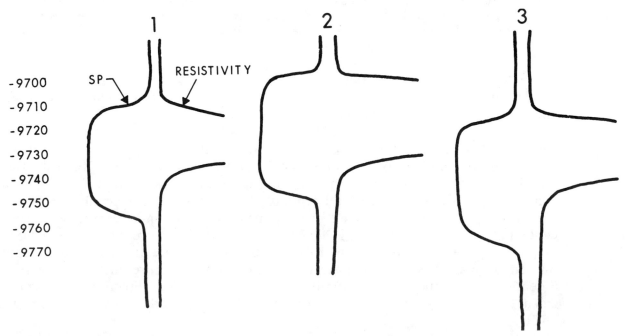

*After Wolff/deWardt

Figure 3-11 Expected lateral uncertainty in a deviated well considering various types of surveys. (Modified from Wolff and de Wardt 1981. Published by permission of the Journal of Petroleum Technology and Gyrodata, Inc.)

Figure 3-12 Different hydrocarbon/water contacts caused by directional survey errors.

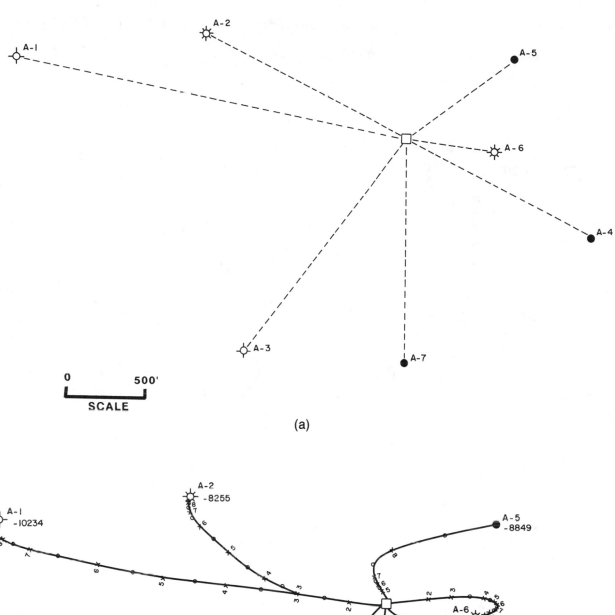

(a)

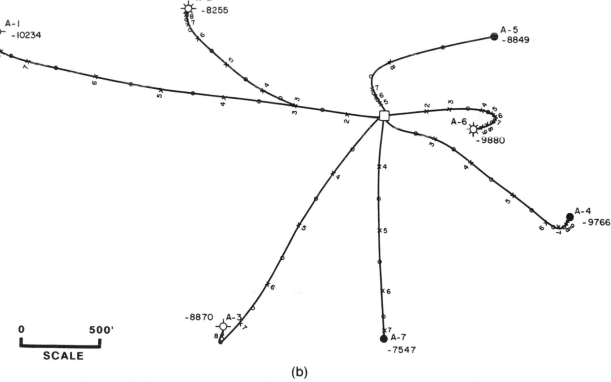

(b)

Figure 3-13   (a) Straight-line method of plotting directional wells in map view. (b) Detailed plot of directional survey data indicating the location and subsea depth of the wellbores along their entire length. Compare this plot to that in Fig. 3-13a.

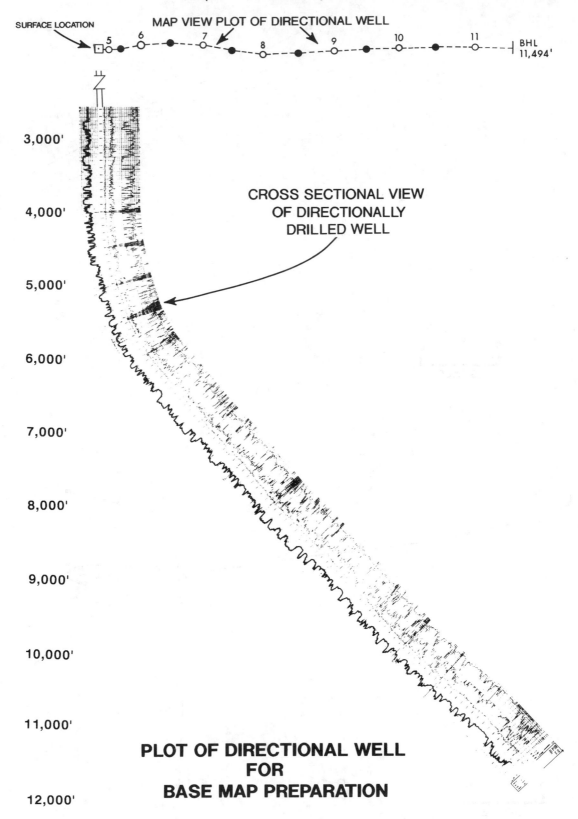

Figure 3-14 Cross-sectional view of a directional well and its detailed map view plot in increments of 500 ft.

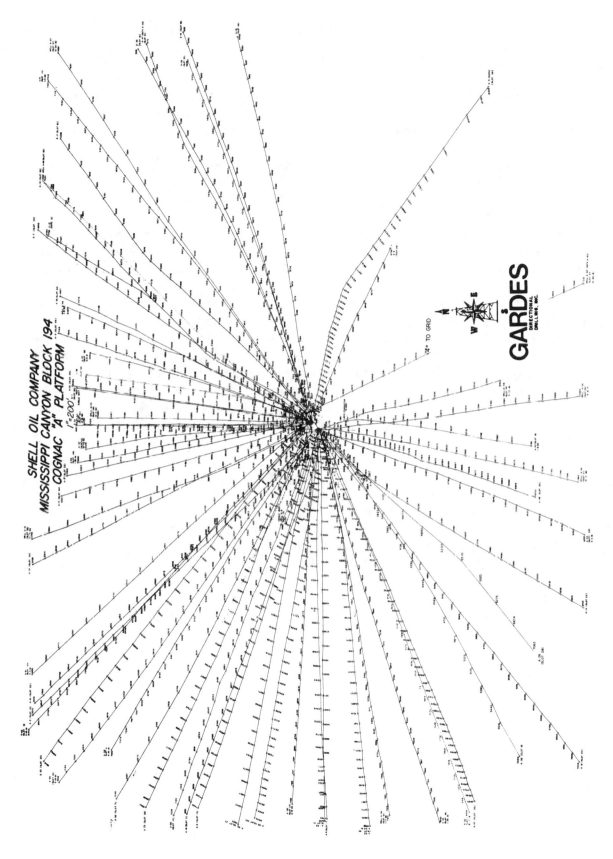

Figure 3-15 Spider directional well plot for the Cognac "A" Platform, Offshore Gulf of Mexico. (Published by permission of Gardes Directional Drilling.)

35

DIRECTIONAL WELL PLOTS

Directional survey data are used to calculate the position and depth of the borehole along its entire length. These data are normally plotted on a base map in one of two ways: (1) as straight lines from the surface to bottom hole location, or (2) as detailed directional plots.

The simplified straight line method of plotting a directional well is shown in Fig. 3-13a. The only data required and plotted on the base map are the surface and bottomhole locations. At times, the measured depth to TD is recorded next to the bottomhole location. A straight dashed line is usually drawn between the surface and bottomhole locations. This directional well plot provides absolutely no information about the position or depth of the wellbore in the subsurface between the surface and bottomhole locations. Such a plot is not helpful in the interpretation, construction, and evaluation of fault, structure, or isopach maps.

When directional survey data are actually plotted to provide detail as to the position and subsea depth of the wellbore throughout its entire length as shown in Fig. 3-13b, the plot has real value. Such a plot provides a visual guide (in map view) to the location and subsea depth of the wellbore anywhere along its path. It saves time in preparing subsurface maps and is extremely helpful in the interpretation, construction, and evaluation of fault, structure, and isopach maps. Later chapters examine several important benefits to fault surface and structure mapping derived from plotting the actual position and subsea depth, at evenly spaced increments (usually 500 ft or 1000 ft) of all directional wells on a base map.

Figure 3-14 illustrates the cross-sectional view of a directionally drilled well and the detailed map view plot of the directional data in 500-ft increments (TVD) along the actual well path. If subsea data are planned for mapping, the directional well data are corrected to subsea before being plotted on the base map.

Figure 3-15 shows the map view well plot for the Cognac "A" Platform in the Mississippi Canyon Block 194 in the US Gulf of Mexico. The platform is the largest multiple well platform in the world with 62 slots. The platform is located in 1000 ft of water and has a total height of 1260 ft. Wells were deviated with high angles, up to 75 deg. Horizontal displacement for the wells, up to 11,500 ft, results in a well pattern that covers over a 2-mi radius from the platform, or a diameter of 4.4 mi. Total cost for the project was approximately *$960 million*.

CHAPTER 4

LOG
CORRELATION
TECHNIQUES

INTRODUCTION

Correlation can be defined as the determination of structural or stratigraphic units that are equivalent in time, age, or stratigraphic position. For the purpose of preparing subsurface maps and cross sections, the two general sources of correlation data are electric wireline logs and seismic sections. In this chapter, we discuss basic procedures for correlating well logs, introduce plans for performing various phases of well log correlation, and present fundamental concepts and techniques for correlating well logs from vertical as well as directionally drilled wells.

Fundamentally, electric well log curves are used to delineate the boundaries of subsurface units for the preparation of a variety of subsurface maps and cross sections (Doveton 1986). These maps and cross sections are used to develop an interpretation of the subsurface for the purpose of exploring and exploiting hydrocarbon reserves.

After the preparation of an accurate well and seismic base map, electric log and seismic correlation work is the next step in the process of conducting a detailed geologic/ geophysical study. No geologic interpretation can be prepared without detailed electric log correlations. **Accurate correlations are paramount for reliable geologic interpretations.**

General Log Measurement Terminology

An understanding of several log depth measurements is important for converting log depths to depths used for mapping. The following is a list of measurements, their abbreviations, and definitions of depth terminology. These terms are illustrated in Fig. 4-1.

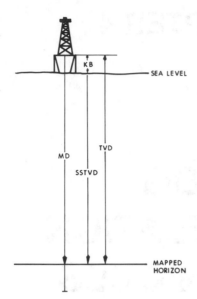

Figure 4-1 Diagram showing general log measurement terminology.

KB = Distance from Kellybushing to sea level.

MD = Measured Depth = Measured distance along the path of a wellbore from the KB to TD (Total Depth of the well) or any correlation point in between.

TVD = True Vertical Depth = Vertical distance from the KB to any point in the subsurface.

SSTVD = Sub-Sea True Vertical Depth = Vertical distance from sea level to any point in the subsurface.

Vertical Wellbore = A well drilled 90 deg to a horizontal reference, usually sea level (also called a straight hole).

The SSTVD measurement is the only depth measurement from a common reference datum, sea level. Therefore, SSTVD is the depth most often used for mapping. Logging depths measured from a vertical or directionally drilled well for mapping are usually corrected to SSTVD. For vertical wells the SSTVD = TVD − KB. The measurements for directionally drilled wells were discussed in Chapter 3.

ELECTRIC LOG CORRELATION PROCEDURES AND GUIDELINES

What is well log correlation? **Electric log correlation is pattern recognition.** It is often debated whether this pattern recognition is more of an art or a science, but we believe both play a part in correlation work. Anyone involved with log correlation must have an understanding of sound geologic principles, including depositional processes and environments, and be familiar with the principles of logging tools and measurements, general reservoir engineering fundamentals, and basic qualitative and quantitative log analyses.

The best way to develop log correlation ability is by actually performing correlation work. A geologist should become more proficient with increased experience in correlation. Proficiency in correlating well logs in one tectonic setting or depositional environment does not always ensure similar competence in other settings. In other words, someone who is an expert at correlating well logs in the US Gulf of Mexico Basin may not be equally competent when working in, for example, the Rocky Mountain Overthrust Belt. Just as it took time to become proficient at correlating logs in the Gulf of Mexico, so too will it take time and familiarity in the new area to become proficient.

When geologists correlate one log to another, they are attempting to match the pattern of curves on one log to the pattern of curves found on the second log. A variety of curves may be represented on a log. For correlation work, it is best to correlate well logs that have the same type of curves; however, this is not always possible. A geologist may be required to correlate logs that have different curves. And at times, even if the logs have the same curves, the character or magnitude of the fluctuations of the curves may be different from one log to the next. Therefore, the correlation work must be independent of the magnitude of the fluctuations and the variety of curves on the individual well logs. Figure 4-2 shows sections from two electric logs. The pattern of curves on Well No. A-1 are very similar to the patterns on Well No. A-2. We can say that these two logs have a high degree of correlation.

The data presented on a well log are representative of the subsurface formations found in the wellbore. A correlated log provides information on the subsurface, such as formation tops and bases, depth and size of faults, lithology, depth to and thickness of hydrocarbon bearing zones, porosity and permeability of productive zones, and depth to unconformities. The information obtained from correlated logs is the raw data used to prepare subsurface maps. These include fault, structure, salt, unconformity, and a variety of isopach maps. Accurate correlation is paramount for reliable geologic interpretations. Subsurface geological maps based on log correlation are only as reliable as the correlations used in their construction. Eventually, a geologist's correlations, right or wrong, are incorporated into the construction of subsurface geological maps. An incorrect correlation can be costly in terms of a dry hole or an unsuccessful workover or recompletion; therefore, it is essential that extreme care be taken in correlating well logs.

In this section, we introduce you to a general correlation procedure, and discuss some guidelines for electric log correlation. The process of correlating logs varies from one individual to the next. As geologists gain experience they modify and eventually establish a correlation procedure which works best for them. If you have no experience in log correlation or want to improve your skills, you can begin by using the procedures and guidelines discussed in this section.

Electric logs are commonly arranged on a work table in one of two ways (Fig. 4-3). The arrangement shown in Fig. 4-3a is preferred over that shown in Fig. 4-3b by most geologists because more log section can be viewed at one time and the logs are easier to slide during correlation.

As a starting point, align the depth scale of the logs and look for correlation as shown in Fig. 4-2. If no correlation is evident, begin to slide one of the logs until a good correlation point is found, and mark it. Continue this process over the entire length of each log until all recognized correlations have been identified. This process may seem relatively easy, but it can be complicated by such factors as stratigraphic thinning, bed dip, faulting, unconformities, lateral facies changes, poor log quality, and directionally drilled wells. There are some basic, universally valid guidelines which are useful in the log correlation process. If followed, these guidelines should improve your correlation efficiency and minimize correlation problems.

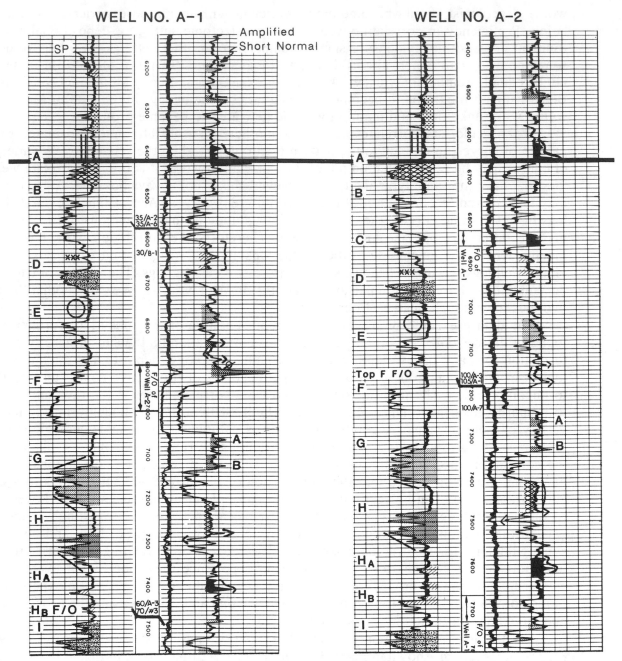

Figure 4-2 Portion of two electric logs illustrating methods of annotating recognizable correlation patterns on well logs.

1. For initial quick-look correlation, review major sandstones using the SP or gamma ray curves.
2. For detailed correlation work, first correlate shale sections.
3. Initially, use the amplified short normal resistivity curve, which usually provides the most reliable shale correlations.
4. Use colored pencils to identify specific correlation points.
5. Always begin correlation at the top of the log, not the middle.
6. Do not force a correlation.
7. In highly faulted areas, first correlate down the log and then correlate up the log.

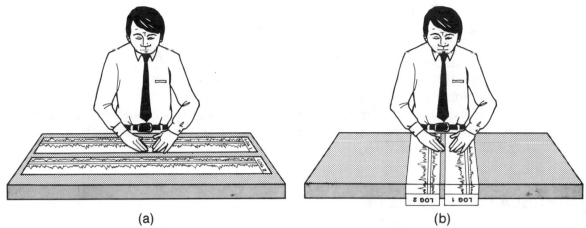

Figure 4-3 (a) Preferred method of arranging logs for correlation. (b) Alternate method of arranging logs for correlation.

After an initial quick look using the SP or gamma ray curves to identify the major sandstones, concentrate your correlation work on shale sections. There are three good reasons for this. First, the clay and mud particles which make up shales are deposited in low-energy regimes. These low-energy environments responsible for shale deposition commonly cover large geographic areas. Therefore, the log curves (sometimes referred to as log signatures) in shales are often highly correlatable from well to well and can be recognized over long distances. Second, prominent sand beds are often not good correlation markers because they frequently exhibit significant variation in thickness and character from well to well and are often laterally discontinuous. Finally, the resistivity curves for the same sand on two well logs being correlated may be different. Variations in fluid content in a sand bed may cause pronounced resistivity differences (i.e., water versus gas).

Individual shale beds exhibit distinctive resistivity characteristics over large areas. Therefore, when all log curves are considered, the amplified short normal resistivity curve provides the most reliable shale correlations. Although all log curves should be used for correlation work, the amplified short normal is five times more sensitive than the short normal, and exhibits patterns that are easier to recognize and correlate from well to well. The amplified short normal is the initial curve used for correlation (Fig. 4-2).

The liberal use of colored pencils is an excellent way to identify and mark correlation patterns on well logs. The correlation patterns might be peaks, valleys, or groups of *wiggles* that are recognizable in many or all of the well logs being correlated (Fig. 4-2). The colored pencils should be erasable in the event that correlations are changed. **Do not mark on original logs.** A blue or blackline copy of the original logs should be used for marking during correlation.

In general, structures become less complicated toward the surface because of several factors. Many faults tend to die upward toward the surface and are either small or non-existent in the upper part of the logs. This makes for easier correlations. Also, in many geologic provinces, especially in soft rock basins, the structural dip, both local and regional, decreases upward. Therefore, beginning correlation at the top of a log is usually easier.

Correlations are not always straightforward and everyone runs into correlation problems from time to time. Often, there is a tendency to force a correlation rather than bypass the problem area until further work is done. This is not good practice. Correlation problems are often due to the presence of faults, high bed dips, unconformities, and facies changes. It is best to pass the problem area and continue the correlation work on the remaining section of the log. Later, when the remainder of the problem log and other logs have been correlated, the questionable correlations can be reviewed again with this new information.

In highly faulted areas it is advantageous to approach a recognized fault cut from two directions. First, correlate down the log to the fault and then correlate up the log to the fault. By taking this approach, determination of the size and depth of the fault in the correlated well will be more accurate (Figs. 4-2 and 4-10). This method is discussed in detail later in this chapter.

CORRELATION TYPE LOG

A Correlation Type Log is defined as a log which exhibits a complete stratigraphic section in a field or regional area of study. The type log should reflect the deepest and thickest stratigraphic section penetrated. Because of faults, unconformities, and variations in stratigraphy affecting the sedimentary section, a correlation type log is often composed of sections from several individual logs and is referred to as a *composite type log*.

Do not confuse a correlation type log with other kinds of type logs such as stratigraphic type logs, composite sand type logs, or show logs. A stratigraphic type log is usually prepared to depict the depositional environments that exist in a particular field or area of study (Fig. 4-4). Although it may include portions of several logs to depict the entire stratigraphic section, it is usually not prepared in the strict sense of a correlation type log. Therefore, it may contain faults or unconformities, and include wells near the crest of the structure which do not represent the thickest sedimentary section.

Composite Sand Type Logs, Pay Logs, or Show Logs are prepared to illustrate the potential sands within a field or area of study that have shows, contain hydrocarbons, or have the potential to be hydrocarbon bearing (Fig. 4-5). These logs are not prepared for use as a correlation aid and therefore are not prepared in the rigid manner of a correlation type log.

When beginning geologic work in a new area of study in which a type log has already been prepared, it is important to carefully review the log to see that it meets with the requirements of a correlation type log. If the type log has an incomplete stratigraphic section, its use will result in correlation errors. The type log must have the complete stratigraphic section if it is to be a useful tool for correlation.

Figure 4-6 shows a cross section through a complex diapiric salt structure. We use this figure to illustrate the procedure for preparing a correlation type log. This structure exhibits a number of complexities, including a salt overhang, several faults, an unconformity, diapiric shale, and stratigraphic thinning and sand pinchouts in the upstructure position near the salt. We will consider each of the four wells that have penetrated the structure and evaluate the applicability of each as a type log.

Well No. 1 is not a good candidate as a type log for several reasons: It is only drilled to a depth of −8700 ft, it crosses a crestal fault, encounters salt at a shallow depth, and does not encounter a complete section. Well No. 2 is drilled off the flank of the structure penetrating a thick nearly complete stratigraphic section. It does, however, cross an unconformity at about −11,300 ft. Well No. 3 is also drilled in a downdip position and penetrates the entire stratigraphic section before encountering diapiric shale near the total depth (TD) of the well. It does, however, cross a fault at about −10,500 ft in the 9100-ft Sand. Well No. 4 drilled in a crestal position is not suitable as a type log because

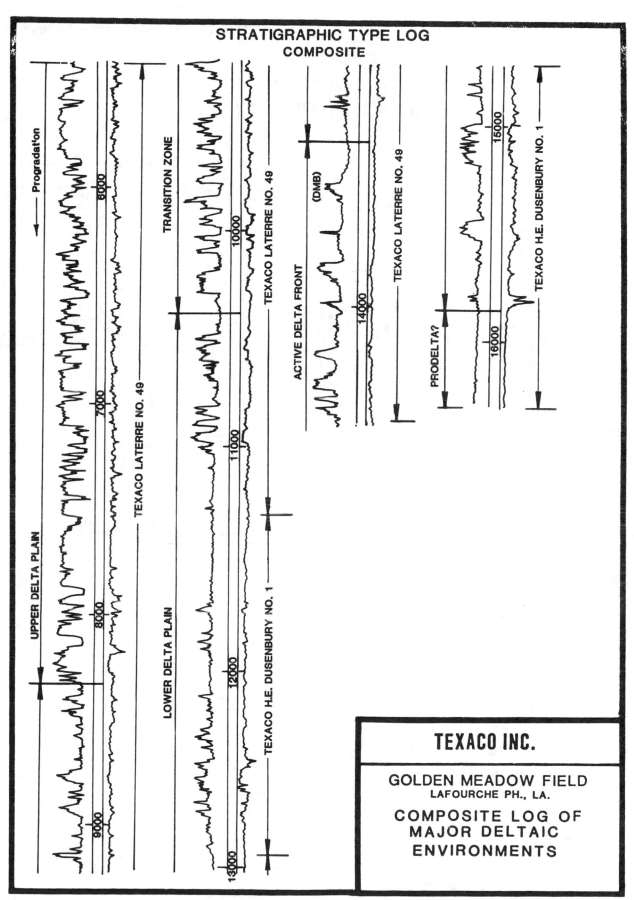

Figure 4-4 Composite stratigraphic type log from Golden Meadow Field, Lafourche Parish, Louisiana. (Published by permission of Texaco USA.)

TYPE LOG
(COMPOSITE)
GOOD HOPE FIELD

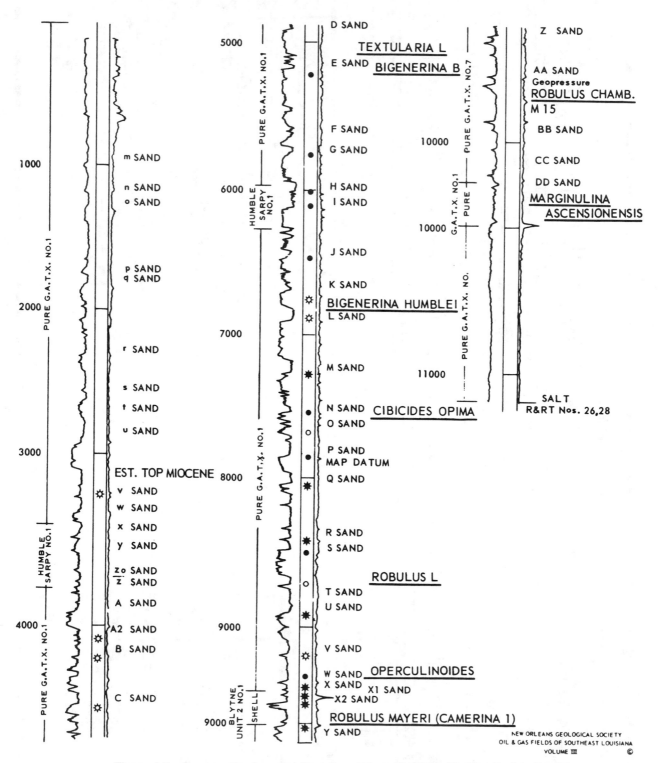

Figure 4-5   Composite show log from Good Hope Field, St. Charles Parish, Louisiana.
(Published by permission of the New Orleans Geological Society.)

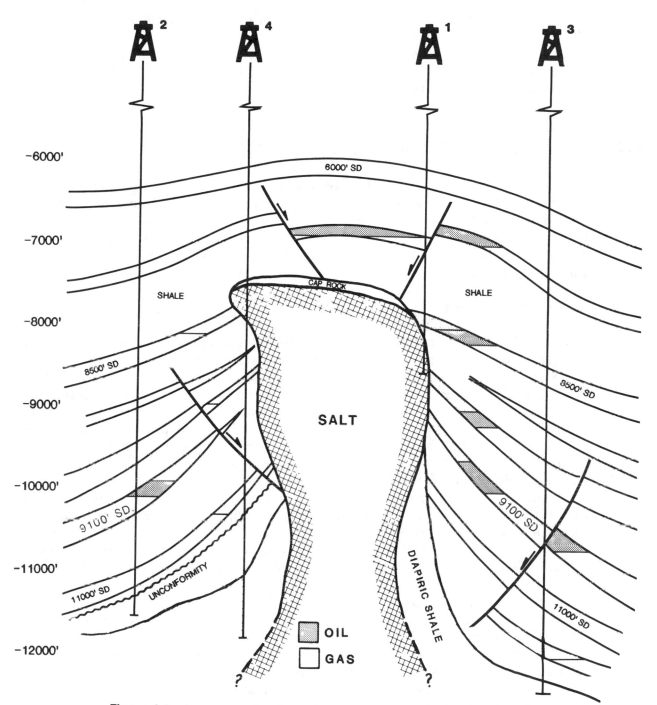

Figure 4-6 A cross section through a complex diapiric salt structure, penetrated by four vertical wells.

it penetrates the salt overhang, encounters a thinner stratigraphic section than that penetrated by Wells No. 2 and 3, crosses a fault, an unconformity, and does not penetrate a complete stratigraphic section.

    For this particular example, the correlation type log must be a composite log including sections from Wells No. 2 and 3. The best type log consists of Well No. 2 from the surface down to a correlation marker just below the 11,000-ft Sand and Well No. 3 from the same marker to total depth (TD) in diapiric shale. This composite type log meets all the requirements in the definition shown earlier and is an excellent standard for all other well log correlation work on this structure.

Normally, faults are not included on a correlation type log. However, if a major decollement, such as a thrust or listric growth fault, serves as the deepest limit of prospective section, it is advisable to place the fault on the type log.

ELECTRIC LOG CORRELATION—VERTICAL WELLS

We begin the discussion of actual correlation work by reviewing electric log correlation in vertical wells. In general, electric log correlation is easier and more straightforward in vertical wells than in wells which are directionally drilled. Later in this chapter, after we discuss the fundamental concepts and techniques of correlation in vertical wells, we review the same concepts and techniques as they apply to directionally drilled (deviated) wells.

Log Correlation Plan

When given the task of correlating logs in a specific field or area of interest, you might ask yourself one of several questions: "Where do I start?" or "Which log do I correlate first, second, third, etc.?" Before starting the log correlation in an area, a general *log correlation plan* needs to be developed. In this section, we illustrate a log correlation plan. This plan provides an answer to the questions asked, and establishes a preferred order in which to correlate electric logs from vertically drilled wells. This correlation plan can be adapted to most geologic settings. For the purpose of illustration, we use a structure map of the 8000-ft Sand on a normally faulted anticlinal structure in an extensional geologic setting (Fig. 4-7).

Figure 4-7 shows a structure contour map on the 8000-ft Sand. The faulted anticlinal structure is the result of a deep-seated salt mass resulting in a structure that becomes more complex in the updip direction.

The following log correlation plan is intended to make correlation work more systematic and easier to conduct, and to result in fewer correlation difficulties.

Step 1. First, prepare a correlation type log. Remember, a correlation type log must show a complete unfaulted interval of sediments representative of the thickest and deepest sedimentary section in the field. For the structure in Fig. 4-7, the wells furthest off structure, such as Wells No. 5 or 7 or a composite of the two, are good candidates for a type log. These wells positioned off the crest of the structure should show the thickest and most complete sedimentary section.

Step 2. A good correlation plan involves the correlation of each well with a minimum of two other wells. To ensure good correlation efficiency over the entire area, the electric log correlation plan should be established to correlate by means of *closed loops*. The correlation plan in Fig. 4-7 illustrates a sequence of closed loops. The recommended order of correlation is represented by "billiard ball" type correlation sequence numbers. Using this procedure, the log correlation work within a loop begins and ends with the same log, eliminating correlation mis-ties and reducing the chance of other correlation errors.

Step 3. First correlate wells expected to exhibit the most complete and thickest stratigraphic section. On a structure such as the one shown in Fig. 4-7, the struc-

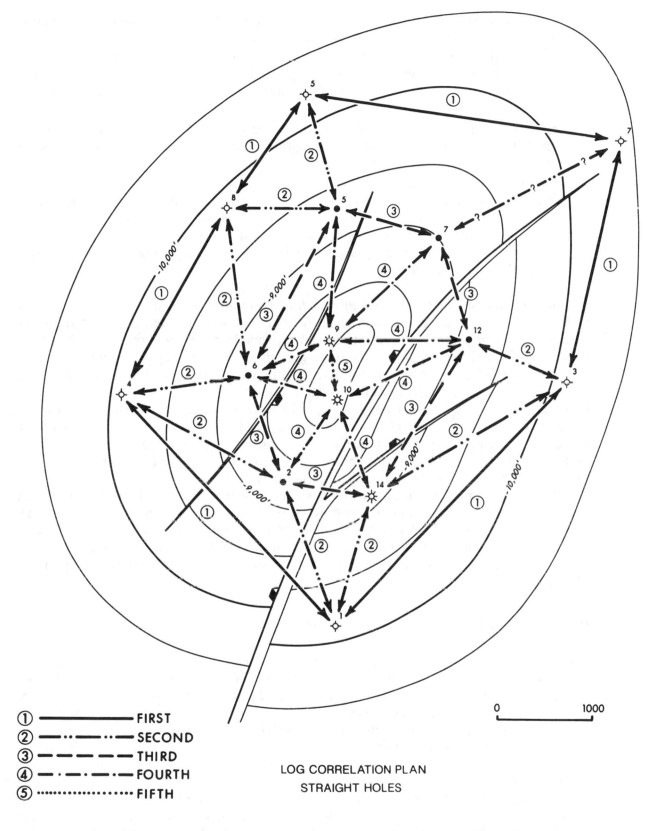

① ——————— FIRST
② —··—··— SECOND
③ — — — — THIRD
④ —·—·—·— FOURTH
⑤ ················· FIFTH

0   1000

LOG CORRELATION PLAN
STRAIGHT HOLES

Figure 4-7 An example of a log correlation plan for vertical wells. Notice that there is a hierarchy in the correlation sequence and that all the wells are correlated in closed loops.

turally lowest wells usually have the thickest section. These include wells represented by the billiard ball type correlation sequence number 1.

Step 4. Continue log correlation progressing from wells in a down-structure position to wells in an up-structure position (see the billiard ball type correlation sequence numbers 2, 3, 4, and 5). These numbers indicate the recommended correlation sequence for the example.

Step 5. Generally, correlate wells located nearest each other. In most cases, closely spaced wells should have a similar stratigraphic section and so correlation is usually easier.

Step 6. In many geologic areas, rapid changes in stratigraphy (particularly changes in thickness) occur over short distances. Where possible, correlate wells anticipated to have a similar stratigraphic interval thickness. In extension or diapiric salt tectonic areas, wells at or near the same structural position usually exhibit similar stratigraphy. For areas involving plunging folds, similar stratigraphy may be exhibited along the plunge of the fold.

In geologic settings involving *growth faults*, some special considerations must be given in preparing a log correlation plan. For our purposes, we define a normal growth fault as a syndepositional fault resulting in an expanded stratigraphic section in the downthrown fault block with displacement on a growth fault increasing with depth.

If a growth fault is present in the area of study, restrict the correlation to wells within one fault block of the growth fault. Keeping in mind that the downthrown block of a major growth fault has an expanded stratigraphic section which can increase the difficulty in correlation, start the correlation in the upthrown block using the plan just outlined. Once the correlations are completed in the upthrown fault block of the major growth fault, carry the correlations, if possible, into the downthrown fault block. Initially, check the correlation in wells located in down-structure positions.

If a significant amount of growth has occurred on the fault, the thickness of the sediments can be so great in the downthrown block that correlation of the section from the upthrown to downthrown blocks may be difficult, if not impossible. In such a case, the best correlations can be achieved by preparing a separate type log for the downthrown block and correlating the downthrown fault block independently from the upthrown block.

Basic Concepts in Electric Log Correlation

Now that we have established a plan of correlation, we shall examine some basic concepts of electric log correlation. Figure 4-8 shows the SP and amplified short normal resistivity curve from the electric logs of two vertical wells. Initial *Quick Look* correlations can be made by reviewing major sands. Sands are the dominant and most obvious feature seen on the SP or gamma ray curves and serve as good quick-look correlations. Because major sand beds frequently exhibit significant variation in thickness and character from well to well and are often laterally discontinuous, however, they are not recommended for detailed electric log correlations.

We suggest that all detailed electric log correlation be undertaken by concentrating on shale sections. We apply this approach to electric log correlation in Fig. 4-8, which shows a log segment from two vertical wells (No. A-1 and No. 3). The SP and amplified short normal curves are shown for each log. There are two major sands seen in each well, labeled the 10,000-ft and 10,300-ft Sands.

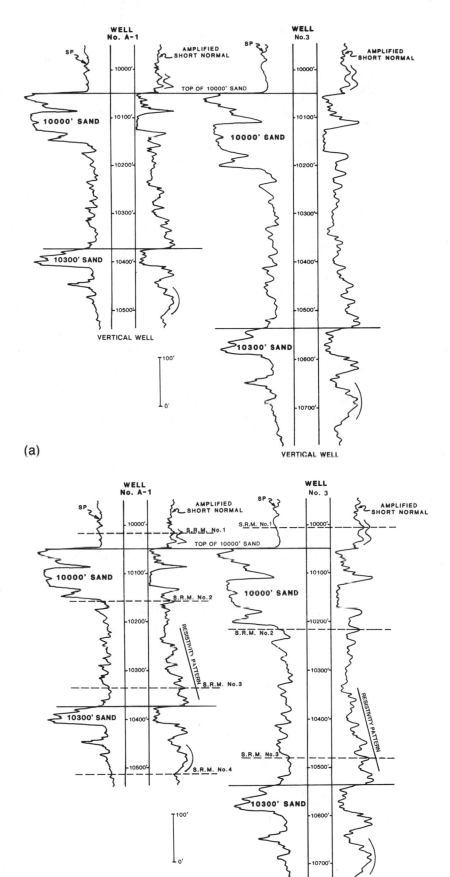

(a)

(b)

Figure 4-8 (a) Correlation of two vertical wells using the major sands as the primary vehicle for correlation. (b) Detailed correlation of the two vertical wells shown in Fig. 4-8a using all the reliable shale and sand correlation markers. (SRM = Shale Resistivity Marker)

First we will review these two logs using the tops of the major sands as the primary vehicle for correlation (Fig. 4-8a). Imagine that we are major sand bed correlators and we are correlating Wells No. A-1 and 3 in Fig. 4-8a. By correlating the sands, we see that the interval from the top of the 10,000-ft Sand to the top of the 10,300-ft Sand is about 325 ft thick in Well No. A-1 and 480 ft in Well No. 3. The interval in Well No. A-1 is short between the two sand tops by 155 ft. This short interval, based on the sand correlations, suggests the possibility of a 155-ft fault in Well No. A-1.

Now we correlate these same two logs, shown in Fig. 4-8b using the guidelines outlined earlier. The guidelines recommend that detailed correlations be conducted in the shale sections using all the electric log curves with an initial emphasis on the amplified short normal resistivity curve. This curve provides the most reliable shale correlations.

Through detailed correlations of the shale sections and the sands, a number of correlation markers are identified on the two logs. These include a series of shale resistivity markers labeled SRM No. 1 through SRM No. 4 (SRM = Shale Resistivity Marker), certain diagnostic resistivity correlation patterns highlighted on the resistivity side of the logs, in addition to the two major sands. All these correlation markers indicate that both log segments have a high degree of correlation and that no fault is present in Well No. A-1.

It appears that the stratigraphic section in Well No. A-1 is uniformly thin relative to Well No. 3. The thickness ratio for the intervals between each of the four shale markers shows a consistency in the stratigraphic thinning in Well No. A-1 when compared to Well No. 3. This uniform thinning supports the idea that although Well No. A-1 is short to No. 3 as a result of stratigraphic thinning, the two logs exhibit correlation.

Faults vs Variations in Stratigraphy

The differentiation between fault cuts and variations in stratigraphic thickness in well log correlation is very important. We stated earlier that reliable interpretations presented on maps and cross sections are bedrocked in accurate correlations. If a stratigraphically thin section is correlated incorrectly as a fault cut, this erroneous fault data will be incorporated into the construction of a fault surface map and later integrated into the structural interpretation. The purpose of this section is to outline procedures which are effective during correlation to help differentiate between faults and variations in stratigraphic thickness.

Fault Cut Determinations.   Now that we have a basic understanding of how shale markers are used to aid in log correlation, look at the log segment from two other electric logs run in vertical wells (Nos. 1 and 3 in Fig. 4-9). By reviewing the logs as major sand correlators, and using the 8600-ft and 9000-ft Sands as the principal correlations, the section in Well No. 3 between the two major sands is 80 ft short and a fault appears possible in the well. With the limited correlation data, the size and location of the fault is uncertain. Also, the correlation of the top of the 9000-ft Sand in Well No. 3 is questionable. Is there a fault in Well No. 3 and is the fault (1) within the shale interval between the base of the 8600-ft Sand and the top of the 9000-ft Sand, (2) at the top of the 9000-ft Sand, or (3) is part of the 9000-ft Sand faulted out? If the fault is at the top of the 9000-ft Sand, the interval from the 8600-ft Sand to the top of the 9000-ft Sand is 80 ft short. If part of the top of the 9000-ft Sand is faulted out, then the interval is short by some amount greater than 80 ft. With the major sand correlation methodology, the nature of the short section in Well No. 3 is not apparent and so we have a correlation problem.

Now we will follow the recommended correlation procedures illustrated in Fig. 4-

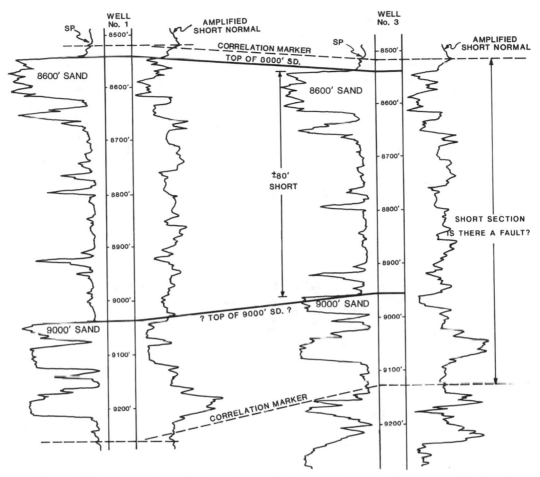

Figure 4-9 Correlation of the major sands in two vertical wells may provide insufficient correlation data to accurately determine the depth and size of a suspected fault.

10. These procedures provide a number of correlation markers, including shale resistivity markers 1 through 7 and specific resistivity characteristics highlighted on the resistivity side of each log. These detailed correlation markers show that the interval between each correlation marker is comparable in both wells except for the short section identified between SRM 5 and SRM 7. Notice that SRM 6 is missing in Well No. 3, as well as the lower segment of the resistivity character highlighted through SRM 6 in Well No. 1. Finally, these detailed shale correlations and the correlation data contained within the 9000-ft Sand indicate that the upper portion of the 9000-ft Sand is also missing in Well No. 3. We have isolated the short section in Well No. 3 to a single specific interval 135 ft thick. **The isolation of this short section to one particular location indicates that the short section is the result of a fault rather than a variation in stratigraphy.** The location of the short section provides the depth of the fault in this well. By measuring the amount of section missing in Well No. 3, we determine the size of the fault (135 ft) by correlation with Well No. 1. The missing section is highlighted in Fig. 4-10. In order to ensure confidence in the fault, Well No. 3 should be correlated with at least one more nearby well.

We refer to a recognized fault in a well as a **fault cut**. For each fault cut there are three important pieces of data that must be obtained for documentation and later use in

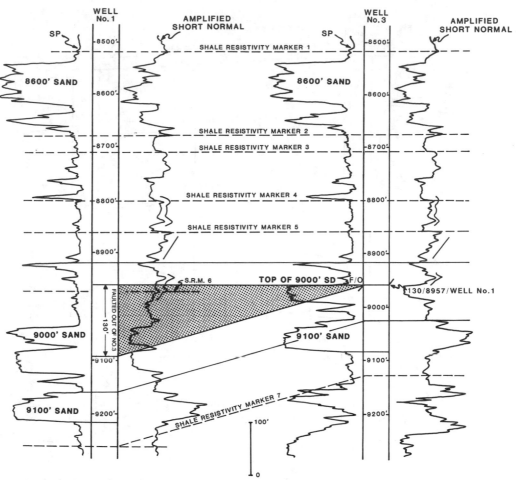

Figure 4-10 Detailed correlation of the two vertical wells shown in Fig. 4-9 using all recognizable correlation markers to determine the depth and size of a fault in Well No. 3. Notice that the top of the 9000-ft Sand and SRM 6 are faulted out of Well No. 3.

mapping: (1) the size of the fault, (2) the log depth of the fault cut, and (3) the well or wells correlated to obtain the fault cut. The fault data (135 ft/8957 ft/Well No. 1) and information regarding the faulted out (F/O) top of the 9000-ft Sand are annotated next to the fault cut symbol on the log. Refer to Fig. 4-10 again for an example of how these data are annotated on a log.

The accuracy of identifying the depth of a fault cut in a well and determining its size is directly related to (1) the detail to which the logs are correlated, (2) the number of logs used for correlation, and (3) variations in stratigraphic thickness seen in the wells. Obviously, the smaller the interval between established correlation markers, the more precise the correlation in pinpointing the depth and size of a fault.

The correlation detail and accuracy required are often dictated by the type of geologic study being conducted. For example, if you are involved in a regional geologic study, pinpointing the depth of a fault within several hundred feet on a well log may be sufficient. Also, you may only be interested in the larger faults (i.e., faults greater than 100 ft). If the study is to be detailed for field development or enhanced recovery, however, it may be necessary to locate the depth of **all recognizable faults** to within ±20 ft. The same variation in accuracy applies to the size of a fault.

Stratigraphic Variations.   Figure 4-11 shows a log segment from Wells No. A-1 and 3. In this section, we use the correlation procedures to establish specific correlation markers to recognize stratigraphic variations so that such thickness changes are not mistaken as a fault cut.

In Fig. 4-11, two correlation markers are identified in each well. Based on these markers, Well No. A-1 is 155 ft short to Well No. 3. Is the short section in Well No. A-1 the result of a fault or variations in stratigraphy? With the limited correlation data shown in the figure, it is impossible to determine why the section in Well No. A-1 is short. We could use the major sands in each well to aid the correlation work, but this added information provides little help in determining the nature of the short section.

So far, we have shown that it is important to identify as many correlation markers as possible, especially in questionable log intervals. **Closely spaced correlations generally improve the accuracy of the correlation, help differentiate between fault cuts and stratigraphic variations, and improve the estimate of the size and depth of identified fault cuts.** Therefore, in order to accurately correlate Wells No. A-1 and 3, additional correlation markers are required.

Figure 4-12 shows the same two logs with additional correlation markers identified. The correlation process is improved with these additional markers. Notice that the shortening in Well No. A-1 is not isolated to one specific interval, but is present in all the

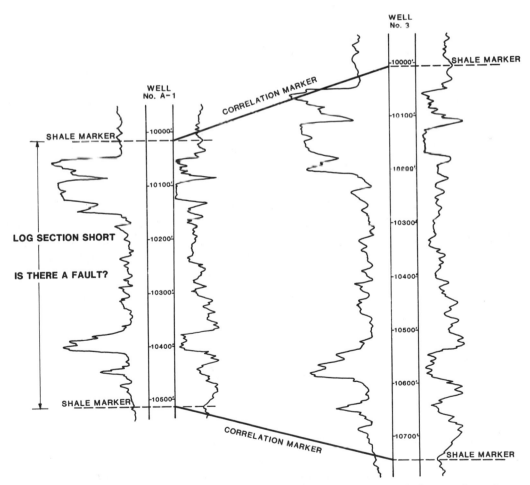

Figure 4-11   Correlation of Wells No. A-1 and 3 using limited correlation markers. Is there a fault in Well No. A-1?

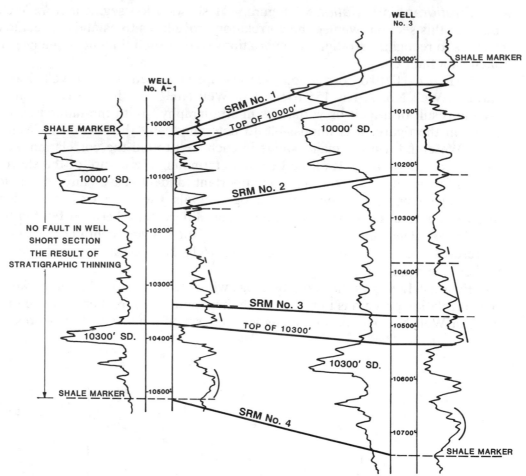

Figure 4-12 Correlation of Wells No. A-1 and 3 using all recognizable correlation markers indicates that there is no fault in Well No. A-1. The short log section is the result of variations in stratigraphy.

intervals in the well between shale resistivity markers 1 and 4. This evidence strongly suggests that the thickness variations in Well No. A-1 are stratigraphic and not the result of faulting. If necessary, interval thickness ratios can be calculated between each correlation marker to provide further evidence to support the conclusion.

$$T_r = \frac{T_s}{T_l} \qquad (4\text{-}1)$$

where

T_r = Thickness ratio

T_s = Interval thickness in short log

T_l = Interval thickness in long log

Pitfalls in Vertical Well Log Correlation

As a final topic on well log correlation in vertical wells, we look at some pitfalls caused by changes in formation dip. Figure 4-13a shows an electric log section from four vertical wells. Using the detailed correlation procedures, we determine that the sand member

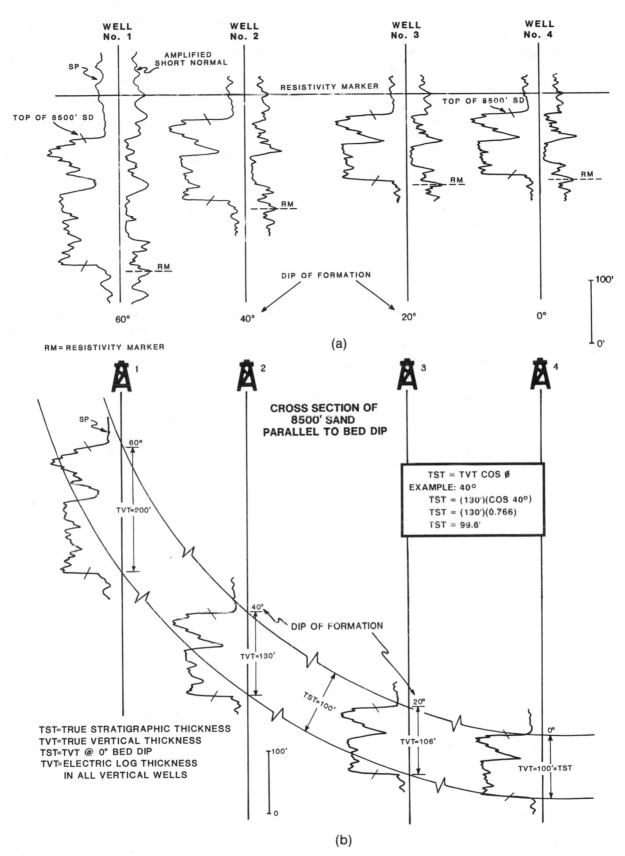

Figure 4-13 (a) Electric log stratigraphic section laid out perpendicular to the strike of the 8500-Ft Sand (parallel to dip). (b) Electric log section showing the relationship of true stratigraphic thickness (TST) to true vertical thickness (TVT) with changing bed dip. The true vertical thickness is that thickness seen in a vertical or straight hole.

shown in each well is that of the 8500-ft Sand. From right to left in the figure, the well logs show an increasing thickness in the 8500-ft Sand from 100 ft in Well No. 4 to 200 ft in Well No. 1. What is the cause of thickness change in this sand—variations in stratigraphic thickness, faulting, or something else?

General structure mapping and dipmeter data show that the bed dip is different in the vicinity of each well: 0 deg in Well No. 4, 20 deg in Well No. 3, 40 deg in Well No. 2, and 60 deg in Well No. 1. All four wells lie in a line perpendicular to bed strike (parallel to bed dip). Analysis of the dip data and logs suggests an increase in thickness of the 8500-ft Sand in the upstructure direction. Normally, however, we expect to see a constant or reduced thickness of a stratigraphic section in the upstructure direction. So, are these thickness changes seen on each log due to faulting, stratigraphic variations, or a geometric problem resulting from changing bed dip?

In Fig. 4-13b, the logs are hung in their true structural position with the dip of the formation at each well location shown. The formation dips and the relationships between true bed or stratigraphic thickness, and log or vertical thickness, are shown on the figure. Notice that even though the log thickness in Well No. 1 is twice that seen in Well No. 4, the stratigraphic thickness is identical at both locations. This example illustrates that caution must be taken when correlating logs on a structure with significant changes in bed dip. Changing bed dip can result in changing log thickness in vertical wells, even though the section is not faulted and the stratigraphic thickness is constant.

To better understand the stratigraphy and growth history of a structure, and resolve some of the geometric problems caused by changes in bed dip, the true stratigraphic thickness (TST) of an interval penetrated by a vertical well can be calculated. The two variables required to calculate TST are the true vertical thickness (TVT) of the section as seen in a vertical well and the bed dip.

$$TST = TVT (\cos \phi) \tag{4-2}$$

where

$$TST = \text{True stratigraphic thickness}$$
$$TVT = \text{True vertical thickness}$$
$$\phi = \text{True bed dip}$$

Logs cannot be correlated in a vacuum. The correlation plan shown earlier illustrated the need to know the structural relationship of logs being correlated. This can be accomplished by having a well log base map that shows the general structure available during log correlation to show the structural position and location of each well log being correlated.

Finally, let's consider a situation as shown in Fig. 4-14. In this case, the sand identified in the two wells has a decreasing stratigraphic thickness in the upstructure direction. We can say that the structure was actively growing during the time of deposition of the sand resulting in stratigraphic thinning toward the crest of the structure. By only log correlation, however, *the vertical log thickness of the sand in each well is exactly the same.* If you recognized the same interval thickness in each well irrespective of structural position, you might make the assumption that since the thicknesses are equal, the structure was not active during deposition of the sand. A review of the wells in cross section in Fig. 4-14 shows that the stratigraphic thickness of the sand actually decreases in the updip direction such that the stratigraphic thickness in Well No. 1 is only one-half

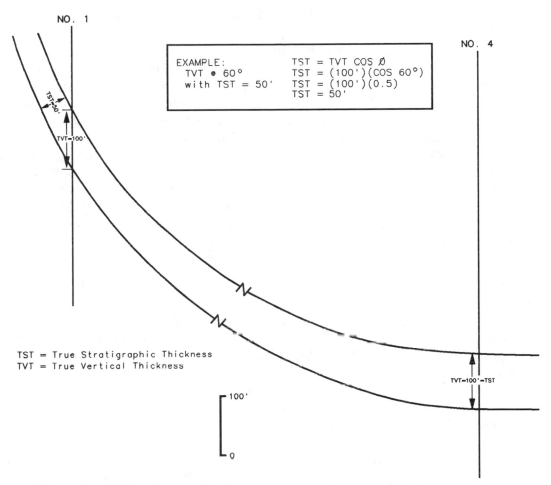

EXAMPLE: TST = TVT COS Ø
  TVT ● 60°   TST = (100')(COS 60°)
  with TST = 50'   TST = (100')(0.5)
                          TST = 50'

TST = True Stratigraphic Thickness
TVT = True Vertical Thickness

Figure 4-14   The cross section shows the effect of changing bed dip on the true
vertical thickness of a stratigraphic section.

the thickness found in Well No. 4. *It is the changing bed dip that causes the vertical log thickness to be the same in each well.* Equation (4-2) can be used to calculate the TST in each well to develop a better understanding of the stratigraphic thicknesses as seen in each well.

These examples show that log thickness in vertical wells varies with changes in formation dip and is equal to the true stratigraphic thickness *only* when the dip of the formation is zero (Fig. 4-14, Well No. 4). They also emphasize that log correlation is not an isolated process. Structural and stratigraphic relationships and geologic knowledge of the area of study must always be kept in mind during correlation. Errors in correlation and incorrect assumptions on such aspects as the local growth history of a structure may be incorporated into the geologic work. Such pitfalls can be prevented if the structural and depositional framework of the area is considered during correlation.

ELECTRIC LOG CORRELATION—DIRECTIONALLY
DRILLED WELLS

In this portion of the chapter, we discuss fundamental concepts and techniques for correlating directionally drilled wells. There are additional complexities in correlation that arise when working with logs from wells deviated from the vertical. We also look at the correlation of vertical wells with directionally drilled wells (often referred to as deviated wells).

What is a directionally drilled well? We discussed earlier that a vertical well is one drilled 90 deg to the horizontal reference, usually sea level. A directionally drilled well can be defined as a well drilled at an angle less than 90 deg to the horizontal reference, as shown in Fig. 4-15. Some general directional well terminology was discussed in Chapter 3. These terms are again illustrated in Fig. 4-15 for ease of reference. Other terminology discussed earlier in this chapter for vertical wells is also applicable to deviated wells.

Most wells drilled offshore and many wells onshore are drilled directionally. The most common well is a simple ramp well (Fig. 4-15a), sometimes called an "L" shaped hole. These wells are deviated to a certain angle which is usually held constant to total depth (TD) of the well. Many wells are drilled with an "S" shaped design. With an "S" shaped hole, the well builds to one angle, maintains this angle to a designated depth and then the angle is lowered again, often going back to vertical (Fig. 4-15b).

Log Correlation Plan

Just as with vertical wells, there must be some system to log correlation of directionally drilled wells. Due to the nature of deviated wells, a good correlation plan is critical to accurate correlations. For this log correlation plan, we once again use the structure map on the 8000-ft Sand on a normally faulted anticlinal structure. The correlation plan outlined here is intended to make correlation systematic, provide a logical method for correlating directionally drilled wells with other directionally drilled wells or with vertical wells, and reduce correlation problems.

> **Step 1.** First, construct a correlation type log. Refer to the section on correlation type logs for the complete definition of a type log. Do not use a deviated well in the construction of a type log because a log from a directionally drilled well does not represent the true vertical stratigraphic section. Wells farthest off structure serve as good type log candidates.
>
> **Step 2.** Correlate all the vertical wells first before correlating the deviated wells since the vertical wells are usually easier to correlate. For the vertical wells, use the same plan outlined in Fig. 4-7.
>
> **Step 3.** Once the vertical wells have been correlated, begin correlating the deviated wells. To begin directional well log correlation, first organize the wells according to their direction of deviation with respect to structural strike. *Deviated wells are classified into one of three groups: (1) wells drilled downdip, (2) wells drilled along strike, and (3) wells drilled updip.*
>
> **Step 4.** Begin correlation of these three groups with the wells drilled generally downdip. First correlate the wells with the least amount of deviation, and where possible correlate in closed loops with each well log correlated with a minimum of two other wells. The wells with the least amount of deviation will have a log section thickness closer to that seen in a vertical well than other wells drilled downdip. Looking at the wells drilled from Platform B in Fig. 4-16, the first directional wells correlated are those represented by a billiard ball type correlation sequence number 1. There are two wells drilled with a minimum downdip deviation (Wells No. B-5 and B-6). These wells can be correlated to each other and then with the straight hole, Well No. B-1, drilled as a vertical well from the platform.

SIMPLE RAMP WELL

"S" SHAPE WELL

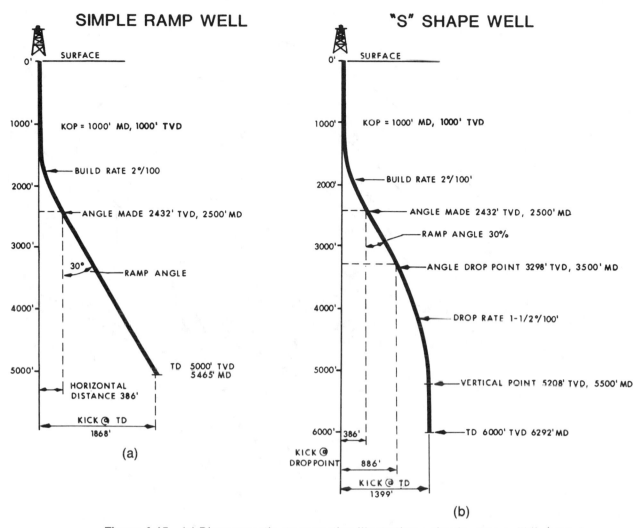

Figure 4-15   (a) Diagrammatic cross section illustrating a simple ramp or "L" shaped well. (b) Diagrammatic cross section illustrating a more complicated "S" shaped well. (Published by permission of Tenneco Oil Company.)

Step 5. Continue correlating wells with increased deviation in the downdip direction. For this example, these are Wells No. B-2 and B-3 indicated by correlation sequence number 2. These two highly deviated wells can be correlated with each other and then with Wells No. B-5 and B-6. Also, the vertical well No. 3 may be used to correlate B-2 and B-3 since it is an offstructure well exhibiting a thick stratigraphic section.

Step 6. When all wells classified as being deviated downdip are correlated, the next group to correlate are those wells deviated along structural strike. From Platform B, Wells No. B-7 and B-9 fall into this category. These wells can be correlated to each other and then with straight hole B-1 to close the loop.

Step 7. Finally, correlate the wells deviated updip. Those wells drilled closest to the crest of the structure usually are complicated by stratigraphic thinning, faulting, and unconformities. The correlation of these wells can be most difficult; therefore, they are normally correlated last when all other correlation information is available. Wells

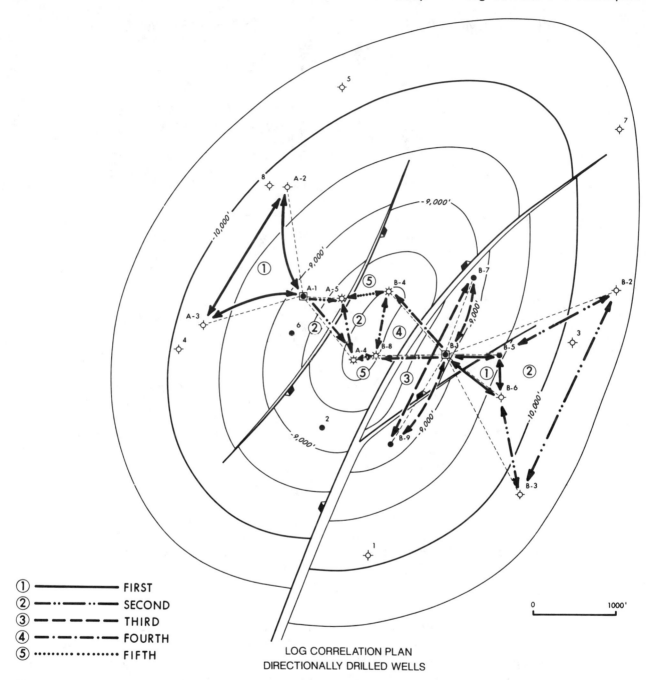

LOG CORRELATION PLAN
DIRECTIONALLY DRILLED WELLS

① —————— FIRST
② —··—··—··— SECOND
③ ——— ——— THIRD
④ —·—·—·— FOURTH
⑤ ·············· FIFTH

Figure 4-16 An example of a log correlation plan for directionally drilled wells. The plan shows the hierarchy of the log correlation sequence and illustrates how to correlate deviated wells in closed loops.

No. B-4 and B-8 drilled from the B Platform fall into this category. They are labeled as correlation sequence number 4.

Step 8. Generally, it is best to correlate wells located nearest each other, especially in areas where significant changes in stratigraphic thickness are probable. Wells nearest each other and approximately in the same structural position usually are expected to have the most comparable interval thicknesses.

Step 9. After correlating the wells from one platform, begin correlation of wells on any additional platforms in the area. In Fig. 4-16, the A Platform wells in the northwest portion of the field should be correlated next. It is not necessary, however, to isolate correlation to a single platform. Often, wells from one platform are drilled in a direction toward another platform. If wells from separate platforms are in close proximity to one another, they should be correlated to each other. Notice that correlation sequence number 5 illustrates the correlation of B-4 with A-5, and B-8 with A-4. Wells No. A-1 and B-1 are straight holes drilled from separate platforms, but since they are located in a similar structural position, they should also be correlated to each other.

The primary focus of this correlation plan is to provide a logical method for correlating all vertical and deviated wells in an area of study. The plan outlined is by no means the only one that can be used. *The important point is to have a plan.* Without one, log correlation becomes a random process often resulting in some type of correlation problem.

Correlation of Vertical and Directionally Drilled Wells

In this section we discuss general procedures for correlating vertical wells with directionally drilled wells. Directional wells have a measured log thickness that can be less than, greater than, or equal to the log thickness in a vertical well drilled through the same stratigraphic section. These different measured log thicknesses result in additional complexities that must be considered when undertaking correlation work using well logs from both vertical and deviated wells.

Now we shall look at the correlation of a vertical well with a deviated well. Figure 4-17 shows a portion of an electric log from vertical Well No. A-1 and the electric log from directionally drilled Well No. A-2. The wells are in close proximity to each other. The detailed electric log correlation (sand and shale sections) for both wells indicates that they have penetrated the same stratigraphic section. Although both wells have a high degree of correlation (see shale resistivity markers 1 through 4), the stratigraphic section in Well No. A-2 is much thicker than the same section seen in Well No. A-1. The log section in Well No. A-1 from SRM 1 to SRM 4 is 490 ft thick. The same section in Well No. A-2 is 735 ft. Earlier in the chapter, in the discussion on vertical wells, we mentioned that a short section in one well with respect to another may be the result of stratigraphic changes or a fault. If the short section is isolated to one particular location, the short section is most likely the result of a fault rather than variations in stratigraphy. Likewise, if the short section is uniformly distributed over a series of intervals, the short section is probably due to stratigraphic variations rather than a fault.

Based on the correlation criteria, the thin section in Well No. A-1 appears to be the result of stratigraphic thinning rather than a fault. In this example, however, we introduce another possible explanation for the shortening. Since Well No. A-2 is directionally drilled, the thickness seen in the well with respect to Well No. A-1 may be completely the result of the wellbore deviation. Figure 4-18 shows vertical Well No. A-1 and deviated Well No. A-2 in its true orientation with respect to the vertical. Well No. A-2 is drilled due west at a deviation angle of 48 deg (48 deg from the vertical). The correlation markers in each well show that the strata are horizontal and the thick section seen in Well No. A-2 is solely the result of wellbore deviation. We have now introduced another complexity in correlation that must be considered when both vertical and deviated wells are present in the area of study.

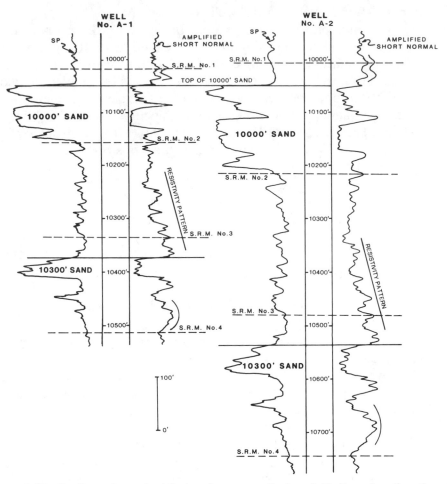

Figure 4-17 Portion of an electric log from a vertical well (A-1) and a directionally drilled well (A-2). The electric log sections show detailed correlations.

There are several procedures that can be used to help correlate a vertical well with a directionally drilled well.

1. Mark the angle of deviation for the directional well on the log at least every 1000 ft. This provides a reminder that the well is deviated and indicates the angle of deviation at 1000-ft intervals on the actual log.

2. To compare interval thicknesses, slide the vertical well log as you correlate from marker to marker. This allows you to compensate during correlation for the expanded or reduced section in the directional well as a result of its deviation.

3. Calculate a thickness ratio for certain correlation intervals of interest to help evaluate whether any short section is the result of faulting, stratigraphic thinning, or just wellbore deviation (Fig. 4-18).

4. If a copy machine with a reduction mode is available, calculate the correction factor required to convert the deviated (stretched) log section to a vertical log section, and then reduce the log by the appropriate reduction factor. Use the reduced log for correlation.

5. **In areas of horizontal beds or low relief**, the measured depth log from a deviated well can be corrected for wellbore deviation and converted into a TVD (true vertical depth) log to use for correlation.

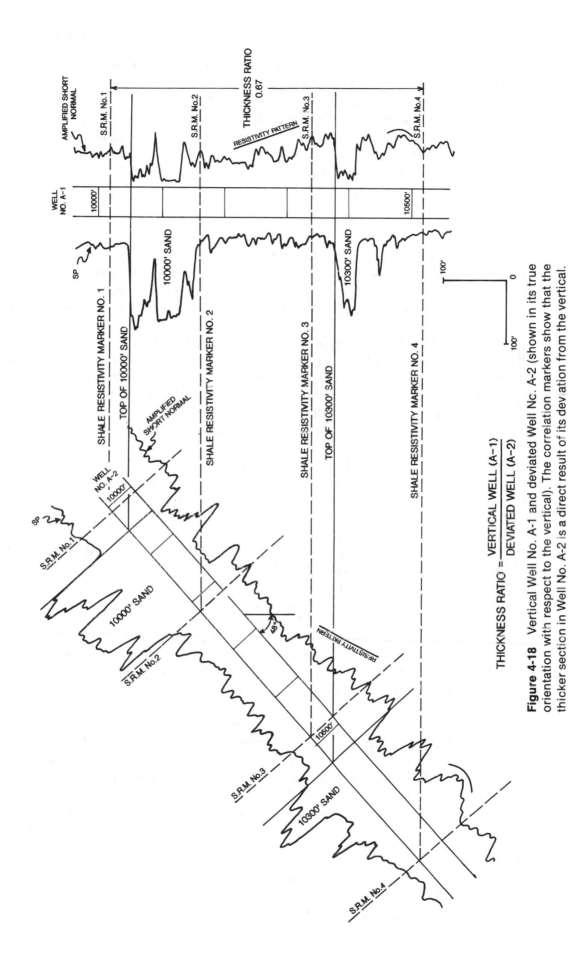

Figure 4-18 Vertical Well No. A-1 and deviated Well No. A-2 (shown in its true orientation with respect to the vertical). The correlation markers show that the thicker section in Well No. A-2 is a direct result of its deviation from the vertical.

$$\text{THICKNESS RATIO} = \frac{\text{VERTICAL WELL (A-1)}}{\text{DEVIATED WELL (A-2)}}$$

6. In areas of steep dip, if dip data are available from a dipmeter log or previously constructed structure maps, these data can be used to convert the deviated log to a TVT (true vertical thickness) log. A TVT log is a log in which the measured thickness has been corrected for wellbore deviation and bed dip to the thickness represented in a vertical well. In areas of steep dip, a TVD log provides little aid, if any, in correlation and can actually cause correlation problems (see section on MLT, TVDT, TVT, and TST).

Estimating the Size of Faults

Earlier in this chapter, we discussed the procedure for estimating the depth and size of a fault cut in a vertical well by correlation with another vertical well. Now we present the method for estimating the depth and size of a fault cut when deviated wells are considered. First, we look at the situation involving an area with horizontal beds.

Horizontal Beds.   We begin with a fault cut in the deviated Well No. A-2 (Fig. 4-19a). By correlation with Well No. A-1, this well is cut by a fault near the 10,000-ft Sand. To determine the location and size of the fault, we correlate the logs in the same manner as previously outlined in this chapter. First, correlate down the logs starting with shale resistivity marker 1. We can say that correlation is lost at points "A" in both wells. Mark this location on the two logs. Next, find a correlation point below this section on the logs, such as shale resistivity marker 4, and correlate up the logs. We now lose correlation in the wells at points "B." By detailed correlation of the shale markers and sands, we have determined that Well No. A-2 is faulted, and the section in Well No. A-1 that is stratigraphically equivalent to the missing section in Well No. A-2 is highlighted in Fig. 4-19a.

The faulted out or missing section in Well No. A-2 is equal to 150 ft by correlation with Well No. A-1. Notice that the base of the 10,000-ft Sand is faulted out in Well No. A-2. This information is annotated on the log along with the size and depth of the fault cut, and the well(s) used to correlate the fault cut.

The size of the fault cut in directional Well No. A-2 is determined by correlation with Well No. A-1, which is a vertical well. In a vertical well, the log thickness and vertical thickness are the same. Since *fault size* is expressed as *the vertical thickness of missing or repeated section*, the vertical thickness of the missing section in Well No. A-2 is 150 ft. The 150 ft represents the size of the fault to use for future structure mapping.

Figure 4-19b is a simplified stratigraphic section showing Wells No. A-1 and A-2 positioned in their true orientation with respect to the vertical. Well No. A-2, which is deviated at 48 deg from the vertical, is pulled apart at the fault cut to show the restoration of the faulted out section. This cross section clearly illustrates that the missing section in Well No. A-2 is equal to the 150 ft of vertical section highlighted in Well No. A-1.

Now consider a fault in vertical Well No. A-1 correlated with deviated Well No. A-2. Well No. A-1 has a fault cut near the base of the 10,000-ft Sand in Fig. 4-20a. Detailed correlation, as shown in the figure, identifies a 225-ft section that is faulted out of Well No. A-1 by correlation with deviated Well No. A-2. The faulted out section is highlighted in the figure. Since the size of a fault cut is determined as *the vertical thickness of missing section*, the estimate of the size of this fault cut in Well No. A-1 (225 ft) based on the deviated log thickness in Well No. A-2, must be corrected to express the size in terms of true vertical thickness.

Figure 4-20b is a stratigraphic section showing Wells No. A-1 and A-2 positioned

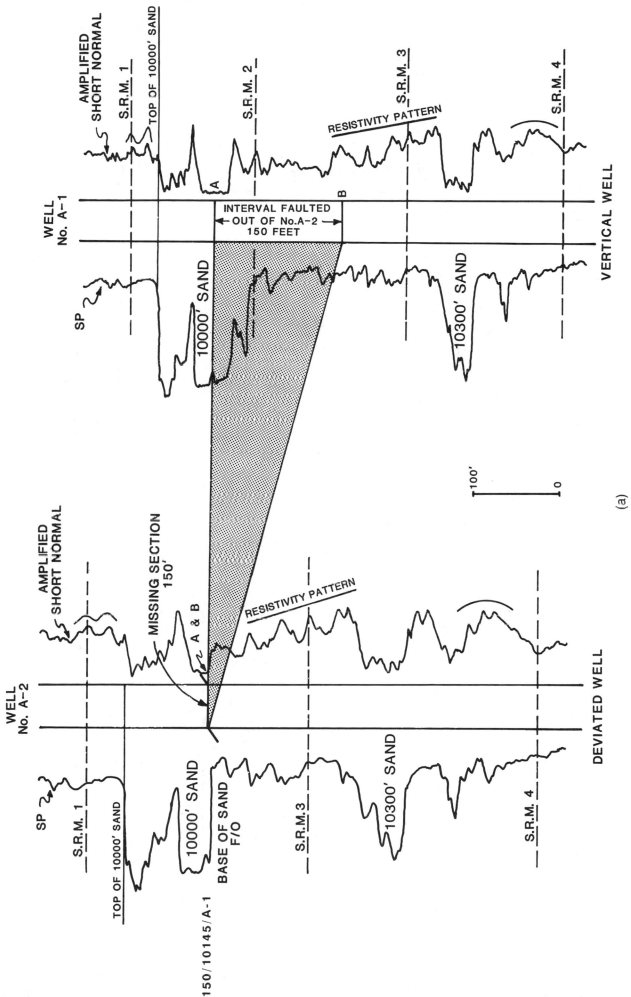

Figure 4-19a Detailed correlation of a deviated well with a vertical well to locate the depth and estimate the size of a fault in the deviated well. The base of the 10,000-ft Sand is faulted out.

(a)

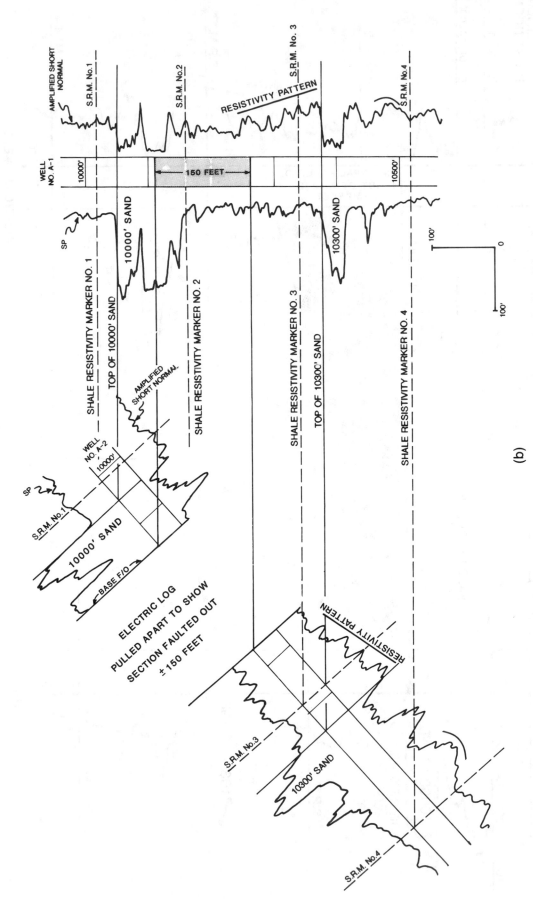

Figure 4-19b The simplified stratigraphic section through Wells No. A-1 and A-2 illustrates that the missing section in Well No. A-2 is equivalent to the vertical section highlighted in Well No. A-1. No thickness correction factor is required in this example.

(b)

in their true orientation to vertical. The log section of Well No. A-1 is pulled apart at the fault cut to show the restoration of the faulted out section. Since we are working in an area with horizontal beds, the correction of the measured log thickness in Well No. A-2 to true vertical thickness is determined by the simple trigonometric solution of a right triangle. The insert in the center of the figure shows that the true vertical thickness of the missing section is equivalent to the opposite side of a right triangle whose hypotenuse is equal to thc log thickness of the missing section in deviated Well No. A-2.

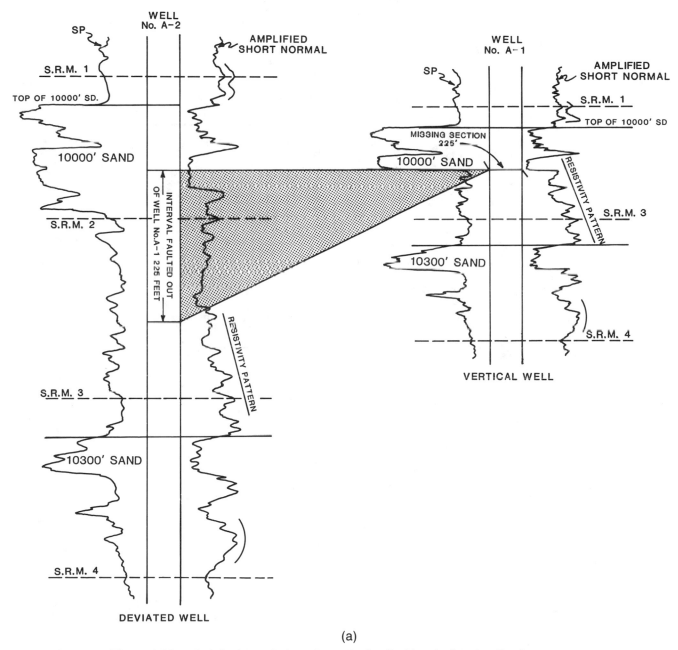

(a)

Figure 4-20a Detailed correlation of a vertical well with a deviated well to locate the depth and estimate the size of a fault in the vertical well.

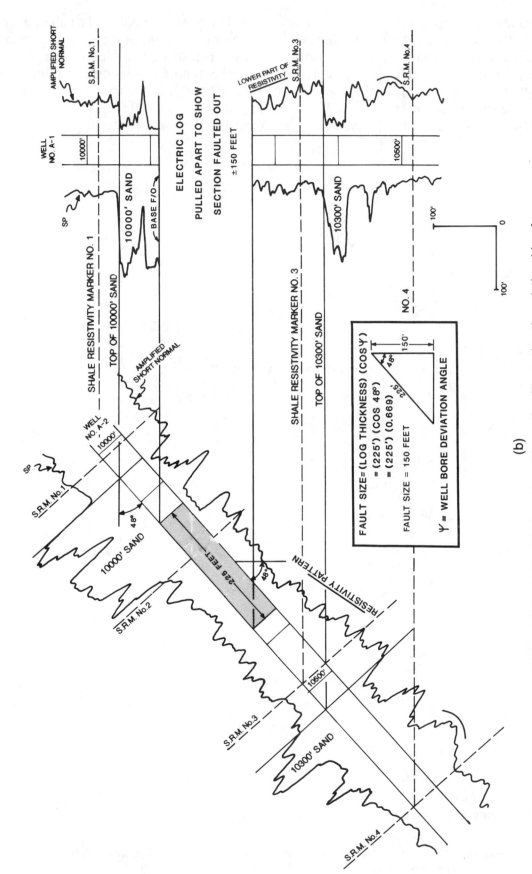

Figure 4-20b A simplified stratigraphic section illustrating the relationship of the missing section in vertical Well No. A-1 to the exaggerated section seen in deviated Well No. A-2. The exaggerated section in Well No. A-2, equal to the missing section in Well No. A-1, must be corrected for wellbore deviation to estimate the size of the fault.

(b)

FAULT SIZE= (LOG THICKNESS) (COS Ɣ)
= (225') (COS 48°)
= (225') (0.669)
FAULT SIZE = 150 FEET

Ɣ = WELL BORE DEVIATION ANGLE

where

$$TVT = (MLT) (\cos \psi)$$

$$TVT = \text{True vertical thickness} \qquad\qquad (4\text{-}3)$$

$$MLT = \text{Measured log thickness in deviated well}$$

$$\psi = \text{Angle of wellbore deviation from vertical}$$

Therefore:

$$TVT = (225 \text{ ft}) (\cos 48°)$$

$$= (225 \text{ ft}) (0.669)$$

$$TVT = 151 \text{ ft}$$

The actual (corrected) size of the fault cut in Well A-1 determined by correlation with deviated Well No. A-2 is 150 ft.

Dipping Beds. The procedure for correlating deviated wells in an area of significant dip is basically the same as those presented thus far in this chapter. **The primary difference occurs in estimating the actual size of a fault cut.** Since the size of a fault cut is defined as the TVT (true vertical thickness) of the missing or repeated section, any fault size determined by correlation with a deviated well exhibiting a measured log thickness must be converted to TVT. In the last section, we defined a simple trigonometric relationship for calculating the correction factor applicable in areas with horizontal beds. When dipping beds are incorporated into the geology, the mathematical correction factor becomes somewhat more complex.

There are several equations available for calculating a correction factor to convert measured log thickness from a deviated well to true vertical thickness. We present two separate methods for computing the correction factor.

Two-Dimensional Correction Factor—Method 1: For the two-dimensional correction factor there are two correction factor equations. The first, called *Type 1*, is used where a deviated well dips in the opposite direction to the bed dip. In other words, the well is deviated in the updip direction (Fig. 4-21a). The second, called *Type 2*, is used when a deviated well dips in the same direction as the bed; the well is deviated in the downdip direction (Fig. 4-21a).

Type 1. The derivation of the Type 1 equation (well-deviated updip) is shown here and illustrated in Fig. 4-21b.

$$\cos \rho = MC/MT \quad (1a) \qquad \cos \phi_a = VE/VT \quad (2a)$$

$$MC = MT \cos \rho \quad (1b) \qquad VE = VT \cos \phi_a \quad (2b)$$

$$MC = VE$$

Therefore, equating 1b and 2b:

$$MT \cos \rho = VT \cos \phi_a$$

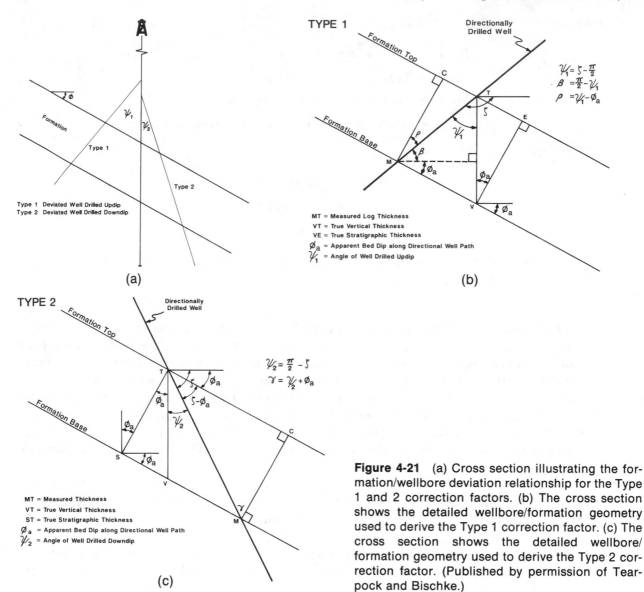

(a)

(b)

(c)

Type 1   Deviated Well Drilled Updip
Type 2   Deviated Well Drilled Downdip

TYPE 1

$\psi_1 = \varsigma - \frac{\pi}{2}$
$\beta = \frac{\pi}{2} - \psi_1$
$\rho = \psi_1 - \phi_a$

MT = Measured Log Thickness
VT = True Vertical Thickness
VE = True Stratigraphic Thickness
ϕ_a = Apparent Bed Dip along Directional Well Path
ψ_1 = Angle of Well Drilled Updip

TYPE 2

$\psi_2 = \frac{\pi}{2} - \varsigma$
$\gamma = \psi_2 + \phi_a$

MT = Measured Thickness
VT = True Vertical Thickness
ST = True Stratigraphic Thickness
ϕ_a = Apparent Bed Dip along Directional Well Path
ψ_2 = Angle of Well Drilled Downdip

Figure 4-21 (a) Cross section illustrating the formation/wellbore deviation relationship for the Type 1 and 2 correction factors. (b) The cross section shows the detailed wellbore/formation geometry used to derive the Type 1 correction factor. (c) The cross section shows the detailed wellbore/ formation geometry used to derive the Type 2 correction factor. (Published by permission of Tearpock and Bischke.)

Rearranging:

$$VT = MT \frac{\cos \rho}{\cos \phi_a}$$

Substituting $\rho = \psi_1 - \phi_a$:

$$VT = MT \frac{\cos (\psi_1 - \phi_a)}{\cos \phi_a} \tag{4-4}$$

Type 2. The derivation of Type 2 (well-deviated down dip) is shown here and illustrated in Fig. 4-21c.

$$\cos \phi_a = ST/TVT \quad (1a) \qquad \cos \gamma = MC/MLT \quad (2a)$$

$$ST = TVT \cos \phi_a \quad (1b) \qquad MC = MLT \cos \gamma \quad (2b)$$

$$MC = ST$$

Therefore, using 1b and 2b:

$$TVT \cos \phi_a = MLT \cos \gamma$$

Rearranging:

$$TVT = MLT \frac{\cos \gamma}{\cos \phi_a}$$

Substituting $\gamma = \psi_2 + \phi_a$:

$$TVT = MLT \frac{\cos (\psi_2 + \phi_a)}{\cos \phi_a} \qquad (4\text{-}5)$$

With Eqs. (4-4) or (4-5), the data required to calculate the correction factor are: (1) ψ = wellbore deviation from the vertical, (2) ϕ_a = apparent bed dip (bed dip in the direction of wellbore deviation), and (3) MLT = measured log thickness in the deviated well. The apparent bed dip is the most difficult data to obtain for these equations. The only source of apparent bed dip is from an already prepared structure map.

Three-Dimensional Correction Factor—Method 2: In this section, we present an exact three-dimensional correction factor equation (After Setchell, 1958). A version of the equation was first presented in 1958 by Setchell (Setchell, J., 1958, A Nomogram for Determining True Stratum Thickness: Shell Trinidad, EP 28884, Abstract in PA Bulletin, No. 127/128, May/June 1958, N.V. DeBataafache Petroleum Maatschappij, The Hague, Production Department, p. 8.) and has been used successfully for over 33 years. We consider this three-dimensional correction factor equation preferable because this equation can be used to calculate the correction factor regardless of the direction of wellbore deviation, and the true dip of the beds is used instead of the apparent dip, which is used in the two-dimensional equations.

To derive the general three-dimensional equation, we introduce a three-dimensional spherical coordinate system (Fig. 4-22) with one axis in the direction of dip (called the *x*-axis), one axis perpendicular to the first and horizontal (called the *y*-axis), and finally, one axis perpendicular to the other two (called the *z-axis*). The origin is the point (T) at which the wellbore first penetrates the bed.

Where

MLT = MEASURED THICKNESS

TVT = VERTICAL THICKNESS

Φ = DIP OF BED

α = Δ AZIMUTH (MIN. ANGLE, WELL AZIMUTH TO BED DIP AZIMUTH

ψ = WELLBORE DRIFT ANGLE

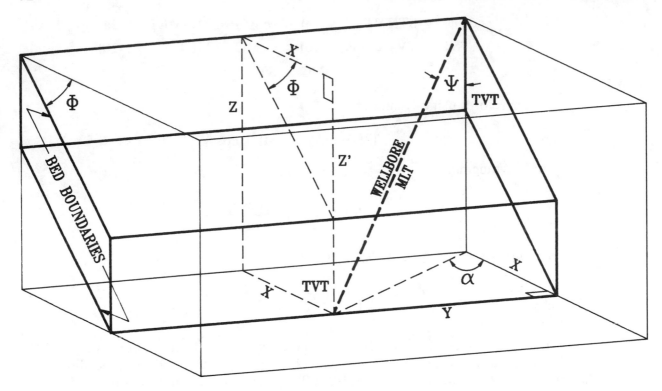

Figure 4-22 A three-dimensional spherical coordinate system with one axis in the direction of dip (x-axis), one axis perpendicular to the first and horizontal (y-axis), and finally, one axis perpendicular to the other two (z-axis). (Published by permission of Tearpock).

Spherical coordinate system equations:

$X = MLT \sin \Psi \cos \alpha$
$Z = MLT \cos \Psi$

Trigonometric definition of Z:

$Z = Z' + TVT \qquad Z' = X \tan \Phi$
$Z = X \tan \Phi + TVT$

Substituting the spherical coordinate definition for Z, X and the alternate trigonometric definition of Z easily yields equation 4-6.

$$Z = MLT \cos \Psi$$

$$X \tan \Phi + TVT = MLT \cos \Psi$$

$$TVT = MLT \cos \Psi - X \tan \Phi \qquad (4\text{-}6)$$

$$TVT = MLT \cos \Psi - MLT \sin \Psi \cos \alpha \tan \Phi$$

$$\text{or}$$

$$TVT = MLT [\cos \Psi - (\sin \Psi \cos \alpha \tan \Phi)]$$

Now apply these methods using the data shown in Fig. 4-23. Wells No. A-1, A-2, and A-3 are drilled from a single location penetrating Bed A of constant thickness dipping at 35 deg due east. Well No. A-1, a straight hole, is cut by a fault that completely faults out Bed A. Well No. A-2 is drilled in a downdip direction with an average deviation angle of 36 deg through Bed A. Well No. A-3 is drilled directly updip with a deviation angle of 35 deg through Bed A. For simplicity, the wells are all assumed to be drilled in a vertical plane parallel to the dip of Bed A.

Through detailed correlation of the three wells, a fault is identified in Well No. A-1. Based on the correlations, Bed A is completely faulted out of the well. Assuming that Wells No. A-2 and A-3 are the only wells available for correlation, the size of the fault cut in Well No. A-1 must be estimated from these deviated wells.

Considering that the size of a fault cut is defined as the true vertical thickness of the missing or repeated section, and the fault in Well No. A-1 completely faults out Bed A, the size of the fault cut in Well No. A-1 is equal to the vertical thickness of Bed A. A review of Fig. 4-23 shows that Bed A has a vertical thickness of 200 ft. When we correlate Well No. A-1 with the two deviated wells, however, we obtain a missing section in terms of deviated log thickness rather than true vertical thickness. Therefore, these deviated log thicknesses must be converted to true vertical thickness to estimate the size

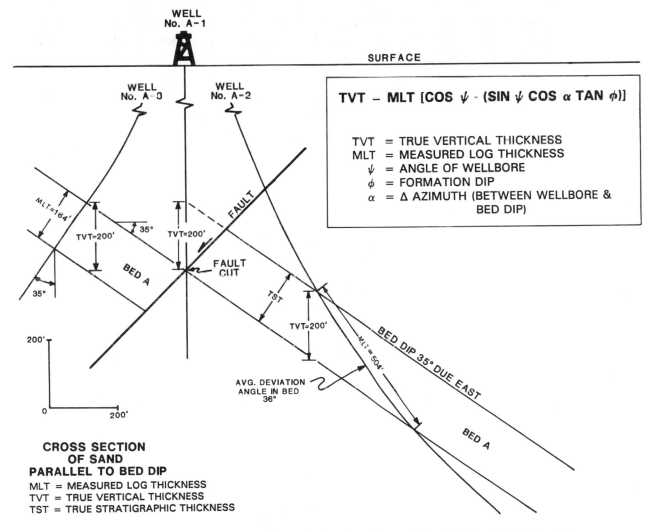

$$TVT = MLT \ [COS \ \psi - (SIN \ \psi \ COS \ \alpha \ TAN \ \phi)]$$

TVT = TRUE VERTICAL THICKNESS
MLT = MEASURED LOG THICKNESS
ψ = ANGLE OF WELLBORE
ϕ = FORMATION DIP
α = Δ AZIMUTH (BETWEEN WELLBORE & BED DIP)

CROSS SECTION OF SAND PARALLEL TO BED DIP
MLT = MEASURED LOG THICKNESS
TVT = TRUE VERTICAL THICKNESS
TST = TRUE STRATIGRAPHIC THICKNESS

Figure 4-23 One vertical and two deviated wells penetrate Bed A. In Well No. A-1, Bed A is faulted out. The cross section shows the relationship of the missing section in Well No. A-1 to the measured log thicknesses in deviated Wells No. A-2 and A-3 that are equivalent to the missing section. Equation shown is used to correct measured log thickness to vertical thickness.

of the fault. The missing section (Bed A) in Well No. A-1 has a logged thickness in Well No. A-2 of 504 ft and a logged thickness in Well No. A-3 of 164 ft. The logged thickness of 504 ft in Well No. A-2 is over two and one-half times greater than the true vertical thickness. The logged thickness of 164 ft in Well No. A-3 is less than the true vertical thickness (0.82).

By log correlation, the size of the fault cut in Well No. A-1 ranges from 164 ft, based on correlation with Well No. A-3, to 504 ft, by correlation with Well No. A-2. What is the actual size of the fault cut? In order to determine the actual size, the log thickness measured in Wells No. A-2 and A-3 must be corrected to true vertical thickness.

TVT Calculation Using Method 1

Type 1—Well Drilled Updip: In order to use Eq. (4-4) to calculate the correction factor, three pieces of data are required: (1) the wellbore deviation angle (ψ_1) which can be obtained from the directional survey of Well No. A-3, (2) the formation dip (ϕ_a) obtained from a completed structure map, and (3) the measured log thickness (MLT) in Well No. A-3 that is equivalent to the missing section in Well A-1.

Data:

$$\psi_1 = 35 \text{ deg}$$

$$\phi_a = 35 \text{ deg}$$

$$\text{MLT} = 164 \text{ ft}$$

$$\text{TVT} = \text{MLT} \frac{\cos(\psi_1 - \phi_a)}{\cos \phi_a}$$

$$= 164' \frac{\cos 0°}{\cos 35°}$$

$$= 164' \frac{1}{0.8192}$$

$$= 164' (1.2207)$$

TVT = 200 ft

Type 2—Well Drilled Downdip: In order to use Eq. (4-5), the same data used for Eq. (4-4) are required.

Data:

$$\psi_2 = 36 \text{ deg}$$

$$\phi_a = 35 \text{ deg}$$

$$\text{MLT} = 504 \text{ ft}$$

$$\text{TVT} = \text{MLT} \frac{\cos(\psi_2 + \phi_a)}{\cos \phi_a}$$

$$= 504' \frac{\cos(36° + 35°)}{\cos 35°}$$

$$= 504' \frac{\cos 71°}{\cos 35°}$$

$$= 504' \frac{0.3256}{0.8192}$$

$$= 504' \, (0.3975)$$

TVT = 200 ft

Since Wells No. A-2 and A-3 are drilled directly downdip and updip respectively, the value for bed dip (ϕ_a) in Eqs. (4-4) and (4-5) is equal to the true bed dip (ϕ). This can be obtained from a structure map or a dipmeter if one was run in the two deviated wells. If the direction of wellbore deviation is not parallel to bed dip, however, then an apparent bed dip in the direction of wellbore deviation must be determined. This apparent bed dip cannot come from a dipmeter log since this log calculates true bed dip. This is one of the main drawbacks to the two-dimensional equations.

TVT Calculation Using Method 2. To solve the general three-dimensional equation, the required data are: (1) wellbore deviation angle (ψ) obtained from a directional survey, (2) wellbore deviation azimuth (α_w) obtained from a directional survey, (3) true bed dip (ϕ) measured from a completed structure map or dipmeter if run in the wellbore, (4) bed dip azimuth (α_a) measured from a completed structure map or obtained from a dipmeter, and (5) measured log thickness (MLT) that is equivalent to the missing section.

TVT for Well No. A-2:
    Data:

$$\psi = 36 \text{ deg}$$

$$\alpha_w = 90 \text{ deg}$$

$$\phi = 35 \text{ deg}$$

$$\alpha_a = 90 \text{ deg}$$

$$\text{MLT} = 504 \text{ ft}$$

$$\alpha = 0 \text{ deg}, \Delta \text{ azimuth}$$

$$\text{TVT} = \text{MLT} \, [\cos \psi - (\sin \psi \cos \alpha \tan \phi)]$$

$$= 504' \, [\cos 36° - (\sin 36° \cos 0° \tan 35°)]$$

$$= 504' \, [0.809 - (0.5878) \, (1) \, (0.70)]$$

$$= 504' \, [0.809 - 0.412]$$

$$\text{TVT} = 504' \, [0.397]$$

$$\textbf{TVT = 200 ft}$$

TVT for Well No. A-3: Data: The data for this calculation are exactly the same as for

Well No. A-2, with two exceptions. The azimuth (α_w) for Well No. A-3 is due west or 270 deg, therefore the Δ azimuth (α) is 180 deg, and ψ is 35 deg.

$$\text{TVT} = \text{MLT} [\cos \psi - (\sin \psi \cos \alpha \tan \phi)]$$

$$= 164' [\cos 35° - (\sin 35° \cos 180° \tan 35°)]$$

$$= 164' [0.8192 - (0.5736)(-1)(0.7002)]$$

$$= 164' [0.8192 - (-0.4016)]$$

$$= 164' [0.8192 + 0.4016]$$

$$\text{TVT} = 164' [1.2208]$$

$$\textbf{TVT} = \textbf{200 ft}$$

Through the use of the two-dimensional and three-dimensional equations, we have successfully calculated the actual size of the fault cut in Well No. A-1. Notice the close agreement between the different equations. The fault cut estimated at 200 ft can now be used for all future fault and structure map integration. The actual procedure for integrating a fault and structure map is detailed in Chapter 8. We have now defined two methods for obtaining the correct size of a fault cut when correlating deviated wells. *Anytime you are correlating in an area with deviated wells and dipping beds, fault sizes must be corrected to true vertical thickness since the missing or repeated section is defined in terms of vertical thickness.* Significant errors in structure mapping can occur if these corrections are not made.

MLT, TVDT, TVT, and TST

The thickness of any given interval on a log is referred to as the *measured log thickness (MLT).* In a vertical well, the MLT for any given interval is equal to the true vertical thickness (TVT) of the interval. We know from the previous discussion in this chapter, however, that the MLT in a directionally drilled well is normally not equal to TVT due to the wellbore deviation and bed dip, in areas of dipping beds.

True vertical depth thickness (TVDT) is defined as the MLT in a deviated well between two specific depth points corrected only for wellbore deviation. *The true vertical thickness (TVT)* is defined as the thickness of an interval measured in the vertical direction. It is the thickness seen in a vertical well. For a directionally drilled well, the TVT can be calculated using the equations introduced in the previous section. *The true stratigraphic thickness (TST)* is defined as the thickness of a given interval measured at a right angle to the bedding surface in a vertical cross section. It can be calculated by multiplying the TVT by the cosine of bed dip.

These various thicknesses are graphically illustrated in Fig. 4-24. In Fig. 4-24a, a well deviated updip at an angle of 50 deg penetrates a sand dipping 35 deg due West. The measured log thickness of the sand is 127 ft. To correct the MLT to TVDT, the

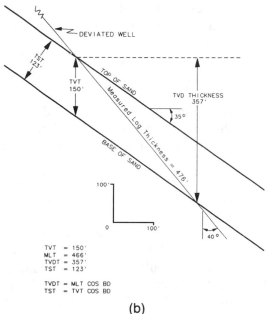

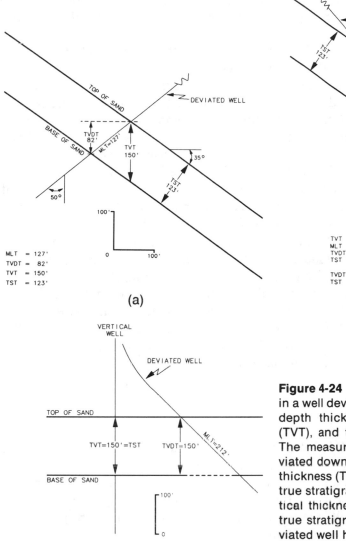

Figure 4-24 (a) The measured log thickness (MLT) in a well deviated updip is compared to true vertical depth thickness (TVDT), true vertical thickness (TVT), and true stratigraphic thickness (TST). (b) The measured log thickness (MLT) in a well deviated downdip is compared to true vertical depth thickness (TVDT), true vertical thickness (TVT), and true stratigraphic thickness (TST). (c) The true vertical thickness, true vertical depth thickness, and true stratigraphic thickness calculated from a deviated well have the same value when the beds are horizontal.

MLT (127 ft) is multiplied by the cosine of the wellbore deviation angle (50 deg). The resultant TVDT is 82 ft.

Figure 4-24b shows the same sand bed penetrated by a well drilled downdip at an angle of 40 deg. The MLT of the sand in the well is 476 ft. Corrected to TVDT, it is 357 ft. The correction factor equations developed in the previous section are used to calculate a TVT of 150 ft for the penetrated sand bed in Fig. 4-24. The TST of the sand calculated by multiplying the TVT (150 ft) by the cosine of the bed dip (35 deg) is 123 ft.

Notice that for the well drilled downdip, the MLT is 3.17 times greater than the TVT, and the TVDT is 2.38 times greater than the TVT. For the well drilled updip, the MLT is less than the TVT, and the TVDT is about one-half the TVT.

The understanding of these various measurements is very important in log correlation and fault cut size estimation. Very often, when a well is directionally drilled, a TVD log is automatically prepared as part of the logging program. The TVD log is then used to aid in correlation, determine the size of fault cuts, and count net sand and pay. Figure 4-24 illustrates that in areas of significant dip, a TVD log may provide little, if any, advantage over the deviated well log for correlation and can actually complicate the size estimation of a fault cut. Observe in Fig. 4-24b that the true vertical depth thickness, which is the thickness seen in a TVD log, is still 2.38 times greater than the true vertical thickness.

Consider a fault in this section which faults out the entire sand. The size of the fault is 150 ft. By correlation with the well deviated downdip, the size of the fault is estimated to be 476 ft; with a TVD log, the estimated size of the fault would be 357 ft. We can conclude that the TVD log did very little to improve the correlation process to estimate the size of the fault. If the data are available, we recommend that a true vertical thickness (TVT) log be prepared. This log can be used to aid in correlation and to estimate the size of a fault cut, since it represents the TVT of the interval logged.

The preparation of a TVD log for the well deviated updip in Fig. 4-24a could actually result in additional correlation problems. Notice that the measured thickness of the sand in the deviated well is 127 ft. Converting this measured log thickness to a TVD log actually reduces the thickness of the interval to 82 ft. This reduced thickness could be mistakenly interpreted as stratigraphic thinning. If the TVD log were used to estimate the size of a fault, it would result in an underestimate. For example, if we again use a fault in this section with a size of 150 ft, by correlation with a TVD log for the well drilled updip, the size of the fault would be estimated at 82 ft, or nearly one-half the actual size.

In areas of horizontal or nearly horizontal beds, the TVDT is equal to or nearly equal to the TVT (Fig. 4-24c) and can be of significant help in correlation and estimating the size of fault cuts. *In areas of dipping beds, however, a TVD log may provide little help and can actually cause additional correlation problems.*

REPEATED SECTION

A repeated section in a well is defined as part of the stratigraphic section appearing twice on a log as the result of a fault, consequently lengthening the section. A repeated section is commonly thought of as a compressional tectonic phenomenon occurring as the result of a reverse fault pushing the stratigraphic section in the hanging wall up and over the same section in the footwall. Figure 4-25 illustrates the geometry required to result in a repeated section due to a reverse fault.

A repeated section can also occur with a normal fault. In this situation, a repeated section requires a specific geometry between a normal fault and a directionally drilled well, or a normal fault and a formation dipping at a steeper angle than the fault. So far, we have looked at vertical and directionally drilled wells penetrating a fault in what is called a **normal sense**; that is, from the downthrown fault block to the upthrown fault block (Fig. 4-26).

Figure 4-27 is a cross section illustrating the geologic and deviated well parameters required to cause a repeated section in a log for a situation involving a normal fault. Geometries as shown in Fig. 4-27 are not uncommon in areas where directionally drilled wells are common. In offshore areas, deviated wells are drilled from a central platform location and are often designed to parallel a known fault (Fig. 4-28). In this way, a well may be drilled to penetrate a series of potential sands upthrown to an important trapping fault. If care is not taken in precisely mapping the fault and drilling the well, it is possible for the well to cross the fault and go downthrown to the fault. In Fig. 4-28, the well deviated toward the salt crosses a normal fault backwards, resulting in a repeated section. Also notice that by crossing the fault, the well does not penetrate productive "Sand B" which is upthrown to the fault, but instead penetrates the nonproductive B Sand downthrown.

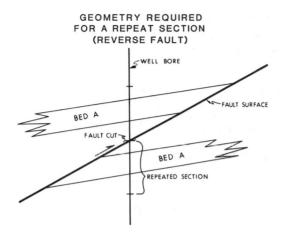

Figure 4-25 The geometry required for a repeated section due to a reverse fault.

Figure 4-29 is an example of a repeated section from a South Louisiana oil and gas field. The amount of wellbore deviation is shown to the right of the log section. Notice that the deviation angle ranges from 40°45′ to 42°15′. The well was drilled to parallel a fault, while staying in the upthrown fault block. Since wellbore deviation is measured from the vertical and fault dip from the horizontal, it is necessary to subtract the hole angle from 90 deg to compare its dip angle to that of the fault. By doing this, the well angle measured from the horizontal, for the section shown, ranges from 47°45′ to 49°15′. Based on the fault surface map, the fault dip is approximately 52 deg. Therefore, the well dip is less than the fault dip, establishing a well/fault geometry similar to that shown in Fig. 4-27. The result of this geometry is the repeated section shown in Fig. 4-29.

Figure 4-30 is another excellent example of a repeated section from East Cameron Block 272 resulting from a normal fault, cut backwards, by a deviated well. Observe that the upthrown fault block is productive of hydrocarbons while the downthrown block is wet.

The identification of a repeated section that has previously gone unrecognized can

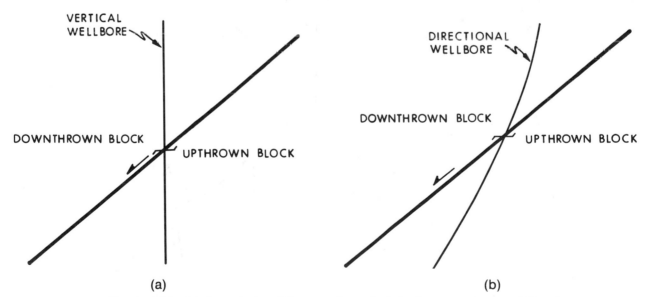

Figure 4-26 (a) A vertical well intersecting a fault in the normal sense (from downthrown to upthrown fault blocks). (b) A deviated well intersecting a fault in the normal sense (from downthrown to upthrown fault blocks).

GEOMETRY REQUIRED
FOR A REPEAT SECTION
(NORMAL FAULT)

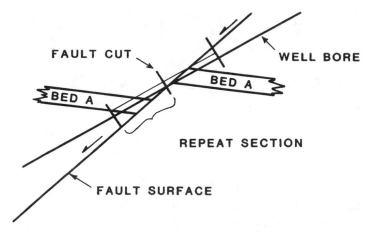

Figure 4-27 The fault and deviated well parameters required for a repeated section resulting from a normal fault.

result in excellent prospect potential. Look at the generic situation shown in Fig. 4-31. In this case, a well was designed to stay upthrown to an important trapping fault. Good control on the fault was not available at the time of drilling and a fault map was not prepared. The well meandered across the fault, resulting in an unrecognized repeated section on the well log. The stratigraphic intervals penetrated upthrown to the fault are productive of hydrocarbons, while the intervals downthrown "A to B" are wet. The repeated section went unrecognized and consequently the intervals penetrated downthrown and wet were condemned as nonproductive.

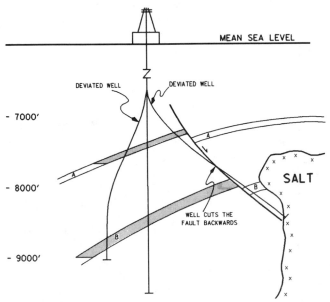

Figure 4-28 The well deviated toward the salt cuts the normal fault backwards (from the upthrown block to the downthrown block) resulting in a repeated section.

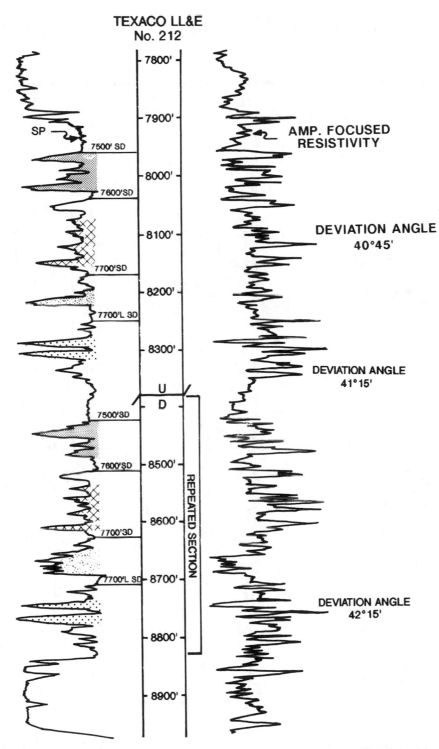

Figure 4-29 Repeated section in the Texaco LL&E Well No. 212. (Published by permission of Texaco, USA.)

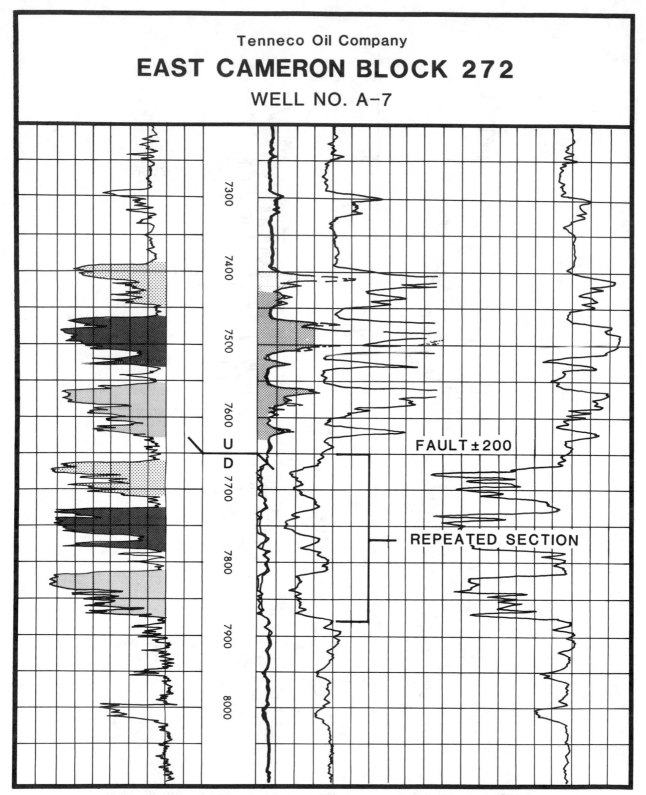

Figure 4-30   Repeated section in the Tenneco Well No. A-7 in East Cameron Block 272, Offshore US Gulf of Mexico. Notice that the upthrown fault block contains hydrocarbons, while the downthrown block is wet. (From Tearpock and Harris 1987. Published by permission of Tenneco Oil Company.)

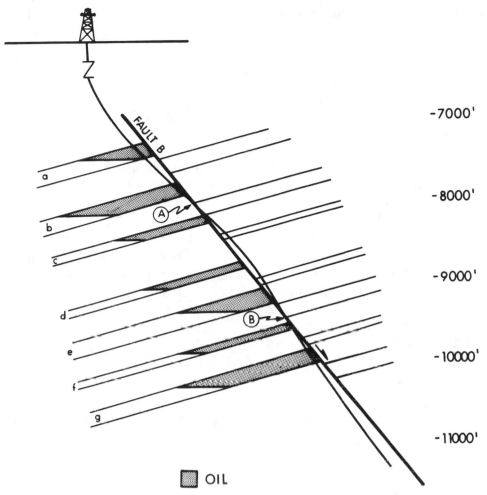

Figure 4-31   The identification of a previously unrecognized repeated section can result in a potential prospect. In this example, Sands C through E are untested upthrown to Fault B.

If an observant geologist recognizes the repeated section and understands that the nonproductive log section represents the well penetration in the downthrown block, a new prospect can be generated in an area previously condemned. In Fig. 4–31, the new prospect includes the series of sands, C through E, that are upthrown to the fault and not seen in the deviated well.

ESTIMATING RESTORED TOPS

At times, a formation that is being mapped may be faulted out of one or more wells in a field. If this situation exists, it is often possible to estimate an upthrown and downthrown restored top for the missing formation in the faulted well(s). *A restored top is defined as an estimated top for a specific marker or formation that is faulted out in a well.* In other words, an upthrown restored top is the estimated top the formation would have had in the upthrown block if the formation had not been faulted out. Likewise, the downthrown restored top is the estimated top the formation would have had in the downthrown fault block had it not been faulted out. These data points should be honored in structure mapping just as any other formation top obtained from well log data. This is important for maintaining the integrity of the structural interpretation. Also, the honoring of restored tops provides additional information which aids in the actual structure contouring.

Vertical Wells

The procedure for determining the upthrown and downthrown restored tops (sometimes referred to as equivalent points) in vertical wells is discussed here and illustrated in Fig. 4-32.

1. Figure 4-32 shows three vertical wells positioned side by side with no horizontal scale. By correlation, we determine that there is a 150-ft fault in Well No. 2 and that the objective formation (C Sand) is faulted out. Well No. 1 is in the upthrown fault block and Well No. 3 is in the downthrown block. To determine the upthrown and downthrown restored tops, these two nearby wells are used.

2. *Downthrown Restored Top for the C Sand.* To estimate the downthrown restored top, Well No. 3, which is in the downthrown fault block, is used to correlate with Well No. 2. Identify a good marker, such as a fieldwide sand or resistivity marker, *above* the faulted out section *in both logs*. In Fig. 4-32, the "B" Sand is used as the marker in the wells.

   Line up the marker on both logs and correlate **down** the logs until the C Sand Top is reached in Well No. 3. The equivalent measured depth reading for the C

ESTIMATING "RESTORED" DATA FOR VERTICAL WELLS HAVING MAP DATUM FAULTED OUT

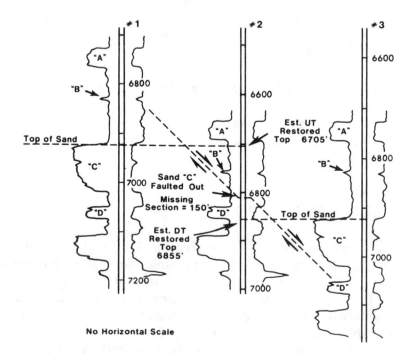

Figure 4-32 The method for estimating restored tops in vertical wells is shown by special correlation of the three vertical wells.

Sand in Well No. 2 is 6855 ft. This represents the estimated depth of the restored C Sand Top in the downthrown block.

3. *Upthrown Restored Top for the C Sand.* To estimate the upthrown restored top, Well No. 1, which is in the upthrown block, is used to correlate with Well No. 2. Identify a good marker in **both wells below** the faulted out section. In this case it is the D Sand top. Line up the D Sand marker on both logs and correlate **up** the logs until the C Sand top is reached in Well No. 1. This equivalent depth reading for the C Sand in Well No. 2 is 6705 ft. This represents the estimated depth of the restored C Sand top in the upthrown block.

The accuracy of the restored top estimates can be checked by subtracting the downthrown restored top from the upthrown restored top and then comparing the difference to the estimated size of the fault. In this case: 6855 ft − 6705 ft = 150 ft. The difference in the restored tops and the size of the fault cut are exactly the same. Therefore, we conclude that the estimated depths for the restored tops are reasonable.

These two new data points should be honored in the structure contouring of the C Sand. If the map is referenced to sea level, these measured log depths must be converted to subsea depths for structure mapping. The specific mapping technique to apply the restored tops in structure mapping is discussed in Chapter 8.

Deviated Wells

In the example shown in Fig. 4-32, all the wells are vertical with no deviation. In many areas where there are numerous deviated wells, the estimation of restored tops requires additional steps.

There are **three separate cases** for log correlation with deviated wells: (1) a fault in a deviated well correlated with a vertical well; (2) a fault in a vertical well correlated with a deviated well; and (3) a fault in a deviated well correlated with another deviated well. In an earlier section, we showed that the size of a fault cut is defined as the vertical thickness of the missing section. Likewise, when estimating restored tops in deviated wells, vertical well log thickness must be used.

Probably the most accurate way to estimate restored tops in deviated wells is to convert the deviated logs to true vertical thickness logs. In order to make a TVT log, bed dip and azimuth data are required in addition to the well deviation angle and well azimuth. If time, costs, or some other factor prohibit the preparation of TVT logs, the restored tops can still be estimated with reasonable accuracy using the deviated well logs.

For case No. 1, in which a fault in a deviated well is correlated with a vertical well, the same basic procedure for estimating the restored top discussed earlier for two vertical wells can be applied as shown in Fig. 4-33 with one variation. Since a fault cut in a well log can be assumed to have zero thickness, it represents one depth for which its subsea equivalent can be calculated. Using a vertical well in the upthrown block, the vertical thickness, in feet, from the fault to the top of the formation in question can be determined. By subtracting this vertical thickness from the subsea depth estimate for the fault cut, the upthrown restored top can be obtained.

We will follow through the procedure for the example shown in Fig. 4-33. A fault is identified in Well No. 3. The size and depth of the fault is 100 ft at −6810 ft. It is recognized that the B Sand is faulted out in the well. We wish to obtain the upthrown and downthrown restored tops for the B Sand. Well No. 1 in the downthrown block and Well No. 2 in the upthrown block are available to estimate the restored tops (Fig. 4-33).

1. *Upthrown Restored Top for the B Sand.* To estimate the upthrown restored top, identify a marker below the B Sand in Well No. 2 that is also in the faulted well. For this example, the lower marker shown in Well No. 2 is right at the fault in Well No. 3. By correlating up the logs, we determine that the B Sand is 80 ft above the lower marker in Well No. 3. Since the marker is right at the fault in Well No. 3, the upthrown restored top for the B Sand in the well is 80 ft above the subsea depth of the fault. The subsea depth of the fault is −6810 ft; therefore, the upthrown restored top is −6730 ft.

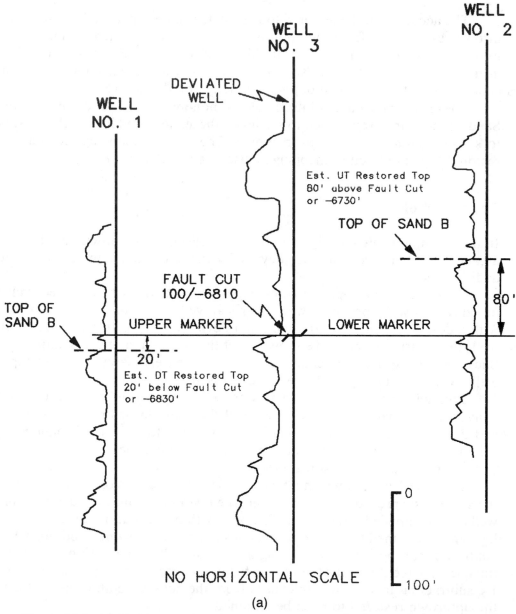

(a)

Figure 4-33 (a) The two vertical wells (No. 1 and 2) and the deviated well (No. 3) illustrate one of the three cases for estimating restored tops in deviated wells. (b) The structural cross section containing the three wells shown in Fig. 4-33a illustrates the method for estimating the restored tops in deviated Well No. 3 and the precise position of those restored tops with respect to the well; directly above (upthrown restored top) and directly below (downthrown restored top) the actual fault cut.

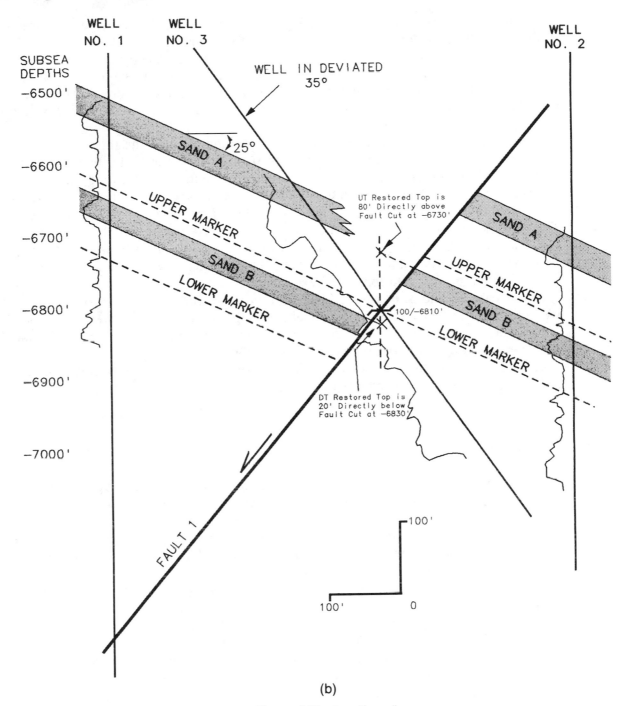

(b)

Figure 4-33 (*continued*)

2. *Downthrown Restored Top for the B Sand.* The same procedure is followed to estimate the downthrown restored top. Using Well No. 1 in the downthrown fault block, identify a marker **above,** but as close to the fault cut as possible. In this case, the upper marker in Well No. 1 is right at the fault cut in Well No. 3. Correlate down the two logs from the upper marker until the top of the B Sand is reached in Well No. 1. This top is 20 ft below the fault in Well No. 3 or at −6830 ft.

The depths for the restored tops were not read directly from faulted Well No. 3, as was done in the vertical well example (Fig. 4-32). Instead, for the upthrown restored top, the vertical distance from the fault to the top of the sand in unfaulted Well No. 2 was **subtracted** from the subsea depth of the fault cut to estimate the upthrown restored top, and likewise, the vertical distance from the fault to the top of the sand in unfaulted Well No. 1 was **added** to the subsea depth of the fault cut to estimate the downthrown restored top. This procedure is required because Well No. 3 is a deviated well with an exaggerated log section.

In Fig. 4-33b, which is a structural cross section through the three wells, an "**X**" is placed 80 ft vertically above and 20 ft vertically below the fault cut in Well No. 3. These X's show the precise position of the upthrown and downthrown restored tops for the B Sand. It is important to understand that the restored tops are located **vertically above and below** the location of the fault cut and not along the path of the deviated well. This understanding is critical when using these restored tops in structure mapping.

The 100-ft difference between the downthrown and upthrown restored tops agrees with the 100 ft estimated for the size of the fault. Therefore, we conclude that the depths for the estimated restored tops are reasonable.

For cases 2 and 3, in which a fault in a vertical well is correlated with a deviated well or a fault in a deviated well is correlated with another deviated well, the procedure is basically the same as for case No. 1 with one exception. For the deviated well, the true vertical thickness from the markers to the faulted formation must be calculated to accurately estimate the restored tops. The measured (deviated) log thickness cannot be used to estimate the restored tops.

UNCONFORMITIES

Unconformities are present in all geologic settings, especially on steeply dipping structures such as salt domes. Unconformities can serve as excellent hydrocarbon traps. Therefore, it is important to recognize unconformities in the subsurface.

There are a host of interrelated versions of the broad term unconformity. Some are primarily erosional, some depositional, and others combinations of both. The subject of unconformities is extensive and beyond the scope of this book. Our coverage of this topic is limited, but it does present important information on the recognition of unconformities during electric log correlation.

An unconformity appears on an electric log as missing section. Since missing section, as a result of an unconformity, can be mistaken for missing section due to a normal fault, care must be taken during correlation so that an unconformity is not mistaken as a fault. In this section we discuss several general guidelines to follow during correlation to recognize an unconformity (Fig. 4-34).

1. Structural dip often is different above and below an unconformity. Dipmeter data can be used to indicate this change in dip. The dip below an unconformity is usually steeper (Fig. 4-34).
2. Missing section as the result of an unconformity can be mistaken for a fault. If missing section is recognized in two or more wells at the same or nearly the same correlative depth, an unconformity should be suspected (Fig. 4-34a).
3. The amount of missing section resulting from an unconformity increases in the up-structure direction. This is illustrated in Fig. 4-34. The missing section in Wells No.

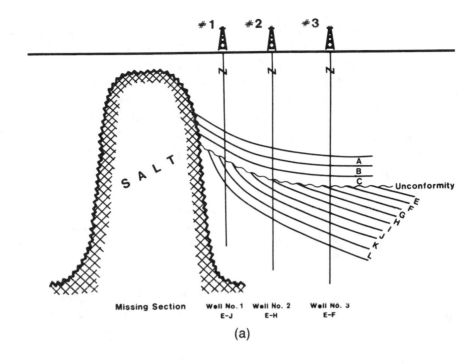

(a)

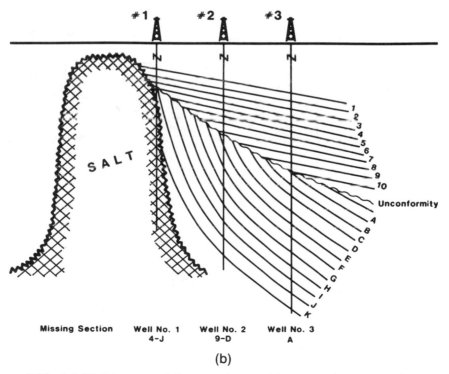

(b)

Figure 4-34 (a) Well log correlation can be used to recognize an angular unconformity. (b) If an onlap sedimentary sequence is deposited above an angular unconformity, certain log correlation guidelines can be used to recognize the unconformity.

1, 2, and 3 increases in the up-structure direction; Well No. 3 has the least amount of section missing and Well No. 1 has the greatest amount missing.

4. The stratigraphic section below an unconformity is truncated in a younger sequence in down-structure wells than in up-structure wells. The sequence gets older in the up-structure direction. The sedimentary sequence just below the unconformity (Fig. 4-34a) in Well No. 3 (the G interval) is younger than the K interval in Well No. 1. To recognize the sequence trend, you must correlate **up** the well logs.

5. If the depositional environment results in an onlap sedimentary sequence (Fig. 4-34b) rather than the sequence shown in Fig. 4-34a, the missing section in logs correlated in the up-structure position increases above and below the unconformity. The sedimentary sequence just above the unconformity is younger in the up-structure direction.

6. Unconformities must be mapped. Since an unconformity can serve as an excellent hydrocarbon trap, a map on the unconformity is vital. In order to identify the intersection of the unconformity and sedimentary sequence below, an unconformity map must be integrated with a structure map. The mapping techniques required are discussed in Chapter 8.

Figure 4-35 illustrates the use of a dipmeter to aid in the identification of an angular unconformity. The figure also shows a dipmeter response to beds affected by a salt dome

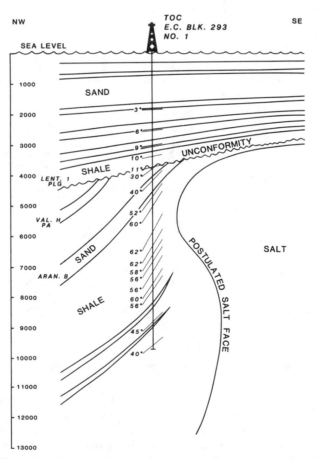

Figure 4-35 Dipmeter data can be used to recognize an unconformity in the subsurface. (Published by permission of Tenneco Oil Company.)

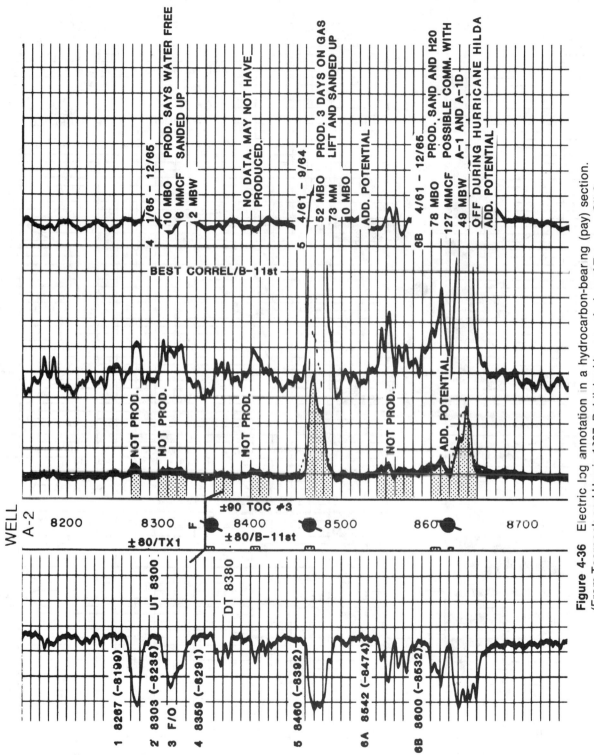

Figure 4-36 Electric log annotation in a hydrocarbon-bearing (pay) section. (From Tearpock and Harris 1987. Published by permission of Tenneco Oil Company.)

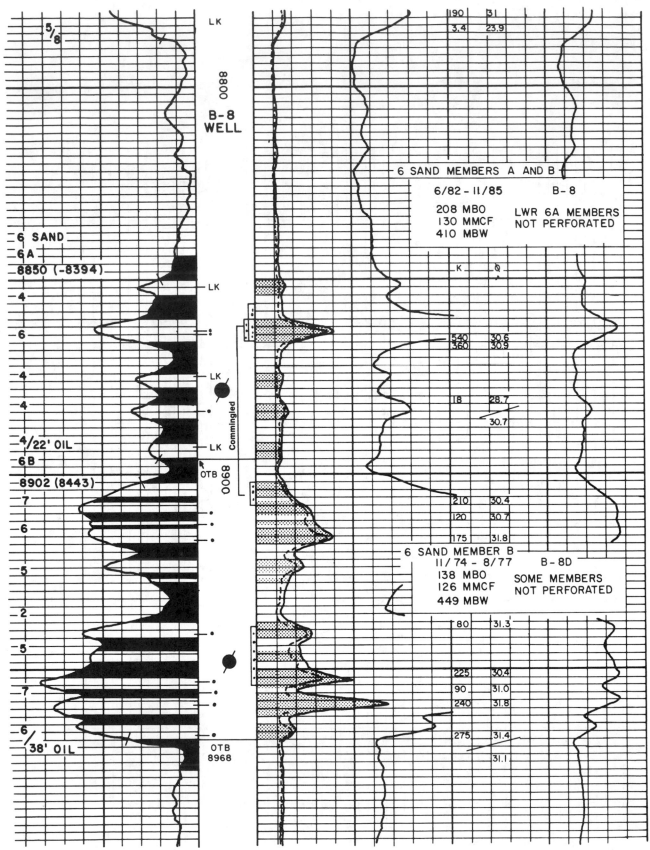

Figure 4-37 Annotation of a detailed "5" log. (Modified from Tearpock and Harris 1987. Published by permission of Tenneco Oil Company.)

with an overhang. The dipmeter reaches a maximum dip of 62 deg at about −6500 ft; thereafter, the dip slightly flattens to 40 deg at TD. This change in dip is in response to the proximity to the salt face. Such dipmeter responses can be used in evaluating a salt dome for possible overhang.

ANNOTATION AND DOCUMENTATION

The importance of good, accurate documentation cannot be overemphasized. The generation of good quality maps depends upon a volume of accurate data. These data include: fault cuts, formation tops and bases, net sand, net pay, and so on. In this section, we illustrate a recommended method of annotating 1- and 5-inch electric logs and documenting the log data.

In Fig. 4-2 we illustrated the importance of marking logs with recognizable symbols and the use of color. These markings are a form of annotation. **They identify your correlations.** The logs in Fig. 4-2 are of intervals that do not contain any hydrocarbon-bearing sands. There are some additional data that should be annotated on a 1- or 5-in. log with recognized pay. Figure 4-36 illustrates the additional data that should be annotated in the pay section on a 1-in. log. The annotation includes the name of each sand, perforation intervals, well status, cumulative production from each interval produced and a note on why each interval went off production, and the measured and subsea depth for the top of each important productive interval. Any intervals that appear productive on the log but have not been produced should be noted, such as Sand 2 and 6A in Fig. 4-36. Finally, any recognized faults must be indicated on the log, such as Fault F at a depth of 8350 ft which faults out the 4 Sand.

The annotation of a detailed 5-in. log is crucial. The annotation of a 5-in. log is shown in Fig. 4-37. The information annotated includes: sand name, measured and subsea depth of the sand tops, perforation intervals and corresponding production, well status, net pay counts, limit of pay (full to base of interval or water contact), and basic core data (at least the porosity and permeability data). Notice that the net pay on the 5-in. log in Fig. 4-37 is assigned per 10-ft intervals on the left side of the log. This annotation is used to support the net pay count and later in preparation of net pay isopach maps.

Finally, the documentation of well data means recording it in some format that can be easily used. There are various types of data sheets available for documenting mapping parameters. These data sheets should be used at all times to document the log correlation data.

CHAPTER 5

INTEGRATION OF GEOPHYSICAL DATA IN SUBSURFACE MAPPING

prepared by Clay Harmon

INTRODUCTION AND PHILOSOPHY

This chapter will discuss the use of geophysical data in subsurface mapping. More specifically, the discussion will center around the use of reflection seismic data both to aid in the visualization of the subsurface geology and to extract useful data to use in the creation of accurate maps. The first section is a very general discussion of the benefits and limitations of using seismic data to construct subsurface maps. The second section is a more detailed discussion of some of the techniques and procedures for integrating seismic data into subsurface geological maps.

Discussion of Objectivity

The discussion in this chapter is intended to benefit the individual who may not be familiar with seismic data, and indeed, may not understand how seismic data are acquired and processed. The orientation of this chapter focuses on practical approaches to using seismic data in the search for hydrocarbon traps. The technical details of seismic acquisition and processing are beyond the scope of this book. These are very important subjects that a working geophysicist must understand. Many mappers who have access to seismic data are not geophysicists. However, it is our intent to illustrate techniques that will make the nongeophysicist comfortable with using these data in the construction of subsurface maps.

This chapter should make it obvious that valuable information is present in seismic data, and that an interpretation that properly integrates both the subsurface geologic data and the seismic data is always more accurate than an interpretation that ignores one of these sources. It will soon become apparent that the discussion has a strong regional orientation. Most of the examples are from the offshore Gulf of Mexico. There are several reasons for this regional bias. Perhaps the most obvious reason is that this region has what is probably a greater abundance of high-quality seismic data than anywhere else in the world. This fact means that (1) it is easier to get good examples from this region than most others, and (2) this region is highly prospective for hydrocarbon prospecting, which increases the likelihood that you may work this region at some point in your career.

This regional orientation does not mean that the techniques outlined are limited in their applicability to the Gulf of Mexico region. In fact, the techniques presented here can be used to establish the three-dimensional geometric validity for subsurface maps in any tectonic environment.

Seismic Data Applied to Subsurface Interpretations

On a fundamental level, seismic data can provide two major benefits. First, seismic data can be acquired over areas that do not have any well data. The map and interpretation can thus be extended with some confidence into areas that have little or no well control. This is an important benefit, especially when one considers that very few wildcat prospects actually have wells already drilled in the immediate area of the prospect. The second

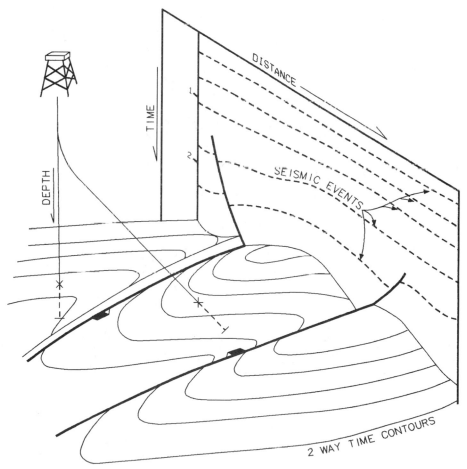

Figure 5-1 Sampling differences between wells and surface seismic. Well data are single points, and correlations of points between wells must be inferred. Seismic events, however, explicitly demonstrate horizon continuity.

benefit is that seismic data can provide explicitly two-dimensional data as opposed to the one-dimensional nature of a wellbore. The two-dimensional character of seismic data as opposed to well data is illustrated in Fig. 5-1. It should be added that in most cases, the two-dimensional appearance of a seismic section is an artifact of the data being reduced to a flat sheet of paper. The data on the line may actually represent a very complex three-dimensional subsurface geologic world!

Where a geologic structure is complex in three dimensions, the most insidious and potentially dangerous pitfall is assuming that all the data on a section represent a planar slice through the earth directly underneath the line. In complex areas, the data on a line may not represent the geologic structure directly underneath the line, as illustrated in Fig. 5-2. Methods for handling some of these effects, called *side-swipe* (Sheriff 1973) are discussed later in the chapter. However, the two-dimensional nature of seismic data often does mean that one less dimension must be inferred in order to construct an accurate subsurface structural interpretation.

Assumptions and Limitations

The techniques outlined in this chapter assume that the data used are properly acquired and processed appropriately, up to and including migration of the data. The techniques also assume that the geology of the subsurface under the line permits the acquisition of

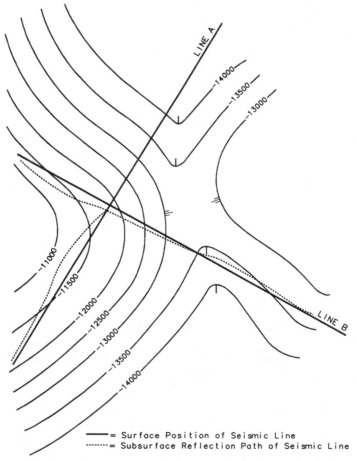

Figure 5-2 Effect of subsurface structure on actual subsurface reflection path of line. Subsurface reflection points do not occur vertically beneath the surface location of the lines when reflected from dipping subsurface horizons.

good quality data. Of course, there are many areas where the subsurface does not co-operate and does not yield good seismic data. Some possible major problems include severe horizontal velocity gradients, high noise areas, or an extremely complex geologic subsurface, all of which may invalidate many assumptions necessary for the acquisition of good data. If these problems are present, they may present a challenge for even the best geophysicist. In such instances, the expertise of a geophysicist may be absolutely necessary to solve the complexities. Before totally discarding the use of geophysics in these areas, you should recognize that if the geology is complex enough to cause these severe problems, it is not going to be a trivial exercise to find the solution on well data alone. Even in these complex areas, seismic data may contain valuable information that can be used in creating a reasonable subsurface interpretation. Examples of the usefulness of seismic data in these complex areas are presented in the section that covers structural balancing.

THE PROCESS

The procedure for making subsurface maps using seismic data is very similar to the sequence of steps used in constructing interpretations from subsurface well log information. The first step is one of data validation, i.e., analysis of what the seismic data represent. Do the seismic data actually have some relationship to the geology in the subsurface? This procedure is similar to the checking one does when a log is first used. In the case of log data, decisions about the validity and meaning of the log response must be made before the data can be used to form an interpretation.

The second step is the actual interpretation of the seismic section. This step is analogous to the correlation of well logs when using just subsurface well log information. Because the validity of the remaining work rests on having an accurate and geologically correct interpretation of the seismic data, this is the most important part of the process. Some aspects of seismic interpretation as they relate to the construction of subsurface maps are covered in this chapter. However, we do not attempt to cover the subject exhaustively. Several excellent books on seismic interpretation are listed at the end of the book and are recommended to those who may be unfamiliar with seismic interpretation techniques. Just as a basic knowledge of well logs is needed to use log data properly, a basic understanding of the reflection seismic method is needed before seismic data can be interpreted correctly.

The third step involves extracting the information from the seismic data and transferring it onto the map so that it can be used effectively. Usually, transferring the data to a map is referred to as *posting*. This procedure is practically identical to that used when recording subsurface well log data. Since seismic sections have a two-dimensional aspect that well log data do not possess, there are some unique aspects to this step when using seismic data. This step also represents the merger of the subsurface and the seismic information. Both types of data should be posted and used to construct the final interpretation. If you don't understand the seismic data and require assistance, experienced interpreters are usually available. There is typically some usable information on even the worst seismic data that can add to the confidence and validity of your final subsurface map. *A valid interpretation should agree with and satisfy all types of information.* If the 3-D data are available, all of it should be used.

An appeal is made to geophysicists as well as geologists. Very often, there are two sets of maps: the *geological map* and the *geophysical map*. A subsurface map should accurately represent the geology and both well logs and seismic data represent physical

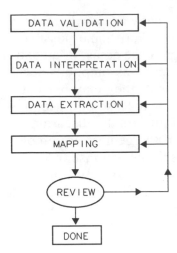

Figure 5-3 Flow diagram of subsurface mapping procedures.

measurements of the subsurface. There is only one configuration of the subsurface, and it is the job of the mapper to create an integrated and reconciled interpretation using all the data available.

The last step is the construction of the subsurface geological maps. This step represents the culmination of all previous work, and in many instances will be the result by which your work is measured. It is only as good as the information it contains, so do not rush to begin this step before the previous steps are completely finished.

In practice, though, this is never the last step. Several iterations of validation, interpretation, posting, and mapping are typically necessary before a satisfactory subsurface map is completed. Figure 5-3 is a conceptual flow diagram of this process. At some point it will become apparent that most of the major questions have been resolved and a satisfactory map has been made. While pride and satisfaction in the result is deserved, always keep in mind that additional data from either drilling or additional seismic shooting will almost always change some of your ideas. These changes may be small, but subsurface mapping can never perfectly represent the structural configuration of the earth. The more your ideas are actually tested, the more obvious it becomes that mapping is both an art and a science. Ideally, you will asymptotically approach the truth as more data become available. The measure of an interpreter is his or her ability to approach the truth quickly with the limited data available.

DATA VALIDATION AND INTERPRETATION

Examining the Seismic Section

The first step toward obtaining the information you need from seismic data is to unfold the line and start the process of deciding what it means. The vast majority of seismic data that are used for subsurface mapping are seismic time sections. Figure 5-4 is a seismic time section over a simply deformed area of the Gulf of Mexico. It is very tempting to think, "This is easy, all of those dark lines are the rock layers, and there is little difficulty picking the fault that dips toward the left part of the section. Since the fault trace on the line is concave upward, this must be one of those common listric faults that everyone writes about."

Without realizing it, you have made assumptions about the data and the geology that may or may not be justified. In many cases, these assumptions are close enough to

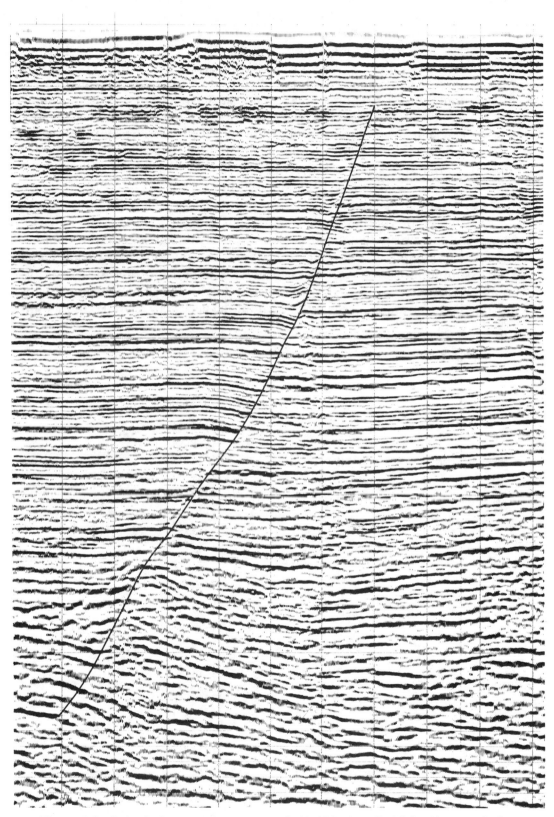

Figure 5-4 Seismic line over faulted area, Gulf of Mexico. (Published by permission of TGS/GECO.)

the truth that it really doesn't matter. In other situations these assumptions, while not completely without merit, may bias your interpretation in a way that may lead you completely down the wrong path.

The first *incorrect assumption* is that each of the reflections represents a discrete layer of rock. A reflection seen on a section may or may not represent a discrete sedimentary boundary. The vertical complexity of the sedimentary sequence and the frequency content of the recorded and processed seismic signal determines the appearance of the seismic wiggles. Figure 5-5 is a *synthetic seismogram* illustrating the relative *"size"* of seismic wiggle in an average velocity Tertiary section such as that in the Gulf of Mexico, in comparison to a well log. It is obvious that the *vertical resolution* of a well log is vastly superior to that of a seismic trace. The seismic wiggle trace is a composite of waveforms from reflections from many boundaries in the subsurface. Figure 5-6 illustrates how a series of interfaces can combine their reflections to produce a simple seismic reflection.

At this point you may be overwhelmed by the potential complexity of the seismic waveform. Let us say that in most cases, it is safe to assume that the *individual reflections represent mappable, isochronous, sedimentary units* (Mitchum and Vail, in Payton 1977). This assumption usually will not sacrifice the integrity of the final map in the least. In areas where there is no radical thinning or thickening of the sedimentary section, it is reasonable to assume that the reflections, at the very least, parallel the sedimentary units.

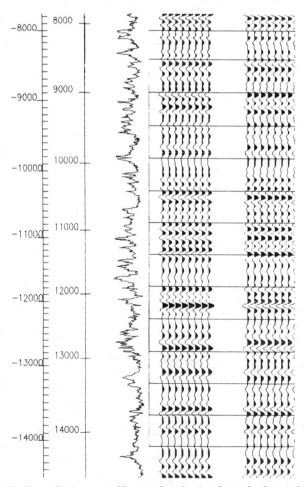

Figure 5-5 Synthetic seismogram illustrating lack of vertical resolution in seismic data.

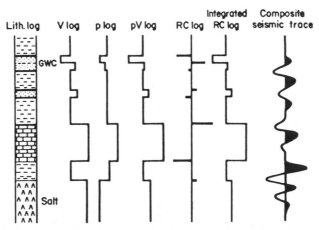

Figure 5-6 Composition of seismic trace from velocity and density contrasts. (After Anstey 1980; and Robinson 1983. Published by permission of IHRDC.)

The exception is noted in the situation when you may be forced to map a horizon that causes no seismic event, but is located in an interval between two diverging reflectors. In some cases, the most likely position of the horizon is not parallel to either reflector. Mapping a non-event is often referred to as *phantoming* (Sheriff 1973).

Keep in mind that the vertical resolution of seismic data will never be as good as that of well log data, but experience has shown that the seismic reflections represent isochronous geologic time surfaces (Payton 1977). This fact makes it possible to map two-dimensional seismic data between wells.

The second assumption, which is shown in Fig. 5-4, is that a curved fault trace seen on a seismic line represents a listric fault. Indeed, the fault shown in the figure is slightly listric, but the reason for stating this is not because of the curved expression of the fault trace on the seismic section, but rather it is the presence of the rollover seen on the seismic time section. A perfectly linear feature in the subsurface may look curved when plotted on a seismic time section—a situation often encountered when plotting direction well paths on seismic sections. You cannot rely on the linearity or nonlinearity of a feature on a seismic time section to be a reliable indicator of its actual geometry in the subsurface without first converting the feature from time to depth.

The most insidious assumption you can make regarding a seismic section is that the section is really just a geologic cross section of the earth directly under the line. What you must always keep in mind is that a seismic section is displayed in two very different dimensions: **space and time, and not space and depth.**

A time section is simply a series of traces displayed next to one another on a piece of paper. The distance between each trace is a physical distance on the paper, and represents a physical distance along the surface of the earth. Therefore, looking horizontally along a section requires only that you understand the scale. A typical full scale section might be 5 in. to the mile along the top of the section. Figure 5-7 shows a typical time section with annotation illustrating the accepted working terminology for the various parts of its display.

It is the *vertical dimension* on seismic time sections that can lead you astray. It sometimes seems reasonable to assume that the vertical dimension also translates directly into a scaleable physical distance. **The vertical section is displayed in two-way time.** It represents the length of time it takes for a seismic signal to travel from the surface, down through the earth, to the reflector, and back to the surface. It would be simple if the seismic velocity field remained constant throughout the earth, but this is **not** the case.

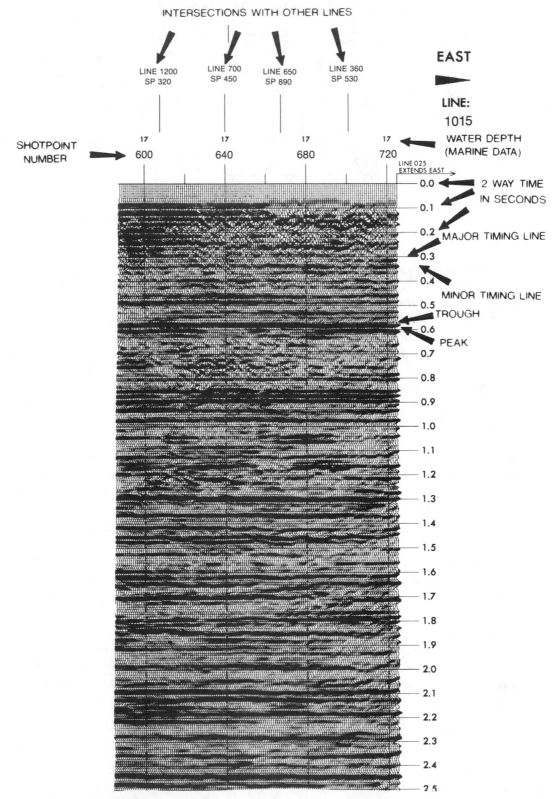

Figure 5-7 Descriptive nomenclature of a typical seismic section. (Published by permission of TGS/GECO.)

Even in *"geologically simple"* areas, there are changes in velocity with depth. In general, the deeper the rock, the *higher* its velocity.

Figure 5-8 is a time-depth table from a well *checkshot* in Pliocene rock, offshore Louisiana. (A checkshot measures the actual time for a surface seismic source to travel to a receiver lowered down a wellbore. This one-way time is converted to two-way time by doubling the times for any given depth.) Underlined in the figure are the depths at 1 sec, 2 sec, and 3 sec. The depth represented by 1 sec of two-way time is 3227 ft. The depth at 2 and 3 sec is 6996 and 11,642 ft, respectively. So in this example, depending on the depth, an incremental 1 sec of two-way time may represent 3227, 3769, or 4646 ft. Therefore, before geometric statements are made from a time section, you must convert the two-way times to depth.

One expensive way to convert the two-way times is to have all the seismic section's depth converted. This is usually not necessary, and in many cases the error in the depth conversion process may exceed the velocity uncertainty inherent in the area you are mapping. An easier method is to convert all time points to depth using a valid checkshot before using them to construct a geometric interpretation.

To demonstrate how different the perspective can become after converting everything to the same dimension, look at Fig. 5-9, which shows a depth-converted fault trace plotted at the same horizontal scale as the seismic line. Does the fault look as curved as the trace on the time section? This example illustrates the effect time sections can have in **distorting the true geometry of a geologic feature.**

	0	10	20	30	40	50	60	70	80	90
0	0	31	63	94	125	157	188	220	251	282
100	314	345	377	408	439	471	502	534	565	597
200	628	659	691	722	754	785	817	849	880	912
300	943	975	1007	1038	1070	1102	1133	1165	1197	1229
400	1260	1292	1324	1356	1388	1420	1452	1483	1515	1547
500	1579	1612	1644	1676	1708	1740	1772	1804	1837	1869
600	1901	1934	1966	1998	2031	2063	2096	2129	2161	2194
700	2226	2259	2292	2325	2358	2390	2423	2456	2489	2522
800	2555	2588	2622	2655	2688	2721	2755	2788	2822	2855
900	2889	2922	2956	2989	3023	3057	3091	3125	3159	3193
1000	3227	3261	3295	3329	3363	3398	3432	3467	3501	3536
1100	3570	3605	3640	3674	3709	3744	3779	3814	3849	3885
1200	3920	3955	3990	4026	4061	4097	4133	4168	4204	4240
1300	4276	4312	4348	4384	4420	4456	4493	4529	4566	4602
1400	4639	4676	4712	4749	4786	4823	4860	4898	4935	4972
1500	5010	5047	5085	5122	5160	5198	5236	5274	5312	5350
1600	5388	5427	5465	5504	5542	5581	5620	5659	5698	5737
1700	5776	5815	5855	5894	5934	5973	6013	6053	6093	6133
1800	6173	6213	6253	6294	6334	6375	6415	6456	6497	6538
1900	6579	6620	6662	6703	6745	6786	6828	6870	6912	6954
2000	6996	7038	7081	7123	7166	7209	7252	7295	7338	7382
2100	7426	7470	7515	7559	7604	7649	7694	7739	7784	7829
2200	7874	7919	7964	8009	8054	8099	8144	8189	8233	8278
2300	8323	8368	8413	8457	8502	8547	8592	8637	8682	8727
2400	8772	8817	8862	8908	8953	8999	9045	9090	9136	9183
2500	9229	9276	9322	9369	9416	9463	9510	9557	9605	9652
2600	9699	9746	9793	9840	9887	9934	9982	10029	10076	10123
2700	10170	10217	10264	10312	10359	10407	10455	10503	10551	10599
2800	10648	10697	10746	10796	10845	10895	10945	10995	11045	11095
2900	11145	11195	11245	11295	11344	11394	11444	11493	11543	11592
3000	11642	11692	11741	11791	11841	11890	11940	11990	12041	12091
3100	12142	12193	12244	12295	12347	12398	12449	12500	12550	12600
3200	12650	12699	12748	12796	12843	12890	12937	12983	13029	13074
3300	13118	13162	13206	13249	13292	13334	13377	13419	13461	13504
3400	13546	13588	13631	13674	13717	13759	13802	13845	13888	13931
3500	13974	14017	14060	14103	14145	14188	14231	14274	14316	14359
3600	14402	14445	14488	14531	14574	14617	14659	14702	14745	14788
3700	14831	14874	14917	14960	15003	15046	15088	15131	15174	15217
3800	15260	15303	15346	15389	15432	15475	15517	15560	15603	15646
3900	15689	15732	15775	15818	15861	15904	15946	15989	16032	16075
4000	16118	16161	16204	16247	16290	16333	16375	16418	16461	16504
4100	16547	16590	16633	16676	16719	16762	16804	16847	16890	16933
4200	16976	17019	17062	17105	17148	17191	17233	17276	17319	17362
4300	17405	17448	17491	17534	17577	17620	17662	17705	17748	17791
4400	17834	17877	17920	17963	18006	18049	18091	18134	18177	18220
4500	18263	18306	18349	18392	18435	18478	18520	18563	18606	18649
4600	18692	18735	18778	18821	18864	18907	18949	18992	19035	19078
4700	19121	19164	19207	19250	19293	19336	19378	19421	19464	19507
4800	19550	19593	19636	19679	19722	19765	19807	19850	19893	19936
4900	19979	20022	20065	20108	20151	20194	20236	20279	20322	20365
5000	20408	20451	20494	20537	20580	20623	20665	20708	20751	20794
5100	20837	20880	20923	20966	21009	21052	21094	21137	21180	21223
5200	21266	21309	21352	21395	21438	21481	21523	21566	21609	21652
5300	21695	21738	21781	21824	21867	21910	21952	21995	22038	22081
5400	22124	22167	22210	22253	22296	22339	22381	22424	22467	22510

Figure 5-8 Time-depth table from Offshore Gulf of Mexico well.

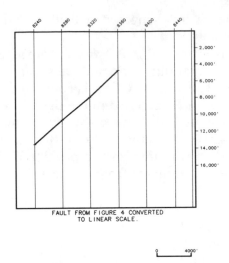

FAULT FROM FIGURE 4 CONVERTED
TO LINEAR SCALE.

0   4000'

Figure 5-9 Depth converted fault trace from seismic line in Fig. 5-4. Notice that the fault does not appear as curved on the depth converted section as does the time section in Fig. 5.4

When considering the listric nature of a fault, you cannot assume that this seismic section orientation is **perpendicular** to the strike of the fault, since fault strike cannot be determined on the basis of one line. Essentially the only statement you can make is that the trace of the fault *on this line appears* to be concave upward. Figure 5-10 shows a hypothetical fault surface and seismic line showing its fault trace on the section. Observe that the fault surface is curved, but the dip of the surface itself maintains a fairly constant angle.

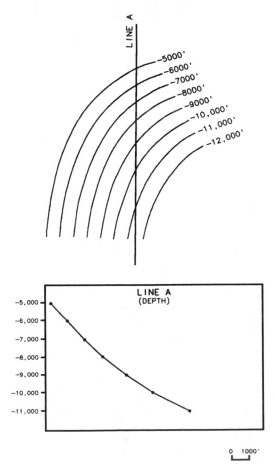

0   1000'

Figure 5-10 Line orientation causing an apparent listric fault on seismic line. The dip on the fault plane is uniform; however, the line orientation makes the fault appear listric, which it is not.

It is enlightening to note that because of the orientation of the line with respect to the fault surface, the trace of the fault on the line *appears* to represent a listric fault, when in fact this is not the case. The easiest mistake to make regarding seismic interpretation is to *infer three-dimensional geology from observations made on two-dimensional seismic data.*

Concepts in Tying Seismic Data

Rationale for Tying Loops.   How do we take into account the three-dimensional nature of the earth when using "two-dimensional" seismic data? (Keep in mind our earlier contention that a seismic section rarely represents a true planar slice through the earth.) The process used to aid in the construction of a valid interpretation is called "*tying*" the lines together. Anyone who has worked with geophysicists has heard the phrases "*tie the data,*" or "*tie the loop.*" What tying the data does for the interpreter is to build a three-dimensional picture of the subsurface. Both subsurface structure contour maps and fault surface contour maps are two-dimensional approximations of three-dimensional geological surfaces. It is logical that two cross sections intersecting a surface (i.e., geologic log cross sections or seismic lines) will show the trace of that surface at the same elevation, at the intersection of the two sections. This is illustrated in Fig. 5-11.

Even though this seems self-evident, the most common error in seismic interpretation is failure to ensure that all geologic surfaces that affect an interpretation have been tied around a loop along the lines. This includes tying the faults from line to line. The only cases when tying faults is impossible are in areas where the fault surfaces are vertical or near vertical, or in areas of complex deformation where the strike lines are poorly imaged. This may seem laborious (and often it is) but the ability to "*tie surfaces*" together by following a laterally continuous seismic event is one of the strongest advantages of seismic data over well information. Well data forces the interpreter to infer a continuous surface from point information, whereas seismic data shows explicit continuity for the horizons being mapped. By tying surfaces, you can eliminate some of the ambiguity that may arise when just using point information from well data. In effect, the act of tying both horizons and faults on a network of lines continually extends the surface, and eliminates a number of possible surface configurations that may arise from the point data in wells.

Figure 5-12 is a set of diagrams illustrating the utility of tying surfaces in order to eliminate this three-dimensional ambiguity. Figure 5-12a is a map showing several fault cuts obtained from well log correlation, along with a possible interpretation of the fault surface. Figure 5-12b shows a different fault surface configuration that satisfies the same arrangement of fault cuts. It is apparent that in some cases the point data from the wells are *insufficient* to uniquely define a geologic surface. Figure 5-12c shows the same area with a set of seismic data, and Fig. 5-12d shows the intersection of these two lines. To satisfy the requirement that *all nonvertical surfaces should tie*, it is easy to see that all but one of the possible fault configurations cannot be justified with the grid of seismic data. Figure 5-12e shows the completed fault surface map with both the well data and seismic data integrated in the fault surface interpretation.

Tying seismic data serves two important purposes. First, it establishes a relationship between the traces of surfaces seen on seismic lines. In other words, by tying the data, we can assure ourselves that a given trace of a geologic surface interpreted on one line is indeed the same surface as interpreted on an intersecting line. The second benefit of tying seismic data is the ability to project the horizon being mapped into areas where *well control may not exist*. This forms the basis for many wildcat prospects. As previously

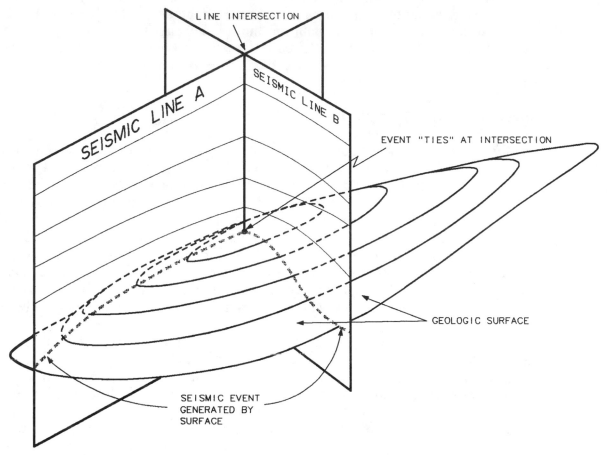

Figure 5-11   Intersection of seismic lines through a subsurface structure. Events must "tie" at line intersections because they represent the same subsurface point.

mentioned, there are very few wildcat prospects that have wells drilled through them. Seismic data allow you to extend a mapped horizon into areas with little subsurface control.

Loop Tying as a Proof of Correctness.   Do seismic interpretations have to tie between the grid of lines? The answer is a qualified yes. Ignoring three-dimensional imaging problems for a moment, we can state that *a valid interpretation must tie to be correct*. All faults and horizons must be related and understood in the framework of a spatial grid of lines. The traces of geologic surfaces as seen on seismic lines (the seismic events) must intersect at the tie points between lines.

Be aware of the tendency, however, to believe that a given interpretation is provably correct because it ties between lines. Any nonvertical surface can be forced to tie between any series of lines. Of course, when it is observed that seismic lines show interpreted horizons crossing the actual seismic events, there is good cause for disbelief in the interpretation. An exception occurs in reverse and thrust faulted terranes where horizons may overlap because of the faulting. More subtle problems can creep into the interpretation during the process of tying faults. Fault traces on seismic lines are sometimes difficult to see, and tying an array of closely spaced faults can present a near impossible task. An insufficiently spaced grid of seismic data in an area of dense faulting can present problems of *aliasing*, particularly if there is little data oriented along the strike of the faults. Figure 5-13 illustrates how aliasing can be a problem in highly faulted areas.

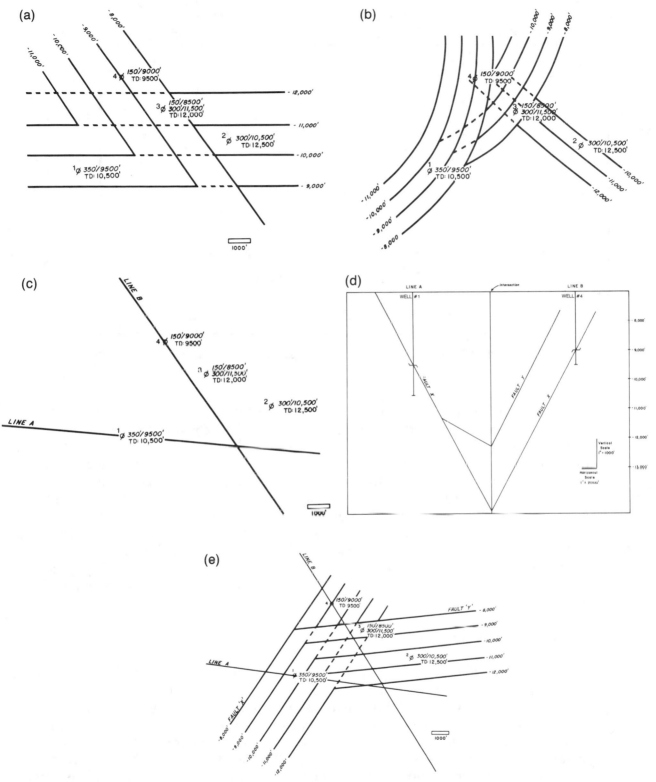

Figure 5-12 (a) Possible interpretation of hypothetical fault cut data observed in a set of wells. (b) Another possible interpretation of the same fault cut data. Multiple fault plane interpretations are possible from this set of data. (c) Hypothetical seismic grid through faulted area. (d) Appearance of hypothetical seismic lines when intersected and "tied." Note that fault traces on lines meet (tie) at the intersection of the two lines. (e) Final fault surface interpretation integrating both seismic and well data.

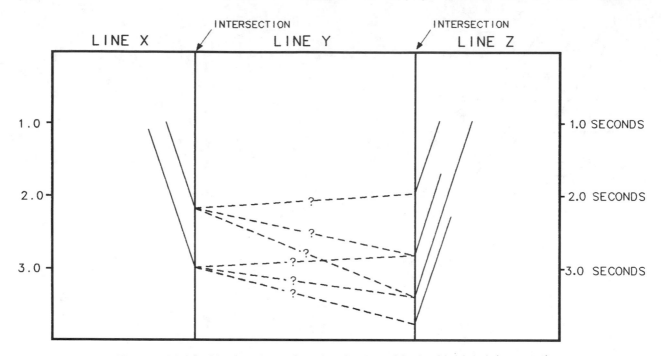

Figure 5-13 Ambiguity when tying closely spaced faults. Multiple interpretations may be possible.

The example in Fig. 5-13 does have a key to help with the tying problem. Notice that the maximum upper limit of two of the faults on line *Z* is the same as the maximum upper limits of two faults on line *X*. It would be a *reasonable first guess* that they are the same pair of faults. This supposition, however, should be made subject to both a believable tie and the construction of a reasonable set of fault surfaces maps.

Perhaps the most useful hint about tying data is to always think in terms of geologic surfaces, and to regard what is seen on seismic lines as merely the trace of this surface (subject to the problems mentioned earlier regarding the vertical dimension being time instead of depth) intersecting the surface of the seismic line. Making geologic interpretations based on one line defies the three-dimensional nature of the earth. Tying the data is a visualization tool that helps create a representation of the subsurface in all its three-dimensional reality.

Procedures in Tying Seismic Data

Contemplating the Data. Although it is tempting to immediately begin mapping, a lot of unnecessary rework can be avoided by first taking the time to arrange your lines into stacks of data with similar orientations. For example, put all your north-south data into one pile arranged consecutively from west to east or east to west. Similarly, put all your east-west oriented data into a separate pile ordered consecutively going north to south or south to north. If your lines are arranged randomly, separate the lines into groups of roughly similar orientations. Then get out your pencil and peruse each line in isolation.

Begin with the stack of lines that is oriented in the same direction as the dip of the predominant geologic features you are mapping. In areas of folds and thrusts, pick lines that are oriented *perpendicularly* to the strike of the fold axes and thrust faults. In listric fault regions, choose lines oriented *perpendicularly* to the strike of the listric faults, and so on. With each line, ask yourself questions about the structure. Which are the dominant faults? Where do the horizons show changes in apparent dip direction? Where are the crests of the highs and the bottoms of the lows? (See Chapter 9 on Long Wavelength Domain Mapping.)

Unfold the next line and put it next to the first line. Look at both of them and ask yourself more questions. Are the same faults present on this line as on the previous line? If a fault is almost certainly the same as one on the previous line, perhaps lightly mark it with the same colored pencil on both lines. Do additional faults appear on this second line? Do the crests of the highs and bottoms of the lows change between the lines? Do the seismic dip rates at similar seismic times differ between the lines? What could be causing these differences? Put away the first line and pull out the third line. Ask yourself the same questions.

Continue this process of looking at two lines at a time until you have looked at all the similarly oriented lines. Make notes about major changes that you observe but cannot adequately explain at the moment. Now follow the same procedure with the other orientations of lines. This may seem somewhat tedious and simple-minded, but we have found that a modest amount of time taken to do this sort of work is well worth the effort. It can save enormous amounts of time spent correcting mistakes made because you went "off the deep end." We have found that when we are "failing" and having problems with an interpretation, doing this little exercise methodically and with a critical and questioning eye can help create an interpretation that is more likely to be correct.

Picking a Reflection to Map.   Seismic profiles (e.g., Fig. 5-4) contain numerous reflections, and it is obvious that it would not be possible or practical to map every event. The interpreter should look critically at the sections and decide which seismic horizons to map. The criteria for choosing an event or horizon to map are usually continuity and event strength. The most recognizable and continuous events will be the easiest to trace through a grid of data. You should pick the strongest and most continuous events to map whenever possible.

Today, most interpreters explicitly or implicitly map on sequence boundaries, because generally these boundaries are the most laterally continuous events on a section. *Sequence boundaries* are geologic unconformities or their correlative conformities and represent approximate isochronous surfaces (Payton 1977). They can generally be located on seismic sections by observing where reflections converge or are truncated against a (usually) strong event.

Seismic sequence analysis is an exciting aspect of the use of geophysics in geologic interpretation. An entire subdiscipline has grown around this area of expertise. Covering this subject in any detail is a book in itself, and therefore we direct you to the references in the bibliography (in particular, Payton 1977; and Berg and Woolverton 1985). You should be aware of the implications of this growing body of knowledge, as it has had and will continue to have a major effect on petroleum exploration.

Annotating the Well Information.   Now that you are confident that you have figured out the geology, mark the position of any wells that intersect the lines. Straight holes require only that you have a reasonably correct checkshot nearby. Directional wells will further require that a directional survey be available. Directional surveys and the projection of deviated wells into a seismic line are covered in Chapters 3 and 6. Remember that you must convert the depth points to their equivalent two-way travel time in order to annotate them correctly on a seismic line.

It is important to remember that any projection is a *compromise*, and will often cause some confusion. Figure 5-14a is an illustration of a projection of a directional well

(a)

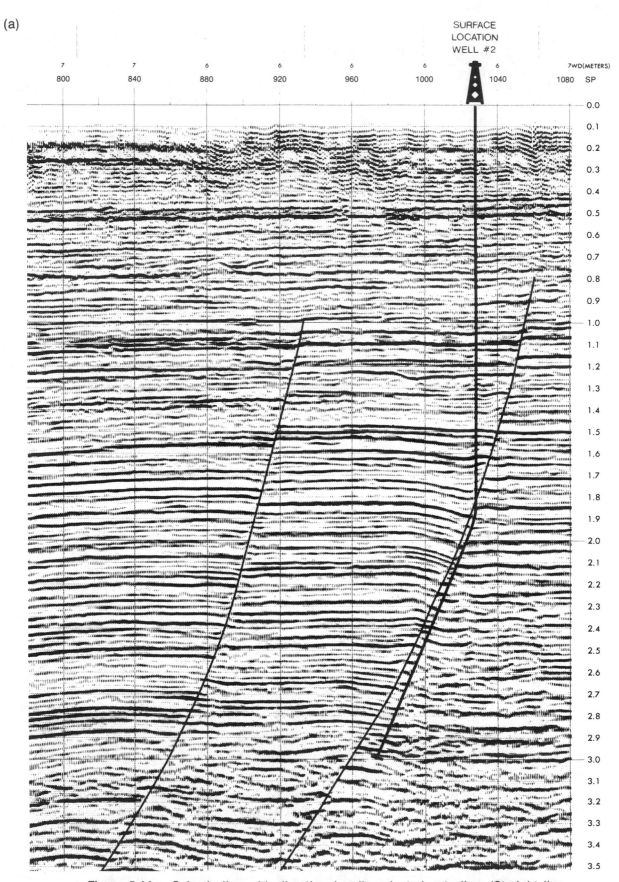

Figure 5-14a Seismic line with directional well projected onto line. (Straight line perpendicular projection used.) (Seismic line published by permission of TGS/GECO.)

110

(b)

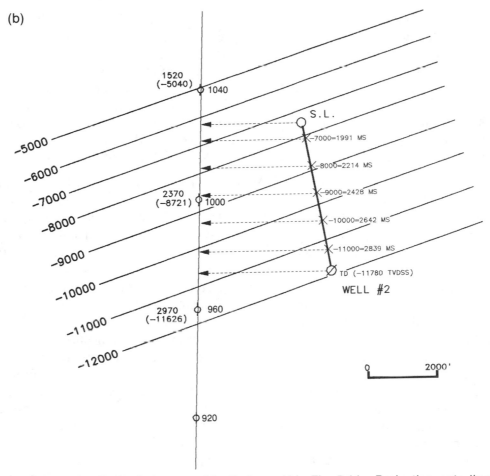

Figure 5-14b Fault plane map of fault observed in Fig. 5-14a. Projection onto line makes well appear to have penetrated the upthrown block. Fault surface and well directional survey clearly indicate that well remained in the downthrown block. Care must be exercised in projecting wells onto seismic lines

onto a seismic line. Notice that the orthogonal projection of the well on the seismic line suggests that the well penetrated the footwall, but the fault surface map (Fig. 5-14b) clearly shows that the well never crossed the fault. This illusion occurs because of the compromises inherent in projecting a three-dimensional entity onto a two-dimensional line.

Once the well position is annotated, the information from the well data, in the form of geologic tops, must be located and marked on the time sections. How do you know where to find the event that corresponds to the geologic horizons? There are basically two methods used to tie the geologic control into the seismic data: (1) using a time/depth function calculated from checkshot data, and (2) tying into the seismic data with a synthetic seismogram.

Tying Well Data to Seismic with Checkshot Information: The simplest and most dangerous method of tying well data to seismic is to use the checkshot data to convert the tops from the log data from depth to time, and post the equivalent horizons on the line at the proper times. The problem with this method is that you never know what kind of assumptions may have been made in the processing of the seismic line to correct to the proper datum. This is why data from different contractors may have static shifts between lines when tied together. Ground "truth" is hard to ascertain in these circumstances.

There is also a temptation to place unwarranted faith in the checkshot data, and to believe it over all other information. We have seen cases where an interpreter has tied a sand occurring in the middle of a 1000-ft shale interval to a no data zone above a distinct event simply because a nearby checkshot indicated such a tie. Particularly when sonic and density information indicate that a sand should generate a strong seismic response, it is likely that the sand ties to the strong event in the middle of a reflection-free zone. This is not to say that pure shale intervals cannot generate strong seismic events, but if a sand is present and fairly close to tying an event on the line, then by all means, correlate the sand to the reflection, and don't be misled by the numbers on the checkshot listing.

Tying Well Data to Seismic with Synthetic Seismograms: Tying the well data into the seismic data with a synthetic seismogram is the preferred method, and will usually give better results. Its usefulness, however, depends almost exclusively on the availability of good quality sonic and bulk density log data from the well. In some areas, these logs are run as a matter of course, while in other regions, they are the exception rather than the rule. Particularly in older basins, there may be a shortage of high-quality sonic and density log data. It is imperative that the interpreter determine the quality of the data used to make the synthetic seismogram. We have seen synthetics made from sonic logs obtained from horribly washed out wellbores that have recorded mostly mud arrivals and cycle skipping. Needless to say, the synthetic seismograms from these logs are useless.

Figure 5-15 shows a typical synthetic seismogram and its tie to a seismic line. This seismogram is shown adjacent to a seismic line going through the actual well location. As you can see, the match is good, though not perfect. The procedure for tying the proper event is to locate the proper horizon on the log plotted next to the synthetic, and draw a horizontal line over to the synthetic trace. Lay the synthetic seismogram over the seismic line at the appropriate location, and shift the synthetic up or down until a good match has been made with the seismic. The horizon line drawn earlier will show where the actual seismic event corresponding to the log horizon is located on the seismic line.

Always be wary of *forcing* yourself to see correlations between the seismic and the synthetic seismogram. When there are problems with either the seismic data or the synthetic data, it may be impossible to make a valid correlation. As a rule of thumb, a shift of the synthetic more than about *one hundred milliseconds* should be highly suspect. Also, if you can turn the synthetic upside down and get an equally good correlation, you

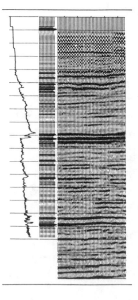

Figure 5-15 Tie of seismic data to synthetic seismogram. Events correlate well from synthetic seismogram to seismic line. (From Badley, 1985, provided by Merlin Profilers, Ltd.)

should be suspicious of the validity of this method for tying well horizons into seismic data.

If applicable, this method of tying in the well data is always preferable over tying into a particular seismic event with just checkshot data, since it will enable you to map with a reasonable degree of certainty on an event caused by the geologic horizon being mapped. This is particularly valuable in areas of rapid thinning and thickening. Figure 5-16 illustrates how an incorrect pick for the mapping horizon can have a profound effect on the depth of the horizon away from the well control. As shown, a small error in the thinner stratigraphic section will cause a much larger error to occur where the stratigraphic section is thicker.

A *vertical seismic profile* (usually abbreviated VSP) derived synthetic trace is also an excellent tool for tying into the seismic data. These tools are just coming into regular use, but the method for using them to tie into the seismic line is the same as that used for synthetic seismograms. Establish a correlation between the traces on the VSP and the seismic, and use the plotted log as a guide for tying into the seismic line. An added benefit to both the synthetic seismogram and the vertical seismic profile is the ability to use these data to analyze the relationship between the lithologic changes in the well and the seismic character generated by these lithologic changes (See Payton 1977 for discussions on using seismic character as a predictive tool for locating and defining lithologic changes).

The subject of tying well information into seismic data is deceptively simple. There is a dangerous tendency to believe that your first tie from a well is the correct one. Make

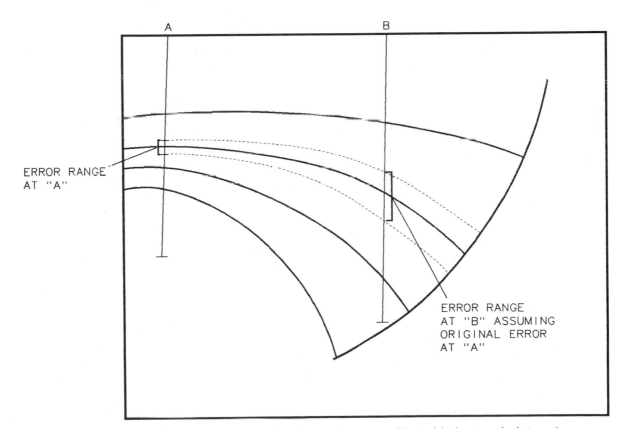

Figure 5-16 Increase in magnitude of error in areas with rapid changes in interval thicknesses. Any uncertainty in tying formation top to seismic event can introduce a greater uncertainty where stratigraphic section thickens away from well control.

sure that it is, because being tied into the wrong horizon can cause you to miss the relationship between a given log response and its correlative seismic response. The only other advice we can give is to avoid a "railroad track" mentality (looking strictly at a narrow 80-millisecond strip of the seismic data), and totally miss what is going on above and below the horizon you are mapping.

Tying the Lines.   If you have followed the procedures so far, you now have stacks of data with a tentative interpretation, and well data annotated on each line. To this point we have not mentioned actually loop tying the data. That is the next step. You will discover very quickly why people do not interpret seismic lines with rapidograph pens. On a good day, about half of what you sketched in preliminarily will be provably wrong when you finish loop tying the data. Again, methodically tying the loops will improve the chances of your finishing the task correctly and in a timely manner. We find it much easier to figure out the most obvious geologic features first, and to tie them together on all the lines. After the large features tie, begin another iteration of tying through the data volume, concentrating on the smaller "second order" features. As the size of the features being tied together decreases, the number of lines required to tie them also decreases, so the work goes more quickly toward the end of the process.

It is important to pick a loop tying scheme that will allow you to make the smallest number of assumptions while carrying your surface around the mapping area. A serious search for the tying path that crosses the fewest faults and that jumps across faults at their area of smallest displacement will more likely be correct.

The initial and most menial task for tying the loops is to post all the intersections of the seismic data on all the lines. Depending on the number of lines that are being tied, this process can take anywhere from a morning to several weeks. Figure 5-17 shows a typical seismic base map with two seismic lines highlighted. The corresponding lines are shown in Figs. 5-18 and 5-19. Line A intersects line B at a location just north of shotpoint 480.

Line intersections are seldom cooperative enough to fall on a *downline* (a downline, shown in Fig. 5-7, merely refers to the dark vertical line printed on the time section at the shotpoint locations also annotated on the maps). Depending on the precision you like to pretend you have, you can either make a rough estimate: "The intersection lies roughly 8 traces past this downline," or use a scale to determine that the intersection is exactly 150 ft, or 1.829 traces to the right of shotpoint 480 on line B. All of the intersections that you intend to tie together must be marked.

The next step is to fold one of the lines at the marked intersection. **This fold must be vertical!** Use a straight edge to ensure that the section is folded vertically. A good method is to always fold the section that has the least dip on the horizons being carried. The reason for this will become apparent later. Align the folded section with the unfolded section at the appropriate intersection. Figure 5-20 shows what this looks like. This is what tying the data is all about.

The first thing you will probably notice is that the lines do not tie perfectly when the intersecting lines are aligned at a common two-way time. Sometimes they will match perfectly, but more likely, they will not match at a common time. At least one section will have to be shifted vertically to establish a good correlation between the lines. Figure 5-21 shows how the two lines have to be shifted relative to one another to "tie" the events. Which line has been shifted? Has line A been shifted down 10 milliseconds (+10 ms), or has line B been shifted up 10 milliseconds (−10 ms)? You now decide which line shows the "real" two-way time to the event.

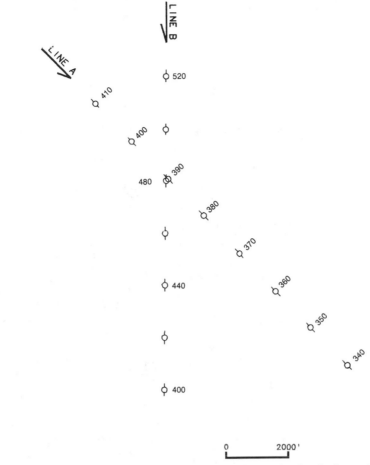

Figure 5-17 Seismic basemap showing positions of seismic lines A and B.

In a mapping project of any complexity, it is necessary to pick a "baseline," so that all the other data may be posted with the appropriate time shifts relative to the baseline. In effect, the procedure for choosing a proper baseline is similar to that for picking a standard correlation log (Chapter 4). When using wells, you are picking the log that best demonstrates the geologic section of the mapping area. When choosing a baseline, pick a line that has as many of the following characteristics as possible.

1. The baseline should cross as many of the other lines as possible. This is required to reduce the number of indirect calculations of static shifts.
2. The baseline should be as close to a dipline as is possible.
3. The baseline should be of very high quality relative to the rest of your data. In short, it must be one of the most believable lines in your collection.

Mis-ties

There are two kinds of mis-ties that must be corrected before posting data on a base-map. There are both *static* and *dynamic* mis-ties. The static mis-ties are reflection-time invariant corrections made to the event times, and are the easiest to recognize and correct. The dynamic mis-ties are corrections that vary with the two-way time of the interval

LINE A

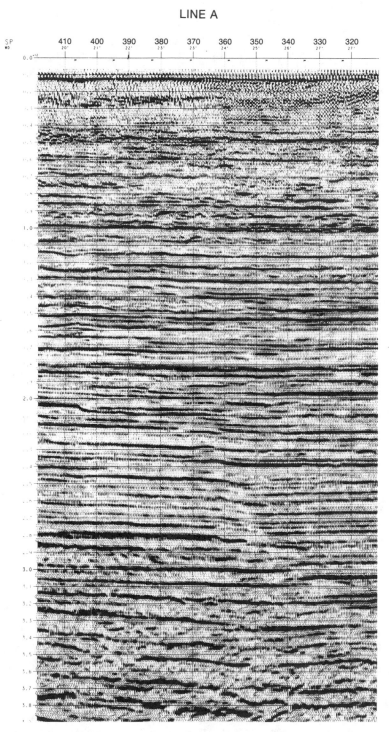

Figure 5-18 Seismic line A. Line shows flat dipping reflectors cut by a small growth fault. (Seismic data published by permission of TGS/GECO.)

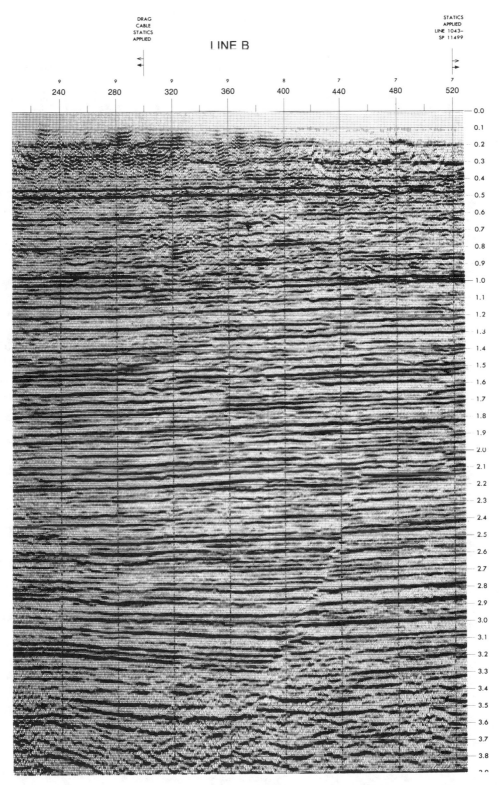

Figure 5-19 Seismic line B. Line shows flat dipping reflectors cut by a small growth fault. (Seismic data published by permission of TGS/GECO.)

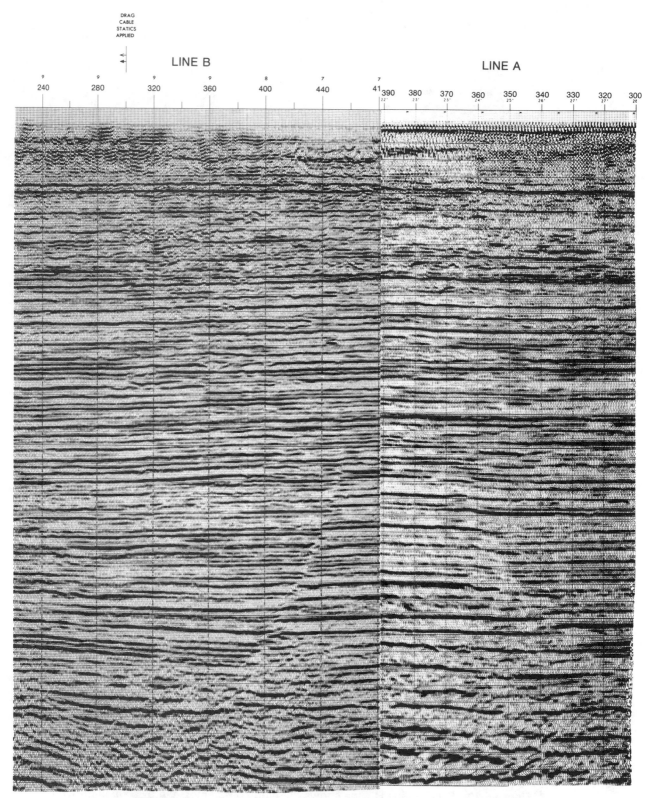

Figure 5-20   Seismic lines A and B intersected with each other. Notice line A appears to be too "shallow" in relation to line B when the timing lines are aligned. (Seismic data published by permission of TGS/GECO.)

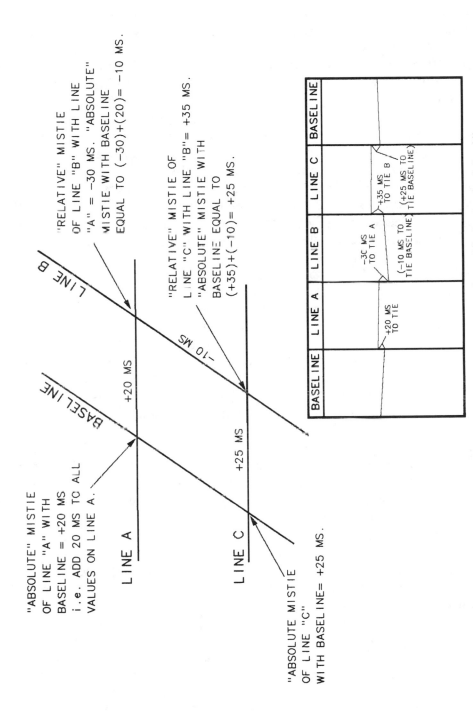

"ABSOLUTE" MISTIE
OF LINE "A" WITH
BASELINE = +20 MS
i.e. ADD 20 MS TO ALL
VALUES ON LINE A.

"ABSOLUTE" MISTIE
OF LINE "C"
WITH BASELINE= +25 MS.

"RELATIVE" MISTIE
OF LINE "B" WITH LINE
"A" = -30 MS. "ABSOLUTE"
MISTIE WITH BASELINE
EQUAL TO (-30)+(20)= -10 MS.

"RELATIVE" MISTIE OF
LINE "C" WITH LINE "B"= +35 MS.
"ABSOLUTE" MISTIE WITH
BASELINE EQUAL TO
(+35)+(-10)= +25 MS.

LINE B

BASELINE

+20 MS

-10 MS

LINE A

+25 MS

LINE C

BASELINE | LINE A | LINE B | LINE C | BASELINE

+20 MS
TO TIE

-30 MS
TO TIE A

(-10 MS TO
TIE BASELINE)

+35 MS
TO TIE B

(+25 MS TO
TIE BASELINE)

Figure 5-21 Keeping track of relative mistie through a grid of seismic data. All absolute misties are in reference to baseline.

being mapped, and are more difficult to correct. Real problems occur when both static and dynamic mis-ties are present in a dataset. The static component must be recognized and corrected first, and the dynamic correction made after the static solution is determined.

Static Mis-ties. Static mis-ties are mis-ties that can be recognized because they cause "bulk" shifts of the intersecting line either up or down to achieve a good correlation (Sheriff 1973). They often occur between datasets of varying vintages and contractors because of different datum corrections and assumptions. Figure 5-20 shows a static mis-tie between two intersecting lines. The easiest way to determine a static mis-tie problem is to search for any shift between flat lying events that are normally present in the shallow part of most basins. In Fig. 5-20, Line A ties line B perfectly with a + 10-ms static shift. Are the times on line A too large or are the times on line B too small? There is no absolute answer. This is the reason for picking a baseline; it establishes a reference or datum. The rest of the lines can then have the time picks for a given event adjusted to the datum established by the baseline. If a line does not directly intersect the baseline, then its relative mis-tie with a line that does intersect the baseline is added together with the value from the line intersecting the baseline. This is harder to describe than to illustrate, so Fig. 5-21 shows an example for keeping track of static mis-tie on lines that do not directly intersect the baseline.

Once a baseline is chosen, annotate the rest of the sections with their respective static mis-tie values relative to the chosen datum. An important point to note about this process is that it should be carried out only when you are tying events that are relatively low dip (probably less than 8–10 deg at most). High dip rates cause an effect called migration mis-tie, which is discussed later. If you are uncertain about whether all of the mis-tie observed at a line intersection is due to static mis-tie, get copies of the unmigrated version of the lines. Unmigrated lines should always tie at their intersections, and any mis-tie observed at the intersection is a static mis-tie. *The mis-tie to annotate in these instances is the mis-tie observed on the unmigrated data.*

Dynamic or Migration Mis-ties. A dynamic or migration mis-tie is one of the more difficult aspects of interpretation of two-dimensional seismic data. The term dynamic mis-tie is a modification of the definition in Sheriff (1973) to apply to interpretation. The spirit of the definition is maintained since it refers to corrections that depend on record time. As noted above, you may encounter varying amounts of mis-tie with different time ranges of events. In general, the shallow, low dip events will tie reasonably well with a static shift. But as the dip increases with increasing depth (greater time on the section), you may notice that events on the dip lines appear to be too deep (too large a time) relative to the intersecting strike line. Figure 5-22 is a sketch of this phenomenon. This mis-tie problem is present because of the limits of two-dimensional seismic data in imaging a three-dimensional subsurface.

To understand this problem, it is first important that you understand what the migration process does to seismic data. Figure 5-23 is a simplified illustration of a two-dimensional seismic line shot in the dip direction. Two simple normal incidence raypaths are drawn on the section from surface positions A and B. By definition, a normal incidence ray will intersect the reflector Z at a right angle. Raypaths drawn to satisfy this condition are A-A' and B-B'. Assume the two-way travel time for a reflection from point A' is 2.0 sec and the two-way travel time for a reflection from point B' is 2.1 sec. Both of these reflection points are actually being recorded at the surface at positions A and B, re-

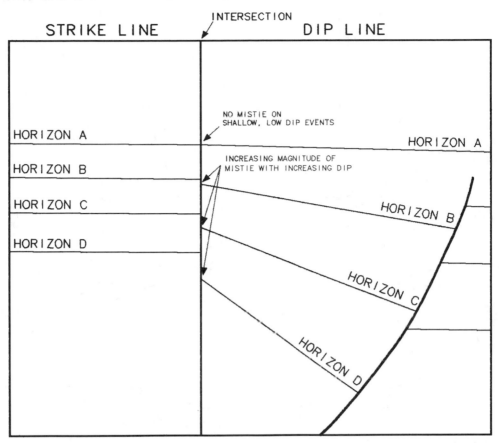

Figure 5-22 Appearance of migrated strike and dip line when intersected. Notice events on strike line are too "high" (too small a time).

spectively. So, on an unmigrated seismic section, the events appear to have the positions shown by dashed line Z'. The seismic line is recording data at a given surface location from a subsurface reflection point updip of the surface location.

Migration is a fairly complex process that corrects the data by moving it to its proper time position relative to the surface location. In other words, after migration, a reflection point is positioned correctly with respect to the surface recording points. Migrating the example would move the event Z' to a position coincident with the actual event in the subsurface. Migration will always *steepen events or reflectors and cause a given event to appear "deeper"* (occur at a larger time) for any given shotpoint when compared to the unmigrated data. Migration of the seismic time data is critically important in getting a reasonably accurate interpretation of the subsurface.

The problem with migration is that it can only fully correct a true dip line. A line that has only a component of the true dip will not be fully corrected, because the two-dimensional migration algorithm cannot move data out of the plane of the line. Figure 5-24 shows the extreme case: a true strike line that is oriented parallel to dip and thus contains no component of dip whatsoever. The data are still coming from updip, but the migration algorithm cannot move data out of the two-dimensional plane of the line. Since the *apparent dip is zero*, the migration algorithm really has no effect on the line. Event Z is positioned on the line at the two-way time of A-A' and B-B', much shallower (less two-way time) than is really the case beneath the surface line A-B. Figure 5-22 shows what happens when a true dip and true strike line are intersected. All of the events on

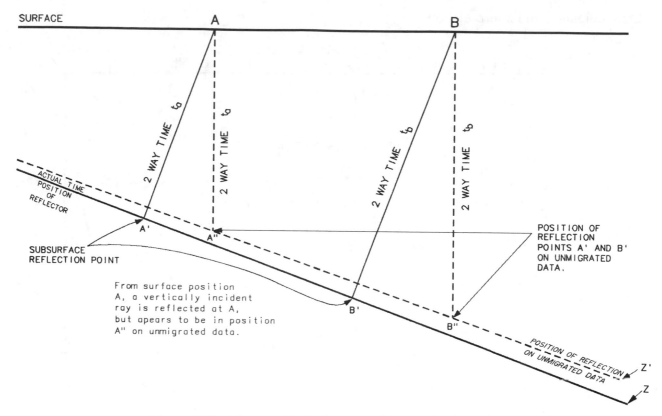

SURFACE

ACTUAL TIME POSITION OF REFLECTOR

2 WAY TIME t_a

2 WAY TIME t_a

2 WAY TIME t_b

2 WAY TIME t_b

SUBSURFACE REFLECTION POINT

POSITION OF REFLECTION POINTS A' AND B' ON UNMIGRATED DATA.

From surface position A, a vertically incident ray is reflected at A, but apears to be in position A" on unmigrated data.

POSITION OF REFLECTION ON UNMIGRATED DATA

Figure 5-23 Diagram illustrating why data need to be migrated. Subsurface reflection points A' and B' are positioned at A" and B" on unmigrated data. Migration moves points A" and B" back to their actual time position A' and B'.

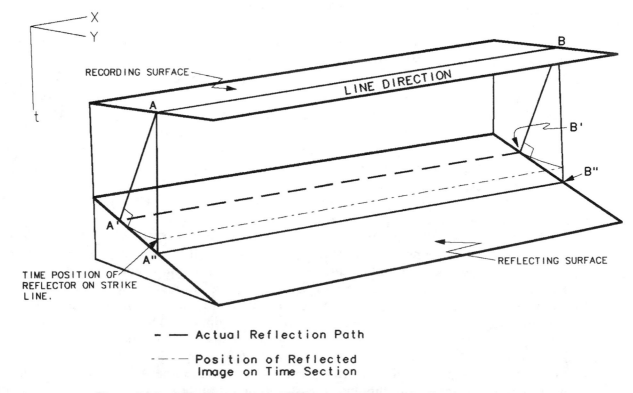

RECORDING SURFACE

LINE DIRECTION

TIME POSITION OF REFLECTOR ON STRIKE LINE.

REFLECTING SURFACE

— — Actual Reflection Path

— · — Position of Reflected Image on Time Section

Figure 5-24 Strike line imaging. Reflection path for strike line is *not* underneath the line. Data is being recorded from updip.

the strike line appear to come in too shallow (too small a time). This is important to remember when tying data. Only in very peculiar circumstances will a strike line event intersect a dip line deeper (greater time) than the corresponding event on the dip line. (This assumes that any static mis-tie has already been taken into account.)

Figure 5-25 illustrates what effect a migration mis-tie can have on a line intersection in areas where there is increasing dip with depth: the mis-tie gets worse and worse in the deeper (greater time) part of the section. In an interesting twist to this problem, notice that the strong event at about 2.6 sec on the strike line is actually "deeper" than its correlative event on the dip line. How can this be possible if the data are coming from updip? Notice that all the events are too shallow on the strike line until about 2.4 sec, when they begin to appear deeper. By tracing horizontally (isotime) from the strike line to the dip line, we can see that the events are deeper because they are being recorded from updip, which happens to be downthrown to a buried growth fault, while the section ties the dip line on the upthrown block. Notice also that the fault intersection point on the strike line is actually 300 ms deeper than a "mechanical" tie would indicate. In growth fault areas, this problem is not uncommon at all, since many growth faults have increasing sediment dips into the nongrowth sediments beneath the fault at depth.

The problem can be handled in two ways. The first and easiest method is to use the strike lines only to tie the events among the dip lines, but ignore them as a valid source of information for drawing maps. In areas where there is already abundant well control and adequate density of dip line coverage, this may be a viable option. These lines can still be used to tie events among the lines, but their actual time values are ignored. The disadvantage of this method is that the strike lines contain information on cross structures which is often critical to development of a trap (Chapter 9).

The best option (and the most time consuming) is to explicitly correct the strike line data by moving the data to their proper position relative to the surface locations. There is an easy graphical way to accomplish this task. Figure 5-26a shows hypothetical dip and strike lines posted on a base map. Figure 5-26b shows an intersection of migrated lines A and B as they would appear on migrated time sections. If you assume that the dip line A is properly migrated, then the data on strike line B at the intersection is actually coming from a position that is updip of the intersection on the dip line.

This concept is easier to illustrate than to explain, and Fig. 5-26b shows an imaginary horizontal line drawn from the event intersection on the strike line to its real reflection point on the dip line. The process for correcting a series of lines is to find this actual reflection point for all the intersections of the strike lines with dip lines and to mark the point on the map from where the event is actually reflected (Point A on Fig. 5-26b). Carried out over a series of intersections, a "corrected" base map can be made that shows the real reflection path of the strike line. The dashed shotpoint map in Fig. 5-26a illustrates how you should "relocate" the strike line updip before posting the data points to use in making your map. Once this corrected base map is made, interpolate the individual shotpoints and post the times from the strike lines at their proper subsurface location. One important point about this technique: the correction is good for only the particular event being interpreted and loop tied. Another event that is perhaps deeper may require another corrected basemap to be constructed for its structure map. In other words, the correction is not fixed, but varies depending on the event being mapped (thus the term *dynamic mis-tie*).

This discussion has been mainly oriented toward tying actual seismic events among lines. Faults must also be tied, and in areas with high sediment dips, the migration mis-tie problem can make mapping fault surfaces extremely difficult. It is vital to remember that if the events are coming from somewhere other than beneath the line, then the

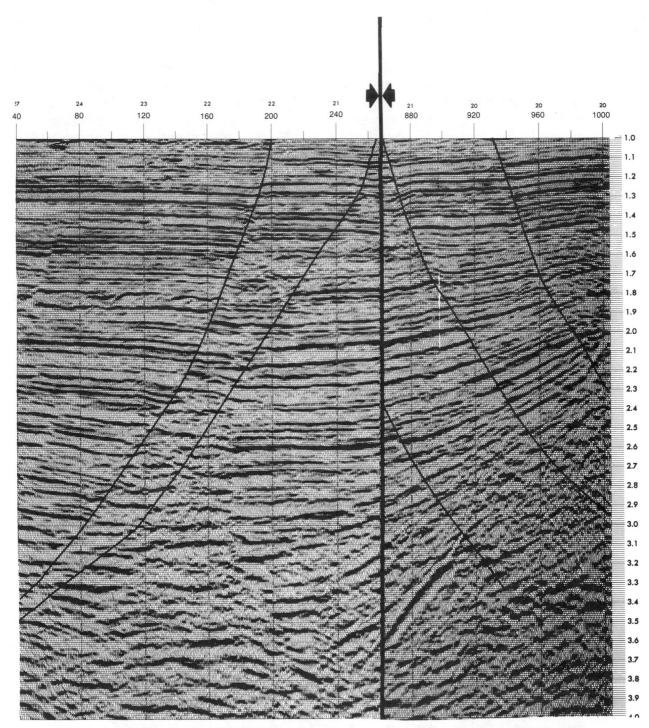

Figure 5-25 An unusual case of migration mis-tie. Strike line events occur earlier than same event on dip lines until approximately 2.4 sec. At the boomer event at 2.6 sec, strike line continues to record updip, however, this is also downthrown to an expanding growth fault! (Seismic lines published by permission of TGS/GECO.)

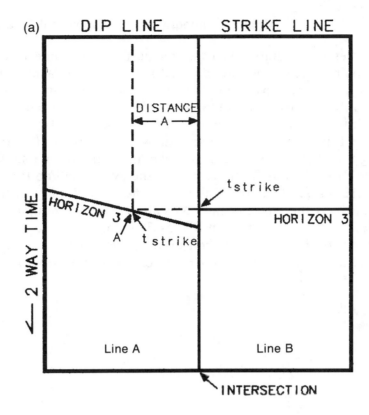

(a)

DIP LINE   STRIKE LINE

DISTANCE
A

2 WAY TIME

HORIZON 3 t_{strike}

A t_{strike} HORIZON 3

Line A   Line B

INTERSECTION

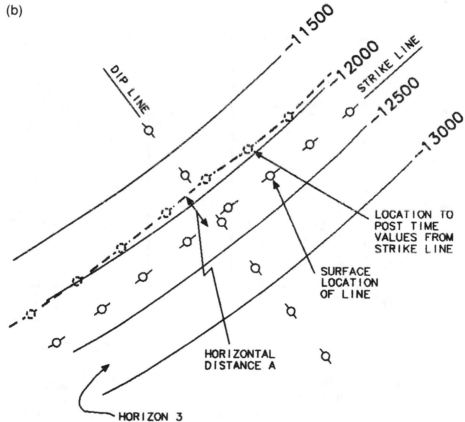

(b)

-11500

-12000

STRIKE LINE

DIP LINE

-12500

-13000

LOCATION TO
POST TIME
VALUES FROM
STRIKE LINE

SURFACE
LOCATION
OF LINE

HORIZONTAL
DISTANCE A

HORIZON 3

Figure 5-26 (a) Practical method for dealing with migration mis-tie involves relocating strike line data points updip as determined from a true dip line. (b) Method for calculating distance strike data must be moved to account for migration mis-tie. Find intersection of time observed on strike line (at line intersection) with the same event on dip line.

discontinuities (i.e., faults) must be coming from the same place as the events. What makes mapping faults especially difficult is the dynamic aspect of the migration mis-tie problem. It is entirely possible to have data points for a given fault surface coming from further and further away from the line, simply because the dip of the section being cut by the fault is increasing with depth. Figure 5-27a shows two seismic lines that intersect each other. Line A is a dip line and shows increasing dip of the seismic events with increasing time (depth). Line B is a strike line with an obvious fault cutting the events on the line. It is obvious from the tie that the events on line B are actually coming from updip of the surface location of the line. The discontinuity in the events is caused by fault X, and thus the image of fault X is also coming from updip of the surface location of the line.The "x" on Line A in Fig. 5-27a is the actual tie point for fault X. Figure 5-27b is an illustration of the method used to construct a corrected fault plane map by moving the times for the fault trace further and further updip with increasing depth of the fault.

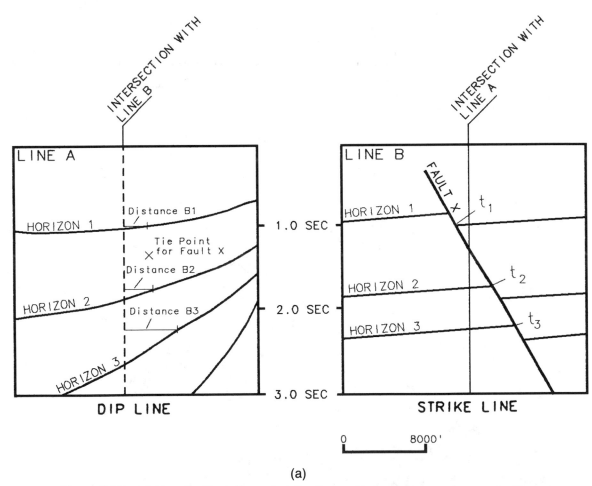

(a)

Figure 5-27   (a) Hypothetical dip and strike lines with fault observed on strike line. Notice the effect that increasing amount of dip with increasing depth on the dip line has on the correlative events on the strike line. (b) Relocation of fault surface points to proper spatial location using method outlined in chapter. Notice that the change in the distance the points must be shifted changes as the record time (depth) increases.

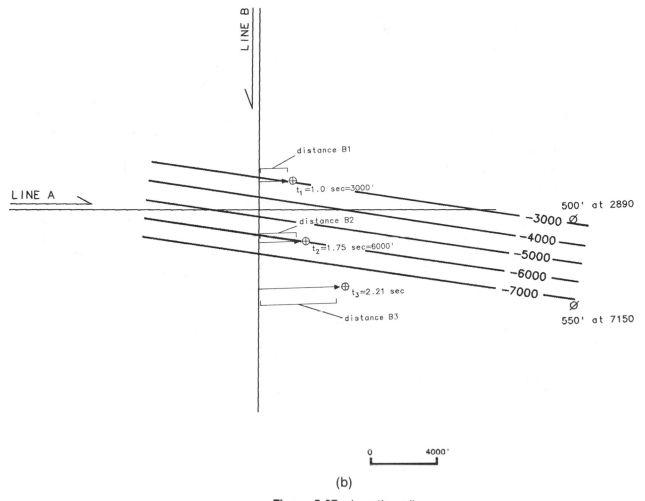

(b)

Figure 5-27 (*continued*)

DATA EXTRACTION

Picking and Posting

After the seismic lines have been interpreted, transfer all the information to a basemap and begin the process of making a subsurface map. As pointed out earlier, the seismic data should be posted along with all of the subsurface information from electric well logs. The mapping process is covered extensively in other chapters, and having seismic data on the map along with the subsurface well data should not affect the techniques used for the actual mapping.

   Types of Data from Seismic.   The most obvious type of data to post from seismic sections are the actual two-way travel time for the event that corresponds to the geologic horizon being mapped. This is analogous to posting formation tops on the map when using well data. The same two-way travel times can be posted for any fault surfaces being mapped. There are several other types of mapping information that can be extracted from the seismic data. One type of information that is extremely useful is the upthrown and downthrown intersection points of the mapping horizon with the surface of a fault. These intersection points have both a vertical datum associated with them, and a spatial location. The significance of these points is that (assuming the data is reasonably high quality) they can be used to position the upthrown and downthrown trace of the fault.

Many interpreters post a solid bar on the map in an easily identifiable color to indicate the fault trace identification. In complex areas, you can assign a unique color code to each fault being tied on the seismic sections and post the trace on a base map in the same color.

Depending on the area, there may be some other useful information that can be posted on the map. If the seismic event being mapped has an amplitude anomaly wherever a hydrocarbon bearing sand is located, then the extent of the amplitude anomaly can be posted. In an area with reasonably continuous sands and no drastic velocity problems, these limits can be used to determine the shape of the contours. In areas adjacent to salt domes, you may be fortunate to have data good enough to identify a salt/sediment interface. This too can be posted and mapped. In areas with stratigraphic discontinuity in the objective sands, you may be able to detect a unique seismic response indicating where the sand is present and where it is not. The extent of the potential reservoir body can thus be mapped.

Extracting the Data. When tying the data around a loop, use a colored pencil to color in the troughs of the seismic data. At each downline or shotpoint marked on the line, mark with a pencil a consistent point on the waveform, perhaps using either the maximum trough (which is easy to see) or the maximum peak.

Posting the Information. Now that the data have been interpreted, they have to be transferred from the seismic lines to the basemap. With an engineer's scale, measure the two-way time in milliseconds to the event being mapped. Next, add or subtract the constant value in milliseconds that represents the amount of static mis-tie between this seismic line and the seismic baseline. Post the two-way time on the map at the actual reflection point for the horizon being mapped. If you have one of your "pseudo" basemaps with all the strike lines repositioned updip, post the seismic events in their true subsurface location. Remember that this location may vary depending on the seismic event that is being mapped. Mark all the intersections of the seismic events with fault surfaces, and post the upthrown and downthrown intersection points. Finally, record any other valuable information and post it on the map.

Converting Time to Depth

You are now at the point where the actual mapping is about to take place. One problem remains: converting the two-way travel time values on your map to depth data. There are several different ways to accomplish this task. As in all the previous discussions, there are both simple and detailed methods. Deciding which method to use depends on the complexity of the time–depth relationship in the area in which you are working. Some areas are so complex that attempting to make a valid map without the assistance of a geophysicist is a mistake. Some areas are so well-behaved in the time–depth relationship that the solution for converting time to depth is trivial.

Brute Conversion with a Time–Depth Table. The first method for converting time to depth is extremely easy. The procedure for converting the values involves looking up the depth values corresponding to the posted time values in a time–depth table that has been generated from checkshot data. There are as many different formats to time–depth tables as there are oil companies, but most bear at least a superficial resemblance to one another. Checkshot data is acquired during the evaluation phase of drilling a well. A checkshot measures the amount of time it takes for the first arrival of a seismic wave to travel from a surface source near the well to a receiver lowered down the wellbore. After the data are acquired, they are usually then interpreted to generate a set of one-way travel times for specific depths in the well.

Conversion of these one-way times to two-way times involves multiplying the time values by two. (The receiver only measures the time it takes for a wave to travel to a given depth, whereas a seismic line measures the time required for a wave to travel to a reflecting horizon, and back to the surface.) The two-way time/depth pairs are then interpolated from the usually rather sparse set of data points to a table with the time/depth pairs calculated at an even time increment. Such a table was shown in Fig. 5-8. A plot of this particular checkshot is shown graphically in a time versus depth graph. The actual times are shown as data points and the line in between the data points is generated by a cubic spline curve fitting routine on a computer.

Once the correct depth for a given two-way travel time has been found, you can post the value beside the time value on the map. Using a contrasting color pencil to post this information is a good method for keeping the time values distinct from the depth values you will actually use to make the map.

Where should this method be used? Any area where there do not appear to be any radical variations in the relationship between seismic time and depth is a good area to use this simple technique. The method for determining whether this is appropriate is simply to use a time depth/table generated from the closest well to the mapping area (or a well in the middle of the area if there is one) to convert all of the times to depth. Then examine your map, looking for obvious discrepancies between the well tops posted on your map and the converted time values. If the depth from a converted time value is 11,200 ft subsea, and a well top at the same location is 10,700 ft subsea, then there is an obvious problem to be addressed.

Recognizing Velocity Problems. If the velocity information does not tie to the wells correctly, there are two possible sources for the differences in the values posted. One possibility often overlooked is that the *wrong horizon* is being interpreted on the seismic line, and thus inappropriate time values are being converted to depth. This problem could arise from an incorrect pick across a fault, or an incorrect tie to a well. The error could be caused by using too large a shift to tie into a synthetic seismogram. Perhaps the synthetic seismogram just isn't very good for tying in this area because of log quality or some other factor. There exists a whole multitude of possible causes similar to these.

Another possibility is that there is a very strong horizontal velocity gradient in the area that causes the time/depth relationship to change laterally. This is often easy to determine if you have the proverbial "lead-pipe cinch" log correlations and seismic event correlations between two wells: You absolutely know that both the correlations and time picks on seismic are correct, and the correlations in the wells are correct. However, they don't agree with one another. In other cases, the problem may not be so easy to identify. If either the well correlations or the seismic event ties are ambiguous, the presence of a gradient may not be obvious because of the difficulty in deciding what interpretation to rely on initially.

Lateral velocity changes can be extremely difficult to manage. It is much beyond the scope of this book to attempt a discussion of the methods of handling velocity gradients. The key question is when to recognize the need for expert help. A simple rule of thumb is this: If the gradient in the area being mapped is severe enough to cause the depth uncertainty for a given two-way time value to exceed the average amount of closure on the features you are mapping, you had better be careful. In particular, if this depth uncertainty can be observed to occur between checkshots in wells that are as physically close as the average dimensions of a prospective closure, the likelihood of correctly mapping a structural closure is pretty small. If this is the problem, seek the assistance of a geophysicist who understands the problems involved in handling velocity gradients.

Accounting for Small Velocity Problems. If the magnitude of velocity changes over your mapped area is not severe, you can often "eyeball" correct a map to account for a gradient. One secret of the working interpreter is to make *ad hoc* adjustments to the map to account for the horizontal velocity gradient. The way this is often done in practice is to find the geologic reason for the velocity differences, and use a different time/depth table on either side of the geologically caused velocity **"boundary."** For example, in the Gulf Coast Tertiary Section, a large growth fault will often have downthrown section with a different velocity field than the upthrown section. In some areas, the seismic velocities may be faster on the downthrown side because of the increased amount of sand in the downthrown block. In other cases, the seismic velocities may be slower in the downthrown block because the thickened section was deposited rapidly and is undercompacted, and slightly overpressured and therefore slower. Whatever the situation, if you can determine the reason for the gradient, you can often adjust the contours to honor the well control.

We have found that the easiest way to handle velocity gradients that do not appear to have definite "boundaries" is to use the following technique, illustrated as an example in Figs. 5-28a through d. A basemap with posted information obtained from both well logs and seismic sections is shown in Fig. 5-28a.

First, prepare a pure time map. Make a map of the isotime contours of the time values, as shown in Fig. 5-28b. Next, determine the average velocity in the depth range being mapped. This is not difficult; simply determine the number of milliseconds of two-way time that the contour interval, in depth, represents at the depth range you are mapping. A typical Tertiary value might be about 22 ms per 100 ft. At each point of well control, use the time map as a guide, and begin contouring in depth, using the distance between each time contour as a rough indicator of the magnitude of dip. Carry the contours about halfway to the next well point, and then start at that well and contour away from that well until you meet the previous contours (Fig. 5-28c). The discrepancy in the depth values can then be adjusted by splitting the difference, and erasing and going back toward the original well points, gradually absorbing the mis-tie in the spacing of the depth contours (Fig. 5-28d).

The points to remember about this technique are: (1) it is quick and dirty, (2) it is appropriate only for minor velocity problems over large areas, and (3) it is useful when the map must be finished quickly. A caveat: Never use this technique when the gradient is severe and is present over a single structure. It is an appropriate and useful technique when you are mapping on a large scale, and need a method of "absorbing" the mis-ties in the synclines between your major closures. Remember that the seismic data just do not have the vertical resolution of well data, in any case. A seismic line sampled at 4 ms, and picked to an accuracy of 10 ms, will give you about 40–60 ft of error in an average

Gulf Coast Tertiary section. Using this technique to account for a 200-ft mis-tie problem does not seem so bad.

The most important point to remember about velocity problems is that it is very easy to exceed your expertise with simplistic solutions. Look critically at your data, and get assistance if it is needed.

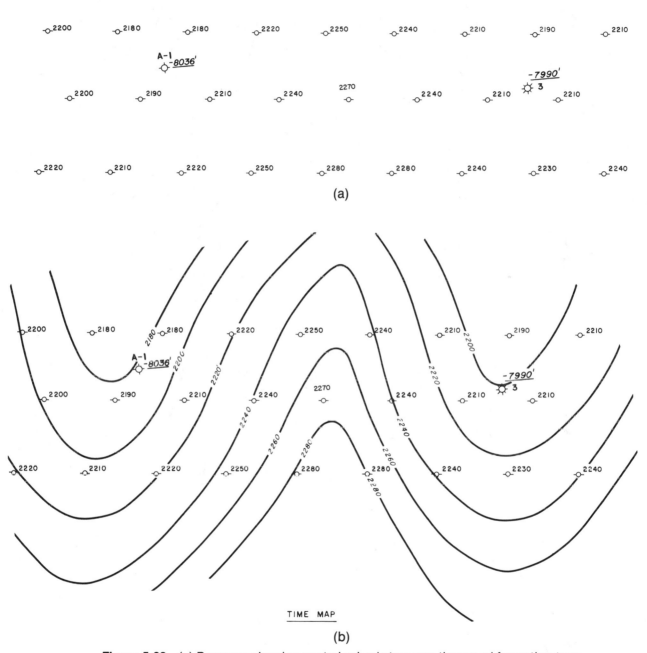

TIME MAP

(b)

Figure 5-28 (a) Basemap showing posted seismic two-way times and formation tops from well data for a mapping horizon. (b) Isotime map of seismic two-way data. (c) Preliminary depth map constructed from each point of control. Contouring begins at each point of well control and moves outward. (d) Final depth map with horizontal velocity gradient "lost" in the syncline between the two plunging noses. The difference in depth values has been adjusted in the contouring of the syncline between the well control points.

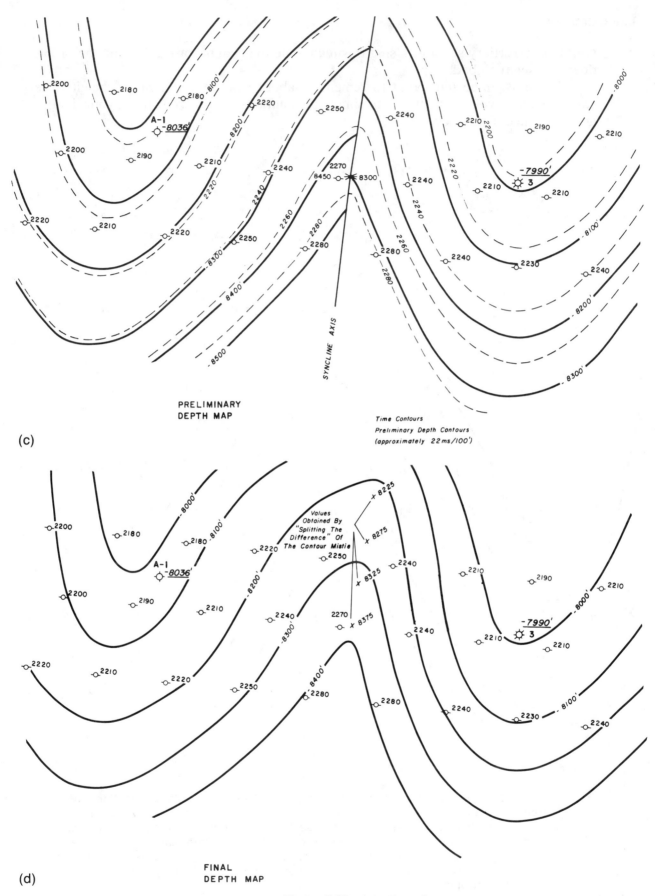

(c)

PRELIMINARY
DEPTH MAP

Time Contours
Preliminary Depth Contours
(approximately 22 ms/100')

(d)

FINAL
DEPTH MAP

Figure 5-28 *(continued)*

SOME FINAL THOUGHTS ON SEISMIC MAPPING

This section is a catchall category about using your common sense and intuition when making maps using seismic data. The main "rule" is to stand back from the work at various stages in the process and ask yourself some hard questions about the work done up to that point. The thoughts are presented in the form of questions that the interpreter must ask himself or herself during the interpretation project.

Have you picked the right event to map? This is meant as more than just a verification to determine if you are tied correctly into a specific horizon. In a more general sense, have you picked the correct event to begin interpreting and tying through your data volume? If there is a choice as to what event to map, always pick the strongest, most laterally continuous event on your data. In many areas, this will often be a seismic sequence boundary. If there is no inherent reason to pick one event in an interval over another, always opt for the one that has the most lateral continuity. Your chances of incorrectly interpreting the correlation across a fault or mis-tying a seismic line are much reduced when mapping these obvious sequence boundaries.

Do your seismic lines make sense? Such geologically unreasonable features as radically thinner downthrown section on a growth fault, or "pencil whipped" horizons that cut across reflectors are warning signs that the interpretation may be suspect. Check the angles of the fault surfaces on your data. Are the angles reasonable for the tectonic regime being mapped? Look for impossibilities on your interpreted section before posting the data on the map. This can save a lot of potential rework. Several minutes of self-criticism each hour can be a very valuable quality control technique for your work.

Has the interpretation taken the path of working from the known to the unknown? It is very easy to become absorbed in solving a few smaller interpretive problems before the main work of understanding the large-scale features has been completed. The best way to work is to solve the obvious, incontestable, problems before tackling the subtleties.

Seismic interpretation and mapping is a process that can only be learned by practice. The more you interpret, the easier it is to recognize the pitfalls and sources of error in your visualization of the subsurface world from seismic data.

CHAPTER 6

CROSS
SECTIONS

INTRODUCTION

A structure contour map depicts the horizontal plan view configuration of a single formation attribute such as structure, thickness, percent porosity, etc. In contrast, a cross section depicts the configuration of many formations as usually viewed in a *vertical plane*. Since a map or cross section alone cannot represent the complete subsurface geologic picture, both must be used to conduct a complete and detailed study.

Geologic cross sections constitute a very important geological exploration and exploitation tool. They are useful in all phases of subsurface geology, as well as in reservoir engineering. Cross sections are used for solving structural and stratigraphic problems in addition to being used as finished illustrations for display or presentation. Used in conjunction with maps, they provide another viewing dimension that is helpful in visualizing a geologic picture in three dimensions.

If a cross section is oriented perpendicular to the strike of the structure, it is termed a *Dip Section*. If the section is oriented parallel to the strike of the structure, it is called a *Strike Section*. Finally, if the orientation is oblique to the structural axis, it is termed an *Oblique Section*.

Various data can be used to construct a vertical cross section. It can be constructed from surface data (dips), electric well log data (formation tops, dips, and fault cuts), seismic data, or entirely from completed subsurface maps. As mentioned in Chapter 1, all available data should be used in the preparation of a subsurface interpretation, whether it is in the form of a structure map or cross section.

PLANNING A CROSS SECTION

Prior to making a cross section, ask yourself a number of questions regarding the planned section construction. The answers to these questions facilitate the preparation of the section and improve its value as an aid to solving problems or illustrating the final geologic picture.

What is the purpose of the cross section? Is the section going to be used as a structural or stratigraphic aid to solving problems? Is it going to be used as a communication device to illustrate the final geologic picture? Will the section show the gross geologic framework or be designed to show significant detail? What sources of data are to be used in the construction? Should the section be prepared true to scale (with the same vertical and horizontal scales) or use an exaggerated scale? What datum is to be used? The answers to these questions provide insight into the planning and preparation of any proposed cross section.

Initially, you must *determine the specific objective* for preparing a cross section. If it is to be prepared to aid in the interpretation of the structural framework, solve problems related to faulting, formation dip, or correlations, then the section required is a structural cross section. If the intent of the section is to solve stratigraphic problems relating to individual sand correlations, permeability barriers, unconformities, facies changes, or changes in depositional environments, then a stratigraphic cross section is needed. A structural or stratigraphic section can be used as a visual aid to communicate or illustrate, as well as to solve problems. The intent will affect the preparation of the section.

The next step in preparing a cross section is to *choose the orientation of the line of section*. The choice is dependent first on the type section you intend to prepare (structure or stratigraphic); second, on the type of geologic structure (i.e., diapiric, extensional, compressional, or wrench); and third, on the data to be used in the section (well logs, structure maps, or surface data).

Finally, *the scales of the proposed section must be selected*. Two separate scales must be considered—the vertical and the horizontal. The scales used are dependent upon the type section being prepared, the actual length of the section, data used, and desired detail. Whenever practical, *use the same horizontal and vertical scales*. However, there may be a special consideration that requires the section to have different vertical and horizontal scales. Often the vertical scale is larger than the horizontal. When this situation occurs, the section is said to have *vertical exaggeration*. Each of the specific conditions that go into the planning and preparation of a line of section are discussed in detail in this chapter.

STRUCTURAL CROSS SECTIONS

Structural cross sections illustrate structural features such as dips, faults, and folds (Silver 1982). They are usually prepared to study structural problems related to subsurface formations, fault geometry, and general correlations. Such problem solving is accomplished by enabling you to visualize the subsurface structure in a vertical plane. Electric well logs (Fig. 6-1) or well log sticks (Fig. 6-2) can be used in the construction of structural cross sections. There are times when electric logs are not available and other data must be used. These data could include drill time logs, core data, and lithologic logs prepared from cuttings descriptions.

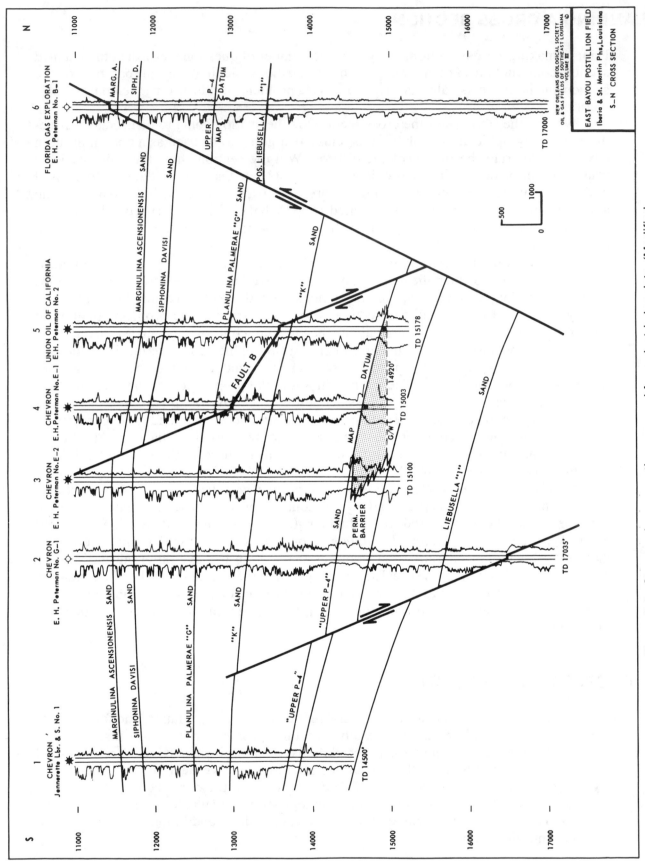

Figure 6-1 Structural cross section prepared from electric log data. (Modified from *Oil and Gas Fields of Southeast Louisiana*, v. 3, 1983. Published by permission of the New Orleans Geological Society.)

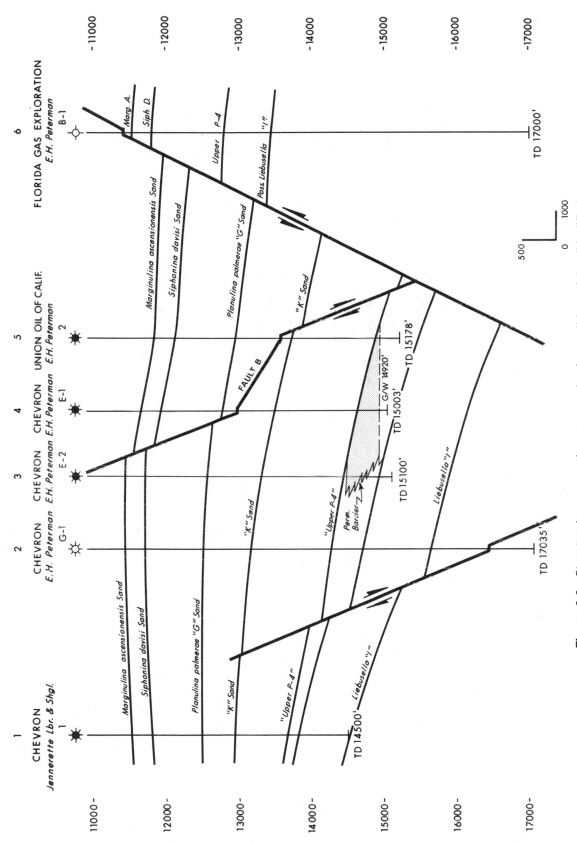

Figure 6-2 Structural cross section prepared using well log sticks. (Modified from Oil and Gas Fie ds of Southeast Louisiana, v. 3, 1983. Published by permission of the New Orleans Geological Society.)

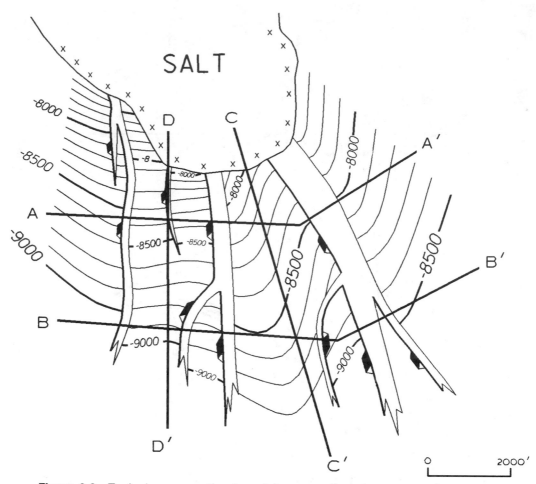

Figure 6-3 Typical cross section layout for a complex piercement salt structure.

Structural cross sections are drawn in the direction of interest. The section can be oriented perpendicular, parallel, or oblique to structural strike. For solving structural problems, it is common for the line of section to be laid out in the dip direction or over the crest of a structure. A line of section perpendicular to the dip of a fault is best for solving fault problems. In a complex area such as that of a faulted piercement salt structure (Fig. 6-3), the evaluation of the structure, salt, and fault geometry may require a number of cross sections to be laid out in the direction of structural dip and perpendicular to fault strike. Each structure and the problems to be solved must be evaluated individually as to the planned direction and the number of sections required to adequately study the geologic feature.

Oil and gas wells do not normally lie in a straight line, as shown in Fig. 6-4a, and so the direction of a planned cross section may not be in a straight line. Instead, line segments between adjacent wells may vary in length and direction, giving the line of section a zig-zag appearance (Fig. 6-4b). With a zig-zag section, whenever the direction of the line of section changes, the apparent dip of formations and faults on the section also changes. An example of such a change can be seen in Fig. 6-1 for Fault B. Notice that the dip of the fault changes from 46 deg between Wells No. 3 and 4 to about 18 deg between Wells No. 4 and 5. This change indicates that the line of section between Wells No. 3 and 4 is perpendicular or nearly perpendicular to the strike of Fault B; but, between Wells No. 4 and 5 the line of section is more parallel to the strike of the fault. These

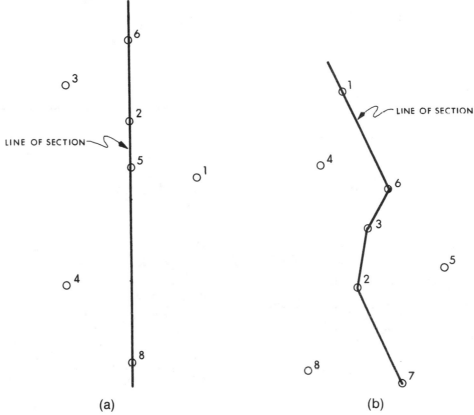

Figure 6-4 (a) A cross section oriented through wells that form a straight line. (b) A zig-zag line of section formed by line segments between adjacent wells.

apparent changes in fault dip are illusions of subsurface geometry caused by the orientation of the line of section. Such illusions must be considered when laying out any line of section in order to prevent confusion to an observer and also to make sure that the section shows what it was prepared to show.

We recommend that structural cross sections be *drawn with the same horizontal and vertical* scales whenever possible. With the same scales, the cross section is prepared true to scale with no vertical or horizontal exaggeration. At times, however, exaggeration is required to permit legible vertical detail. The effects of vertical exaggeration are discussed in detail later in this chapter.

Electric Log Sections

When preparing cross sections with electric logs, certain procedures are helpful in maximizing the usefulness of the section.

1. The logs that are chosen for a section must fit the scale of the cross section. The logs may need to be enlarged or reduced to make the scale equal that of the cross section.
2. If possible, use the same vertical and horizontal scale. This is usually convenient for field studies, however, regional or semi-regional sections may require exaggerated vertical scales to get them down to a manageable size.
3. Data must be legible. Log heading data, correlations, depths, etc., may have to be posted after the section is laid out in order for them to be legible.

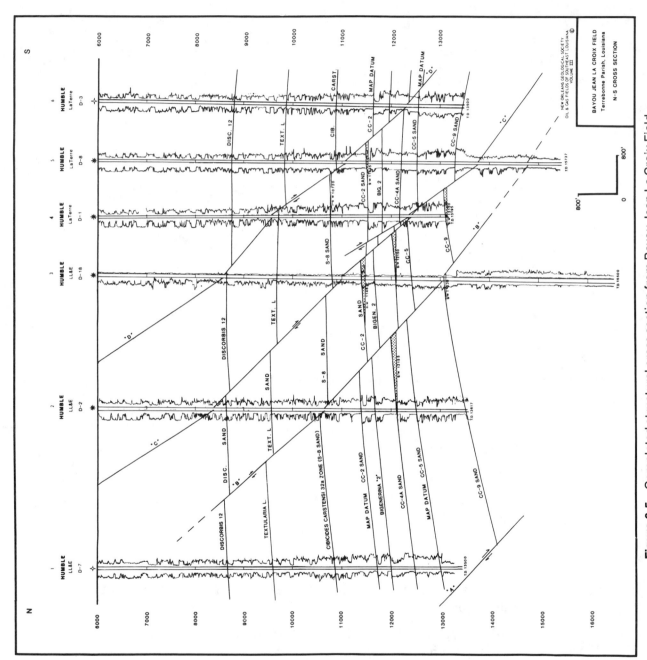

Figure 6-5 Completed structural cross section from Bayou Jean La Croix Field, Terrebonne Parish, Louisiana. (Reprinted from *Oil and Gas Fields of Southeast Louisiana*, v. 3, 1983. Published by permission of the New Orleans Geological Society.)

How are electric logs mounted on a cross section? The procedure usually begins with the preparation of a film positive for each of the original logs to be used in the section. If required, the logs are reduced to the appropriate vertical scale to accommodate the section. The film positives are normally taped on coordinate grid cross section paper with the correct horizontal spacing between wells. It is good practice to plot your line of section on a well base map or structure map (Fig. 6-13) so the spacing between well logs for the section can be measured directly from the base map. For a structural section the logs are normally hung with sea level as the referenced datum. Therefore, the measured log depths must be converted to subsea depths for vertical position on the section.

Finally, an ozalid or xerographic print of the cross section base is made and you are now ready to begin the cross section interpretation. We recommend that all recognized correlation markers, as well as fault cuts from each well, be indicated on the well logs used in the section. The fault cuts can be connected in each well to reflect the proposed fault interpretation; and lastly, lines can be drawn from well to well connecting the correlation markers developing your interpretation (Fig. 6-5). Straight line sections are usually used to initially evaluate a structure (Fig. 6-8a). They portray the dip in straight line segments, so such features as variations in dip between wells will not be apparent. Although such a section does not represent the true attitude of the structure, during the initial phases of a study it provides significant information on the general structure, fault geometry, and correlations.

Cross sections usually go through several stages of revisions, with each such revision improving the accuracy and reasonableness of the interpretation. Remember that the final cross-sectional interpretation must agree with the completed geologic maps, be geologically reasonable, have three-dimensional geometric validity, and conform to the structural style of the area. If possible, cross sections should be retrodeformable or structurally balanced (see Chapter 9).

Figure 6-5 is a completed structural cross section from Bayou Jean La Croix Field in Terrebonne Parish, Louisiana. It is a good example of what is called a "*Finished Illustration*" (Langstaff and Morrill 1981) structural cross section prepared from electric well logs (1 in. = 100 ft) and finished structure maps on the various horizons shown. The section is geologically reasonable for the tectonic setting, illustrating the subsurface structural geology including fault geometry, correlations, and areas of hydrocarbon accumulation. The section was prepared true to scale with the same horizontal and vertical scales shown graphically in the lower right-hand corner of the section.

Stick Sections

An alternative to electric logs is the use of log sticks in the preparation of a cross section. A *stick* is defined as a vertical or deviated line that represents an electric log. A stick section has several advantages over the electric log section, including simplicity, clarity, and ease of construction (Lock 1989).

Since sticks do not show any correlation data (formation correlations or fault cuts), it is necessary to record the depth of all pertinent correlations and fault cuts obtained from the actual electric logs. Stick sections are often used to solve structural problems because of their simplicity and lack of clutter. A typical stick structure cross section is shown in Fig. 6-2. This is the same cross section shown in Fig. 6-1, which uses the actual electric well logs in its construction.

STRATIGRAPHIC CROSS SECTIONS

Problems related to changing stratigraphy require a stratigraphic cross section. They are drawn to illustrate stratigraphic correlations, unconformities, permeability barriers, stratigraphic thickness changes, facies changes, and other stratigraphic characteristics. Many of the comments made about structural sections also apply to stratigraphic sections. Vertical and horizontal scales must be assigned, the line of section laid out based upon the intent of the section, a datum chosen, and the logs prepared to place on the cross section. The datum for a stratigraphic cross section is normally chosen as some stratigraphic marker with the section set up so that the chosen datum is horizontal. By using a horizontal datum, the distorting effects of structure (folds and faults) are eliminated. This is equivalent to unfolding and unfaulting the strata. Figure 6-6 is an example of two stratigraphic cross sections from southwest Kansas/northwest Oklahoma, each hung on a specific stratigraphic datum. These sections were prepared as part of a detailed study to evaluate the complexities of the structural and stratigraphic factors controlling the trapping of hydrocarbons in the Morrowan Sandstones (G. W. Mannhard and D. A. Busch 1974).

In preparing structural cross sections, recall that you had the choice of using the actual electric well logs or stick representations of the logs. For stratigraphic cross sections, the actual well logs must be used since the work involves solving problems related to the actual lithology. Changes in lithology from well to well can only be evaluated with real log data.

In the preparation of a stratigraphic section, the choice of an appropriate marker bed to use as the datum is extremely important. The choice of a poor datum can pose problems such as incorrectly illustrating the original configuration of the formations or sands under study. In other words, a poorly chosen datum may, for example, incorrectly depict an actual channel sand as a bar.

Remember, one of the primary objectives in laying out a stratigraphic section is to reconstruct the sand geometry at the time of deposition or shortly thereafter. When working in areas of predominant sand/shale deposition, keep in mind that sands and shales compact to differing degrees. The effects of differential compaction as the result of sediment burial are recorded on the electric logs. If you have a good idea of the environment of deposition for the sands being evaluated, the choice of the datum may be relatively easy. If the environments are in question, however, you should ask yourself which marker is most likely to have been close to horizontal at the time of deposition.

Figure 6-7 illustrates three cross sections of the Pennsylvanian age Anvil Rock Sandstone using three different coal seams as the chosen reference datum. In the upper section, the No. 7 coal seam above the sand is chosen as the datum. With this reconstruction, the cross section does not show the effects of draping over the sand. In the middle section, the top of the Anvil Sandstone is used as the datum. By using this datum, the sand may give the appearance of a channel fill sand regardless of the sand's actual original configuration. The bottom section using the No. 5 coal seam below the sand is the best choice in this particular situation, because this last section depicts the Anvil Sandstone as a channel fill sand that shows the effects of differential compaction. The channel sand interpretation is also supported by the truncation of the 5A and 6 coal seams against the channel. If the intent was to reflect the geometry at the time the Anvil Rock Sandstone was deposited and the channel was active, the middle section hung on the sandstone itself would come closest to showing this geometry.

We emphasize that care must always be taken when choosing the stratigraphic datum for a stratigraphic cross section. There are no actual rules of thumb to apply when choos-

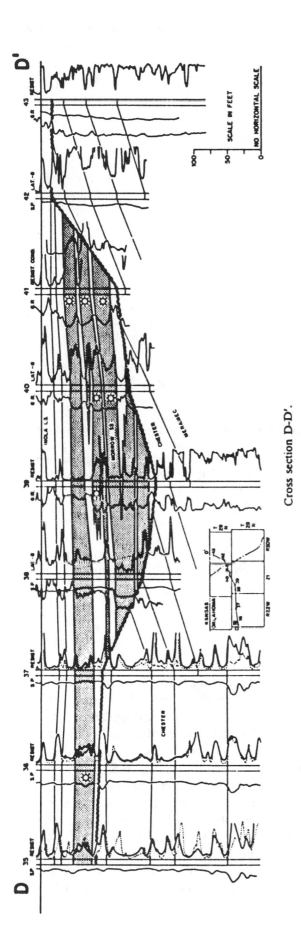

Cross section D-D'.

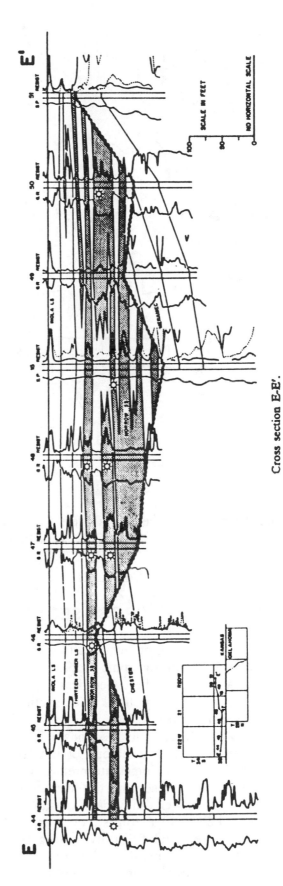

Cross section E-E'.

Figure 6-6 Two stratigraphic correlation sections prepared to evaluate stratigraphic complexities in the Morrowan Sandstones. (From Mannhard and Busch 1974. Published by permission of the American Association of Petroleum Geologists.)

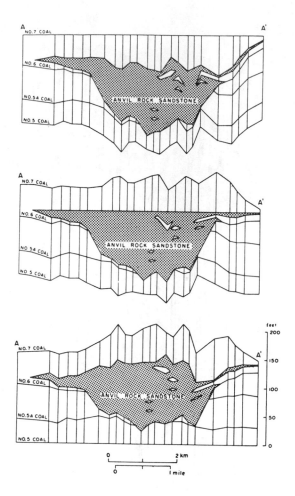

Figure 6-7 The choice of a reference datum is very important in the construction of a stratigraphic cross section. Three stratigraphic cross-sectional interpretations of the Anvil Rock Sandstone are shown based on three different reference data. (Potter 1963.)

ing the datum; however, answers to the initial questions presented earlier prior to constructing a section often can serve as a guide in your choice of the best datum.

PROBLEM-SOLVING CROSS SECTIONS

Cross sections can be very useful from the very beginning of a project in helping to solve structural and stratigraphic problems. We call such sections *Problem-Solving Cross Sections*. As stated earlier, electrical logs are required if the problem is stratigraphic. If the basic problem is structural, stick sections may be useful in helping solve fault and structural geometry problems.

Two different stick sections using the same data are shown in Fig. 6-8. Figure 6-8a is an example of a problem-solving cross section utilizing the straight line method of illustrating the geologic interpretation. With this method, straight lines are drawn from well to well representing the formation and fault correlations. Obviously, the straight lines between wells may not illustrate the true geologic picture, such as changes in bed dip between control points; however, in the initial stage of a geologic study, this type of section is very helpful in evaluating alternate correlations, and fault and structural interpretations. Remember, the interpretation must be geologically reasonable and have three-dimensional geometric validity.

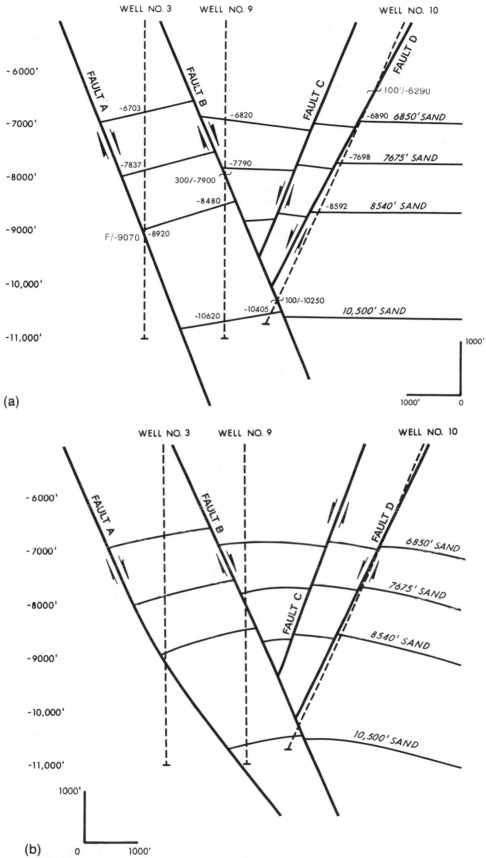

Figure 6-8 (a) Straight line problem-solving cross section. (b) Finished illustration cross section constructed from completed fault and structure maps (compare this with Fig. 6-8a)

FINISHED ILLUSTRATION (SHOW) CROSS SECTIONS

A *Finished Illustration* or *Show Cross Section* illustrates the final interpretation. It is constructed after all the fault and structure maps have been prepared. Such a section reflects the final interpretation and is used to complement the fault and structure maps. Finished illustration cross sections also serve as a visual aid to communicate and present the final geologic interpretation.

Figure 6-8b shows a completed finished illustration cross section for the initial interpretation shown in Fig. 6-8a. Observe that the correlations between wells reflect the true geometry of the formation and faults as opposed to the initial straight line interpretation.

Figures 6-9a and 6-9b present the details of constructing a finished illustration (or show) cross section from electric well log control and completed subsurface maps (in this case a structure and fault map). The section is a dip section crossing a normally faulted anticlinal structure. The figures depict the data used in constructing the 6850-ft Sand and Faults A, B, C, and D on the finished illustration cross section. The final illustration cross section constructed from all the fault and structure maps is shown in Fig. 6-8b.

We can review Figs. 6-9a and 6-9b to illustrate the actual procedure for constructing this type cross section. The upper part of Fig. 6-9a is a structure contour map on the 6,850-ft Sand. The line of section, drawn on the structure map, starts at the downthrown trace of Fault A, continues through Wells No. 3, 9, and 10, and terminates at the 7100-ft contour line upthrown to Fault D.

It is always a good idea to review the planning and preparation procedures presented earlier before beginning any cross section until you feel that you are thoroughly familiar with them. The first task is to prepare the cross section coordinate grid paper and then begin plotting all available data points. Since all the data are acquired from completed maps and accepted log correlations, it is not necessary to place the actual electric logs on the section; however, this is a matter of personal preference. The first points to plot are those for the 6850-ft Sand. Starting with Well No. 3, the first data point to place on the section is the top of the sand in this well, which is at a depth of −6703 ft. This point, shown at Well No. 3 as location A on the structure map, is plotted at the appropriate depth on the log stick for Well No. 3 on the cross section (position A). The second data point is the intersection of the section line and the −6700-ft structure contour line shown as location B on the structure map, 100 ft south of Well No. 3. This −6700-ft subsea depth point is plotted on the cross section 100 ft from Well No. 3 (position B). The third data point is the intersection of the section line with the −6600-ft contour line as shown at location C on the structure map. This point is 725 ft southeast of Well No. 3 and is plotted on the cross section at this location.

What is the fourth data point? The fourth data point is the intersection of the section line and the upthrown trace of Fault B shown as location D on the structure map. This intersection is at an estimated subsea depth of −6585 ft and measures 900 ft from Well No. 3. This point is plotted on the cross section at position D.

This procedure is continued for each measurable data point along the line of section on the structure map. Even though there are no wells north of Well No. 3, the section extends to Fault A using all contour lines as data points. It should be noted that measurable data points include such items as gas/oil contacts, oil/water contacts, and upthrown and downthrown fault traces. You will also notice that there is an obvious structural crest about 150 ft south of Well No. 9. The crestal position and its subsea depth have been estimated and incorporated into the cross section. Remember that this section

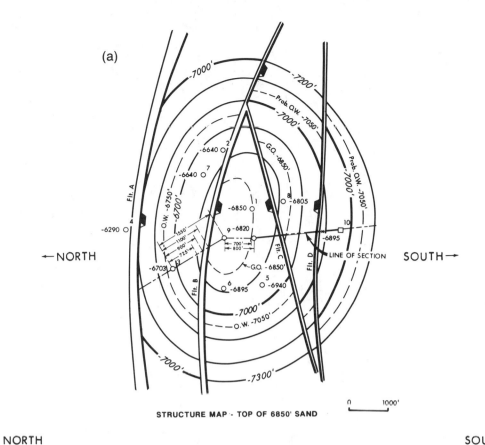

(a)

STRUCTURE MAP - TOP OF 6850' SAND

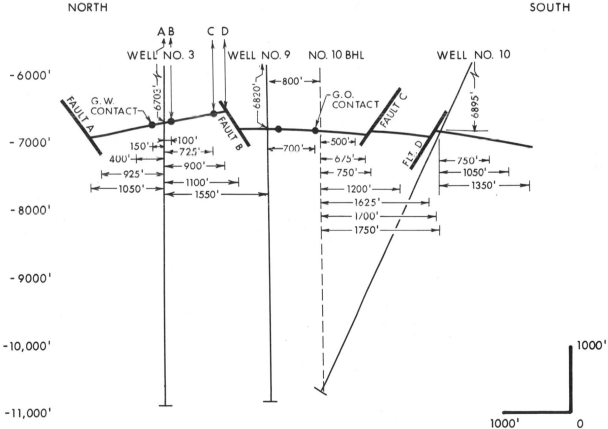

Figure 6-9 (a-upper) Structure map on the top of the 6850-ft Sand. Line of cross section shown on map. (a-lower) details for construction of a finished illustration section for the 6850-ft Sand. (b-upper) Fault surface map for Fault B. Line of section shown on map. (b-lower) Detailed construction for Fault B for incorporation into finished illustration cross section.

(b)

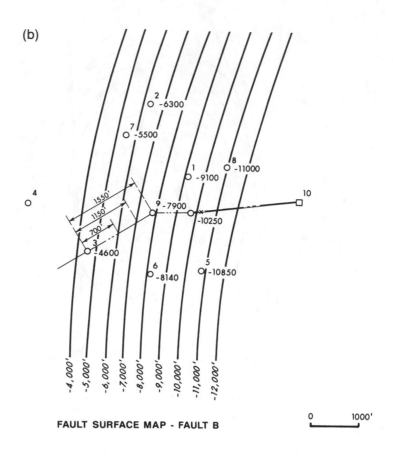

FAULT SURFACE MAP - FAULT B

0   1000'

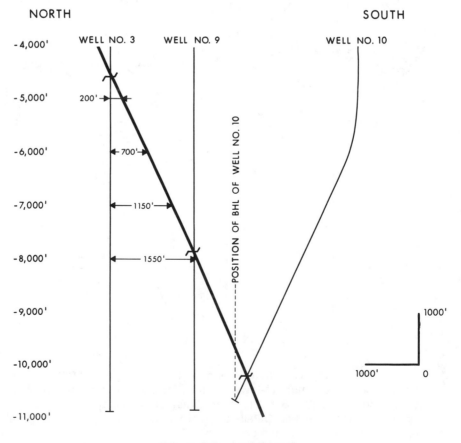

NORTH   SOUTH

-4,000'   WELL NO. 3   WELL NO. 9   WELL NO. 10

-5,000'   200'

-6,000'   700'

-7,000'   1150'

-8,000'   1550'

-9,000'

-10,000'   1000'

-11,000'   1000'   0

Figure 6-9 (*continued*)

is being constructed after the structure maps are in their final stages of completion and all data shown on the maps should be used in the construction of the cross section.

Each point for this cross section is identified on the structure map along the section line and plotted on the cross section. When all the data points are plotted and connected with a correlation line, a detailed interpretation of the 6850-ft Sand is illustrated in cross-sectional view (lower portion of Fig. 6-9a). This procedure should be completed for all the sands planned for the cross section.

Turning to Fig. 6-9b, we see a contour map for Fault B with the line of section drawn on the map (upper portion of the figure). Following the same procedure outlined for the 6850-ft Sand, the location of each fault data point is identified on the fault surface map and plotted on the cross section in the lower part of the figure. The same procedure is followed for Faults A, C, and D.

We now have completed the plotting of all the available data points for the 6850-ft Sand and all the faults for the finished illustration cross section shown in Fig. 6-8b. The interpretations of the other sands and faults were constructed using the same procedure just outlined. The result is a detailed cross section representing a very accurate picture of the structural interpretation of the formations and faults in a vertical plane.

In summary, a finished illustration cross section is important for several reasons. It serves as a visual aid for presenting a completed geologic picture and can also be helpful in identifying mapping problems that might otherwise go undetected, particularly in fields with closely spaced productive sands. An accurately detailed cross section might indicate any number of mapping "busts" such as areas where thickness compatibility between horizons has not been honored. The cross section in Fig. 6-10, prepared to evaluate cross fault drainage, is an example of a detailed finished illustration cross section constructed almost exclusively from final structure maps for each of the horizons shown on the section. The section shows that the interpretation for each horizon, the faults, and the intersection of the faults with the individual horizons is geologically reasonable. There are, however, a few areas on the cross section which appear to indicate some minor geologic busts, which are highlighted by asterisks. Although none of these problems appear to be very serious, the fact that they are visible shows the sensitivity of this detailed cross section to interpretation error.

When mapping several very closely spaced horizons, it is not too difficult to cross contours from one horizon to the next, if extreme care is not taken during mapping. It is good practice to underlie the structure map currently being prepared with one already constructed on a formation immediately above or below. Such practice ensures compatibility in the structural interpretation for a series of horizons being mapped and prevents major mapping busts. This topic is discussed in detail in Chapter 8, but it is introduced here to show how detailed cross sections are used to identify structure contouring errors.

At times it is even possible to cross structure contours from two mapped horizons that are actually hundreds of feet apart, resulting in a mapping bust. Figure 6-11 illustrates just such a situation. Figure 6-11a shows a structure map on two different horizons. The prospective stratigraphic section for an exploratory play lies between these mapped horizons (P-5 and P-7) which are separated vertically by nearly 800 ft of stratigraphic section in the off flank position labeled "T" on the structure maps. A cursory review of the structure maps indicates that they appear geologically reasonable.

Based on a study of the growth activity of a salt dome to the west, it was evident that the salt was actively growing during the time of deposition of the prospective section. So the section between the P-5 and P-7 horizons should thin in the up-structure position to the west, and it does. The stratigraphic thinning was incorporated into the structural picture; but, because care was not taken during the mapping process, the two structure

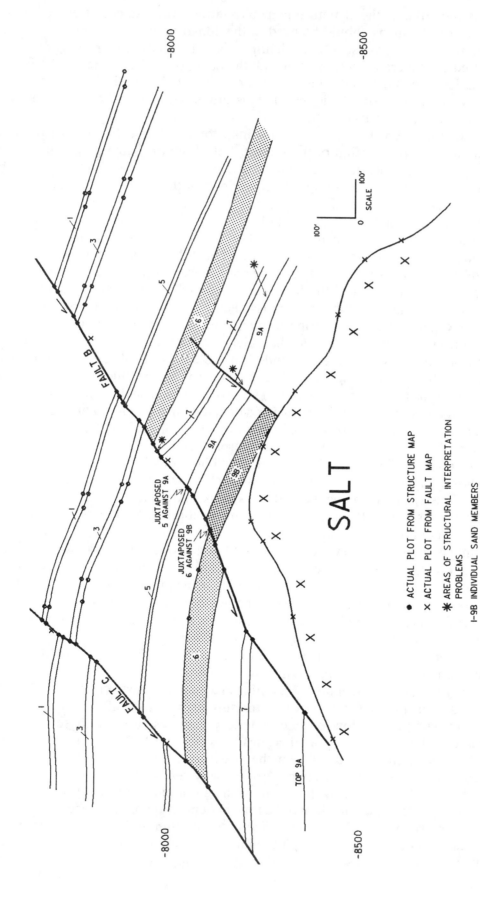

Figure 6-10 A detailed finished illustration cross section constructed almost exclusively from completed maps, including fault maps, structure maps on the top and base of the seven horizons shown, and a salt map. Notice that the detailed section indicates that the 6 Sand downthrown to Fault B is in juxtaposition with the 9B Sand upthrown to Fault B. Cross fault drainage was evident from production data and confirmed by this detailed section. The original scale of the cross section was 1 in. = 100 ft; therefore, detail on the order of tens of feet could be incorporated into the section.

150

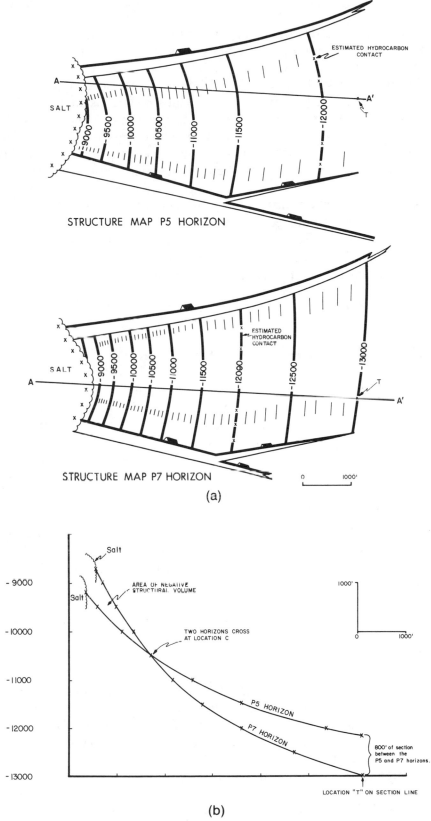

STRUCTURE MAP P5 HORIZON

STRUCTURE MAP P7 HORIZON

(a)

(b)

Figure 6-11 (a) Completed structure maps on the P5 and P7 horizons. These two horizons bracket a prospective section trapped by salt to the west and faults to the north and south. (b) Structure cross section prepared from the completed structure maps on the P5 and P7 horizons. The cross section clearly indicates that the completed structure maps were not constructed correctly.

maps portray a picture of the stratigraphic section and the two mapped horizons, P-5 and P-7, not only thinning but actually crossing up-structure. The detailed cross section of the P-5 and P-7 horizons in Fig. 6-11b shows the effect of this mapping bust very clearly. Notice that the two mapped horizons intersect at the location marked C, and then cross, creating a negative structural volume which is an *impossible* geologic situation. This geological mapping bust is due to carelessness in not using the technique of underlaying the completed P-5 structure map when constructing the P-7 map. Such a mistake can put a question in the minds of management as to the reliability of the work and jeopardize what might otherwise be a great exploration prospect. The preparation of a quick cross section would have caught this mapping bust.

A precise finished illustration cross section can identify many mapping problems providing the time to correct the work before the final interpretation and maps are prepared. Also, these cross sections are excellent aids for reviewing the possibility of juxtaposed sands across a fault (Fig. 6-10). Studies of such sand occurrences are important in evaluating the possibility for cross fault drainage from one fault block to another (Smith 1980).

CORRELATION SECTIONS

In this segment of the chapter, we discuss a special type of cross section called a *Correlation Section*. This section, which is a special type of stratigraphic cross section, is primarily used as a detailed correlation aid. There are several important guidelines helpful in preparing this type of section:

1. Choose a stratigraphic datum that best serves the intended purpose of the correlation cross section (see Fig. 6-7). A reliable shale marker is usually a good choice for a stratigraphic datum (Sneider et al. 1977).
2. Limit the section to a short vertical log interval in order to show significant correlation detail.
3. Position the logs as close as possible with no horizontal scale to include as many logs as needed in the section.

A correlation section can serve as an excellent correlation aid in defining the lateral and vertical continuity of sands within a specific area or reservoir. These sections can also be used as good prospecting tools to evaluate and illustrate the potential for hydrocarbons.

Figure 6-12 shows the layout for a typical correlation section. The actual electric log showing all the curves can be used; however, to reduce clutter and provide sufficient space for the inclusion of many logs, the SP and amplified resistivity curves are often sufficient. As mentioned in Chapter 4, the amplified resistivity curve is usually the most helpful for correlation work.

In preparing a correlation section, follow the same correlation procedures outlined earlier in Chapter 4. Since a correlation section is a type of stratigraphic section, hang it on a well defined datum such as the one shown in Fig. 6-12a. Next, correlate the shale sections using all correlatable shale markers indicated by the SP and resistivity curves. Figure 6-12a shows the shale correlations for this section. Observe the parallel to semi-parallel trend of the shale markers which indicates a fairly uniform stratigraphic thickness.

Now that the shale markers have been established, we can begin the sand corre-

CORRELATION SECTION

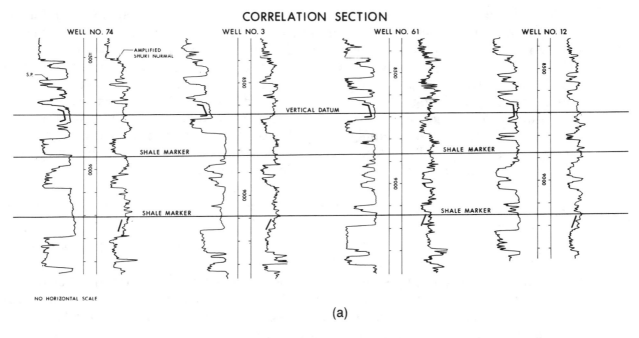

(a)

CORRELATION SECTION

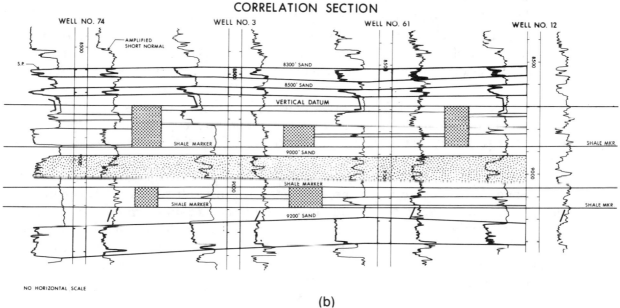

(b)

Figure 6-12 (a) Correlation section is first laid out by hanging the logs on a reference datum and correlating all recognizable shale markers. (b) Completed correlation section shows the lateral and vertical continuity (or lack of continuity) of the individual sands seen in each well.

lations as shown in Fig. 6-12b. As can be seen, the sands are not as vertically and laterally continuous and uniform as the shale sections. Lateral changes in depositional environment can cause sudden variations in sand thicknesses even within an area of limited lateral extent. Because the deposition of clays and muds which make up the shales are usually deposited in quiet waters over extremely large areas, rapid changes in stratigraphic thicknesses in shales is not common. Such extensive and consistent deposition of shales usually provides for good lateral correlation continuity from well to well.

The 8300-ft, 8500-ft, and 9200-ft Sands are separated vertically from the other sands, but they appear laterally continuous from well to well. The crosshatched pattern for the 9000-ft Sand is representative of the entire gross sand package. Within this overall gross sand package there are several distinct sand members in each well. Based on this correlation section, it is speculative whether there is vertical or lateral continuity of individual sand members from well to well within the gross sand package. As for the other individual sands shown in each log, they appear to be laterally discontinuous from well to well, indicating rapid changes in depositional environments and laterally limited sand bodies. This type of information regarding the continuity of sands is most important to the development geologist and reservoir engineer. The layout of one such section or a number of sections can often aid in making critical decisions, such as the following.

1. Which well or wells must be perforated to maximize the drainage efficiency within a reservoir?
2. Which sand interval or intervals should be perforated within a single wellbore to optimize hydrocarbon recovery?
3. Is a specific reservoir competitive with an adjoining lease operator? In other words, can the operator on an adjacent lease drain your reserves by producing his wells because of possible lateral continuity of sands across the lease? If so, what action must be taken to protect your reserves?
4. Are any additional development wells required to optimize field production?
5. Can remaining reserves, identified in an abandoned well, be recovered with other existing wellbores? By this we mean, is there continuity of the hydrocarbon sand from the location of the abandoned well to a well capable of being completed?

We mentioned earlier that correlation sections can be used as an excellent exploitation tool. Figure 6-13 shows part of a geologic and engineering prospect package developed to justify the drilling of a development oil well into what was considered a depleted/nonproducible oil reservoir.

Figure 6-13a is a structure map on the top of a prospective sand called the "9300-Foot Sand." An oil reservoir is present upthrown to Fault A. Six wells penetrated the oil reservoir with Wells No. 4 and 10 having produced minor amounts of oil and gas. Due to the minimum amount of oil production from Well No. 10 and the minimum gas production and pressure decline in the No. 4 well, the reservoir was considered depleted. Further study, however, with the use of detailed maps, log correlations, perforation data, and production data, and the correlation section A-A' (Fig. 6-13c), revealed that the reservoir consisted of three distinct sands: (1) a small highly calcareous fringe complex, such as that seen in Wells No. 7 and 10; (2) a major cut and fill channel sand seen in Wells No. 3 and 4; and (3) an upper transgressive sand member separated from the fringe and channel sands by a shale break. This transgressive sand member is seen in all wells.

The correlation section (Fig. 6-13c) shows that the oil production in Well No. 10 was from the small fringe complex, and the gas production in Well No. 4 from the upper transgressive sand member. Reserves calculated volumetrically, from net sand and net hydrocarbon isopach maps, when compared to the actual oil and gas production, suggested that the fringe complex and transgressive sand are separate members not in communication with the main channel sand (see delineation of major channel sand on the structure map Fig. 6-13a, shown as a permeability barrier, and on the net sand map Fig. 6-13b, highlighted as the limit of major channel sand). Therefore, significant reserves remain to be produced in this channel sand, which can be recovered by a recompletion in Well No. 3, if possible, or through the drilling of a new well. This prospect used the

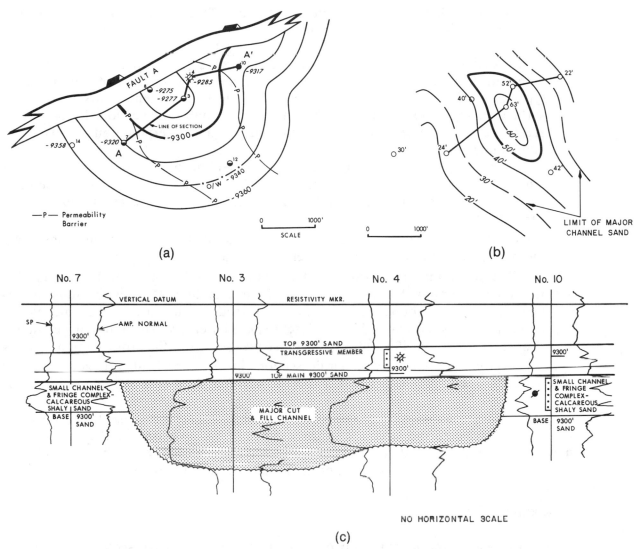

Figure 6-13 (a) Structure map on the top of the 9300-ft (hydrocarbon-bearing) Sand trapped upthrown to Fault A. (b) Net sand isopach map of the 9300-ft Sand delineating the limit of good quality major sand development. (c) A detailed correlation section through the 9300-ft Sand Reservoir trapped upthrown to Fault A. The section clearly illustrates that the upper transgressive member, the fringe complex, and the channel sand are separate and distinct members of the 9300-ft Sand package. Production data indicate minor oil production from Well No. 10 and that the transgressive member perforated in Well No. 4 pressure depleted (it is a closed system separated from the main 9300-ft Sand by a continuous shale break).

correlation section as an integral part of the prospecting process, as well as a final illustration to present the idea to management.

CROSS SECTION DESIGN

In this section, we discuss the specific procedures for laying out cross sections. We review the layout of sections for four different tectonic settings: (1) extensional (normally faulted), (2) diapiric salt, (3) compressional (reverse or thrust faulted), and (4) wrench fault settings.

Before reviewing the section design for the four different tectonic settings, we summarize some cross section design guidelines.

1. Cross sections can be run from well to well in a zig-zag pattern or as a straight line with data from each well not on the line projected into the line of section. Both types of sections have inherent problems. Zig-zag sections tend to distort the subsurface geology due to changes in the strike direction of the section. If you treat a zig-zag section as a series of two-well straight sections, interpretation problems can be minimized.

    Straight line sections may require the projection of well data over long distances, resulting in well projection problems. If you understand the various methods and limitations of projecting well data, however, straight line sections can be used very effectively.

2. Deviated wells can be included in a cross section if the line of section runs along the plan view path of the deviated well (Fig. 6-15).

3. If a line of section being laid out intersects two closely spaced wells, include the well that penetrates the deepest section, if the total depths are significantly important. Spacing may not be the critical factor; instead, similar structural geometry may be critical. In such cases, if two closely spaced wells reflect different geometry, choose the well that illustrates the geometry expected in the cross section.

4. When preparing both strike and dip sections in the same field, it is good practice to tie the sections together with a specific well found on both sections (Fig. 6-14).

Use common sense in laying out all sections. Remember, a cross section is intended to help you visualize the structure in three dimensions, give you another perspective view of the structure, and serve as an aid in solving a variety of problems related to general correlations, fault geometry, or the structural interpretation.

Extensional Structures

Figure 6-14 shows the layout for two dip sections and one strike section for a typical extensional structure. These dip sections, which are perpendicular to the strike of the faults, provide the best information for studying the faults. For single fault systems, these sections can be balanced to help develop the best structural interpretation. Sections involving bifurcating or compensating faults are difficult to balance because there is out-of-the-plane motion, which is often difficult to account for in balancing (see Chapter 9).

    Initially, the cross sections can be used as problem-solving sections to help delineate the fault and structural geometry for the area under study. At the initial stages of the geologic study, the straight line method of construction for faults, formations, salt, unconformities, and other features is recommended. During the later stages of mapping, the sections can be upgraded or revised to help develop and illustrate the final interpretation. Also, cross sections such as section A-A' in Fig. 6-14 can be useful in resolving possible correlation problems in deviated wells.

Diapiric Salt Structures

Diapiric salt structures are in general structurally complex and therefore often require the layout of a number of sections to develop a reasonable geologic interpretation. For an example of the cross section design for a diapiric salt structure, we use the piercement

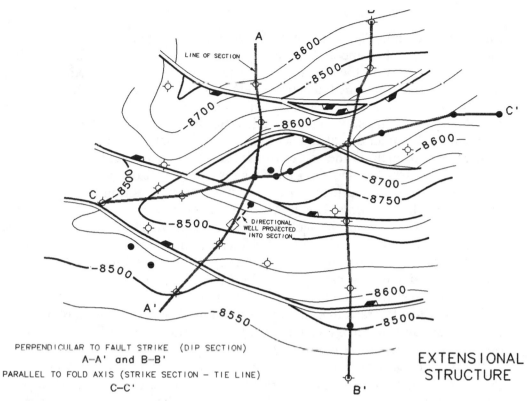

PERPENDICULAR TO FAULT STRIKE (DIP SECTION)
A–A' and B–B'
PARALLEL TO FOLD AXIS (STRIKE SECTION – TIE LINE)
C–C'

EXTENSIONAL
STRUCTURE

Figure 6-14 Typical layout of cross sections in an extensional tectonic setting. (Map was computer drafted.)

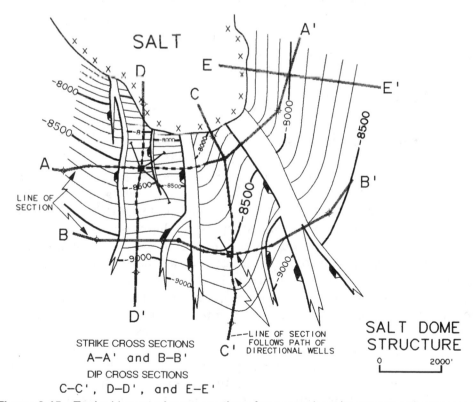

STRIKE CROSS SECTIONS
A–A' and B–B'
DIP CROSS SECTIONS
C–C', D–D', and E–E'

SALT DOME
STRUCTURE

0   2000'

Figure 6-15 Typical layout of cross sections for a complex piercement salt structure including both straight and vertical wells. (Map was computer drafted.)

salt dome shown in Fig. 6-15. A typical cross section layout for a piercement salt structure must be designed to incorporate both straight and deviated wells since both types are commonly drilled on these structures. There are two basic directions for the cross sections: (1) Strike Sections (sometimes referred to as peripheral sections)—these sections parallel or semi-parallel structural strike, and (2) Dip Sections (sometimes refered to as radial sections)—these sections parallel structural dip. With the presence of salt structures, we recommend that dip sections be constructed to continue past the last updip well control to include the salt in the section (see section C-C' in Fig. 6-15).

For the structure in Fig. 6-15, we show the layout of two strike and three dip cross sections. Due to the nature of the structure and the position of the wells, there is only one straight line section (E-E'). All other sections have a zig-zag pattern. The actual procedure for laying out the sections is the same as that outlined in the previous section. Observe how each line of section that includes a deviated well follows the path of that deviated well from the surface to total depth. The portion of the line of section that follows a deviated well path is dashed on the figure for clarity.

Once the structural interpretation is complete, the sections can be upgraded to serve as displays to illustrate the final interpretation. The structure maps represent the horizontal view and the cross sections show the vertical view of the three-dimensional geometry of the structural interpretation.

Look at sections A-A' and B-B' in Fig. 6-15. With these sections, the general fault geometry in the southern portion of this field can be studied because the sections are laid out perpendicular to the strike direction of the faults. They can also be used to study the juxtapositioning of productive sands to evaluate the possibility of cross fault drainage. The cross sections C-C', D-D', and E-E', which are dip sections, provide information on the correlations, sediment/salt interface, growth characteristics of the structure, and data on the downdip extent of any hydrocarbon accumulations.

Finally, notice that section E-E' is laid out with no well control. This section may be very important in evaluating the structural interpretation developed for the eastern flank of the field. Such a section is constructed using the detailed illustration cross section techniques discussed earlier. It is constructed solely from completed structure, fault, and salt maps.

Compressional Structures

The most common hydrocarbon trap in compressional areas is found in hanging wall anticlines which form as a direct result of thrust faulting and include such structures as fault propagated folds, fault bend folds, and duplexes (Chapter 9). The anticlines commonly exhibit an asymmetry with steep frontal limbs and elongated longitudinal axes (B-axis) perpendicular to transport direction. In order to study the internal geometry of these plunging compressional folds, cross sections are normally laid out perpendicular to the B-axis or parallel to transport direction.

Figure 6-16 shows a structure map on top of the Upper Triassic Nugget Sandstone in the Painter Reservoir and East Painter Reservoir Fields. Cross sections A-A' and B-B' shown on the map are laid out parallel to transport direction and perpendicular to the strike direction of the thrust faults. In compressional settings it is best to construct straight line sections and plunge project well data into the line of section to preclude the possibility of physically moving structural geometry within the fold. Such a section will also provide the best information for studying the thrust faults and balancing the structural interpretation.

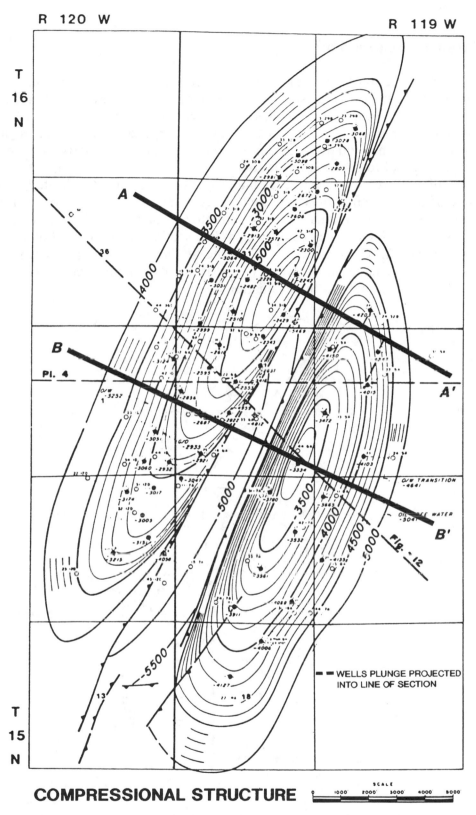

COMPRESSIONAL STRUCTURE

SECTIONS A–A' AND B–B' ARE PARALLEL TO TRANSPORT DIRECTION AND
⊥ TO STRIKE OF THRUST FAULTS.

Figure 6-16 Cross section layout for a typical compressional structure. Structure map is on the Upper Nugget Sandstone in the Painter Reservoir and East Painter Reservoir Fields, Wyoming. (Modified from Lamerson 1982. Published by permission of the Rocky Mountain Association of Geologists.)

Wrench Fault Structures

The cross section layout for associated wrench fault structures is basically the same as discussed in the previous section. Wrench fault systems are often associated with faulted or unfaulted elongated anticlines. If we wish to study the geometry of an associated anticline, dip cross sections must be laid out perpendicular to the elongated fold axis. It is also recommended to tie the dip sections with at least one strike section laid out parallel or subparallel to the axis of the fold.

Figure 6-17 shows a wrench fault system with associated faulted anticlines on either side of the wrench fault. Cross sections A-A' and B-B' are laid out perpendicular to the fold axes. These sections can be constructed as zig-zag sections in which the section line passes through each well or as straight line sections with the well data projected into the line of section. Cross section C-C' is the tie line to Sections A-A' and B-B'. If internal faults such as those shown in Fig. 6-17 exist on the structure, cross sections perpendicular to the fault can help in the evaluation.

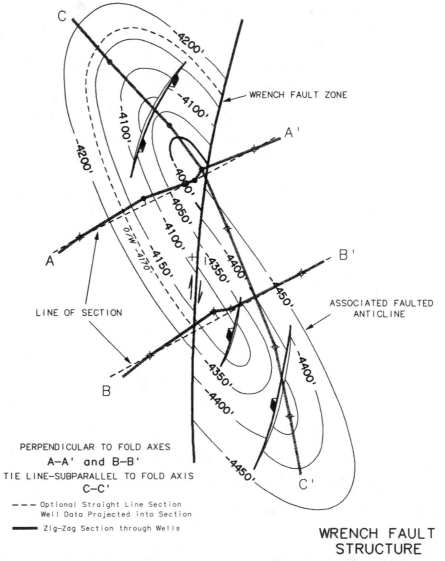

Figure 6-17 Cross section layout for a wrench fault system with associated faulted anticlines.

VERTICAL EXAGGERATION

Earlier in this chapter, we mentioned that whenever possible, cross sections should be constructed using the same horizontal and vertical scales. Special considerations may require that a section be prepared with different (exaggerated) scales, particularly when constructing large regional or semi-regional sections. Often it is the vertical scale that is exaggerated in the construction of a cross section. Vertical exaggeration can be incorporated into both structural and stratigraphic cross sections. Keep in mind that with the use of a vertically exaggerated scale comes various types of *distortion*, such as that of bed thicknesses and dip angles of formations and faults.

The degree of vertical exaggeration is defined as:

$$V_E = \frac{H_L}{V_L} \tag{6-1}$$

where

V_E = Vertical exaggeration

V_L = Length of unit distance on the vertical scale

H_L = Length of unit distance on the horizontal scale

Say, for example, that you wish to prepare a cross section with a horizontal scale of 1 in. = 10,000 ft and a vertical scale of 1 in. = 2000 ft. By using Eq. (6-1), the vertical exaggeration for this cross section is:

$$V_E = \frac{10,000 \text{ ft}}{2000 \text{ ft}}$$

$$V_E = 5$$

Likewise, if a map has a horizontal scale of 1 in. = 4000 ft and you wish to prepare a cross section with a vertical exaggeration of "4," Eq. (6-1) can be rearranged to determine the vertical scale required for this exaggeration.

$$V_L = \frac{H_L}{V_E}$$

Therefore:

$$V_L = \frac{4000 \text{ ft}}{4}$$

$$V_L = 1 \text{ in.} = 1000 \text{ ft}$$

We mentioned at the beginning of this section that there are certain times when a cross section with a vertical exaggeration is required; and in fact, it may have some advantages over a cross section with equal scales. Several situations that may require a cross section with a vertical exaggeration are: (1) the preparation of a cross section in an area of low structural relief, (2) the construction of a section in an area of gently dipping beds, (3) a section that would be unreasonably long with equal scales, or (4) the need for extensive vertical detail. Consideration must also be given to the cost and size limitations of available reproduction equipment. Figure 6-18 shows two cross sections across the Uinta Basin, Colorado. The upper section has a vertical exaggeration of 12; the lower section is true scale with the same horizontal and vertical scales. Notice how much detail can be shown on the upper vertically exaggerated cross section as compared to the true scale cross section.

When a cross section is prepared with vertical exaggeration, the angle of dip of beds, faults, or any other line is exaggerated. The exaggerated dip angle is not simply the product of true dip multiplied by exaggeration. Equation (6-2) below defines the relationship between true dip and the exaggerated dip (see Dennison 1968; and Langstaff and Morrill 1981).

$$V_E = \frac{\text{Tan [exaggerated dip } (\delta_E)]}{\text{Tan [true dip } (\delta)]}$$

Therefore:

$$\text{Tan [exaggerated dip } (\delta_E)] = (V_E) \text{ Tan [true dip } (\delta)] \qquad (6\text{-}2)$$

Using Eq. (6-2), the exaggerated dip for any cross section can be calculated if the true dip and vertical exaggeration are known. Likewise, if the exaggerated dip and vertical exaggeration for a formation or fault are known, the true dip can be calculated. Figure 6-19 is a graphic representation of Eq. (6-2) from Langstaff and Morrill (1981). We can look at an example to illustrate the use of the graph. Referring once again to Fig. 6-18, the Dakota Formation has a dip of 50 deg at location A and the cross section has a vertical exaggeration of 12. By entering the graph on the X-axis at 50 deg, representing the exaggerated dip of 50 deg and entering the Y-axis at 12, representing the vertical exaggeration, the intersection of the two lines generated from these data points indicates a true dip of 5.7 deg. You can see by this example that there can be significant exaggeration of dip on a cross section with a large vertical exaggeration.

As pointed out earlier, formation thicknesses are also affected by vertical exaggeration. If we consider that the exaggeration in the vertical direction on a cross section is a function of Eq. (6-1), it becomes apparent that interval thickness in cross section varies as a function of the exaggerated dip. Figure 6-20 shows two cross sections. Figure 6-20a is plotted true to scale and shows a formation of constant thicknesses "T." Figure 6-20b shows the effect on the formation thickness with a vertical exaggeration of two (2). Notice that in the area of low dip the thickness of the interval increased by a factor of two (2). In the area of steep dip, the interval thickness has been attenuated as a result of the exaggerated scale. It is interesting to note, however, that even though the apparent true bed thickness (stratigraphic thickness) is attenuated in the area of steep dip, the vertical thickness is still increased by a factor of two (2). This effect can also be seen in Fig. 6-18 across the Uinta Basin. In the area of Rangely Dome, we can see the effect of attenuation of the formation on the flanks of the dome and the apparent thickening of the formation over the crest of the structure and in the synclines on each side of the dome. In the preparation and review of cross sections with vertical exaggeration, be sure to keep such thickness changes in mind.

The decision to use vertical exaggeration in the construction of a cross section must be based upon the scale of the section and its intended use. Employed wisely, vertical exaggeration can be a very useful tool. Finally, we emphasize that the use of both a

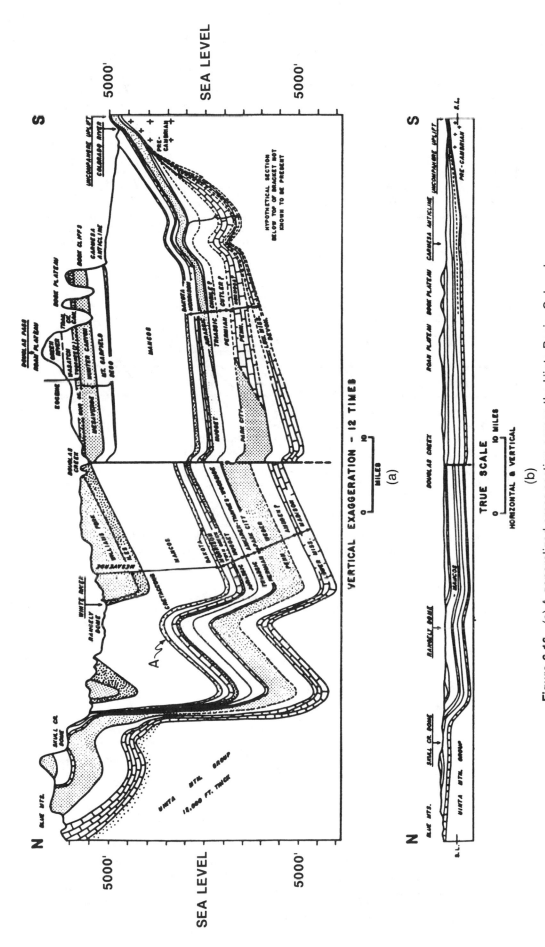

Figure 6-18 (a) A generalized cross section across the Uinta Basin, Colorado, showing the effect of vertical exaggeration (12 times). (b) Same general cross section constructed to true scale. (Modified from Suter 1947. Published by permission of the American Association of Petroleum Geologists.)

163

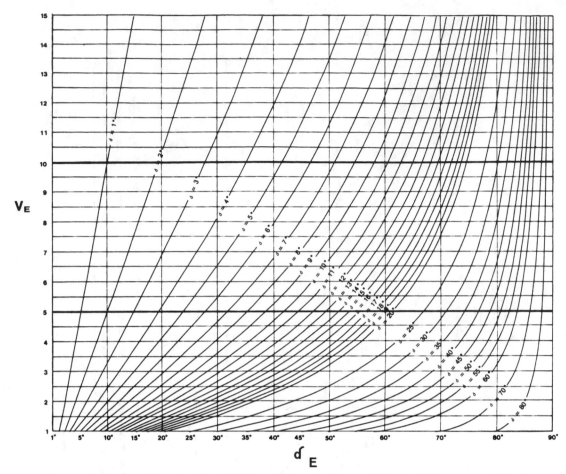

Figure 6-19 Solutions to Eq. (6-2) for different values of V_E (vertical exaggeration) and δ (true dip). True dip occurs where $V_E = 1$. (From Langstaff and Morrill 1981. Published by permission of IHRDC Publishers.)

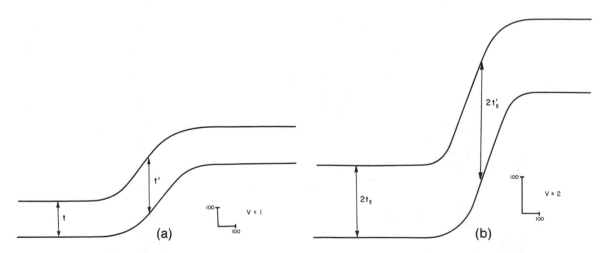

Figure 6-20 (a) is constructed to true scale. (b) shows apparent attenuation due to vertical exaggeration. Attenuation is greatest where true dip is steepest. Notice in the area of steep dip that, although the stratigraphic thickness is attenuated, the vertical thickness is twice the vertical thickness of that shown in (a) which is drawn with a true scale. (Modified from Langstaff and Morrill 1981. Published by permission of IHRDC Publishers.)

vertical and horizontal graphic scale is a must for all constructed cross sections, whether they are true scale or incorporate some type of exaggeration.

PROJECTION OF WELLS

It is best to construct straight cross sections directly through wells; however, for various reasons this may not be possible, so it sometimes becomes necessary to project a well into a line of section. There are several ways in which a well can be projected into a line of section (Fig. 6-21). These include:

1. *Plunge projection*
2. *Strike projection*
3. *Up or down dip projection*
4. *Normal to the section line projection (Minimum distance method)*
5. *Parallel to fault projection*

All these methods can be used to project well data into a cross section. The best method to use, however, depends upon various factors, including: (1) the structural style of the area being worked, (2) the orientation of the section to the axis of the structure, (3) the horizontal distance of a well from the line of section, (4) whether or not there are faults on the section line, and (5) the general dip of the structure. The most commonly used methods for projecting wells into a line of section are the *plunge and strike* projections.

Plunge Projection

In structural settings such as compressional fold and thrust belts involving plunging structures, the projection of well data into a line of section along the B-axis of the fold in the hanging wall, either up- or down-plunge, is the preferred method. Figure 6-22 is a map

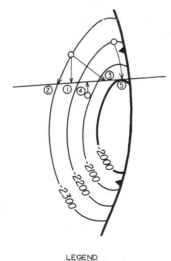

LEGEND

1. Plunge Projection
2. Strike Projection
3. Updip Projection
4. Normal to Section Projection
5. Parallel to Fault Projection

Figure 6-21 The five most common methods of projecting well data into a line of section include: (1) Plunge, (2) Strike, (3) Up or downdip, (4) Normal to the section line (minimum distance), and (5) Parallel to fault. (Modified from W. Brown 1984. Published by permission of author and the American Association of Petroleum Geologists.)

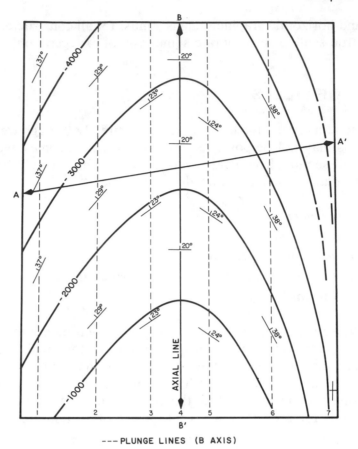

Figure 6-22 Map view of a plunging cylindrical fold. The longitudinal fold axis (B-axis) is shown as B-B'. The dashed lines represent plunge lines which parallel the B-axis. (Modified from W. Brown 1984. Published by permission of author and the American Association of Petroleum Geologists.)

view of a cylindrical fold. The longitudinal or B-axis of the fold is shown as B-B' (W. Brown 1984a), which is coincident with the axial line. The dashed lines represent plunge lines (fold elements) which are, by definition, parallel to the B-axis in a cylindrical fold. Notice that the dip rates are the same along each individual fold element, including the axial line. This is true only as long as the plunge rate is constant for the portion of the fold depicted.

One of the main objectives in plunge-projecting well data into a line of section is to preclude the possibility of physically moving **structural geometry** within the fold. By definition, a cylindrical fold is one in which all fold elements are parallel to one another, and to the B-axis, which is parallel to the direction of plunge. A good way to visualize this is to equate a cylindrical fold to a cylinder of pipe. The edges of the pipe represent fold elements, and are parallel to one another for the entire length of the pipe. Thus, the shape of the pipe's cross section is the same at each end, and everywhere in between. Similarly, a cylindrical fold contains fold elements which are everywhere parallel, and the cross section geometry is the same throughout the length of the cylindrical portion of the fold. Therefore, projection of fold geometry into cross section A-A' (Fig. 6-22) along a line that is not parallel to the plunge lines (i.e., sides of the pipe) would result in the attempted **physical movement of geometry to an improper position within the fold** (i.e., projection of a 38-deg dip from plunge line No. 6 to plunge line No. 5, which has only 24 deg of dip).

We shall now consider the application of plunge projection. Figure 6-23 is a structure

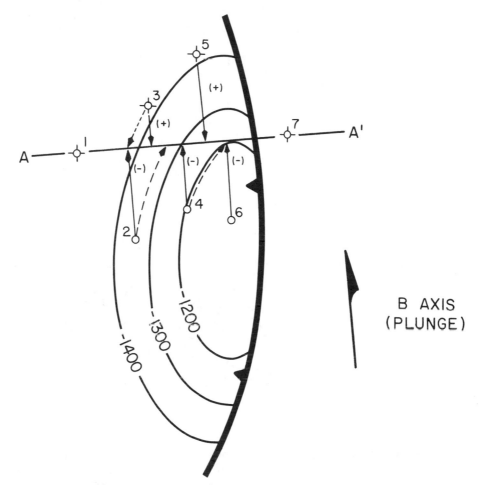

Figure 6-23 Structure map of a plunging cylindrical fold cut by a west dipping reverse fault. (From W. Brown 1984. Published by permission of author and the American Association of Petroleum Geologists.)

map of a plunging cylindrical fold, cut by a reverse fault. We want to construct cross section A-A' as shown on the structure map. Since only Wells No. 1 and 7 actually lie on the line of section, and we wish to include more data than that available from the two wells, we have no choice but to project the additional wells into the line of section.

Well No. 1 is drilled off the flank of the structure and Well No. 7 is in the footwall in the opposite fault block. Wells No. 2 through 6 can be projected into the line of section in several different ways: most commonly either parallel to structural strike (dashed arrow lines), or parallel to plunge (solid arrow lines). In the former case, Wells No. 2 through 5 are projected into A-A' along structural strike (parallel to the contours), and Well No. 6 cannot be projected at all. Figure 6-24, the cross section obtained by projecting along structural strike, is obviously very confusing, and results in an **unacceptable interpretation** of the fault shape, as well as the relationship of hanging wall and footwall rocks.

The direction of plunge of the B-axis is indicated on the structure map in Fig. 6-23 and therefore the direction in which the wells must be plunge-projected. The B-axial direction can be obtained by analysis of any form of dip data. The solid arrows show the planned projection path of the wells into the section. Next to each projection is a "+" or "−" sign. These signs indicate whether the structural elevation for a specific formation or structural marker (fault, dip rate, etc.) in the wells must be adjusted posi-

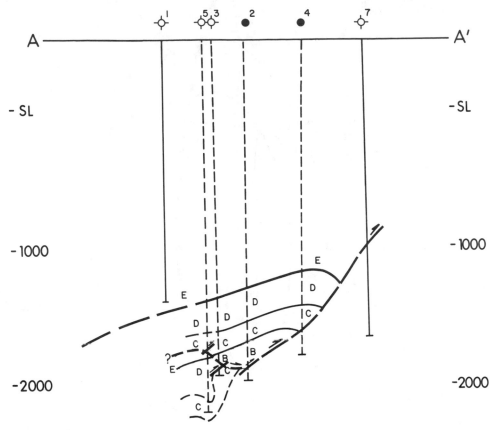

Figure 6-24 Structural interpretation along cross section A-A′ (Fig. 6-23) based on strike projecting well data into the line of section. The strike projected data results in an unacceptable interpretation. (Modified from W. Brown 1984. Published by permission of author and the American Association of Petroleum Geologists.)

tively or negatively for the direction and rate of plunge over the distance projected into the section line. For example, Well No. 2 has a subsea depth for the E Horizon of −1340 ft. When we project this well along plunge, the top of the E Horizon intersects the cross section at a subsea depth of −1410 ft. Therefore, the top must be adjusted downward by 70 ft to correct for the change in elevation. This correction in elevation can actually be calculated with trigonometry by using the angle of plunge and the length of the projection from the well site to the line of section. Each well must be corrected in this manner, with (+) values indicating elevations raised because the well is projected up-plunge, and (−) values indicating elevations lowered because the well is projected down-plunge.

Figure 6-25 is the completed cross section using all the data from the wells projected into the line of section parallel to plunge, or the B-axis (in the hanging wall). In order to differentiate between the wells that actually lie on the section line from those that have been projected, the wellbore stick for each projected well is dashed. Notice also that the numerical sequence (1 through 7) is now in proper order.

In conclusion, in the case of plunging folds, the plunge-projection method offers the most accurate means of projection of structural geometry for the hanging wall block into a cross section. This projection may be either up- or down-plunge, or both, and structural elevations likewise are adjusted up or down. The projections are made in order to ascertain the true cross-sectional shape of the structure.

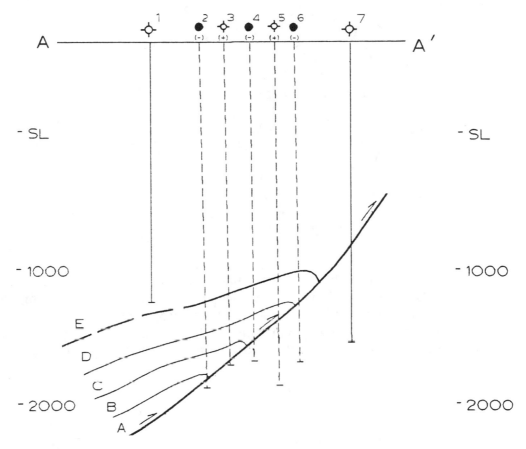

Figure 6-25 Structural interpretation along cross section A-A' (Fig. 6-23) based on the well data projected into the line of section parallel to plunge of the B-axis. The projection of data along plunge provides the most accurate method of projecting the fold geometry into the cross section. (Modified from W. Brown 1984. Published by permission of author and the American Association of Petroleum Geologists.)

Strike Projection

If you arc mapping in a geologic setting involving vertical tectonics, such as diapiric uplift, or in a tectonic area with nonplunging folds, projection along strike is often more beneficial than other types of projections. In the areas of extensional or vertical tectonics, such as the Gulf of Mexico Basin, the dip rate on structures is often constant in a direction parallel to the contours, or along structural strike (Fig. 6-26). Therefore, in order to preserve structural geometry, as well as stratigraphic relationships (syndepositional structures), wells should be projected along strike.

We first look at an example of projecting along strike in an area without faulting (Fig. 6-27). In Fig. 6-27a, Well No. 2 is projected along strike into the line of Section A-A'. In cross section, the top for the M Horizon projects into the correct structural position on the section and also projects at the proper rate of dip. Using this method, the projection does not require any correction factors for elevation or dip rate. In Fig. 6-27b, Well No. 2 is projected along strike and again projects correctly into the line of section B-B'. We can conclude that in this type of structural situation, projection along strike is preferred.

Now we shall look at some examples of projecting a well along strike in an area cut by faults (Fig. 6-28). Figure 6-28a is a portion of a structure map showing a west dipping

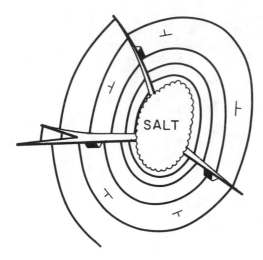

Figure 6-26 Nonplunging diapiric salt structure exhibiting constant dip along structural strike.

structure cut by an east dipping normal fault. East-west cross section A-A′ is plotted on the map. Wells No. 1, 2, and 4 lie on the section line. Since data from Well No. 3 is important to the interpretation, the well must be projected into the line of section. Considering the specific geologic conditions, Well No. 3 is strike projected into the A-A′ cross section, as shown below the structure map in the figure. The top of the horizon

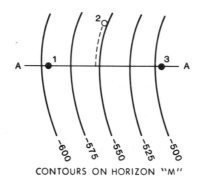

CONTOURS ON HORIZON "M"

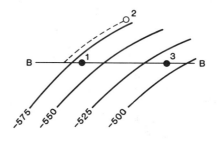

CONTOURS ON HORIZON "M"

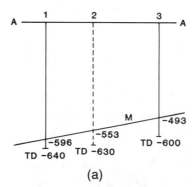

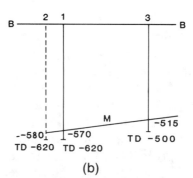

(a)   (b)

Figure 6-27a (a) Projection of well data along strike. Upper figure is a portion of a structure contour map. The lower figure is cross section A-A′ illustrating the projected data for Well No. 2. (b) Projection of well data along strike. Upper figure is a portion of a structure contour map. The lower figure is cross section B-B′ illustrating the projected data for Well No. 2. (Published by permission of Tenneco Oil Company.)

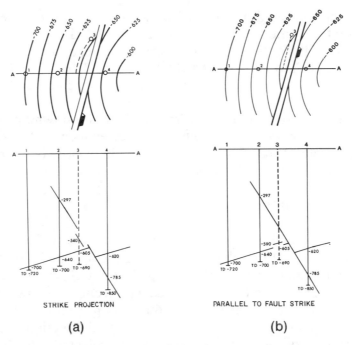

Figure 6-28 (a) Strike projection of well data in an area cut by a normal fault. The strike direction of the fault does not parallel the strike of the formation. Upper figure is the structure contour map. In the lower figure, cross section A-A' illustrates that the formation top for the mapped horizon fits the cross section: however, the fault cut in Well No. 3 is too deep and must be adjusted. (b) Parallel to fault strike projection. The well data are projected parallel to the strike of the fault surface. The fault cut projects correctly into the line of section, but the formation top must be adjusted. (Published by permission of Tenneco Oil Company.)

projects into the section in the correct structural position and maintains the correct structural dip, but the depth of the fault cut in the projected well is incorrect and must be adjusted. This adjustment is required because as Well No. 3 is projected from its actual position into the line of section, the distance from the well to the fault changes.

You might already have asked yourself the question, "If the well is projected parallel to the fault, how will the fault and horizon top project into the line?" Figure 6-28b depicts this situation. Well No. 3 is projected into the section line parallel to the strike of the fault. With this projection the fault cut fits the section correctly, but the horizon top is projected too shallow, requiring an elevation adjustment from −590 ft to −605 ft.

When projecting wells along strike in an area cut by faults, either the formation tops or the fault cuts usually will require an elevation adjustment. Therefore, we caution that projected wells often have limited use in a cross section and can, at times, cause more confusion than clarity. If a well must be projected into a line of section, be sure to identify the projected electric log or wellbore stick with a dashed line and clearly note any required corrections or adjustments.

Other Types of Projections

Figure 6-29 illustrates two other type projections: one along dip and one normal to the line of section. Well No. 2, in Fig. 6-29, is projected normal or at a right angle to the section line. This is sometimes referred to as the minimum distance method (Brown

1984a). Using this method for well projection, the top of the formation does not fall correctly on the section. Also, it is possible that the dip of the formation may be different at the actual well location than at its projected location on the line of section, since the well is projected into a higher structural position. This can add confusion to the section and may also require adjustments in the stratigraphic interval thicknesses in the projected well.

The second projection shown in Fig. 6-29 is a downdip projection of Well No. 4. With this projection the formation top is projected incorrectly. Also, a problem may exist with the projected formation dip rate, since the well is being projected into a downdip position. Neither method is recommended for projecting well data into a cross section.

Projection of Deviated Wells

If a directionally drilled well is to be included in a cross section, it is best to have the line of section follow the plan view path of the deviated well (Fig. 6-15). If the line of section is not coincident with the well path, the well will have to be projected into the cross section. The procedure for such data projection is shown in Fig. 6-30, which shows the map view of directional Well No. 6 and its projection into the line of section A-A'. The deviated well path is almost coincident with the section line, but not exactly. Therefore, directional survey data points must be projected into the line of section. This is accomplished by first determining the approximate direction of formation strike either from a structure map or seismic data and then projecting each data point parallel to the formation strike direction as shown in the figure.

If the electric log is represented by a stick, each directional survey data point is plotted individually in cross section and then the points are connected with a smooth curved line representing the electric log in stick form. Notice in Fig. 6-30 that the projected data points for the 4000-ft, 5000-ft, 6000-ft, and 7000-ft true vertical depths are not the same distance from the surface location as the actual depths plotted along the directional

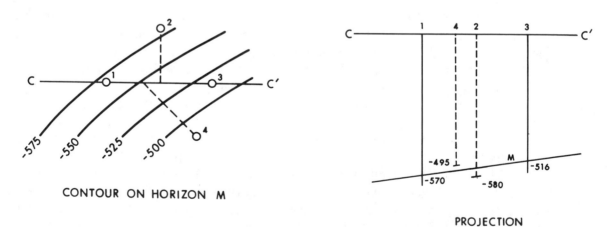

CONTOUR ON HORIZON M

PROJECTION

Well No. 2 – Normal to Section Projection.
Well No. 4 – Downdip Projection.

Figure 6-29 Projection of well data into a line of section using the normal to section and downdip projection methods. Neither method is recommended for projecting well data. (Published by permission of Tenneco Oil Company.)

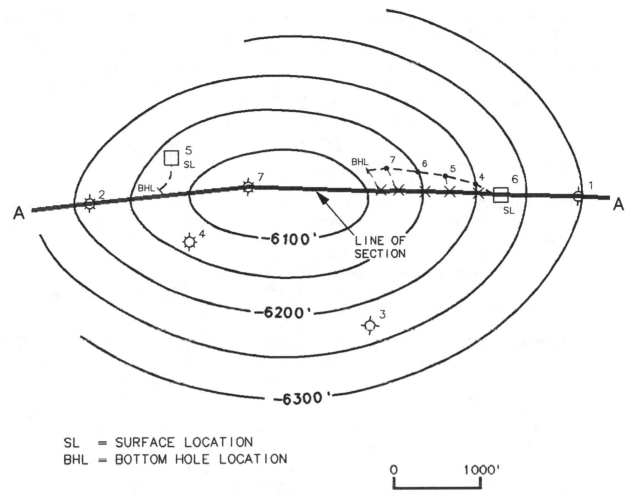

SL = SURFACE LOCATION
BHL = BOTTOM HOLE LOCATION

0   1000'

Figure 6-30 Structure contour map with line of section A-A'. Deviated Well No. 6 is projected into the line of section parallel to formation strike.

path. This correction must be made when the individual directional survey points are plotted in cross section. If each survey data point is projected correctly and plotted correctly in cross section, the projected directional wellbore stick will depict the true representation of the directionally drilled well in cross section.

When the actual electric log for a deviated well is used to project into a cross section, the log must be cut at various locations to retain true vertical or subsea depth, whichever is used. In Fig. 6-30, we see that the projected depth points plotted on the map are not the same distance from the surface location as the actual plotted depths. Therefore, adjustments must be made to the log in order to include it in the cross section. These adjustments are made by cutting out small sections of the log at various depths, as needed, to maintain true vertical or subsea depth on the section after projection.

Figure 6-31 shows a deviated well that is cut at various depths to retain its true depth in cross section after projection. It is best to delete sections of the log that do not contain any important correlations. If possible, shale sections should be used to make these adjustments.

Figure 6-32 is a north-south structural cross section through West Cameron Block 192 Field, Offshore Louisiana. The section is composed from six straight and six directionally drilled wells. The structure map on the U Sand is placed as an insert on the cross section. Notice that the line of section has a zig-zag pattern from north to south. It appears that this zig-zag pattern was used to accommodate directionally drilled wells. In other words, the path of the section line changes direction to parallel as closely as possible the path of the directional wells. By doing so, the projection of data from each deviated well into the line of section is minimized, thereby providing the most accurate representation of the directionally drilled well data in the cross section.

Projecting a Well into a Seismic Line

A well path projected into a seismic line (Fig. 5-14a and b) can often cause more confusion than clarification to an observer unfamiliar with the projection or details of the geologic structure. Therefore, care must be taken to properly project a well into a seismic section and clearly label the projected well data.

The techniques outlined for projecting a well into a cross section are also valid for projecting well data into a seismic section. The following are a few additional suggestions that may help when projecting a well into a seismic section.

1. Be certain that the velocity function used to convert the depths to time for plotting the well is the most accurate one for the area. Incorrect velocity information can cause some serious errors when projecting a well, particularly if it involves a directionally drilled well. Also remember that a seismic section **is not** a geologic cross section, and that the apparent angle of a well on a seismic section may not resemble the angle of the well on the equivalent geologic cross section. It is also important to know both the horizontal and vertical scales of the seismic section being used.

2. Consider dashing all projected wells posted on a line. If the well path actually crosses the plane of the line, mark the point clearly with a different color solid line or a horizontal dash across the well path.

3. Remember that when a well is projected along structural strike, many of the faults may appear to intersect the projected well path at different depths than the actual intersections in the well. Be ready to field questions when a line like this is used for presentation purposes.

4. Too often the **minimum distance** projection method is used for projecting well data into a seismic section. Whenever possible, do not use this method because the data frequently will be projected incorrectly, resulting in a conflict between the well and seismic data (Fig. 5-14a and b).

5. It is not good practice to project well data over very long distances into a seismic section. Remember, the geometry of any structure changes laterally even over short horizontal distances. Projections of well data over long distances can and often do cause great confusion in the interpreted seismic section.

CROSS SECTION CONSTRUCTION ACROSS FAULTS

In Chapter 7 we discuss the importance of the fault component *vertical separation* and define the missing or repeated section in a wellbore, resulting from a fault, as being equal to this fault component. Therefore, when constructing a cross section with faults, the

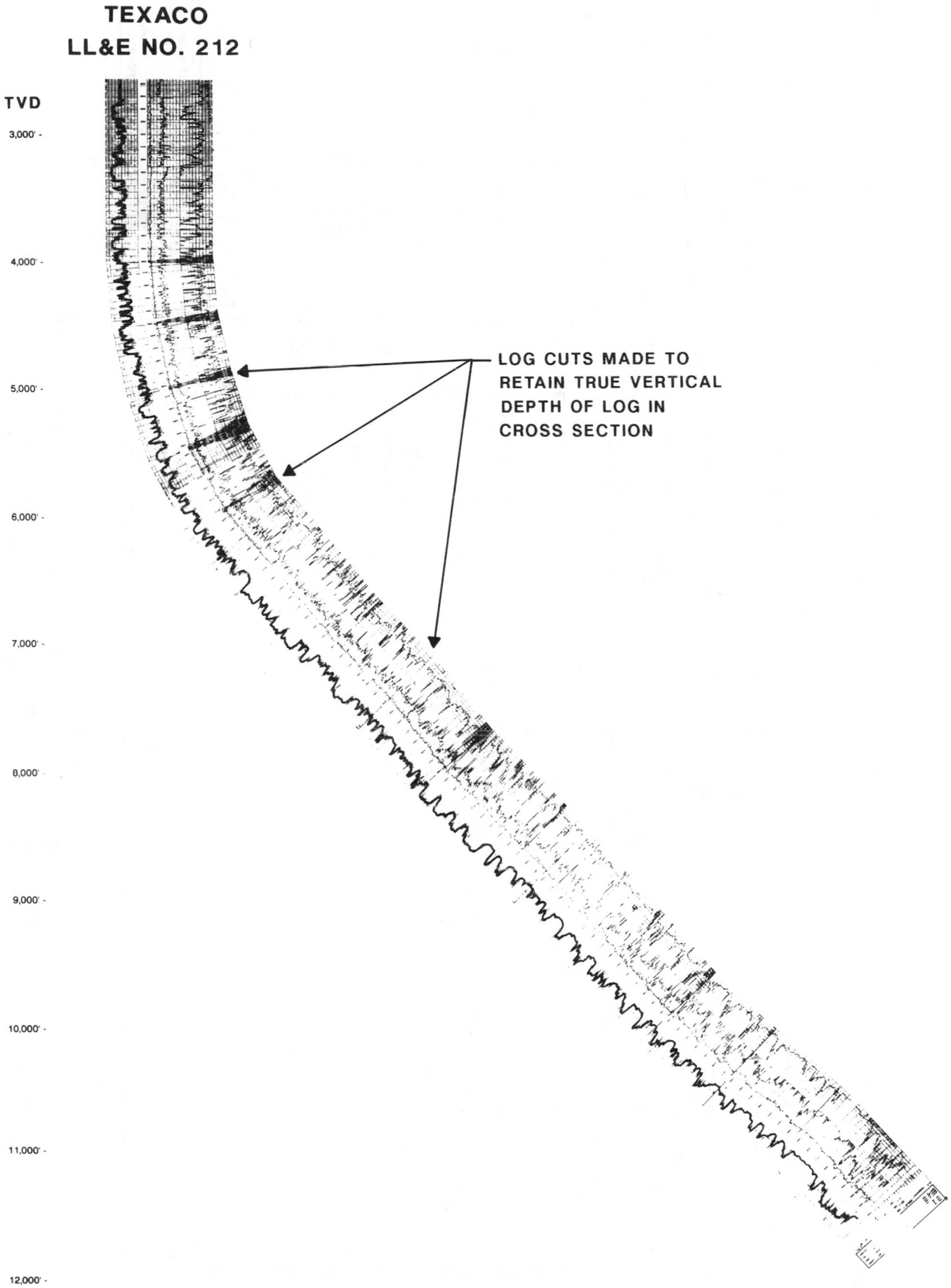

**TEXACO
LL&E NO. 212**

TVD

3,000' -

4,000' -

5,000' -

6,000' -

7,000' -

8,000' -

9,000' -

10,000' -

11,000' -

12,000' -

LOG CUTS MADE TO
RETAIN TRUE VERTICAL
DEPTH OF LOG IN
CROSS SECTION

Figure 6-31 The film of the log for Well No. 212 is cut at various depths to retain true vertical depth of the entire log in cross section.

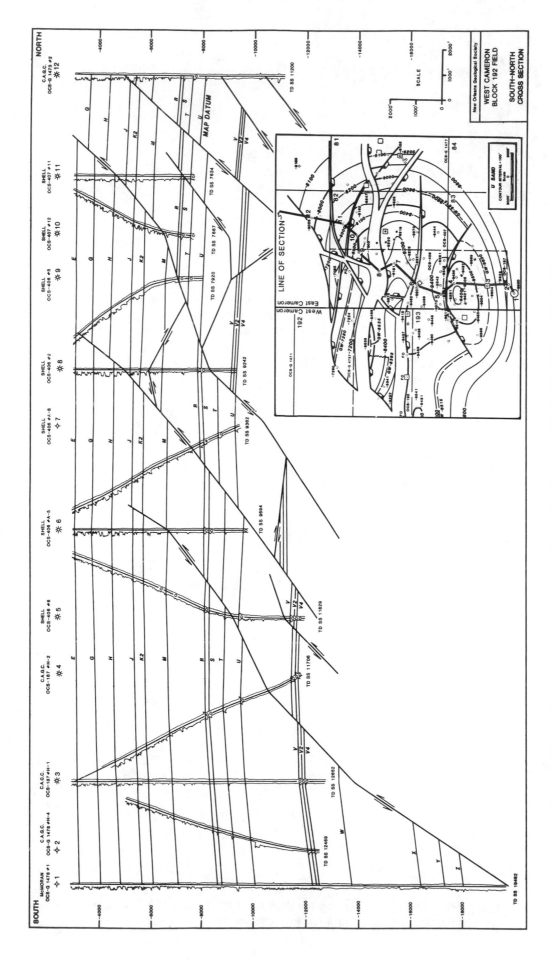

Figure 6-32 North-south structural cross section in West Cameron Block 192, Offshore Gulf of Mexico. The section line (see structure map insert) has a zig-zag pattern to accommodate the directionally drilled wells, thus minimizing the projection of well data into the line. (From *Offshore Louisiana Oil and Gas Fields*, v. II, 1988. Published by permission of the New Orleans Geological Society.)

vertical separation must be used to correctly displace a formation, bed, or marker from one fault block to another. In this section, the method for using the vertical separation in the preparation of a cross section containing a fault is discussed in detail. The concept is basic and the correct technique of construction very simple; yet the technique is often misused. Before reading this section, it may be necessary to become familiar with the fault *component vertical separation* and the difference between this component and fault *throw*. These subjects are covered in the beginning of Chapter 7, particularly in the sections entitled **Definition of Fault Displacement** and **Fault Data Determined from Well Logs**.

If the missing section in an electric log is mistakenly used as throw in the construction of a cross section involving a fault, chances are the displacement of beds, formations, or markers across the fault will be incorrect. The values for vertical separation and throw are the same only for a geologic structure with a vertical fault or a situation in which a fault cuts through horizontal beds. With increasing formation dip, however, the values are different and can be significantly so with steeply dipping beds. Therefore, the bed or formation displacement error in a cross section, using the missing section as throw, will usually be greater with increasing structural dip.

Figure 6-33 graphically depicts the correct method for using vertical separation (missing section) to obtain the proper bed displacement across a fault. The incorrect method using throw as the missing section is also shown for comparison. In this example, assume there is sufficient control from well data or a structure map in the downthrown fault block to establish the dip of the formation (Bed A) and its intersection with Fault 1 in the downthrown block at −7100 ft, as shown in Fig. 6-33a. The missing section for Fault 1 determined by log correlation is 300 ft. With this information, the next step is to determine where Bed A intersects the fault in the upthrown fault block. This is accomplished by correctly plotting the 300 ft of missing section across the fault into the opposite fault block on the cross section as vertical separation.

First, look at the missing section plotted incorrectly as throw. *Throw* is defined as the difference in vertical depth between the fault intersection with the bed in one fault block and the fault intersection with the same bed in the opposing fault block determined in a direction perpendicular to the strike of the fault. Applying this definition to the cross section in Fig. 6-33b, which is perpendicular to the strike of Fault 1, the depth of intersection of Bed A with the fault in the upthrown fault block is −6800 ft.

Vertical separation is defined as the measured vertical distance between a bed projected from one fault block across the fault to a point where the projection is vertically over or under the same bed in the opposite block. *The missing or repeated section observed in a wellbore as the result of a fault is vertical separation and not throw.* Applying this definition to the cross section in Fig. 6-33c, the dip of Bed A is projected from the downthrown block through the fault, as if the fault were not there, to the upthrown block until it is 300 ft vertically under the fault. The depth at this point (−6900 ft) is the upthrown intersection of Bed A with the fault.

There is a 100-ft vertical difference between the two methods in the estimated depth of the upthrown intersection of Bed A with the fault. Incorrectly using missing section as throw places the upthrown block 100 ft too high. This mistake could be costly, particularly if a well were planned for the upthrown block based on the incorrect cross-sectional interpretation using vertical separation incorrectly as throw.

Because this technique is so important, we will review it with two other geologic examples, one involving a normal fault and the other, a reverse fault. Figure 6-34 shows cross section A-A′ and an insert enlargement of a portion of the section. The missing

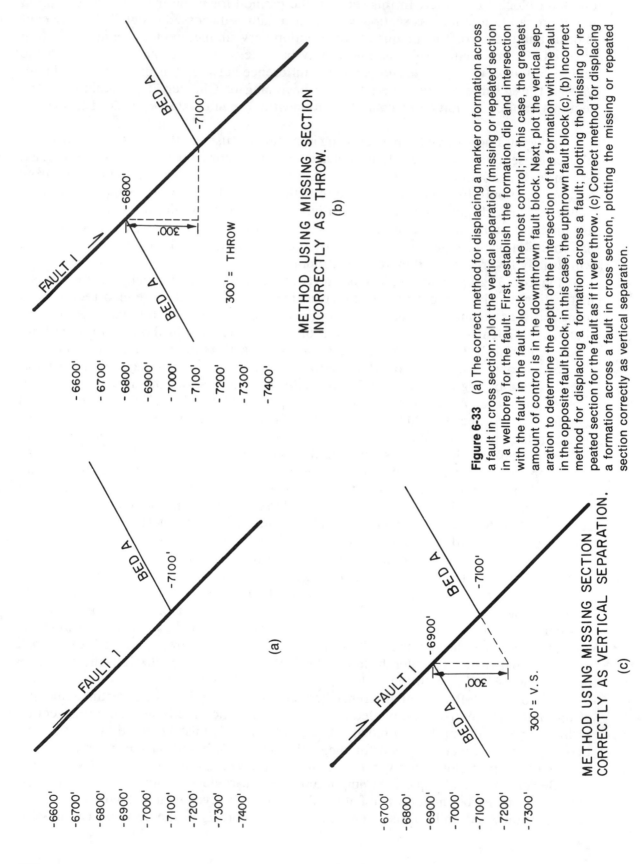

Figure 6-33 (a) The correct method for displacing a marker or formation across a fault in cross section: plot the vertical separation (missing or repeated section in a wellbore) for the fault. First, establish the formation dip and intersection with the fault in the fault block with the most control; in this case, the greatest amount of control is in the downthrown fault block. Next, plot the vertical separation to determine the depth of the intersection of the formation with the fault in the opposite fault block, in this case, the upthrown fault block (c). (b) Incorrect method for displacing a formation across a fault; plotting the missing or repeated section for the fault as if it were throw. (c) Correct method for displacing a formation across a fault in cross section, plotting the missing or repeated section correctly as vertical separation.

section in Wells No. 3, 5, and 2 for the normal Fault 1 is 400 ft. Since the missing section is equal to vertical separation, we use this fault component value to prepare the cross section construction across Fault 1.

The attitude of the 6000-ft Sand in the upthrown block of Fault 1, in cross section, is established by the available well control in that block and a completed structure map. Well No. 5 is close enough to the intersection of the 6000-ft Sand with the fault to establish the depth at which this intersection occurs. Once this depth has been established, we must determine the displaced position (depth) at which the sand intersects the fault in the downthrown block. The depth chosen again depends upon your understanding of missing section. Remember, the missing section is not equal to throw in areas involving dipping beds. Therefore, if the mistake is made of using 300 ft of fault displacement as throw, the estimated depth of the intersection of the sand with the fault in the downthrown block would be incorrect. In this case, shown in Fig. 6-34, the intersection would be too shallow by approximately 75 ft, shown in the enlarged insert. You must use the definition of vertical separation to project the dip of the 6000-ft Sand from the upthrown block, through the fault as if the fault were not there, into the downthrown block until the projection is 300 ft vertically over the fault. This point, labeled I, is the correct intersection of the sand with the fault in the downthrown block. Notice that the projection of the 6000-ft Sand from the upthrown to downthrown blocks is curved to follow the actual changing dip of the sand.

We now look at one example involving a reverse fault. Figure 6-35 shows cross section A-A' and an enlarged insert of a portion of the cross section. The repeated section resulting from Fault 1 as correlated with all surrounding well control is 400 ft, which is the vertical separation of the fault.

Assume for this example that because of more well control, such as Wells No. 3 and 5, the attitude of the 5000-ft Sand and its intersection depth and location with Fault 1 is first established in the hanging wall fault block. The task now is to determine the displaced position of the sand in the footwall and elevation of the bed intersection with the fault. With reverse fault geometry, the technique is actually easier and can be used to establish two control points for the 5000-ft Sand in the opposing fault block (in this case the footwall). Working with the enlarged portion of the figure, follow along the sand in the hanging wall until the bed is at a point that is 400 ft vertically above the fault surface. This point, labeled I, is the intersection of the sand with the fault in the footwall.

A second point of control for the 5000-ft Sand can be established by measuring 400 ft vertically down from the intersection of the 5000-ft Sand with the fault in the hanging wall. This point, labeled I', corresponds to the depth to the top of the sand in the footwall at the point vertically under the intersection of the sand with the fault in the hanging wall.

As an academic exercise, estimate the intersection of the 5000-ft Sand with the fault in the footwall using the incorrect assumption that the 400 ft of repeated section is a measurement of throw. As a second exercise, estimate the size of the actual fault throw considering the correct geometry shown in cross section in Fig. 6-35.

It is important to note again that the preciseness of the projection of bed or formation dip, through a fault, determines the accuracy of the bed or formation displacement across the fault. If the bed dip is changing in the area where the projection is made, the projection must follow this changing bed dip (observe again the curved projection of the 6000-ft Sand in the enlargement insert in Fig. 6-34). In other words, the projection of a bed or formation across a fault need not be a straight line. If the formation dip is not constant across the fault, then the projection will not have a constant dip since it must follow the changing bed dip to accurately represent the vertical separation.

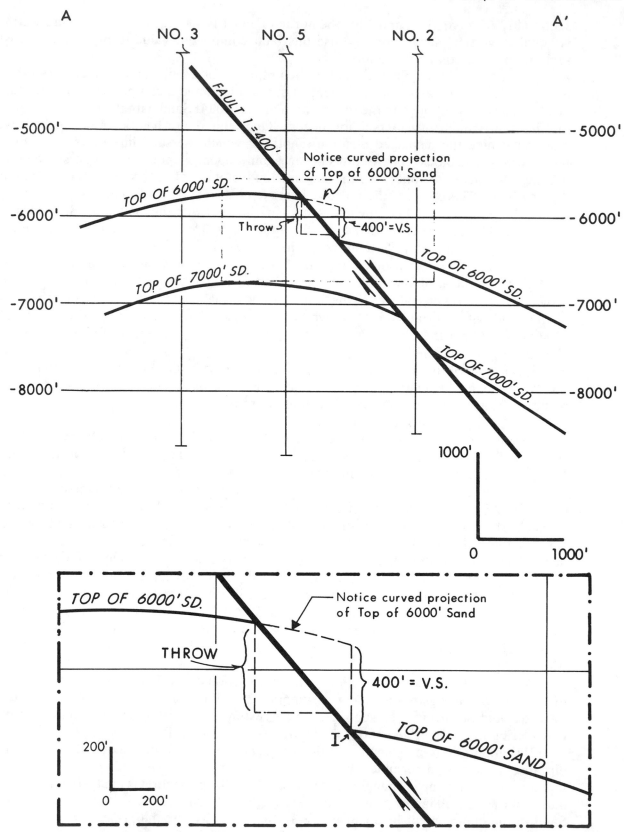

Figure 6-34 An illustration using the correct method for displacing a formation across a normal fault.

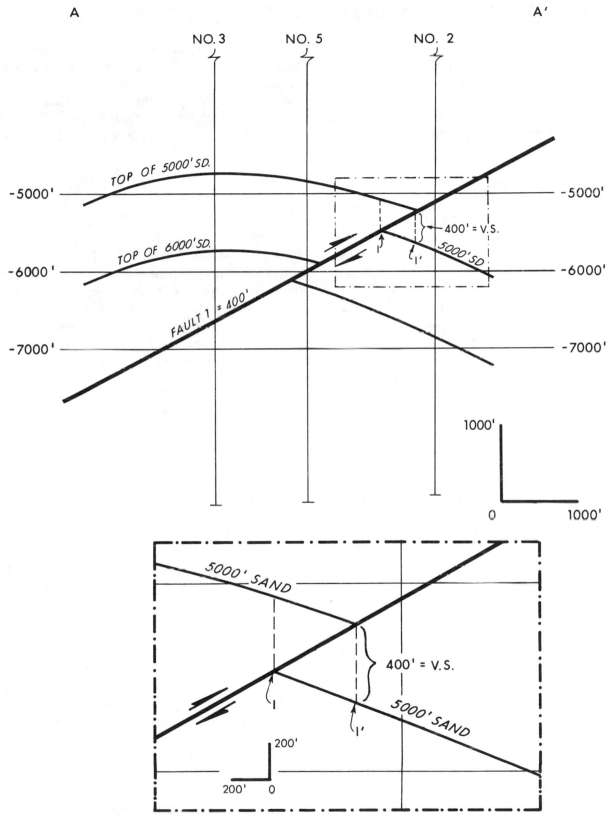

Figure 6-35 An illustration using the correct method for displacing a formation across a reverse fault.

THREE-DIMENSIONAL VIEWS

In petroleum exploration and exploitation, a map or cross section alone cannot always represent the complete subsurface geologic picture, because each is limited to two dimensions. Since neither a map nor cross section presents data in three dimensions, a detailed study may require the use of additional techniques that aid in visualizing the geology in three dimensions. Such additional aids include *log maps, stick or fence diagrams, isometric projections, three-dimensional structural models, and block diagrams*.

Log Maps

Log maps are the simplest means of combining plan view information with some vertical dimensional data. Figures 6-36 and 6-37 illustrate two different types of log maps. In Fig. 6-36, electric log sections and data charts for the Middle Cruse Miocene Sand Member from Trinidad, BWI are superimposed on a structure map of the same sand member. This type map allows you to quickly review the vertical sand conditions and producing characteristics of the sand in each well with respect to the well's structural position within the reservoir. Notice that only the SP and the amplified short normal resistivity curves were used in the log sections. By looking at this map, you can very quickly see the complexity of the one or more sand members making up this reservoir and the relationship of sand quality to structural position and well location.

Figure 6-37 is another type of log map in which the SP curve for each electric log is placed next to the appropriate well location in an attempt to visualize the three-dimensional paleo-geographic conditions during the time of sand deposition. In this field, the sands appear to be of two types: (1) a major cut and fill channel sand, and (2) a very poor quality, highly calcareous fringe complex. The calcareous nature of the sand was determined from cores taken in a number of wells.

If the actual structure map for the 10,500-ft Sand was superimposed on this log map in a manner similar to that shown in Fig. 6-36, this additional step would improve the evaluation of the sand conditions with respect to the localization of the major channel sand and calcareous fringe complex sand in comparison to the structural position of the individual wells, production characteristics of the local reservoirs, and hydrocarbon trapping conditions. Such information can lead to the identification of overlooked workovers and recompletions, in addition to possible development drilling locations.

Fence Diagrams

Fence diagrams, also called *panel diagrams*, consist of a three-dimensional network of cross sections drawn in two dimensions. They are designed to illustrate the areal relationship among several wells that are located in close proximity to each other. Fence diagrams can either be structural or stratigraphic. Figure 6-38 is a typical example of a fence diagram in which actual electric log segments have been traced for use at each well location. Fence diagrams are often prepared using only log sticks, although log sticks do not provide the correlation detail usually needed for detailed work. Fence diagrams are also usually diagrammatic because they seldom have a common horizontal and vertical scale.

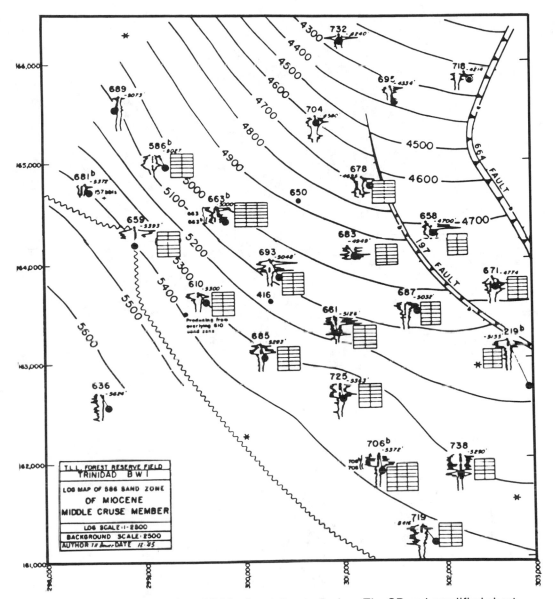

Figure 6-36   Log map of the Middle Cruse Sand member. The SP and amplified short normal log curves are superimposed on the structure map for the sand member. The relationship of sand quality to structure position within the reservoir can quickly be evaluated. (Modified after Bower 1947. Published by permission of the American Association of Petroleum Geologists.)

Fence Diagram Construction.   The construction of all fence or panel diagrams begins with a well base map. The plane of the base map represents the chosen datum plane and each well location is taken to be the point where the well intersects the datum plane. A vertical line is drawn at each well control point and the well data are either hand plotted along this line, or a film of the actual log curves, reduced to the proper scale, is attached to the line. For example, in Fig. 6-38, the SP and resistivity curves were hand traced from reductions of the actual log. Figure 6-39 is a fence diagram constructed using the actual reduced log curves.

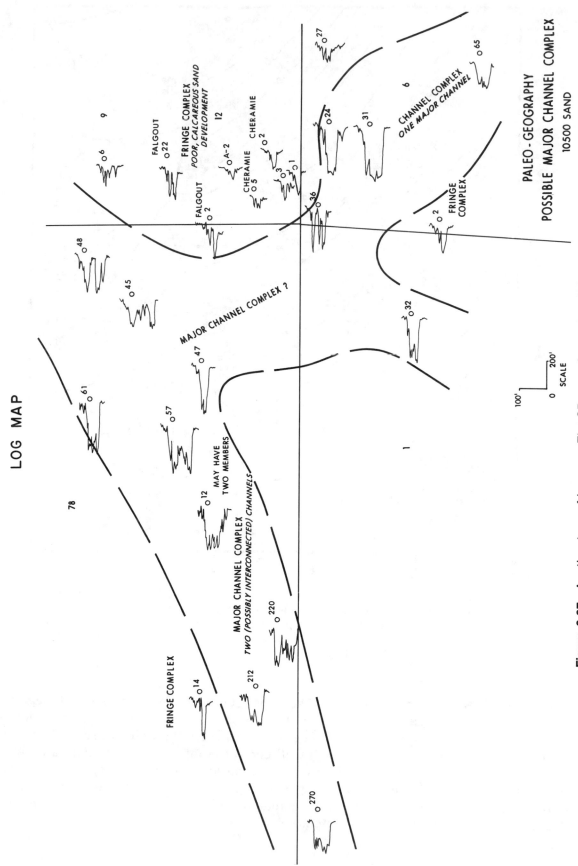

Figure 6-37 Another type of log map. The SP curve for each well is placed at the surface location for the well in an attempt to visualize the three-dimensional characteristics of the 10,500-ft Sand. (Published by permission of Texaco USA).

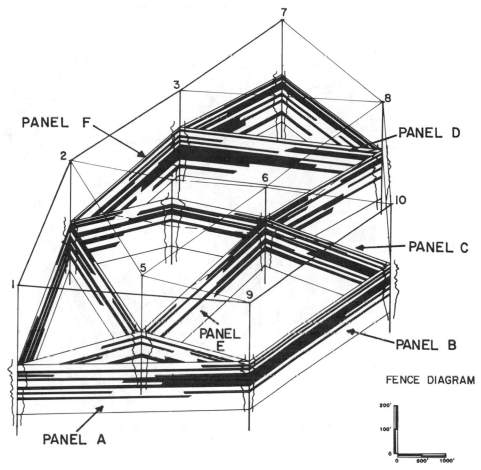

PANEL F

PANEL D

PANEL C

PANEL E

PANEL B

FENCE DIAGRAM

PANEL A

Figure 6-38 Layout of a typical stratigraphic fence diagram in which the actual electric log segment for each well was traced at the well location. (Modified from Boeckelman, unpublished. Published by permission of the author.)

In the preparation of a structural fence diagram, the plane of the map is the chosen datum which can be sea level or any chosen elevation above or below. For example, the chosen datum for the fence diagram in Fig. 6-40 is −1500 ft, which is a subsea datum equivalent to 1500 ft below sea level. The log sections or hand-traced curves are hung on the vertical line at each well location referenced to the chosen datum. The structural profile of the formation or bed(s) on the section is drawn by initially connecting the major correlation markers from each well. The number of correlation markers connected depends upon the detail desired for the diagram.

It is important to try to minimize the interference of one panel with another to reduce the effect of masking data. It is a good idea to begin to complete the panels starting with the east-west sections first, and in particular, the panels nearest the lower or front end of the map (see panels A, B, and C in Fig. 6-38). The northwest-southeast, or northeast-southwest panels are completed next up to a point where they intersect the east-west panels from the front or where they disappear behind these panels from the rear, such as panels E and F in Fig. 6-38. Once all the correlation markers have been connected on all the panels and the desired labeling added, the diagram is complete. If dips are most prominent in the north-south direction, an isometric projection may be more helpful, or

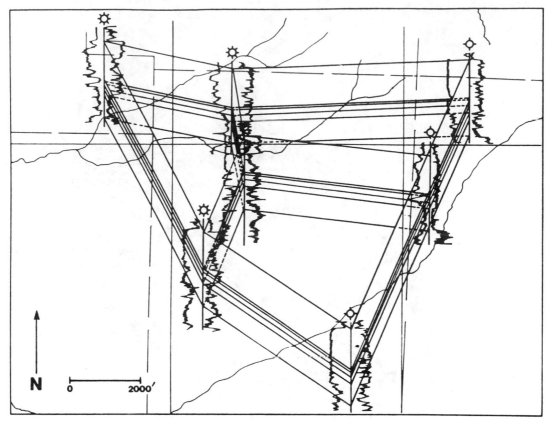

Figure 6-39 Layout of a stratigraphic fence diagram using the actual reduced log curves. (From Langstaff and Morrill 1981. Published by permission of the IHRDC.)

the diagram may be drawn with north toward the right instead of toward the top of the map, as is more common.

For the construction of a stratigraphic fence diagram, the plane of the map is considered to be either the bedding plane or an unconformity that is the chosen datum horizon. Just as with the structure fence diagrams, a vertical line is drawn at each well

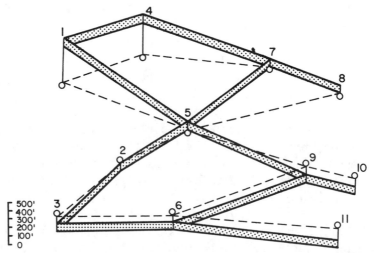

Figure 6-40 Structural fence diagram: reference plane (dashed line) is 1500 ft below sea level. (From Bishop 1960. Published by permission of author.)

location. Since most stratigraphic fence diagrams are prepared to illustrate correlations, facies changes, permeability barriers, sediment wedging, or some other stratigraphic feature, the actual log sections are normally required. Connect all the correlation markers to complete the fence diagram (Figs. 6-38 or 6-39).

Isometric Projections

Isometric projections are sometimes used to emphasize certain stratigraphic variations that might not be apparent on the conventionally oriented fence diagram. In an isometric projection, the map base is shown as if it were turned at an angle and tilted toward the front, which transforms the projection from orthogonal to nonorthogonal axes (Langstaff and Morrill 1981). Figure 6-41 shows the difference between a standard fence diagram and its transfer into an isometric projection. Notice that the rectangular base map becomes a parallelogram in the isometric projection.

The base map used in Fig. 6-41 has a coordinate grid system with both north-south and east-west scales provided. Any isometric projection will have distortion in any direction except in the directions of the original grid system. In other words, all lines parallel to the original north-south and east-west axes remain parallel in the isometric projection. Therefore, any data points to be measured from the base map must be measured along one or both of these coordinated grids and by doing so, they can be measured accurately. Measured distances in any direction except parallel to the original grid coordinates will be distorted. They may be greater or less than the actual distance. It is very important, therefore, to always include some type of coordinate grid system on all map bases that are planned as isometric projections.

Three-Dimensional Reservoir Analysis Model

One of the most accurate and detailed types of projections is what we call a *Three-Dimensional Reservoir Analysis Model*—abbreviated **RAM** (Wayne Boeckelman, unpublished). Using the basic rules of descriptive geometry, we can construct a three-dimensional model of a structure from which true length and true slope can be measured. This is accomplished by constructing the folding plane at 90 deg to dip (Fig. 6-42).

Construction Procedure. The construction of a three-dimensional reservoir model must begin with a completed structure map such as the one shown in Fig. 6-43a. Choose a grid that is 90 deg to the dip of the structure (Fig. 6-43b) so that true length and true slope can be plotted and measured from the model. Be sure that grid lines, which later become the three-dimensional panels, go through all the wells on the structure map, as shown in Fig. 6-43b, to ensure the use of all the geologic control. In this example, there are six wells to include in the grid. Notice, however, that there will eventually be more control points than just those from the six well locations. The model shows that there are actually 18 grid line intersections in addition to those through the six wells. Each of these intersections serves as a data control point. This means that there are a total of 24 geologic control points for the construction of this example model. You can immediately see that one of the benefits of this type model is to provide a more accurate interpretation because of the large number of control points.

Once the grid system is laid out, extend a line vertically straight down from each

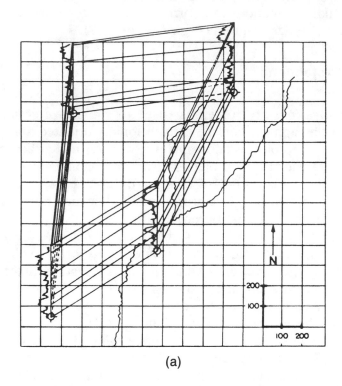

(a)

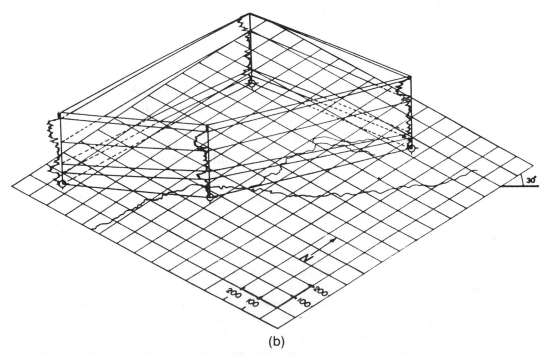

(b)

Figure 6-41 Comparison of (a) a standard stratigraphic fence diagram to (b) an isometric projection. (From Langstaff and Morrill 1981. Published by permission of the
IHRDC.)

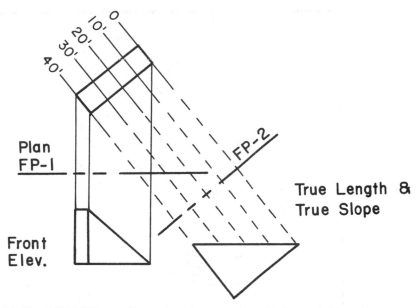

Figure 6-42 A RAM (three-dimensional reservoir analysis model) can be constructed to measure true length and slope by constructing the folding plane at 90 deg to dip. (From Boeckelman, unpublished. Published by permission of author.)

well and grid intersection as shown in Fig. 6-43c. As with all the other types of three-dimensional diagrams, a datum must be chosen. For this example we chose −9100 ft as the datum, so all data are referenced from this datum. Once the datum is chosen, we are ready to plot data for the specific stratigraphic interval to be studied. In this model, we are going to study the 9200-ft Sand.

We begin with a review of how to place the log or hand-drawn curves on the vertical line with reference to the datum. Since the intent is to evaluate the entire 9200-foot Sand package, the representation of this sand package must be placed on the model, and in doing so there are several choices. The most accurate method is to make a reduced film of the actual log section to accommodate the scale of the model and hang the log section on the structural top of the 9200-ft Sand at each well location. As mentioned before, it is best to use a limited number of log curves to reduce clutter on the model. The second choice is to hand trace the reduced log curves on the vertical line at each well location. The choice is more a matter of preference.

Our interest, in this case, is to hang the section on the top of the 9200-ft Sand in order to study and review the entire 9200-ft Sand package. At Well No. 3, for example, the top of the 9200-ft Sand is at a depth of −9150 ft. Since the referenced datum is −9100 ft, the first depth point is 50 ft below the datum. The log or hand-drawn curve(s) is positioned so that the top of the sand is 50 ft below the reference grid at Well No. 3, as shown in Fig. 6-43d. For Well No. 1 the top of the sand is at a depth of −9450 ft, which is 350 ft below the referenced grid datum of −9100 ft. Therefore, the top of the sand represented on the actual log or hand-drawn curve(s) is placed 350 ft below the grid at the Well No. 1 location. The depth below the referenced datum is calculated for each well and the appropriate log curve(s) is posted at that location.

Now what about the 18 other grid intersections? Using the grid overlay on the structure map of the 9200-ft Sand (Fig. 6-43b), determine the structural elevation for the top of the formation at each grid intersection, calculate the difference between the formation top and the reference depth, and post the structural top at the appropriate depth

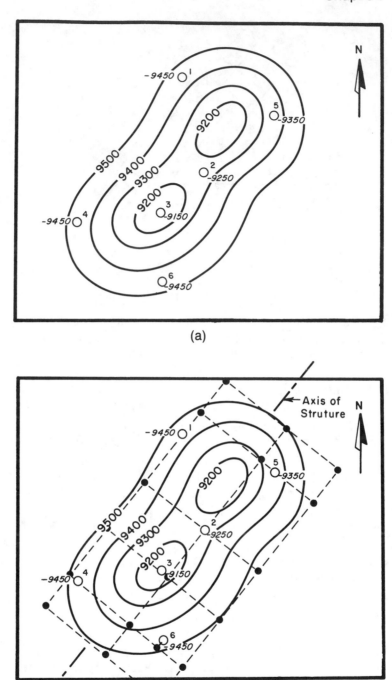

(a)

(b)

Figure 6-43 (a) Completed structure map on the top of the 9200-ft Sand. (b) RAM grid positioned at 90 deg to dip. Grid lines must intersect each well to ensure the use of all geologic data. (c) Choose a datum and construct lines from the datum vertically straight down from each well and grid location. (d) Post the elevation for the top of the 9200-ft Sand at each grid intersection and connect the points with straight lines representing the top of the 9200-ft Sand. (e) Overlay the grid system on the base of sand map and post the base of the sand at each grid intersection and each well location. (f) Begin detailed construction of the sand and shale data for each panel starting at each well location, working outward in all panel directions. (From Boeckelman, unpublished. Published by permission of author.)

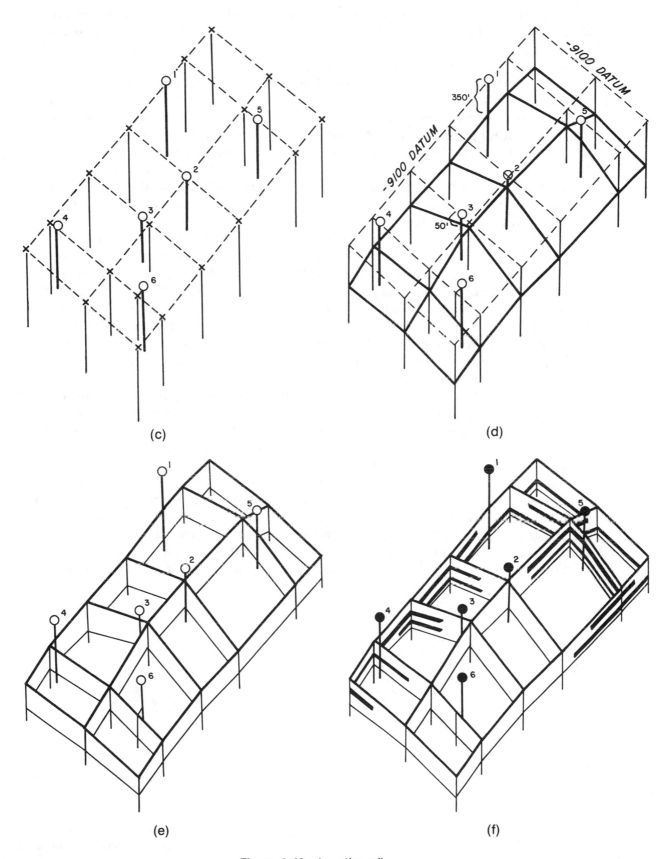

(c)

(d)

(e)

(f)

Figure 6-43 (*continued*)

on the vertical line at each separate grid intersection. After posting the elevation for the structural top of the 9200-ft Sand at each of the 24 grid intersections, connect the points with straight lines as shown in Fig. 6-43d. These connected lines represent the top of the 9200-ft Sand on the three-dimensional model.

The control points connected with straight lines may not accurately represent the attitude of the sand top, since the true top of the sand has curvature between many control points. The process can be improved further by posting the depth values for all structure contour/grid intersections and connecting these depths. Approximately 30 depth control points for contouring the top of the sand are added using this technique. With these additional control points, the top of the sand can be contoured to more closely represent the true attitude of the sand. Also, grid lines can be added at strategic locations to refine the RAM as desired. The amount of detail depends upon the individual interpreter and the intent of the reservoir model.

From the log sections for the six well locations, the base of the sand can be recognized. But what about the base of sand at the other 18 grid intersections? We must now repeat the process outlined above for the structural base of the sand. Overlay the grid over a structure map on the base of the sand and determine the structural elevation of the base of the sand at each grid intersection and mark this base on each vertical intersection line. Connect all the depth points for the base of the sand and the three-dimensional panels for the 9200-ft Sand are complete, as shown in Fig. 6-43e. Notice that all the panels are at right angles to one another, so very little data on any particular panel are hidden by other panels.

Finally, using the net sand and shale data from each of the electric logs, begin construction of the model panel by panel. Start at each well location and work outward in all panel directions to prepare a complete interpretation of the sand/shale distribution for the 9200-ft Sand. This sand/shale work can be done in a general way or detailed based upon each foot of analyzed electric log (Fig. 6-43f).

This three-dimensional reservoir analysis model (RAM) can be used to very accurately evaluate the distribution of sand and shale, and aid in developing a depositional model for the overall sand package as well as the individual members. If hydrocarbons are present, water levels can be plotted and an entire color scheme developed to represent sand, shale, hydrocarbons, and water. There are advantages to this model not associated with other types of cross sections, fence diagrams (Fig. 6-44), or isometric projections.

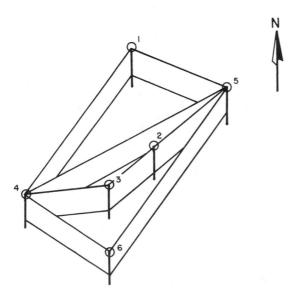

Figure 6-44 Typical fence diagram layout for the 9200-ft Sand. (From Boeckelman, unpublished. Published by permission of author.)

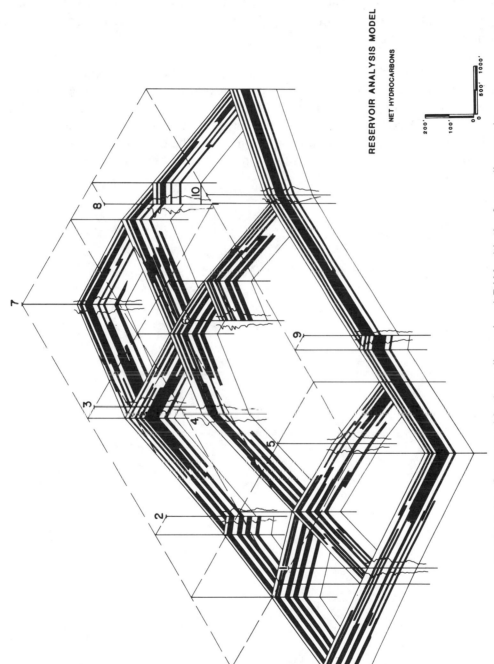

Figure 6-45 Compare this three-dimensional RAM with the fence diagram in Fig. 6-38, prepared from the same well data.

RESERVOIR ANALYSIS MODEL

NET HYDROCARBONS

1. True length and slope measurements can be made.

2. More panels are available for study, providing more detailed and accurate analysis.

3. Because of the right angle grid system, less data are obscured in this model as compared to other types of panel diagrams.

4. By measuring total net sand at each grid intersection and well location, it is possible to construct a more accurate net sand and net hydrocarbon isopach map.

5. The overall appearance of the model is more pleasing to the eye and is easier to understand and follow.

Figure 6-45 is another example of a completed three-dimensional reservoir model. Compare this model with the fence diagram in Fig. 6-38, which is constructed from the same well data. Notice how clearly the structure is represented in the three-dimensional model (Fig. 6-45) compared to the fence diagram (Fig. 6-38). Also observe how much more data are presented with minimal obstruction in the three-dimensional model. Finally, true length and slope can be measured on the model, which cannot be done with the fence diagram.

CHAPTER 7

FAULT
MAPS

INTRODUCTION

Faulted structures play a very significant role in the trapping of hydrocarbons; therefore, it is imperative that anyone involved in the exploration for or exploitation of hydrocarbons use correct and accurate subsurface mapping techniques for the preparation of fault and integrated structure maps. A reasonable structural interpretation, in most faulted areas, must begin with an accurate fault interpretation resulting from the construction of fault surface maps and the proper integration of these fault maps with structure maps. We refer to constructed fault maps as **fault surface maps** rather than the more commonly used term "fault plane maps," since most fault surfaces are not true planes.

The data required to construct a fault surface map are obtained from the correlation of well logs and interpretation of seismic sections. In Chapter 4, we presented the methods and procedures for recognizing a fault in a well log and determining its size for mapping. In this chapter, we discuss the importance of mapping faults and present the methods for constructing fault maps with fault cut data acquired from electric well logs and seismic sections.

The preparation of accurate fault surface maps requires three-dimensional thinking and a good understanding of the structural style of the area being mapped. When we make reference to the understanding of structural style, we are referring to that specific assemblage of geologic structures common to a particular petroleum province (Fig. 7-1). In order to prepare geologically reasonable maps, one must be familiar with the tectonic setting being worked and the fault and structural patterns expected. Although many of the basic concepts and techniques for mapping faults are universally valid, their con-

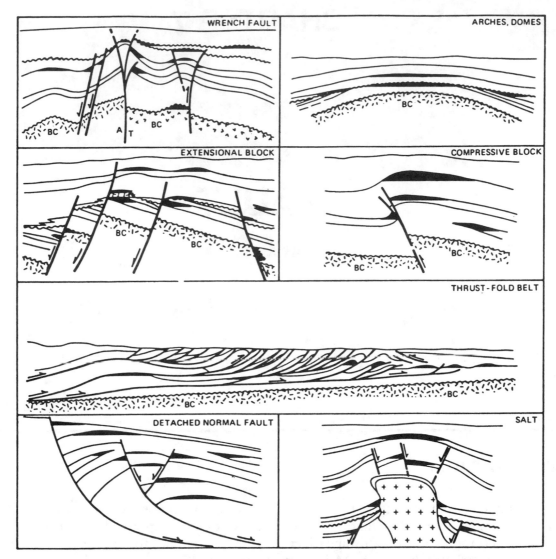

Figure 7-1 Schematic diagrams of hydrocarbon traps associated with various structural styles. BC, basement complex: T, displacement toward viewer: A, away from viewer. (From Harding and Lowell 1979. Published by permission of the American Association of Petroleum Geologists. Salt-related closures modified from *Salt Domes* by Michel T. Halbouty. Copyright 1979 by Gulf Publishing Company, Houston TX. Used with permission. All rights reserved.)

struction and application often depend upon the kinds of geologic structures being mapped.

Detailed discussion of structural geology or structural styles is beyond the scope of this text, although some aspects of this subject are presented in several chapters, in particular Chapter 9 on structural balancing. Since this is an advanced level text with the primary focus on mapping techniques, we make the assumption that you have a general understanding of the principles of structural geology as outlined in such texts as Billings (1972), Harding and Lowell (1979), and Suppe (1985).

Faulted structures can be simple or complex. To provide the best structural interpretation with the available data, the integrity of the structure must be shown to be sound and geologically reasonable. To provide the most accurate and sound geological interpretation in faulted areas, the construction of fault surface maps as a fundamental part of the geologic work is absolutely necessary. The integration of these fault maps with structure maps is also essential to support the structural interpretation and to prepare accurate maps. The integration of fault and structure maps is discussed in detail in Chapter 8.

Too often, geologic maps and cross sections are prepared without giving much consideration to the three-dimensional geometric validity of the interpretation. Testing the validity of geologic interpretations is discussed in some detail in Chapter 9 under structural balancing; but it needs to be stated here that the proper construction of fault surface maps and their correct structural integration can go a long way toward providing three-dimensional validity or consistency to any interpretation.

It is not sufficient to rely solely *on what the well logs are indicating or what is seen on the seismic section.* Cross sections and seismic sections can in themselves misrepresent true subsurface relationships by the simple nature of their orientation, as well as by other factors. A good understanding of three-dimensional geometry is essential to any attempt at reconstructing a subsurface picture.

No hydrocarbons have ever been trapped by a fault trace. The trapping mechanism is the fault surface itself. Therefore, the mapping of the surfaces of all important faults is an integral part of any subsurface interpretation, particularly in areas involving multiple faults which can result in extremely complicated structural relationships. Attempting to reconstruct a complicated structure by using isolated fault cut data from electric well logs or seismic sections without the benefit of reasonable fault surface maps and their integration with various structural horizons can provide erroneous geologic interpretations. We often take short cuts in our preparation of subsurface maps. Such short cuts include failure to construct fault maps, the preparation of a structure map on only one horizon, or one line (seismic section) interpretations. In this chapter and in Chapter 8, we show how such short cuts can often lead to structural interpretations that are misleading, unreasonable, and at times costly.

The basic concepts and techniques discussed in this chapter apply to the use of data obtained from both vertical and deviated wells, in addition to data from seismic sections. The discussions, illustrations, and practice problems deal principally with extensional and compressional faulting that reflect mainly dip-slip movement, but the methods are broadly applicable to all styles of faulting.

FAULT TERMINOLOGY

"Probably no portion of geologic literature has a more confused terminology than that dealing with faults." This is a profound statement that is as applicable today as it was when made by H. W. Straley in 1932. A literature search on the subject of fault nomenclature or terminology shows that as far back as the turn of the century there was great inconsistency in the use of fault component terminology.

In 1908, the Council of the Geological Society of America appointed the "Committee on the Nomenclature of Faults." This committee was charged with establishing proper fault nomenclature. Since that time, there have been numerous papers on the subject of fault components and their related terminology, usage, and nomenclature; however, despite these numerous publications, many earth scientists are still confused when it comes to the definitions and usage of various fault component terms.

The correct understanding and usage of fault terminology with respect to certain fault components is essential to the preparation of correct subsurface fault and structure maps. Therefore, in this section, we define the fault components that are important to subsurface mapping in the petroleum industry. There are many other fault component terms that will not be discussed, not because they are less important, but because they do not apply to the mapping techniques that are presented in this text. Figure 7-2 graphically defines the fault components of interest to us.

Definitions

Vertical Separation (AE) is the vertical component of bed displacement. It is measured as the vertical distance between a line or plane (such as a formation top) projected from one fault block across a fault to a point where the projection is vertically over or under the same line or plane in the opposite fault block. It is that separation, seen in vertical wellbores, shafts, and vertical cross sections (Dennis 1972).

Throw (AC) is the vertical component of dip slip. It is the difference in vertical depth between the fault intersection with a line or plane (such as a formation top) in one fault block and the fault intersection with the same line or plane in the opposing fault block, *determined in a direction perpendicular to the strike of the fault.*

Heave (BC) is the horizontal component of dip slip. It is determined in a direction perpendicular to the strike of the fault.

Fault Cut is the vertical thickness of the stratigraphic section missing or repeated in a wellbore as a direct result of a normal or reverse fault cutting through the section. The fault cut, also referred to as the missing or repeated section in a wellbore, is determined by correlation of an electric log from one well with other electric logs from surrounding wells as presented in Chapter 4. **Technically, the fault cut is a representation of vertical separation.**

Notice in Fig. 7-2 that the value for the throw of the fault is not equal to the value

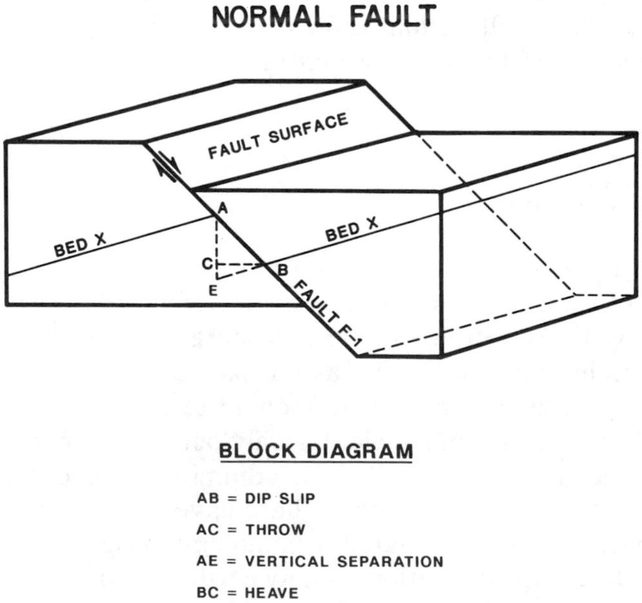

NORMAL FAULT

BLOCK DIAGRAM

AB = DIP SLIP

AC = THROW

AE = VERTICAL SEPARATION

BC = HEAVE

Figure 7-2 Block diagram of Bed X displaced by a normal fault, illustrating four different fault components. The front panel is perpendicular to the strike of Fault F-1. (Modified from Tearpock and Harris 1987. Published by permission of Tenneco Oil Company.)

for the vertical separation. The fact that they are different components of a fault is of significance when fault cut data obtained from well logs are used for integrating faults into a structural interpretation (Tearpock and Bischke 1990). These terms, throw and vertical separation, are commonly misunderstood, and more importantly, often misused in the preparation of subsurface fault and structure maps.

Because the understanding of these terms is so important to the correct construction of subsurface fault and structure maps, we attempt to clarify the issue without causing more confusion than currently exists. We discuss the terms in a general way and review them with respect to subsurface mapping techniques. For a complete review of the subject of fault nomenclature, refer to the selected bibliography at the end of the book.

DEFINITION OF FAULT DISPLACEMENT

We apply a definition to the word displacement similar to that given by Reid et al. (1913). It is here applied to the relative movement of the two sides of a fault, measured in any direction, when the direction is specified, or to the change in position of a bed, formation marker, etc., caused by fault movement. There are two ways to estimate displacement resulting from a fault. The first is the *actual relative displacement* of the two sides of a fault, and the other is the *apparent relative displacement*.

SLIP is the term used to describe the *Actual Relative Displacement* of a fault (Hill 1959). It is defined as the measurement of the distance of the actual relative motion between two formerly adjacent points on opposite sides of a fault, *measured on the fault surface*. **SEPARATION** is the term used to describe the *Apparent Relative Displacement* of a fault (Hill 1959). It is defined as the distance, measured in any specified direction, between two parts of a displaced surface on opposite sides of a fault. Separation concerns apparent movement of a fault with respect to a reference horizon cut by the fault (Dennison 1968). Separations are measurable, while slip is usually calculated. **Throw and heave are slip terms, and vertical separation is a separation term.**

Numerous authors (including Reid et al. 1913; Hill 1947; Crowell 1959; Dennis 1972; Tearpock and Harris 1987; and others) have emphasized the importance of distinguishing between fault components related to slip and those components related to separation. **Fault displacement in terms of slip (actual relative displacement) is, in most cases, impossible to measure in the subsurface.** This is so because there are no subsurface sources of data from which these measurements can be made. In general, the only descriptive information available from subsurface sources on the displacement of a fault is information about separation (apparent relative displacement). **Too often, however, the separation information is carelessly used as though it were slip, which causes serious mapping problems.** The two terms that have been carelessly misused the most are **throw** and **vertical separation**. Straley (1932) pointed out that there are at least three independent usages for the terms throw and heave in the literature. Mason, Crowell, Straley, and others have recommended the elimination or restricted use of the terms throw and heave.

The correct usage of fault component terms by all earth scientists may never be achieved, but we must clearly define the fault component terms throw, heave, vertical separation, and fault gap in order to construct subsurface geologic maps correctly. Mapping errors occur when these fault components are confused, and the subsequent mapping errors often are significant.

We again emphasize that the actual slip (measurements of throw and heave) cannot routinely be measured with conventional subsurface sources of data and are therefore not mappable. Throw and heave cannot be seen or measured in well logs; and, with seismic sections, slip measurements require each interpreted seismic line to be oriented perpendicular to the strike of the fault. Separation components, on the other hand, are measurable fault components with conventional subsurface data and therefore are mappable. They can be measured in a vertical shaft, an electric log from a wellbore, or on a seismic section, **regardless of orientation with respect to fault strike**. Of the various separation components, vertical separation is the most important parameter for constructing subsurface fault and structure maps (Tearpock and Harris 1987).

We cannot leave this subject with the idea that slip-related fault components are not important. Fault slip and related components are important, particularly in mining operations. Slip generally can be determined in mines where the actual fault surface is visible. In fact, the usage of the terms throw and heave originally came from the coal fields of Great Britain where the strata are nearly horizontal or the faults are strike faults (Reid et al. 1913). These terms, as mentioned earlier, have found their way into the petroleum industry and are probably here to stay. However, they have limited practical application in subsurface petroleum mapping. In addition, they cause confusion and significant mapping errors. We discuss their application and relationship to separation components later in this chapter and again in Chapters 8 and 9.

MATHEMATICAL RELATIONSHIP OF THROW TO VERTICAL SEPARATION

Throw is directly related to fault slip, for cross sections drawn perpendicular to fault strike (or parallel to maximum dip), as shown in Fig. 7-3.

$$\text{Throw} = AC = AB \sin \theta$$

where

$$AB = \text{fault slip}$$

$$\theta = \text{fault dip}$$

Thus, as the fault changes dip, the value of throw for a given amount of slip (AB) must also change. We must emphasize at this point that throw (which is related to fault dip and displacement) cannot be directly measured from electric well logs. We do, however, present methods which enable you to calculate the amount of throw, if desired, knowing the vertical separation and other properties, such as fault and bed dips. However, *normally throw does not enter into proper subsurface mapping techniques* (Tearpock and Bischke 1990).

The vertical separation (AE) in Fig. 7-2 or 7-3 is defined as the distance that a bed has been vertically displaced during faulting (Hill 1947). This distance is of primary importance to us, as the vertical separation is recorded within and is directly measurable on electric well logs. To illustrate this point, consider the following hypothetical example. Assume that a structure exists which contains beds that dip uniformly to the west (Fig. 7-4). The SP from two wells drilled into these beds is shown on the figure. The dashed line in Fig. 7-4 represents a future normal fault. This future fault will displace the beds in such a manner that the upper portion of Well No. 1 in the downthrown block is placed

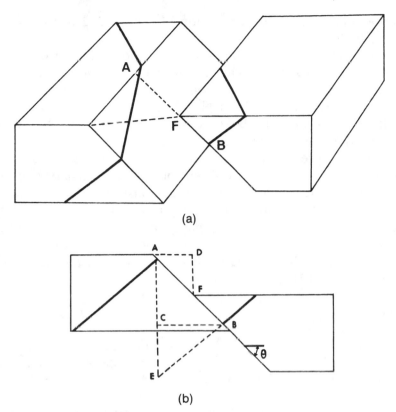

(a)

(b)

Figure 7-3   (a) Block diagram showing displaced surface by normal fault. (b) Vertical cross section perpendicular to strike of fault; along same plane as front of hanging wall block of block diagram. DF is vertical slip; AD is horizontal dip slip; AC is the throw; CB is the heave; AE is the vertical separation. DF, AC, and AE are all used by some geologists as throw. Θ is dip of fault. (Modified from Billings 1972. Published by permission of Prentice-Hall, Inc.)

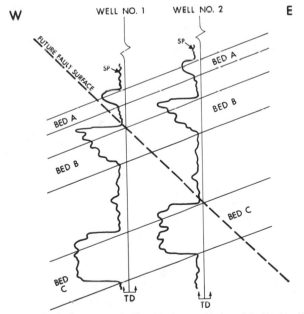

Figure 7-4   Hypothetical example. Unfaulted structure with beds dipping uniformly to the west. Dashed line shows location of future normal fault. (Published by permission of Tearpock and Bischke.)

in juxtaposition with the lower portion of Well No. 2 in the upthrown fault block, as shown in Fig. 7-5.

The geometric configuration produces the following observations. As the upper portion of Well No. 1 is displaced above the lower portion of Well No. 2, the top of Bed B in the hanging wall is brought into contact with the Top of Bed C in the footwall. Therefore, the missing section in Well No. 1 of the hanging wall includes the stratigraphic section from the top of Bed B to the top of Bed C. Inspection of the electric well logs now reveals that the missing section as the result of the fault cut in Well No. 1 (in the hanging wall) is represented by the coarsening upward sand sequence and the lower shale section present in the hanging wall portion of Well No. 2 (shaded section in Fig. 7-5).

This hypothetical example clearly demonstrates that the throw of the fault is not equal to the missing section in the hanging wall portion of Well No. 1. However, the missing section is equal to vertical separation as defined in Figs. 7-2 and 7-3. *We therefore have shown that electric well logs record vertical separation and not throw, and that throw does not directly enter into subsurface mapping techniques* (Tearpock and Harris 1987; Tearpock and Bischke 1990).

Quantitative Relationship

Vertical separation can be related to throw utilizing the relationships contained in Fig. 7-6 (Tearpock and Bischke 1990). Performing some trigonometry and using the Law of Sines, the following relationship is developed.

$$AE \sin(\pi/2 - \phi) = AB \sin(\phi - \theta)$$

Substituting

$$AB = \frac{AC}{\sin \theta}$$

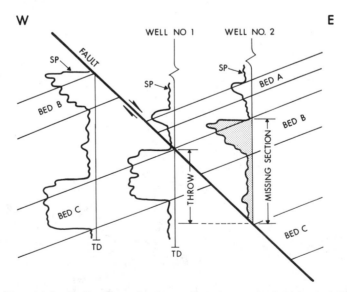

Figure 7-5 Normal fault displaces beds so that the upper portion of Well No. 1 in the hanging wall fault block is juxtaposed with the lower portion of Well No. 2 in the footwall. Missing section in Well No. 1 by correlation with Well No. 2 is highlighted on the SP curve for Well No. 2. The missing section is equal to the vertical separation and not throw. (Published by permission of Tearpock and Bischke.)

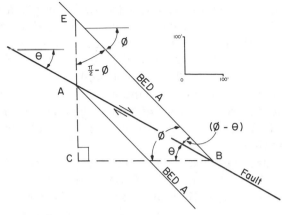

175' = Repeat Section by Log Correlation
AC = Throw
AE = Vertical Separation
AB = Dip Slip Displacement

Figure 7-6 Using trigonometry and the law of sines, the relationship of vertical separation to throw is shown graphically in this figure and mathematically in Eq. (7-1). (Published by permission of Tearpock and Bischke.)

yields

$$\frac{AE}{AC} = \frac{\sin(\phi - \theta)}{[\sin\theta \cdot \sin(\pi/2 - \phi)]}$$

Utilizing trigonometric identities yields

$$\frac{AE}{AC} = \left| \frac{\tan\phi}{\tan\theta} - 1 \right| \qquad\qquad (7\text{-}1)$$

where

AE/AC is taken relative to the absolute value
ϕ = bed dip
θ = fault dip
ϕ and θ are taken clockwise from 0 deg to 180 deg.

Equation (7-1) has significant application in regard to the evaluation of subsurface structure maps (Tearpock and Bischke 1990). The equation gives us the ability to check completed structure maps for accuracy of construction when the missing or repeated section is used to prepare an integrated subsurface structure map (see Contouring Faulted Subsurfaces, Chapter 8, for definition of *integrated structure map*). In Chapters 8 and 9 we discuss the application of Eq. (7-1) to test the validity of structure maps constructed using well log or seismic data, and present a method for analyzing the magnitude of error if mapped incorrectly.

FAULT DATA DETERMINED FROM WELL LOGS

As a standard practice in reviewing subsurface structure maps with faults, two questions should be asked. (1) What fault data were used to estimate the size of the fault or faults in the preparation of the structure maps? (2) What technique was used to contour across the faults? If the mapper responds that throw was mapped across the faults and the source of the fault throw data was subsurface well logs or seismic sections, a review of the maps can easily be made to determine whether the use of the word throw is simply a verbal substitution for vertical separation or if the mapper actually used the vertical separation data incorrectly as if the data were throw. If fault cut data from electric logs were used as throw for a fault in the construction of an integrated subsurface structure map, the map will probably be incorrect and require revision. The use of well log fault cut data as *throw* in mapping across faults on structure maps is an incorrect technique. **Herein lies one of the most basic problems with the construction of many subsurface structure maps**—a misunderstanding of what fault data are actually obtained from electric well logs for use in subsurface structure mapping.

Determination of fault cut data begins with well log correlations. Although there have been numerous publications on fault component terminology covering the subject of throw and vertical separation, we were not able to find one figure that diagrammatically illustrates the geometric relationship of missing or repeated section, as seen on an electric log from a vertical wellbore, with respect to the fault components throw and vertical separation. Figure 7-3, from Billings (1972), illustrates a vertical cross section perpendicular to the strike of the fault showing such fault components as throw, heave, and vertical separation. Notice that the figure shows at least three separate fault components defined by various geologists as throw, demonstrating the confusion that surrounds the use of these fault component terms. Although this figure correctly illustrates the difference between throw and vertical separation, it does not relate this geometry to what would be seen in a wellbore.

Since fault cut data are ultimately derived from well log correlation, it is very important to understand that the thickness of missing or repeated section as the result of a fault, measured in an electric log by correlation with surrounding electric logs, is actually a measurement of the fault component vertical separation. The cross section in Fig. 7-7 diagrammatically shows the geometric relationship between the missing section in a wellbore and the vertical separation of the fault. The east-west structural cross section, which is perpendicular to the strike of the fault, shows two beds (A and B) which have been faulted (displaced) by the normal Fault F-1. This normal fault cuts Well No. 1 at −5230 ft and is dipping at an angle of 45 deg to the east. The formation is dipping at 30 deg to the west. By correlation with Well No. 2, the missing section in Well No. 1 is determined to be 100 ft and the fault is shown to have entirely faulted out Sand Bed "A" in Well No. 1. As shown in the cross section, the throw of Fault F-1 is represented by the vertical line AC, which is equal to 63 ft. The vertical separation, represented by the vertical line AE, is equal to 100 ft. Thus, by the correlation of Well No. 1 with Well No. 2 in Fig. 7-7, we have diagrammatically shown that the missing section obtained for the fault cut in Well No. 1 is not throw nor equal to throw, but is a measurement of the fault component *vertical separation*. With this particular set of conditions, the throw of the fault is only 63% of the vertical separation. As shown in Fig. 7-7, there can be a significant difference between the throw and vertical separation of a fault. If mapped incorrectly, this difference can result in significant error in an integrated structure map.

For the most part, an understanding of fault terms and their application to subsurface mapping comes from academic studies and company-sponsored training programs. Textbook discussions on faults frequently use very simplistic examples showing faults cutting horizontal beds. These examples using horizontal beds lead to the misconception that throw and vertical separation are the same fault component or have the same value. Throw and vertical separation, however, have the same value in only three specific circumstances: (1) when the strata being faulted are horizontal, (2) when the fault is vertical, or (3) in a cross section perpendicular to the strike of the fault, when the fault strike is at a right angle to the strike of the strata. In this latter situation, the strata will have an

NORMAL FAULT

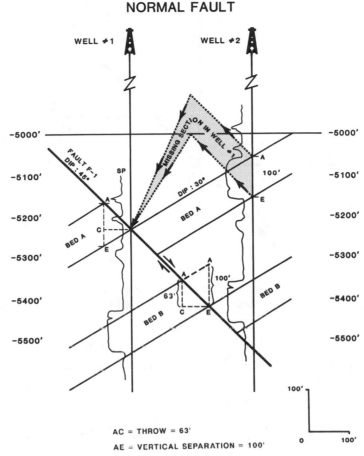

Figure 7-7 Diagrammatic cross section illustrates the geometric relationship between the missing section in a wellbore and the vertical separation of a fault. The 63 ft for the value of throw was calculated mathematically using Eq. (7-1). (Published by permission of Tearpock.)

apparent dip of zero degrees. Figure 7-8 shows the situation involving horizontal beds where the values for throw and vertical separation are the same. Many textbook examples illustrate faults cutting a stratigraphic section with no formation dip (horizontal beds), such as that shown in Fig. 7-8. In these cases, the values for throw and vertical separation are the same and are equal to the missing or repeated section in a wellbore. The use of models where faults cut dipping beds can eliminate the misconception that missing section is always throw, an idea that leads to the preparation of incorrect maps.

The first set of conditions from our list is discussed in greater detail here because it is the most common situation presented in textbooks and the one that has resulted in more misunderstanding of the fault component terminology than any other. Because of exposure only to simplistic examples using horizontal beds (where throw and vertical separation have the same value), many of us have failed to recognize that the values for these two different fault components vary from each other depending upon the attitude of the formation. This major misunderstanding can result in numerous mapping errors in a structural interpretation.

Vertical separation is directly measurable in well logs and is a measure of the missing section caused by a fault, which is valid regardless of the apparent attitude of the formation. Throw and heave are dependent fault variables which change with variations in the apparent attitude of the fault and formation. For most petroleum-related mapping,

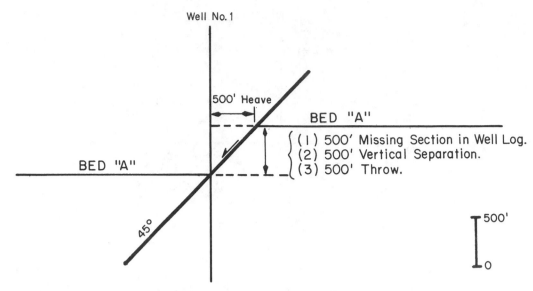

Figure 7-8 The values for throw and vertical separation are the same when the disrupted beds are horizontal. (Published by permission of Tearpock.)

the estimates for throw and heave have only academic value, and can be measured only in a cross section or seismic section that is oriented perpendicular to the strike of the fault, or on a structure map after the map has been completed using fault cut data correctly as vertical separation to construct a technically and structurally reasonable map. By using Eq. (7-1), however, the measured throw across a fault on a completed structure map can be used to check the accuracy of the map. This is discussed in detail in Chapter 8.

FAULT SURFACE MAP CONSTRUCTION

A fault surface map is a type of contour map. It differs from a structure contour map in that the contours are on the surface of a fault rather than on some stratigraphic marker, formation, or bed. The contouring of a faulted horizon presents numerous complex problems, both in the contouring of key horizons and fault surfaces. Although the contouring of key horizons may be the main objective in a mapping project, the contouring of the fault or faults provides essential information about the geology being studied.

In an area where a fault serves as the boundary limit of a hydrocarbon reservoir, the trapping mechanism is the fault surface itself. Reconstruction of the geologic picture involves the integration of all fault surface maps with several key structural horizons. Therefore, construction of an accurate fault surface map for each important fault, using all available data, is required.

Earlier in this chapter we mentioned that the preparation of accurate fault surface maps requires three-dimensional thinking and a good understanding of the regional tectonics being studied. This is so because each tectonic setting has its own characteristic patterns of faulting. For example, most of the faults in the Gulf of Mexico Basin are normal faults typically downthrown to the basin, and although they may strike in any direction, the preferred strike direction is roughly parallel to the present or historical coastline. Regional knowledge is very important in developing a geologic interpretation, comparing alternative geologic solutions and generating final subsurface maps that are geologically reasonable.

Note that we use the term fault *surface* map or just fault map instead of the more common usage of fault *plane* map. Fault surfaces tend to differ from true mathematical planes. They may increase or decrease in dip with depth, as well as change strike direction reflecting a sinuous or angular appearance which may trend in a specific direction or represent an arcuate shape. Fault surfaces are therefore rarely perfect planes: they may be listric, antilistric or even deformed. However, on a very localized or field scale, some faults may appear planar and can be mapped as such. Some of the basic fault examples in this textook represent idealized data used to present and teach a specific technique. In these cases, the fault examples are often simplified as true planes.

The construction of fault surface maps has numerous benefits in the interpretation and understanding of the development of faulted structures. Fault surface maps can:

1. aid in solving three-dimensional structural problems;
2. define the location of a fault in space in both the horizontal and vertical dimensions;
3. be integrated with structure maps to delineate accurately the
   a. upthrown and downthrown fault traces
   b. fault gap or overlap
   c. hydrocarbon reservoir limits;
4. help delineate complex fault patterns;
5. eliminate the distortion of a fault as seen on a cross section with zig-zag well spacing;
6. be used to estimate the dip and strike of a fault at any location along the fault;
7. aid in the designing of well plans, particularly for directionally drilled wells; and
8. help identify prospects that otherwise might be overlooked.

In the subsurface, faults can be recognized in one of three ways: (1) through the correlation and interpretation of electric logs, (2) by the interpretation of seismic sections, and (3) by inference. In this section, we discuss the use of electric logs to obtain fault data required to construct a fault map. The fault information begins with electric log fault cut data points. These fault cut data points, which result from the intersection of a drilled well with a fault surface, establish the actual presence of faulting (refer to Chapter 4). For normal faults, the fault cut is usually represented by a loss of stratigraphic section and by a repeat of section for reverse faults, as shown in Fig. 7-9. There are exceptions to this generalization, however, such as the special case of steeply dipping beds cut by a normal fault resulting in a repeated section shown in Fig. 7-9(c).

For a normal fault, any key horizon encountered in a well above the fault cut is in the downthrown fault block; and any marker encountered below the fault cut is in the upthrown fault block [Figs. 7-9(a) and 7-9(c)]. For a reverse fault, any datum encountered in a well above the fault cut is in the upthrown fault block; a bed encountered below the fault cut is in the downthrown block [Fig. 7-9(b)]. This relationship between the fault surface cut point and any particular horizon thus indicates whether a well is in the upthrown or downthrown fault block for any particular horizon being mapped. This relationship is further discussed in Chapter 8.

For each fault cut point, two values are required for use in the interpretation and construction of a fault map: (1) an estimate of the size of the fault cut, which we define as the vertical separation, and (2) an estimate of the subsea depth of the fault cut in the well. In Fig. 7-10 the recognized fault cut data in Well No. 2 are clearly marked to indicate the size of the fault cut (150 ft), the depth of the fault cut (a measured depth of 7280 ft),

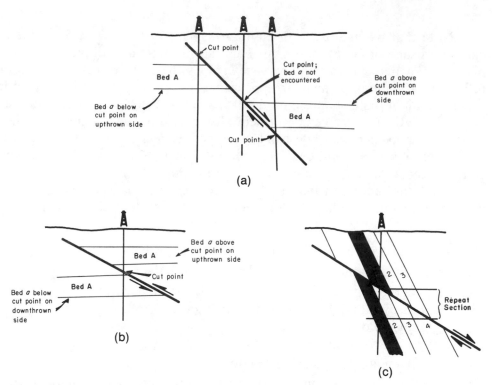

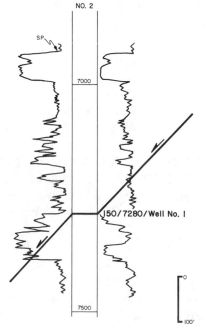

Figure 7-9 (a) Normal fault resulting in a missing section. (b) Reverse fault resulting in a repeated section. (c) Normal fault resulting in a repeated section. In example (c), the beds are dipping at a steeper angle than the fault. (Modified from Bishop 1960. Published by permission of author.)

and the well(s) used in the correlation (Well No. 1). If you are still not sure of how to identify faults in electric well logs, refer again to Chapter 4 on electric log correlation.

Fault cut data from at least three wells, not in a straight line, are required to accurately contour a fault surface in the vicinity of the well control. If you are familiar enough with the area or if data from one or more seismic sections are available, however,

Figure 7-10 Fault cut information is documented on the electric log. It indicates the size and depth of the cut and the well(s) used for correlation.

then an accurate fault map may be constructed with data from just one well. Obviously, the more fault cut data available the better the interpretation of the fault surface. Fault maps also can be prepared from seismic data alone if the coverage is sufficient. This topic is presented later in this chapter.

Contouring Guidelines

In preparing fault surface maps, certain general guidelines should be followed. If a sufficient number of fault cuts are available, the fault surface can be contoured in the same way as the elevations of a key horizon (Reiter 1947). Inasmuch as faults result from breaks rather than bends in the strata, they pose some special problems in contouring. The rules differ from the general rules for contouring in that angular relationships may exist between two intersecting fault surfaces or between a fault surface and a bedding plane. The general guidelines for contouring a fault surface are as follows:

1. Contours of a fault surface may be open ended. They do not have to close upon themselves.
2. Changes in either fault strike or dip are assumed to be gradual unless evidence indicates otherwise (cross faulting). An exception to this guideline might occur in the case of mapping a reverse fault ramp and flat surface (see Chapter 9), however even in these cases the changes in dip are often more gradual than abrupt.
3. Changes in fault strike for normal faults are usually represented by smooth curves rather than by sharp angles. There are exceptions to this guideline, such as deformed fault surfaces, the effect of cross structures, and ramp and flat thrust faults.
4. Changes in dip are generally shown as smooth curves rather than plane segments intersecting at a sharp angle. Again, there are exceptions to this guideline, such as those listed in guideline 3.
5. Use the interpretive method of contouring outlined in Chapter 2 for preparing fault surface maps.
6. Several fault surfaces are commonly contoured on a single base map. Contours of individual faults may merge inasmuch as faults intersect one another in nature. Note: When constructing compensating, bifurcating, or intersecting faults, denote the lines of termination, bifurcation, or intersection on the fault maps.
7. Fault surface maps must be geologically reasonable for the area being mapped.
8. Fault surface maps are normally contoured with a 500-ft or 1000-ft vertical contour interval, since fault surfaces are usually relatively steep. Thrust faults may dip at a low angle, and therefore, a smaller contour interval may be appropriate.
9. The contours of a fault surface join those of a given datum surface at points of intersection between the fault and the key horizon contours at the same elevation.
10. Faults commonly dip at different angles from a key bed being mapped, and consequently, only a portion of the fault surface contours intersect with those of a given datum (see Chapter 8).
11. The fault surface contours cut beds above and below a given datum unless the fault is extremely limited in vertical extent.

Fault map construction actually involves some subjective interpretation; thus, the more data available for mapping, the less uncertainty in the interpretation. As each of us has a different idea of the geologic picture of an area being worked, it is possible in

areas with limited well or seismic control to construct a fault map with several different
interpretations using a single or similar set of fault cut data. Such a situation is shown
in Fig. 7-11. Figures 7-11a–c show three different fault surface map interpretations for
the same Fault "A" in the Kings Bayou Field, Cameron Parish, Louisiana. The three
different interpretations are based on the same well control. These maps, prepared in
the mid-1960s, were probably constructed with little, if any seismic control to help in the
fault interpretation, and with limited well control. Our intent here is not to determine
which interpretation is more correct, but to review the maps in terms of their technical
construction and accuracy. All three fault maps may be geologically reasonable for the
tectonic setting and incorporate correct construction techniques. However, after reading
the next section of this chapter, Fault Surface Map Construction Techniques, return to
Fig. 7-11c and see if you can identify any problems with this fault surface map interpretation.

   Fault maps, like many other subsurface geological maps, tend to change with time
as new well and seismic data become available. Therefore, a fault surface interpretation
is never complete until the last well is drilled and all the seismic data to be shot have
been shot and interpreted.

Fault Surface Map Construction Techniques

We shall begin with a relatively simple fault contour map involving a single fault illustrated
in Fig. 7-12. There are 18 vertical wells in this field example from which fault cut data
have been obtained (Fig. 7-12a). First, the size and depth for each fault cut are posted
next to the appropriate well in which the cut was observed, as shown in the figure. The
most common way of posting these data are first to indicate the size of the fault and then
the subsea depth of the fault cut (i.e., 300'/−4240'). The minus sign in front of the depth
number indicates that the fault cut is below sea level.

   The well control in this example is located in three separate areas on the base map.
Seven wells are located in the western portion of the map, three in the central portion,
and eight to the east. As discussed under the general contouring rules, begin contouring
in the area or areas of **maximum** control. In this case, we first contour the eastern area,
where there are eight wells, followed by the western area, with seven points of control,
and finally, the central area with three fault cuts. Use a free-hand style of contouring or
proportional ten-point dividers to initially establish the contour spacing for the map.

   Figure 7-12b shows the contours established for the fault in the three areas of well
control. Based on the fault cut data we assume that the three fault segments are parts
of the same fault, so the final step is to extend the contours into the area of no control
and connect the fault segments. There are two possible ways to do this. One method is
to extend the contours from each segment toward one another as straight lines until they
intersect. This method, illustrated in Fig. 7-12c, appears to be a more unreasonable or
unlikely interpretation. The second and preferred method is to use an interpretive form
of contouring (Fig. 7-12d). With this method, some geologic license is used in the interpretation to reflect the expected geometry of the fault surface in this tectonic setting.
We gradually change the strike direction of the fault connecting the adjacent segments
with a smooth curve.

   Now that the fault map is complete, we can estimate the dip of the fault at any
location. At a depth of around 5000 ft subsea, the dip of the fault is 65 deg, decreasing
to 55 deg at 8500 ft subsea, and finally flattening to about 40 deg between −10,000 ft
and −11,000 ft. This type fault shape is very common for growth faults (see section on
Growth Faults in this chapter). The fault in Fig. 7-12 is contoured as a listric (curvilinear

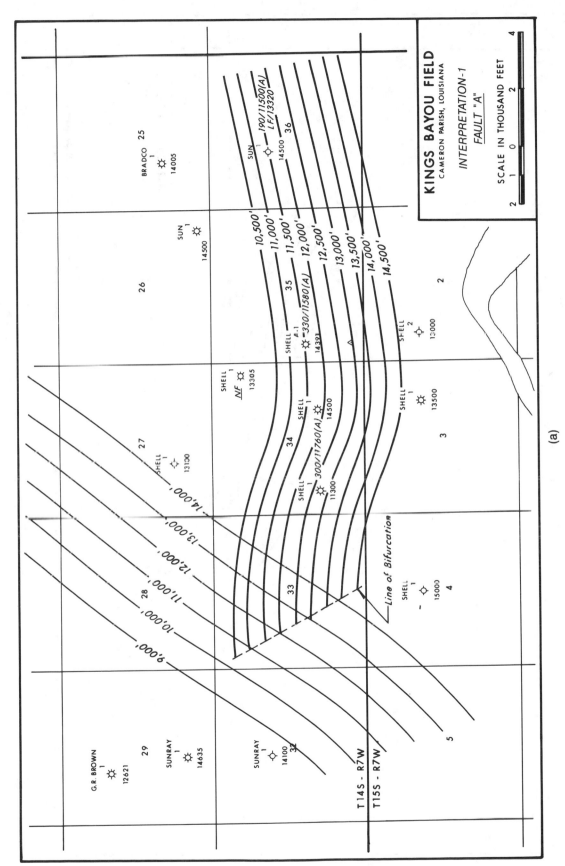

Figure 7-11 (a) Fault interpretation 1. (b) Fault interpretation 2. (c) Fault interpretation 3, Kings Bayou Field. (From Louisiana Department of Conservation.)

(a)

KINGS BAYOU FIELD
CAMERON PARISH, LOUISIANA

INTERPRETATION-1
FAULT "A"

SCALE IN THOUSAND FEET

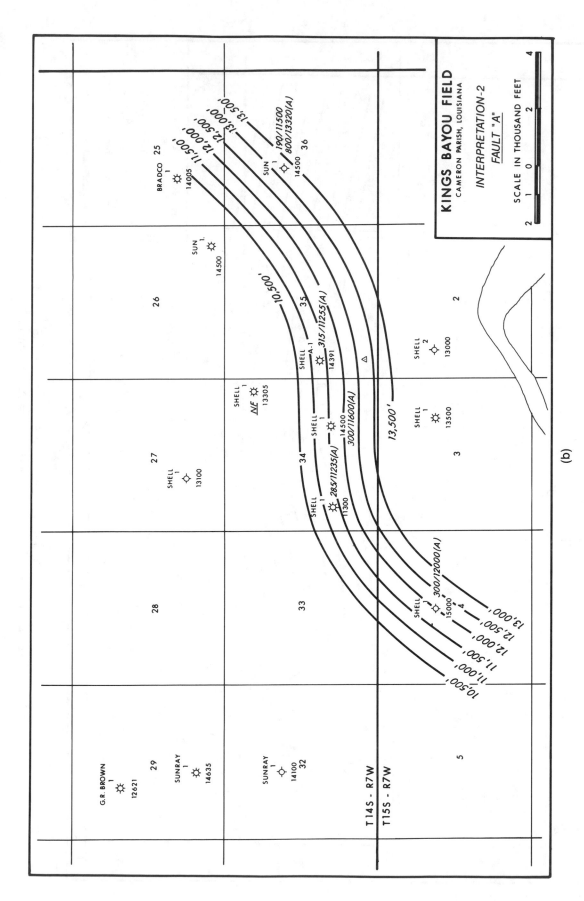

Figure 7-11 (continued)

(b)

212

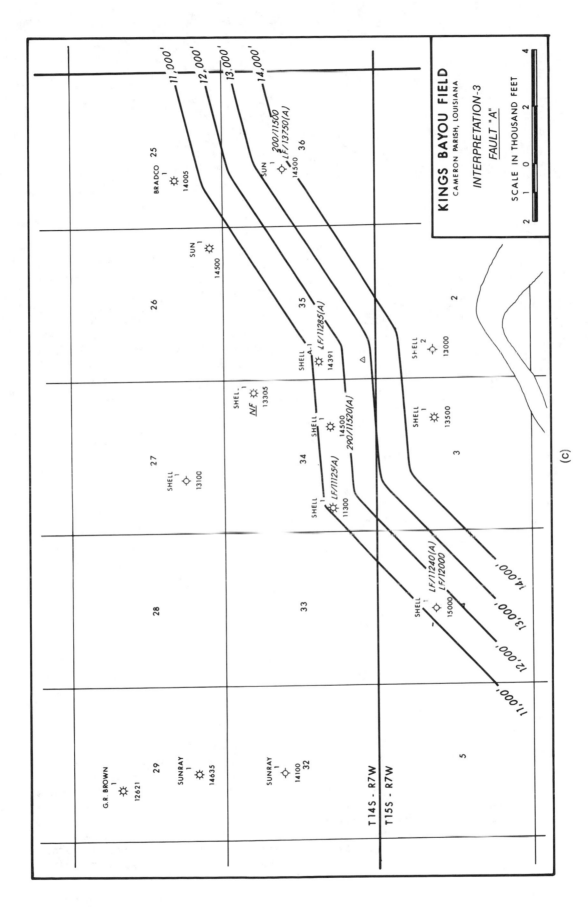

Figure 7-11 *(continued)*

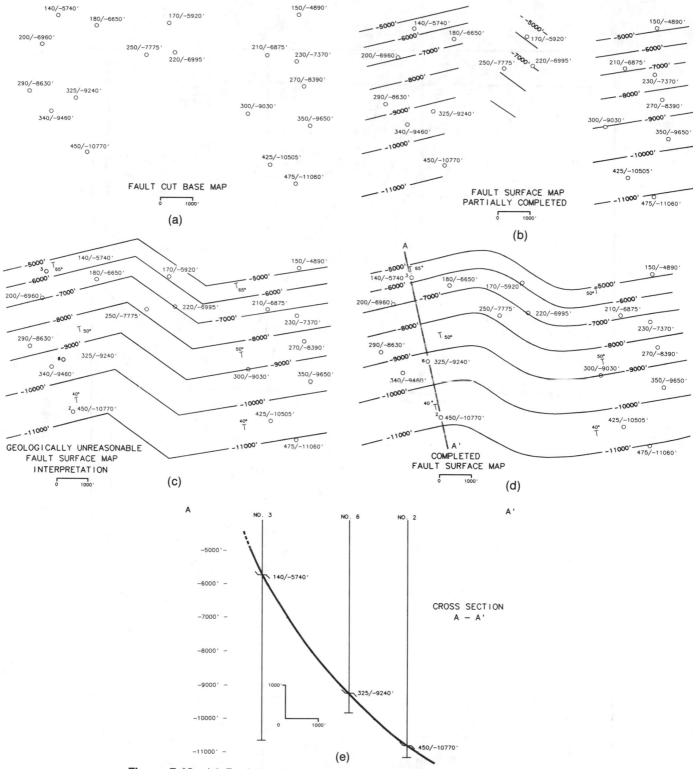

Figure 7-12 (a) Fault cut base map showing the size and depth of the fault cut in each well. (b) Fault contours established in three areas of well control. Each fault segment is part of the same fault. (c) Unrealistic fault surface interpretation results from connecting each fault segment (Fig. 7-12b) with straight lines. Mechanical approach to contouring. (d) Completed fault surface map using the interpretive form of contouring to reflect the expected geometry of the fault surface. (e) Cross section A-A′ passes directly through Wells No. 3, 6, and 2 and is laid out perpendicular to fault strike. Using an interpretative approach to contouring results in a gradual, rather than abrupt, change in fault dip with depth.

concave upward surface) growth fault; that is, a fault whose dip decreases with depth, while the vertical separation or size increases.

Figure 7-12e shows a strike cross section (A-A') laid out in a northwest-southeast direction perpendicular to the strike of the fault. Three wells lie on the section with fault cut data for each well posted on the cross section. An interpretive method of contouring was used to contour the fault surface with depth. This is the preferred method which provides the most reasonable interpretation. Other methods could have been used, including the mechanical contouring method, in which the dip rate is constant between each two well control points but changes at each well. The equal-spaced contouring method also could have been used. Both methods provide a less reasonable interpretation of the fault surface.

In developing a final fault surface map interpretation, keep in mind that a fault need not remain constant in strike direction, dip, or size over the entire fault. Along strike, the size of a fault may increase, decrease, or remain constant; the strike direction may actually change. The size of a fault might increase with depth, die upsection, or even decrease with depth. A fault may die laterally, transfer displacement to other faults or folds, combine with other faults, or terminate against another fault. In areas of salt diapirism, a fault may terminate against salt, cut through it, or even be distorted by the salt intrusion. We again emphasize that a good interpretation of a fault surface must have three-dimensional validity, comply with the tectonic characteristics of the region being mapped, and use correct mapping techniques in its construction.

Figures 7-13 and 7-14 are examples of completed fault surface maps. Figure 7-13 is the fault surface for Fault "V" in Bayou Villars Field, St. Charles Parish, Louisiana. The correlation data from over 20 wells were used in this fault surface map interpretation, which is contoured with a 500-ft contour spacing. The fault is a growth fault downthrown to the north, with a displacement of 130 ft at a true vertical depth (TVD) of 2430 ft increasing to about 600 ft at 8000 ft (TVD) in the deepest penetrated section. The contour spacing, which widens toward the north, indicates that the fault flattens with depth. Also

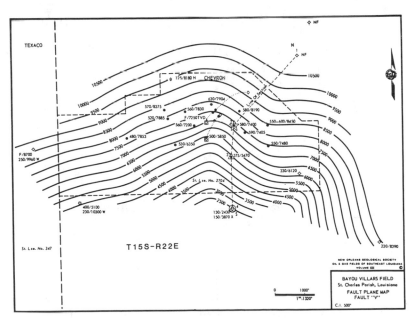

Figure 7-13 Completed fault surface map for Fault V in the Bayou Villars Field, St. Charles Parish, Louisiana. The fault is a down-to-the-north growth fault. (Published by permission of the New Orleans Geological Society.)

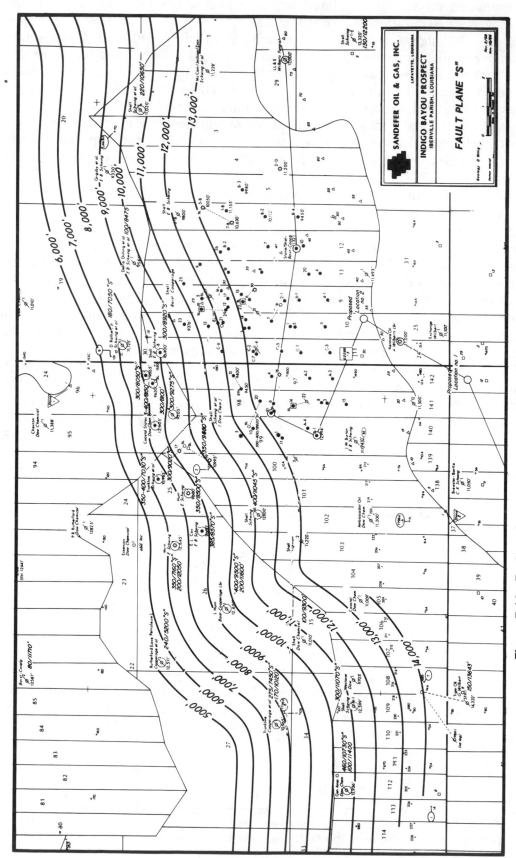

Figure 7-14 Fault surface map on Fault "S" at Indigo Bayou, Iberville Parish, Louisiana. Vertical separation varies slightly from well to well, ranging in size from 220 ft to 400 ft. The fault shows very little change in dip with depth. (Published by permission of Sandefer Oil and Gas, Inc.)

216

notice that changes in strike direction are represented by smooth curved contours rather than angular bends. This conforms to the interpretive contouring guideline outlined earlier. In the shallowest wells, the dip rate is measured at about 63 deg, decreasing to about 50 deg at around 9000 ft. Oil is trapped upthrown to this down-to-the-north growth fault in sands of the Middle Miocene age.

The fault map in Fig. 7-14 is that of the "S" Fault in the Indigo Bayou area, Iberville Parish, Louisiana. Although there is some variation in the size of this fault, for the most part it appears to be a post-depositional fault with a vertical separation ranging from 220 ft to 400 ft. The fault exhibits little, if any, growth with depth. This fault surface map is contoured in accordance with the guidelines and rules outlined in this section.

TYPES OF FAULT PATTERNS

Extensional Faulting

Normal faulting may be defined as motion along a dipping fault surface on which the hanging wall has gone down relative to the footwall. Normal faults often exist as a set with parallel strikes but opposing dips, referred to as a conjugate fault system. Commonly, each fault has a different amount of slip, with the fault having the major displacement called the **master fault**, and the fault with the relatively minor displacement called an *antithetic or compensating fault*. Normal faults are typically steeply dipping; however, the dip of normal faults may range from almost horizontal to vertical, with dips between 55 to 70 deg most common. In South Louisiana, for example, the average fault dip has empirically been determined to be 51 deg (Ocamb 1961). Normal faulting usually results in a missing stratigraphic section in electric well logs and a gap on structure maps between the formation in the upthrown (footwall) and downthrown (hanging wall) fault blocks. This was illustrated in Figs. 7-2 and 7-7.

Extensional basins associated with salt tectonics can have very complex fault patterns. The maximum principal stress axis in extensional basins is vertical, resulting in normal faulting with initial dips of about 60 deg or greater due to extension over the dome. Salt intrusions, commonly called *salt domes*, are often associated with the formation of crestal grabens, as well as radial and peripheral faulting. Antithetic faults, also referred to as compensating faults, are common in extensional areas.

In addition to single normal faults (structural and growth), there are three principal patterns of normal fault intersections and terminations common in areas of extensional tectonics. These patterns, illustrated in Figs. 7-15, 7-18, and 7-21, are: (1) bifurcating, (2) compensating, and (3) intersecting. For each of the fault patterns discussed, a fault surface map, one or more cross sections, and a block model are provided to explain and illustrate the pattern.

Bifurcating Fault Pattern. A *bifurcating fault pattern* results from two normal faults that dip in the same general direction, as shown in Fig. 7-15. The strike direction of each fault is such that the two faults merge laterally in the subsurface and continue on as one fault. The line along which the two faults **merge** is called the **line of intersection** or **bifurcation**. The total vertical separation of the fault across the line of bifurcation must be conserved. This means that the vertical separation of the single fault, where only one fault exists, is equal to or nearly equal to the sum of the vertical separations of the two faults, where two faults are present. The contoured fault map in Fig. 7-15a shows two

(a)

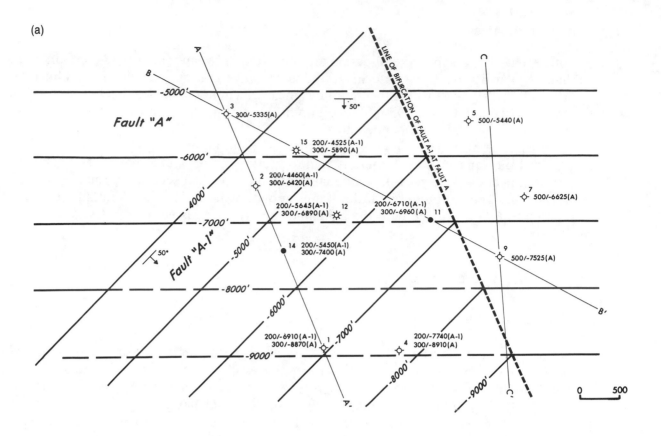

(b)

CROSS SECTION A - A'

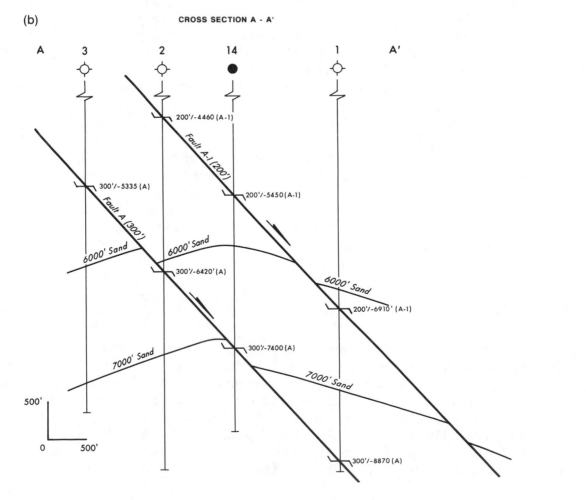

Figure 7-15

intersecting fault surfaces dipping in the same general direction. This interpretation was made using fault cuts from 11 wells, the contouring guidelines, and an understanding of the geologic setting.

Using Fig. 7-15, we review the fault pattern in detail and illustrate the specific characteristics that classify this as a bifurcating fault pattern. Fault A (F-A) is striking east-west and dipping 50 deg to the south. Fault A-1 is striking northeast-southwest with a dip rate of 50 deg to the southeast. The two faults are dipping in the same general direction.

Fault A-1 merges with Fault A where the two faults intersect, as indicated by the dashed line called *the line of bifurcation*. There are two faults present west of this line. Fault A has a vertical separation of 300 ft and the vertical separation for Fault A-1 is 200 ft. East of the line of intersection only one fault (F-A) exists, with a vertical separation of 500 ft. These vertical separation values across the line of bifurcation satisfy the conservation of vertical separation, also referred to as the additive property of faults (see Chapter 8). Notice that the contours for Fault A are dashed west of the line of intersection, indicating that the contour values are deeper than those for Fault A-1. This is good drafting practice, which helps reduce confusion on maps when more than one fault surface is constructed on the same base.

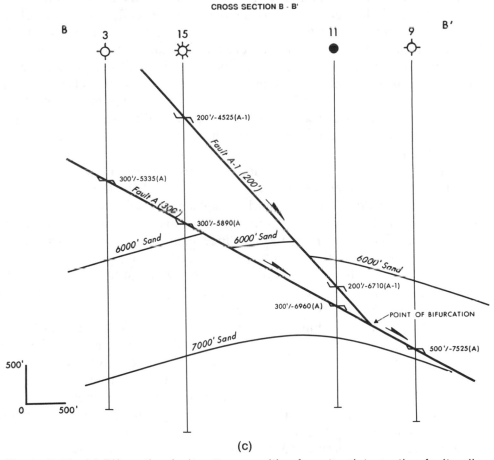

(c)

Figure 7-15 (a) Bifurcating fault pattern resulting from two intersecting faults, dipping in the same general direction. Line of bifurcation indicates where the two faults intersect. (b) Cross section A-A' bisects Faults A and A-1 in such a way that the two faults do not appear on the cross section as bifurcating faults, but instead appear as two parallel dipping faults. (c) Cross section B-B' is laid out almost perpendicular to Fault A-1 and at an oblique angle to Fault A. In cross section, the faults appear to merge with depth rather than to bifurcate laterally.

Figures 7-15b and 7-15c are two cross sections with a different orientation to the two fault surfaces for each line of section. Remember, cross sections used in conjunction with maps provide another viewing dimension that can be helpful in visualizing the geologic picture and solving structural problems. The orientation of the section line, however, is very important. Chosen incorrectly, the line of section can be more confusing then informative.

In the two cross sections through the bifurcating fault pattern, the fault geometry appears different in each section. In cross section A-A' (Fig. 7-15b), the fault pattern does not appear to be bifurcating. Instead, the two faults appear as parallel faults. Is this real or an optical illusion as a result of the line of section? In Fig. 7-15c showing the B-B' cross section, Fault A-1 appears to merge with Fault A with depth. Real or illusion? Although the two cross sections are geologically and technically correct, they can pose problems for those unfamiliar with the geology of the area. When laying out a cross section, be sure to consider the purpose of the cross section and your audience (Chapter 6).

The vertical separation values for the faults have been incorporated into each cross section to represent correctly the offset of the 6000-ft and 7000-ft Sands by Faults A and A-1. The term **bed offset** means that the bed has been displaced by the fault and the displacement is defined in terms of vertical separation. Earlier in this chapter, we showed that the missing section in a wellbore as the direct result of a fault is the measurement of the displacement in terms of vertical separation. Since this understanding is very important when preparing cross sections, we detailed the procedure for using vertical separation in the preparation of cross sections in Chapter 6. Therefore, in the two cross sections in Fig. 7-15b and 7-15c, the offset for the beds is constructed using the vertical separation from the wellbore fault cut data. We cannot over emphasize that wellbore fault cut data are not throw; therefore, we cannot construct a cross section using the fault cut data as throw.

Figure 7-16 is a block diagram of a bifurcating fault pattern. A review of Figs. 7-16(a) and (b) illustrates the geologic development of the fault pattern and individual fault blocks. Fault 1 develops first with a displacement of 100 ft downdropping Fault Block B by an amount equal to the displacement of the fault (100 ft). Next, Fault 1A develops also with a displacement of 100 ft. Since Fault 1A stops or terminates at Fault 1, the surface of Fault 1 west of the intersection must accommodate the 100 ft of displacement from Fault 1A. It does so by moving downward an additional 100 ft. Therefore, the net displacement of Fault 1, west of the intersection of the two faults, is 200 ft.

Figure 7-17 is an example of a bifurcating fault pattern from Golden Meadow Field in Lafourche Parish, Louisiana. The fault system shown is that of two faults dipping in the same general direction and merging laterally as indicated by the line of bifurcation. The sum of the vertical separations in the area where two faults are present, east of the line of bifurcation, is equal to or nearly equal to the vertical separation of the one fault west of the intersection in the area where one fault is present.

Fault L dipping to the north-northeast is contoured between −9500 ft and −14,000 ft. The available well control east of the intersection of the two faults indicates that the size of Fault L ranges from 65 ft to 105 ft. Fault K, which dips to the north, is contoured between −10,500 ft and −14,000 ft. Based on the well control, the size of this fault appears to be about 120 ft. The sum of the vertical separations for Faults K and L east of the line of bifurcation is ±185 ft to 225 ft; west of the bifurcation line, where both faults have laterally merged into one, the vertical separation of the Fault L is 165 ft to 225 ft. The nearly equal values for the vertical separation on both sides of the fault

(a)

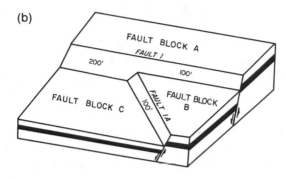

(b)

Figure 7-16 Block models show the development of a bifurcating fault pattern. (a) Fault 1 develops with a vertical separation of 100 ft. (b) Fault 1A develops (vertical separation of 100 ft) and terminates against Fault A. Fault A, west of the intersection of the two faults, accommodates the 100 ft of added displacement by Fault 1A by moving downward an additional 100 ft.

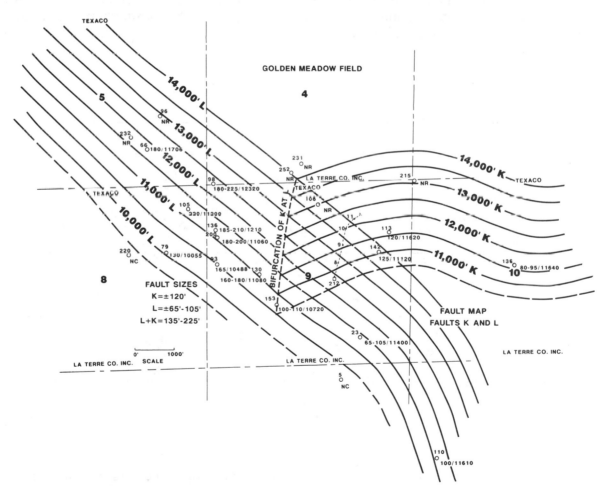

Figure 7-17 Example of a bifurcating fault system from Golden Meadow Field, Lafourche Parish, Louisiana. Note the conservation of vertical separation on either side of the line of bifurcation. (Published by permission of Texaco, USA.)

intersection show that the vertical separation across the line of bifurcation has generally been conserved.

Compensating Fault Pattern. A *compensating fault pattern* results from two normal faults dipping in opposite directions toward one another (Fig. 7-18) with the acute angle between the strike direction of the two faults being less than 90 deg. At the point of intersection, one of the two faults terminates against the other. Considering the additive property of faults, the vertical separation of the single fault where only one fault exists is equal or nearly equal to the sum of the vertical separations of the two faults in the area where both faults exist.

We can look in detail at the fault surface map for the compensating fault pattern shown in Fig. 7-18a. There are 11 wells from which the fault cut data were obtained. Based on these fault cuts, the general guidelines presented earlier, and an understanding of the expected fault surface geometry in this setting, two intersecting fault surfaces were contoured as shown in the figure.

The completed fault map illustrates the specific characteristics that classify this as a compensating fault pattern. Notice that Fault A is striking east-west and dipping to the south with a dip rate of 50 deg and Fault B is striking northeast-southwest and dipping at 50 deg to the northwest. The two faults are dipping in opposite directions toward one another. Fault B terminates against Fault A, where the two fault surfaces intersect at equal subsea elevations, as indicated by a dashed line referred to as the *line of termination*.

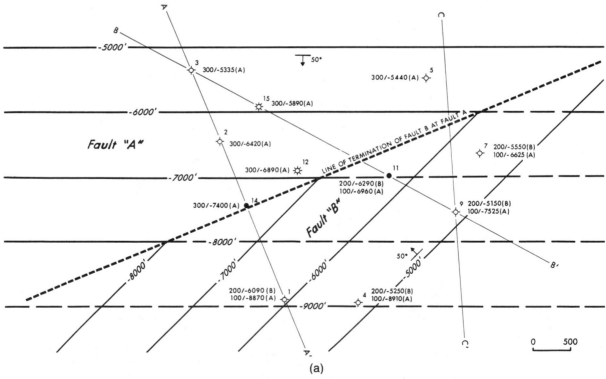

(a)

Figure 7-18 (a) Compensating fault pattern resulting from two intersecting faults dipping in opposite directions. Line of termination of Fault B at Fault A indicates the intersection of the two faults. (b) Cross section A-A' illustrates the termination of Fault B at Fault A. West of the intersection, Fault A has a vertical separation of 300 ft, while east of the intersection Fault A is 100 ft and Fault B is 200 ft. The vertical separation is conserved across the intersection.

CROSS SECTION A - A'

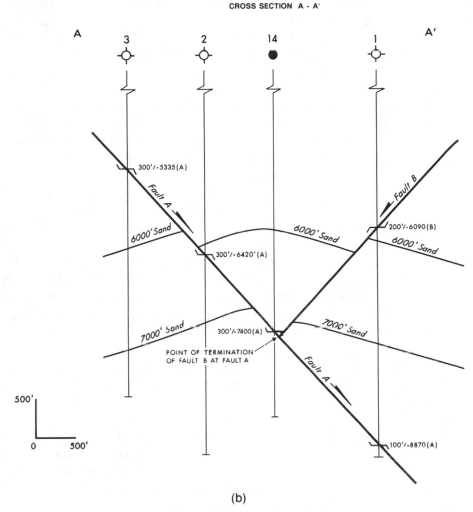

(b)

Figure 7-18 (*continued*)

South of this line there are two faults (F-A and F-B). Fault A has a vertical separation of 100 ft, and the vertical separation of Fault B is 200 ft. North of the termination line, only F-A is present with a vertical separation of 300 ft. These displacement values satisfy the conservation of vertical separation. Therefore, we say that Fault B is compensating with respect to Fault A.

Figure 7-18b is a northwest-southeast stick cross section A-A' shown in plan view on the fault contour map in Fig. 7-18a. The fault cuts from Wells No. 3, 2, 14, and 1, which lie directly on the cross section, are posted on the section. Fault B terminates against Fault A at a depth of −7460 ft, which corresponds to the point on the fault map (Fig. 7-18a) where the termination line for Fault B intersects the cross section. In the area where Fault A and Fault B are present, Fault A has a vertical separation of 100 ft, and the vertical separation of Fault B is 200 ft. This is shown in Well No. 1 by the fault cut point for F-A of 100 ft at −8870 ft and 200 ft at −6090 ft for F-B. North of the termination of Fault B, only Fault A is present with a vertical separation of 300 ft shown in the fault cuts in Wells No. 2, 3, and 14. In Well No. 14, for example, the 300-ft fault cut is at a depth of −7400 ft. The vertical separation values for the faults have been incorporated into the cross section to correctly represent the offset of the 6000-ft and 7000-ft Sands by Faults A and B.

Figure 7-19 is a block diagram of a compensating fault pattern. At times, to explain the final fault displacement for this type of fault pattern, you may hear it explained as follows: "Think of the system in this way. West of the fault intersection, Fault 1 has a size of 200 ft. Since Fault 2 is 100 ft, it *takes away* 100 ft of displacement from Fault 1 east of the intersection of the two faults, leaving 100 ft of displacement left for Fault 1." This explanation may provide you with a visual understanding of the size of each fault on both sides of the termination line, but technically it is incorrect and can lead to confusion. *One fault cannot take displacement away from another fault.*

Looking at Fig. 7-20, think of the geologic development of the fault pattern and individual fault blocks in the way they were formed. Fault 1 develops first with a displacement of 100 ft downdropping Fault Block B by an amount equal to the displacement (Fig. 7-20a). Next, Fault 2 develops, also with a displacement of 100 ft, creating Fault Block C and downdropping it by 100 ft. Since Fault 2 terminates at Fault 1, the surface of Fault 1 west of the intersection must accommodate the 100 ft of displacement created by Fault 2 by moving down an additional 100 ft (Fig. 7-20b). Therefore, the net displacement of Fault 1 west of the intersection of the two faults is 200 ft. Some geologists refer to such movement as a reactivation of the older fault surface by the younger fault. If we check for conservation of vertical separation, we see that the 100 ft for Fault 2 plus the initial 100 ft for Fault 1 east of the fault intersection equal the final 200 ft of vertical separation for Fault 1 west of the intersection of the two faults.

Intersecting Fault Pattern.   So far, we have discussed two types of fault patterns in which one fault merges or terminates against the other fault at their intersection. Now we look at the intersecting fault pattern, which results from two faults (*normal or reverse*) dipping in such a manner as to intersect in the subsurface; unlike the bifurcating pattern, in which the two faults merge, or the compensating pattern, in which one fault terminates against the other, both faults continue downward past the depth of intersection. Neither fault stops or merges against the other. The geometric relationship of intersecting faults is very difficult to visualize; block models can help illustrate this pattern in three dimensions.

Because of the complexity of this fault pattern, a correct interpretation is rarely achieved, even in areas of adequate well and seismic control (Dickinson 1954). When considering the three fault patterns discussed in this section, the intersecting fault pattern presents the most complexities and the solutions are not at all straightforward. With limited available data, a decision must be made whether the intersecting faults formed contemporaneously (Horsfield 1980) or are of two different ages. Without good seismic control, such as a three-dimensional survey, it is often difficult if not impossible to determine if the faults formed contemporaneously or at different times.

If the conclusion is that the faults are of two different ages, the next step is to determine which fault formed first and which was second. Such conclusions impact the

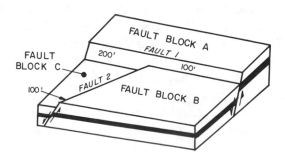

Figure 7-19 Block model of a compensating fault system.

(a)

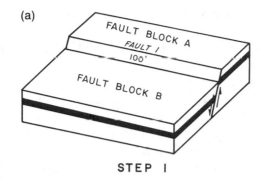

STEP 1

(b)

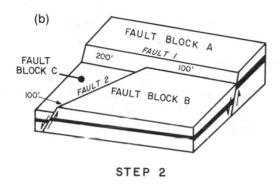

STEP 2

Figure 7-20 Block models show the development of a compensating fault pattern. (a) Fault 1 develops with a vertical separation of 100 ft. (b) Fault 2 develops (vertical separation of 100 ft) and terminates against Fault 1. The portion of Fault 1 west of the intersection of the two faults accommodates the additional 100 ft of displacement by Fault 2 by moving downward 100 ft.

construction of the fault surface maps, as well as completed structure maps. Even with today's technology, rarely are there sufficient data to make such decisions.

Because of the complexities and uncertainties surrounding this fault pattern, *we recommend that the intersecting faults be mapped as if there is no offset of one fault by the other.* This assumption does result in some error around the intersection of the faults and integrated structure maps; but it does save time, and the actual interpretation may be impossible to determine from data available. This **compromise** introduces less error than an incorrect guess as to the age of faulting. This subject is further discussed in Chapters 8 and 9.

For the intersecting fault example shown in Fig. 7-21a, we make one of two assumptions: (1) the faults are contemporaneous, or (2) the relative age of the faults is unknown. With either assumption, we construct a fault map with both faults meeting at their intersection and continuing downward, unaffected past their intersection. The fault maps were prepared using fault cut data from the 11 wells shown on the map. Fault A is striking east-west and dipping to the south at 50 deg and Fault B is striking northeast-southwest with a dip rate of 50 deg to the northwest. The intersection of the two faults is shown as a dashed line. After intersection, the two faults continue downward with no change in vertical separation, since these values are not affected by the intersection of the faults as in the cases for the bifurcating and compensating fault patterns.

Around the area of fault intersection, a *chaotic zone* may exist in which both the sediments and fault surfaces are disrupted. However, well control and seismic data rarely can identify such disruption, and therefore, the mapping around these intersections is inaccurate. This must be kept in mind when mapping horizons affected by intersecting faults.

The cross section A-A' shown in Fig. 7-21b illustrates the simplified (or compromised) method of preparing the fault maps as if the two faults were contemporaneous. Although neither fault surface is offset on the fault map, when the fault map is integrated with a structure map, all four fault traces may be offset. This is covered in detail in Chapter 8.

Figure 7-22 is a block diagram of an intersecting fault pattern resulting from two different ages of faulting. Fault 1 developed first, followed by Fault 2. Notice in the figure that Fault 2 cuts through Fault 1, displacing the older fault surface and causing a gap in this displaced surface. Since both faults carry to depth, no change in displacement occurs for either fault after they intersect.

An interpreted seismic line from a three-dimensional survey over an offshore Gulf of Mexico field is shown in Fig. 7-23. It shows very clearly an intersecting fault pattern in which both faults appear to pass through one another as if the faulting were contemporaneous, similar to the example used in Fig. 7-21. Intersecting faults dipping in opposite directions, as shown in the figure, are given the special name **intersecting horst-graben** faults. Looking at the figure, we can see how this pattern gets its name. Above the fault intersection, the faults form a central graben block; below the intersection, a horst block develops, thus the name horst-graben fault system.

In areas where there are significant seismic data (three-dimensional data), it is sometimes possible to determine the relative ages of the faults. If this is possible, fault surface maps can be constructed for both faults showing displacement of the older fault by the younger one. Integration of these fault maps with structure maps results in a more accurate picture of the fault intersection.

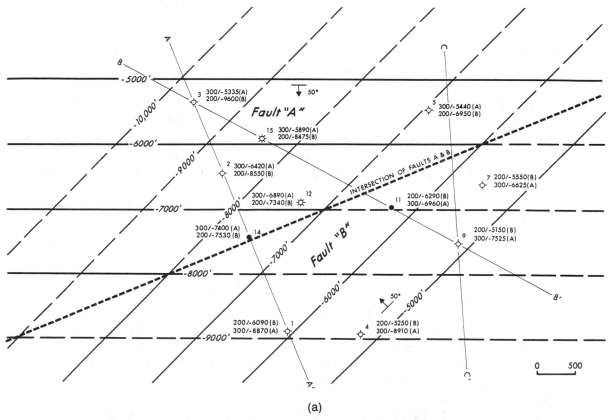

(a)

Figure 7-21 (a) Intersecting fault pattern resulting from two faults dipping in opposite directions. Unlike the compensating fault pattern, both faults continue downward after they intersect. This fault map was prepared using the simplified method of assuming neither fault surface is offset. (b) Cross section A-A' illustrates this intersecting fault pattern. Observe that the faults form a central "graben" block above the intersection and a central "horst" block below the intersection.

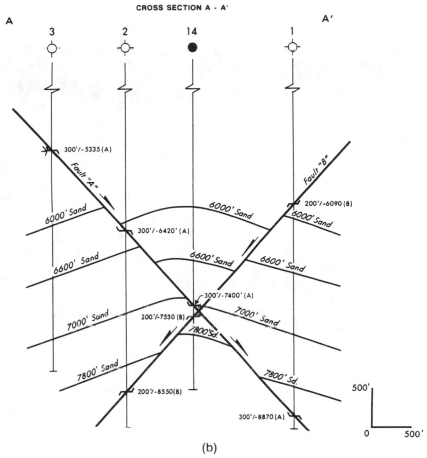

CROSS SECTION A - A'

(b)

Figure 7-21   (continued)

One final note on the different fault patterns; these patterns can be very complex, involving numerous faults in a single area. Also, a fault need not remain as one pattern over its lateral extent. In other words, a fault which is part of a bifurcating fault pattern in one area can be part of an intersecting or compensating pattern in another.

Combined Vertical Separation.   The term *combined vertical separation* applies

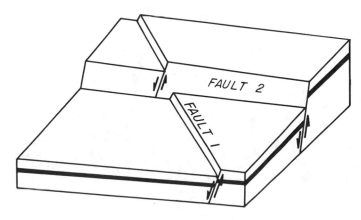

BLOCK DIAGRAM
Figure 7-22   Block model illustrates an intersecting fault pattern.

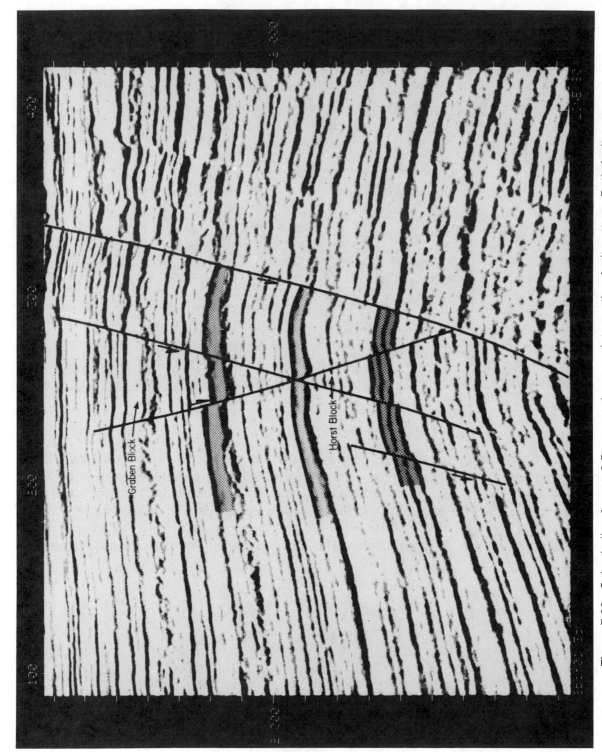

Figure 7-23 Seismic line from a 3-D survey shows an intersecting fault pattern. Both faults appear to have formed contemporaneously since neither fault is offset by the other. Notice that the east dipping fault intersects and terminates against a second west dipping fault, forming a compensating fault pattern. (Modified from Tearpock and Harris 1987. Published with permission of Tenneco Oil Company.)

228

to the relationship which results when two faults of different ages intersect. The zone of combined vertical separation applies to that segment of the intersecting (younger) fault which lies between the offset limbs of the intersected (older) fault. This zone is referred to as the "Zone of Combined Throw" by Dickinson (1954); however, his use of the word throw is a verbal substitution for vertical separation.

Figures 7-24a and 7-24b illustrate a sequence of faulting involving two normal faults which results in a combined vertical separation. If a well penetrates the area of combined vertical separation, only one of the two intersecting fault surfaces will be crossed (*only one fault cut in the well*), but the interval shortening or missing section will be equal to the sum of the fault cuts for both faults or 100 ft. The example in Fig. 7-24 shows two intersecting faults of different ages dipping in opposite directions. The initial Fault F-1 has 50 ft of displacement. The younger Fault F-2, which also has 50 ft of displacement, has displaced Fault F-1 in a manner similar to a fault displacing a formation or bed. A review of the stratigraphic section in the area affected by both faults shows a vertical shortening of 100 ft, which is equal to the combined sizes of Faults 1 and 2.

Figure 7-25 is a fault surface map for Fault J in the Golden Meadow Field, Lafourche Parish, Louisiana. This fault is the north dipping component of an intersecting fault system composed of two faults of different ages dipping in opposite directions. The north dipping fault has a vertical separation of about 80 ft, and the vertical separation of the south dipping Fault D is about 150 ft. Notice along the line of fault intersection the fault cuts in Wells No. 2 and 109 are of unusually large size (235–250 ft) compared to the size of Faults D and J. These two larger fault cuts are the result of a combined vertical separation.

When working in an extensional area of complex or intersecting faults where an unusually large fault cut is present in one or more wells, keep the idea of combined vertical separation in mind. An unusually large cut could be the result of a new, previously

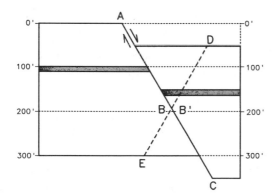
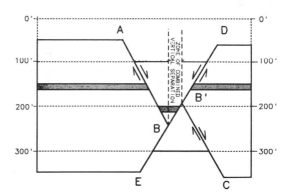

Figure 7-24 Working model illustrating the zone of combined vertical separation (zone of combined fault cut) which develops from two intersecting normal faults.

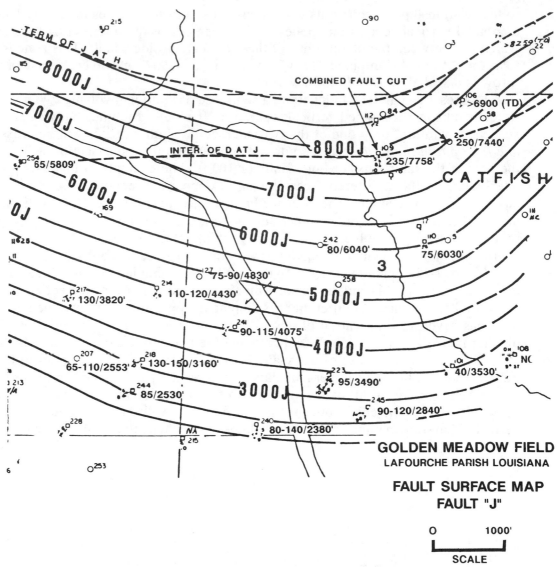

Figure 7-25 Wells No. 2 and 109 have combined fault cuts as the result of intersecting Faults D and J, Golden Meadow Field, Lafourche Parish, Louisiana. (Published by permission of Texaco, USA.)

unrecognized fault, a bifurcating or compensating fault pattern, a buried fault, or a combined vertical separation resulting from intersecting faults of different ages.

The effects of intersecting normal and reverse faults are detailed in Figs. 7-26a through 7-26d. The upper part of each figure shows a structure contour map depicting the interruption of a dipping horizon "O" by various combinations of intersecting normal and reverse faults. The cross section in the lower part of each figure shows the structural effects of the intersecting faults on interval "O–P" which has a constant thickness defined as "c." Although the size of the fault cut in the zone of combined vertical separation is a function of many variables including formation dip, fault dip, vertical separation, and the relative movements of the individual faults, it is usually equal to the algebraic sum of the vertical separation of both faults.

When the intersecting fault is normal (Figs. 7-26a and 7-26b), the vertical separation of the intersecting fault in the zone of combined vertical separation is equal to the algebraic sum of individual vertical separations, that is, it is equal to $(-b)$ plus $(+a)$. The figures illustrate that when both faults are normal, the vertical separation across the intersected fault in the zone of combined vertical separation will also be normal. When the intersected fault is reverse, however, the vertical separation will be normal only when the intersecting fault has the greater vertical separation and will be reverse when the intersected reverse fault has the greater vertical separation (Dickinson 1954).

Figures 7-26c and 7-26d show the resulting fault geometry when the intersecting fault is reverse. In these two cases, the vertical separation across the intersecting fault in the zone of combined vertical separation is equal to the algebraic difference between the vertical separations of the intersecting and intersected faults, that is, it equals $(+b)$ minus $(+a)$. The vertical separation across the zone of combined vertical separation will be normal when the intersected fault is reverse and has a vertical separation greater than that of the intersecting fault. When the intersected fault is reverse and has a vertical separation smaller than that of the intersecting fault, or when it is normal, the vertical separation across the zone of combined vertical separation is always reverse for a reverse intersecting fault. Unlike the geometry involving normal faults where only one fault is seen in the zone of combined vertical separation, the geometry in Figs. 7-26c and 7-26d results in *three faults*.

Compressional Faulting

Reverse faulting is defined as motion along a dipping fault surface on which the hanging wall block rises relative to the footwall block. End members of the reverse fault spectrum consist of vertical and horizontal fault surfaces, as is the case with normal faults. An idealized reverse fault is shown in Fig. 7-27. As in Fig. 7-2, for normal faults, this figure graphically defines the common fault components used in mapping. Depending upon relative amounts of fault and formation dips, normal faults usually omit section, while reverse faults usually repeat section in electric logs. The terms normal and reverse are in wide use to indicate these conditions. However, we should recognize the origin of section omission and repetition, and in doing so, use the genetic terms extensional and compressional faulting. The previous section of this chapter dealt with the practical aspects of extensional faulting. In this section we discuss a few principles and practical aspects of compressional faulting.

In compressional areas, the determination of fault cut data begins with well log correlations just as it does in extensional areas. For a reverse fault, the repeated section recognized on an electric log by correlation with surrounding well logs is the *vertical separation* (Tearpock and Harris 1987). Figure 7-28 diagrammatically represents the determination of fault cut data by the correlation of well logs. The east-west structural cross section perpendicular to the strike of the fault shows two beds which have been cut by a reverse fault (F-1). The reverse fault cuts Well No. 1 at -5300 ft and is dipping at an angle of 35 deg to the east, while the formation is dipping at 30 deg to the west. By correlation with Well No. 2, the repeated section in Well No. 1 is determined to be equal to 100 ft and is shown to have completely repeated Sand Bed "A" in Well No. 1. As shown in the cross section, the throw of Fault 1 is represented by the vertical line AC, which is equal to 55 ft. Vertical separation, represented by the vertical line AE, is equal to the repeated section, which is 100 ft. This correlation example demonstrates that the repeated section obtained for the fault cut due to a reverse fault is equal to vertical separation, which is similar to that shown in Fig. 7-7 for a normal fault.

Figure 7-29 is another reverse fault example used to illustrate the representation of a repeated section by log correlation. The figure shows a cross section consisting of two

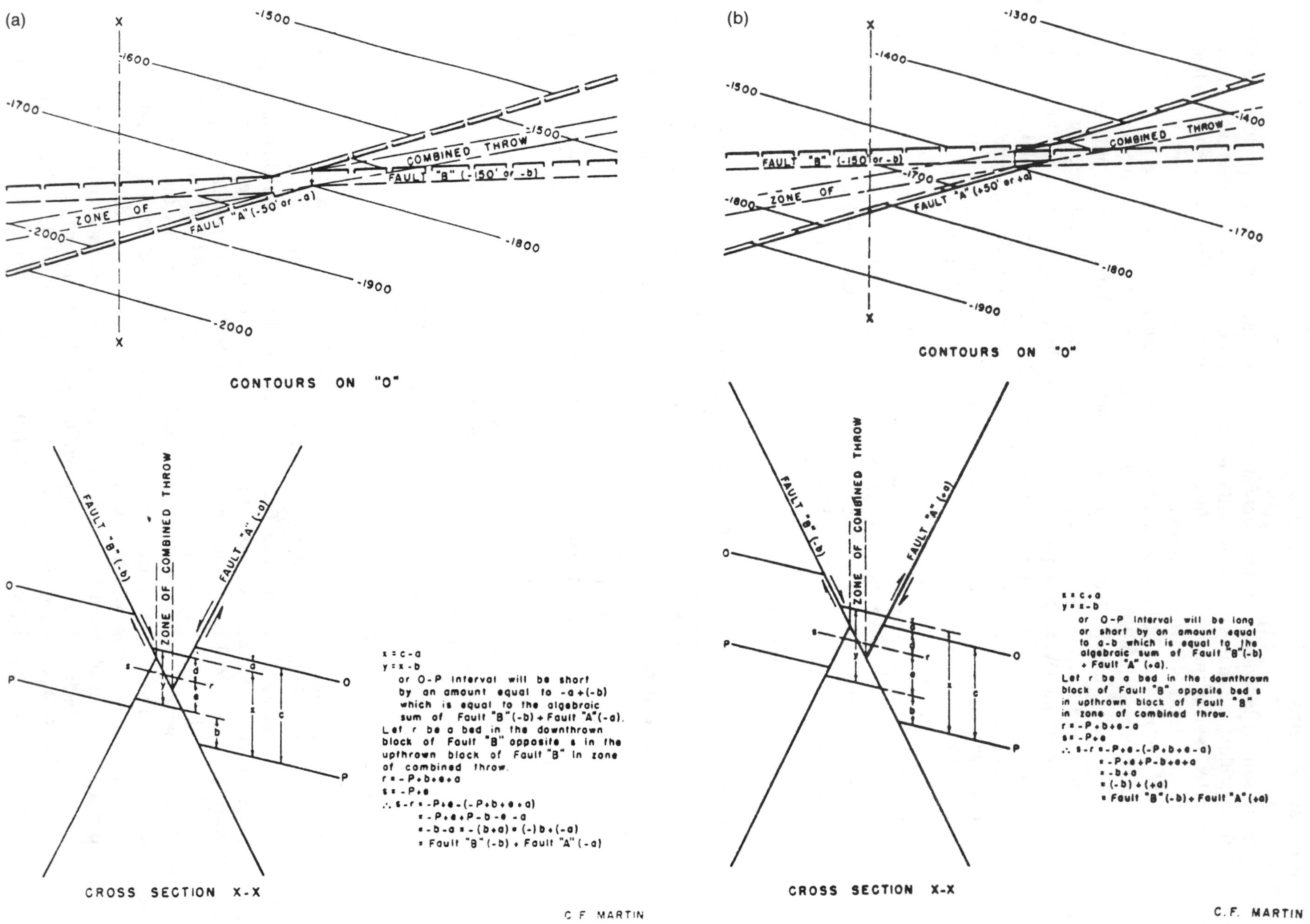

Figure 7-26

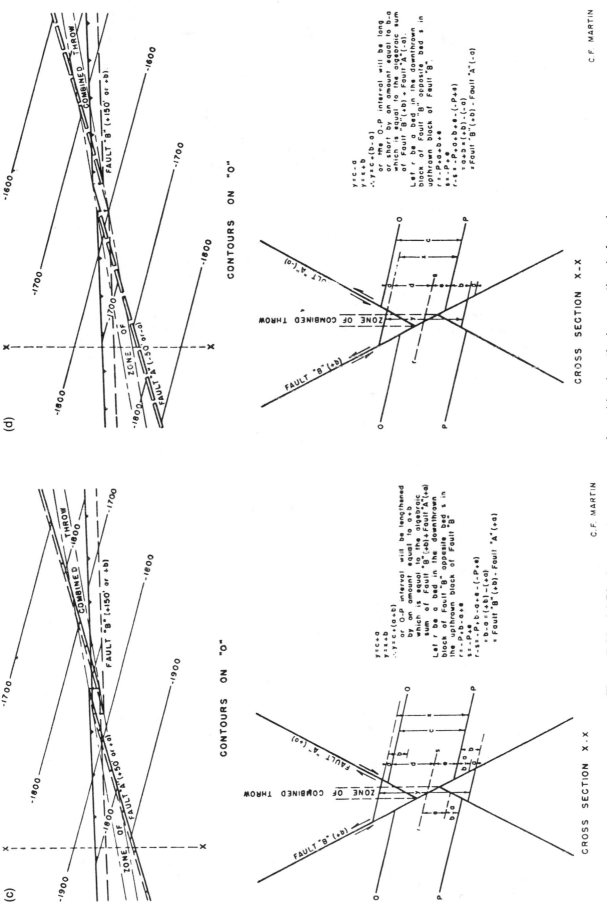

Figure 7-26 (a) Displacement across zone of combined vertical separation (referred to by Dickinson as comb ned throw) for intersecting normal faults. (b) Displacement across zone of combined vertical separation for a reverse fault intersected by a normal fault. (c) Displacement across zone of combined vertical separation for intersecting reverse faults. (d) Displacement across zone of combined vertical separation for a normal fault intersected by a reverse fault. (Dickinson 1954. Published by permission of the American Association of Petroleum Geologists.)

REVERSE FAULT

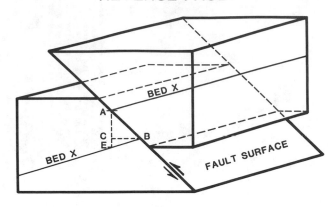

BLOCK DIAGRAM

AB = DIP SLIP   AE = VERTICAL SEPARATION

AC = THROW   BC = HEAVE

Figure 7-27 Block diagram of Bed X displaced by a reverse fault, showing four different fault components. The front panel is perpendicular to the strike of the fault. (Modified from Tearpock and Harris 1987. Published by permission of Tenneco Oil Company.)

REVERSE FAULT

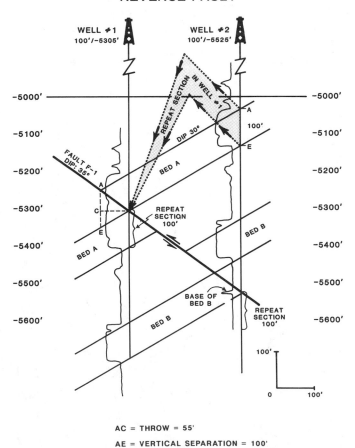

AC = THROW = 55'

AE = VERTICAL SEPARATION = 100'

Figure 7-28 Diagrammatic cross section illustrating the geometric relationship between the repeated section in a wellbore and the vertical separation of a fault. (Published by permission of Tearpock.)

wells, each of which has intersected reverse Fault F-1. The fault and beds are dipping in the same direction. By log correlation, we see that the repeated sections in the wells are equal to 100 ft, which is the vertical separation (AE). Unlike the first example in Fig. 7-28 in which the throw was about one-half the size of the vertical separation, here a measurement of throw reveals it to be over twice as large as the vertical separation. We once again diagrammatically show that the throw of Fault F-1 cannot be measured by log correlation and also that the throw is not equal to the repeated section.

Vertical separation is directly measurable in well logs and seismic sections. It has a fixed value for any particular fault cut which is valid regardless of the apparent attitude of the formation or fault. Depending upon the attitude of a fault or formation, throw can be less than, equal to, or greater than the vertical separation. These geometric relationships are vital to understanding the subsurface and its depiction in map form.

Horizontal shortening of section is one component of compressional faulting, while vertical motion is another. We do not attempt to determine which is the main driving force in mountain building. We can say that crustal shortening has been documented by drilling in areas such as the Rocky Mountains and many other regions of the world in the form of compressional faults and folds.

Thrust faults are dip-slip faults in which the hanging wall has moved up relative to

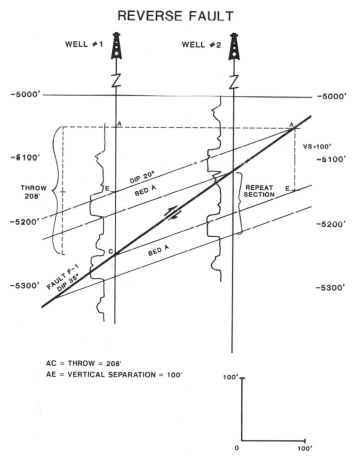

Figure 7-29 Diagrammatic cross section illustrating the relationship between a repeated section in a wellbore and the vertical separation when the beds are dipping in the same direction as the fault. Compare the relationships in this figure to those shown in Fig. 7-28.

the footwall. Thrust faults generally dip less than 30 deg during active slip and commonly dip between 10 and 20 deg at their time of formation (Suppe 1985). Many thrust faults lie along bedding planes over part of their length.

There are two primary settings in which thrust faults develop: (1) compressional plate boundaries, and (2) secondary faulting in response to folding. The foreland of fold-and-thrust belts have been the most extensively studied areas of thrust faulting because of significant potential for both petroleum and coal reserves. There are two basic types of fold-and-thrust belts: (1) those dominated by thrust faulting such as the southern Appalachians and the Cordilleran belt in western Canada and the United States, and (2) fold-dominated belts such as the central Appalachians and Jura mountains of Switzerland (Suppe 1985).

Individual thrust faults commonly cut up the stratigraphic section in a sequence of discrete crosscutting segments called *ramps and flats*. The thrust fault separates the deformed hanging wall structure from the substrate along a bedding plane zone called a decollement (meaning ''unglue'' in French). Thrust and reverse faults, as with normal faults, can bifurcate in the ramp areas with total displacement distributed among the many fault splays, as well as absorbed in folds. In many cases the hanging walls

CONTOURS ON REVERSE FAULT 1

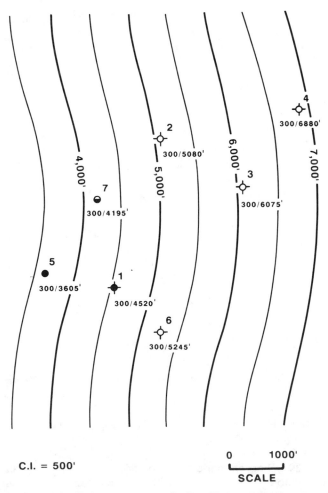

C.I. = 500'

SCALE

Figure 7-30 Fault surface for a reverse fault with a vertical separation (repeated section) of 300 ft.

of thrust faults are folded; this results from bending of the fault blocks as they slip over nonplanar fault surfaces. These folds are called fault bend folds (Suppe 1985). The best-known fault bend folds are those associated with the stepping up of thrust faults from lower to higher decollement surfaces.

There are three names commonly applied to compressional faults: (1) high angle thrust faults usually referred to as reverse faults (steeply dipping), (2) low angle thrust faults (moderately dipping, sometimes defined as less than 45 deg of dip), and (3) overthrust faults (very shallow dip, usually with long horizontal transport).

How do we map the surface of a reverse fault? Most of the fault surface map and cross section techniques already described for extensional faulting apply to compressional fault mapping. For the most part the differences lie in the complexities of the compressional fault patterns and the limit of good quality well and seismic data. Probably more than in any other tectonic setting, we must have a thorough understanding of the structural styles at play and be able to think and visualize in three dimensions.

Single Compressional Faults.   Figure 7-30 is an example of a single compressional

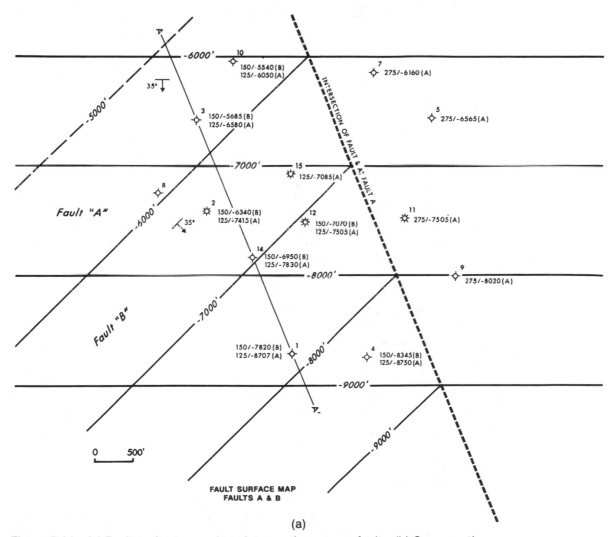

(a)

Figure 7-31   (a) Fault surface map of two intersecting reverse faults. (b) Cross section A-A' bisects the two faults at an orientation which makes the two intersecting reverse faults appear to be parallel.

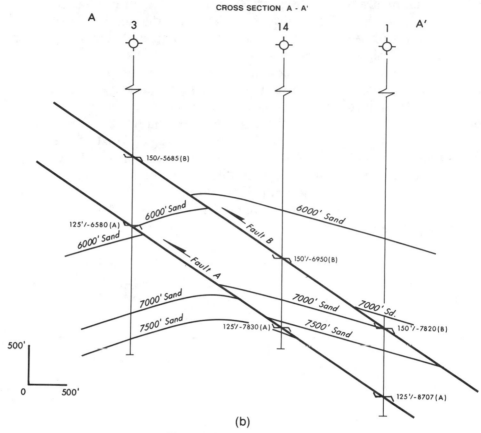

CROSS SECTION A - A'

(b)

Figure 7-31 (*continued*)

reverse fault surface map. The interpreted fault map is based on data from seven wells. The techniques used to construct this map are the same as those used for normal faults. Notice that the depth for each fault cut is expressed as a positive number (300 ft/5245 ft). This means that these fault cuts are above sea level. The cut size of 300 ft refers to the thickness of the repeated section.

Intersecting Compressional Faults. A compressional fault pattern showing two reverse faults dipping in the same general direction is shown in Fig. 7-31a. The data from 12 wells, the general guidelines presented earlier, and an understanding of the compressional tectonic setting were used to construct the fault surface map interpretation for the two reverse faults A and B. Fault A is striking east-west and dipping to the south at 35 deg, and Fault B is striking northeast-southwest and dipping 35 deg to the southeast. The two faults merge laterally as indicated by the dashed line of intersection. We can see by reviewing the fault map that the vertical separation is conserved across the line of intersection; this means that the combined vertical separations for Fault A (125 ft) and Fault B (150 ft) west of this line equal the vertical separation of Fault A (275 ft) east of the intersection.

The change in displacement for these reverse faults across the intersection line is the result of their intersection. Do not mistake the change in displacement in this example with changes in fault displacement resulting from its transfer from one structure to another. In other words, this example is not related to what Dahlstrom (1969) refers to as the compensatory mechanism for reverse faults (thrusts) wherein fault displacement which is diminishing in one fault is replaced by an echelon fault whose displacement is

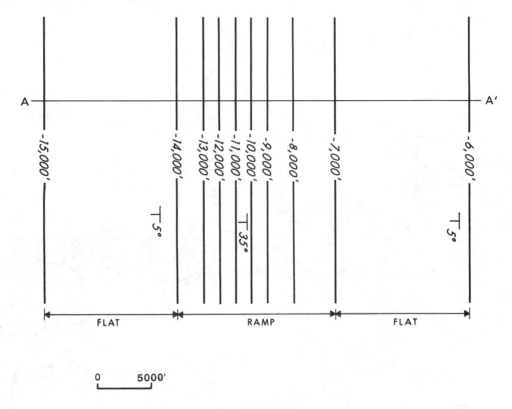

RAMP & FLAT THRUST FAULT SURFACE MAP

(a)

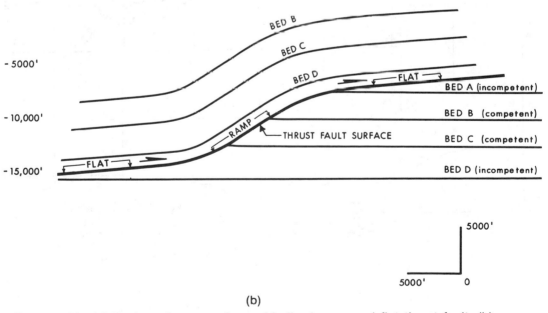

(b)

Figure 7-32 (a) Fault surface map for an idealized ramp and flat thrust fault. (b) Diagrammatic cross section, parallel to the transport direction of the thrust fault shown in Fig. 7-32a.

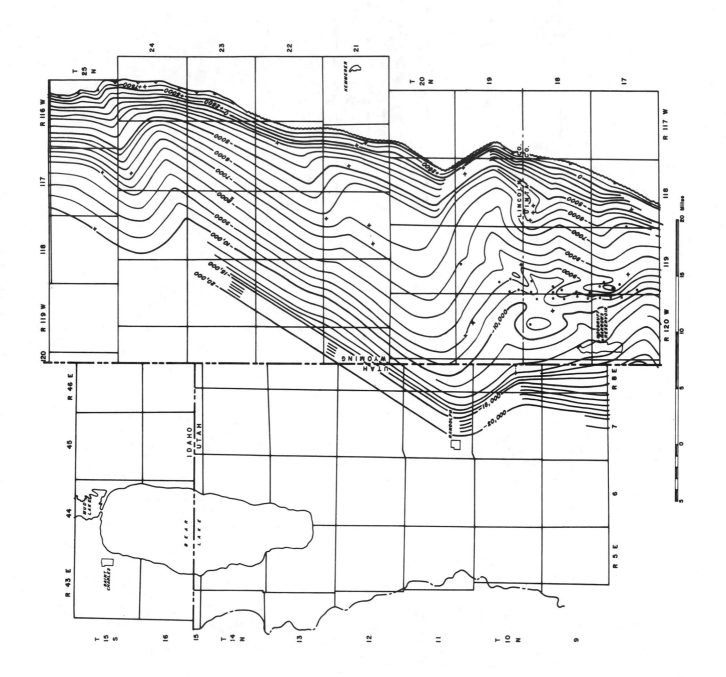

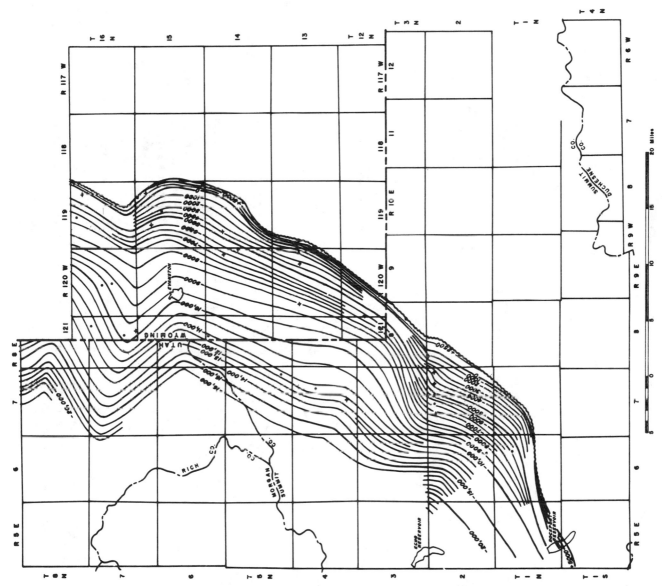

Figure 7-33 Fault surface map of the Absaroka Thrust Fault in the Fossil Basin of Utah and Wyoming. Note the changes in fault dip from east to west. (From Lamerson 1982. Published by permission of the Rocky Mountain Association of Geologists.)

241

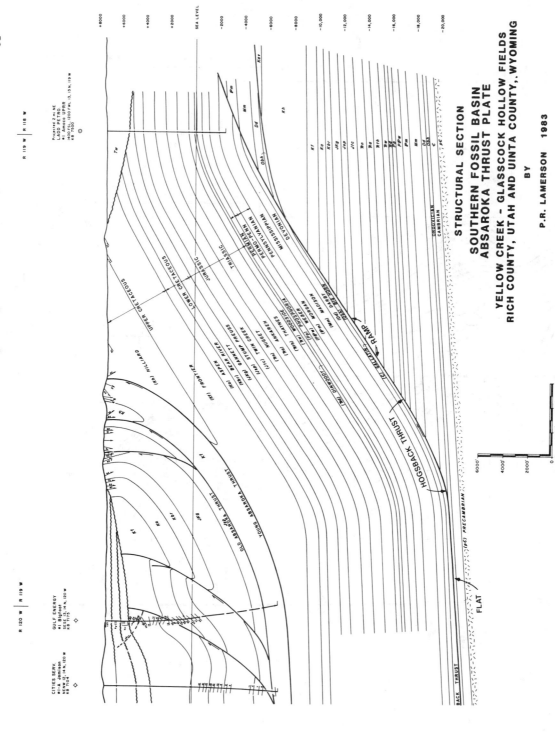

Figure 7-34 Cross section illustrating the geometry of the Hogsback ramp and flat thrust fault. (Lamerson 1982. Published by permission of the Rocky Mountain Association of Geologists.)

242

increasing. Those displacement changes occurring in a *transfer zone* result from all the associated faults being rooted in a common sole fault. This type of faulting and change in displacement is discussed in Chapter 9.

Figure 7-31b shows cross section A-A'. The displacement for each fault has been incorporated into the offset of each sand shown on the cross section. Remember, just as in the extensional faulting examples, the offset of the beds, by the faults, *refers to the displacement in terms of vertical separation.* Therefore, the upthrown and downthrown intersections for each bed with Faults A and B are constructed with the vertical separation technique used to displace a formation correctly from one fault block to another in cross section. This is the same technique used to construct the cross sections for the extensional and compressional fault patterns which was introduced in Chapter 6.

Ramp and Flat Thrust Faults. Ramp and flat fault surfaces develop such that the fault surface moves subparallel in incompetent beds and ramps to higher structural levels in competent rock, and finally when the ramp intersects another incompetent rock, the ramp may transform again into another flat. This geometry is illustrated in Fig. 7-32. Figure 7-32a is a fault surface map of an idealized ramp and flat thrust fault. As with all fault surface maps, the interpretive method of contouring is recommended. This gives you the geologic license to construct the fault surface map to reflect the expected geometry in the area of study.

Figure 7-32b is cross section A-A', which shows the fault surface and the disrupted beds. Notice how the beds in the hanging wall are thrust up and over the same beds in the footwall. The geometry of these faults is detailed in Chapter 9.

A fault surface map of the Absaroka Thrust fault in the Fossil Basin of Utah and Wyoming is shown in Fig. 7-33. Twenty oil and gas fields have been found in the hanging wall of the Absaroka Thrust fault in the southern Fossil Basin. The fault surface map in general reflects a steep ramp to the west from about −20,000 ft to −12,000 ft, a flattening or gentle rise in the central portion of the fault map between −12,000 ft and −5000 ft, and then another steepening ramp to the east (Lamerson 1982).

A portion of the Hogsback ramp and flat thrust fault is shown in cross-sectional view in Fig. 7-34. This cross section clearly illustrates the features of the ramp and flat portions of this type thrust fault. Observe that the flat parallels the Cambrian rocks which serve as the zone of weakness for the basal detachment zone. As the Hogsback thrust fault cuts up-section across Paleozoic and Mesozoic rocks in the footwall, it creates a ramp with west dip on the hanging wall rocks of the thrust between 20 deg and 30 deg (Lamerson 1982).

Fault surface mapping is not commonly done in compressional areas, although there is *no reason* why it should not be used, and indeed it can aid understanding of complex thrust fault geometry. Frequently, thrust faults do not act as seals, and commonly the anticlines that form in the hanging walls act as the primary trapping mechanism. These structural rollovers are often sufficient for trap delineation with little, if any, fault interaction. We strongly believe, however, that the mapping of faults is an integral part of any geologic study regardless of the tectonic setting and can significantly add to the understanding of the geology.

FAULT DATA DETERMINED FROM SEISMIC INFORMATION

Up to this point, most of the information reviewed on fault mapping has related primarily to fault cut data determined from well logs and the use of these data. Seismic data can also provide valuable fault information to aid the mapping process (Chapter 5). This section covers some basic techniques for integration of fault data determined from seismic sections with data obtained from well control.

First, we make a few assumptions with regard to the seismic data being used to assist in mapping fault surfaces. These assumptions are as follows.

1. The data being used are of *reasonable quality*, and have been correctly processed up to and including migration of the data.
2. *The two-way time-to-depth conversion is known* (i.e., a reliable velocity function exists) and does not vary significantly across the area being mapped.
3. The lines have been *interpreted correctly*, with all faults "tied" at line intersections. In areas of steep dip, this may not be possible because of the migration mis-tie between strike and dip lines. In these cases you may have to rely mostly on the dip-oriented lines. This sort of problem makes fault surface mapping a very helpful tool in identifying faults on successive dip lines (Chapter 5).

If these assumptions are met, seismic data can provide **three** very useful types of fault information which aids in the integration of well-derived fault data to construct a final fault surface map interpretation. First, seismic data offers a method of establishing a fault identification (ID). Second, it provides the location of a fault surface in space, and third, the amount of vertical separation for a fault can be measured.

The tying process provides us with a method of establishing a fault *ID*. In short, by tying a fault on a series of lines, we can be sure that the fault seen on various lines is formed by the same fault surface. A fault interpretation that does not tie between lines cannot be correct. Merely tying the faults, however, *does not* ensure that the fault interpretation is valid, since any nonvertical surface can be tied between a series of lines. To establish validity, a fault surface map must be constructed to confirm that both the dip and changes in strike of fault contours are geologically reasonable.

Seismic data offer a method of visualizing a fault interpretation through a series of lines. By tying the fault traces from these lines, we can be assured that a given fault seen on a line at one end of a survey is the same fault seen at the other end. A schematic illustration of this is shown in Fig. 7-35.

Seismic data give us a *location of the fault surface in space*, and unlike well data, data from seismic sections are over a series of points along the profile of the line. This enables us to map fault surfaces over a greater area, and in areas where there is no well control.

It is important to remember that accuracy of points plotted from seismic data is very dependent upon the accuracy (appropriateness) of the velocity function used to convert seismic time to a depth value. Figure 7-36 illustrates what influence a velocity function can have on dipping surfaces. The figure shows two positions for a proposed directional well plotted on a seismic line based on two different velocity functions. Both functions are possibly valid for this prospect. As the illustration shows, the use of the slow velocity function places the directional well in the upthrown fault block; however, the fast function places the well in the downthrown fault block. The upthrown block is the target block to test horizons 1 and 2. The use of the wrong velocity would cause the exploratory well to be drilled in the wrong fault block. In this particular situation, since the correct velocity was not known, a straight hole design was chosen rather than risk missing the prospective fault block with a directional well. The same problem can apply to fault interpretations. In some cases, there may be some uncertainty as to the two-way seismic time corresponding to a particular depth. If the position of a fault is especially critical, error bars

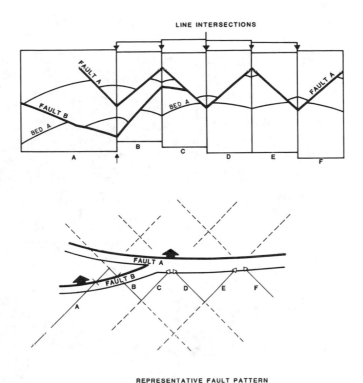

Figure 7-35   Illustration of the usefulness of fault tying in propagating fault surface over a series of seismic lines. (Prepared by C. Harmon. From Tearpock and Harris 1987. Published by permission of Tenneco Oil Company.)

can be placed around a posted point that will graphically illustrate spatial uncertainty for a particular datum.

The third type of seismic information can provide is a ***measurement of the vertical separation*** of a fault. There is often disagreement between geologists and geophysicists with regard to the size of a fault estimated from well log correlation as compared to that estimated from a seismic section. In many cases, the difference between what the geophysicist is calling displacement and what the geologist is calling displacement is the result of a misunderstanding of fault component terminology. If we assume the vertical resolution limitation of seismic data is known, then the answer to this discrepancy is that two different values for displacement are being measured. As discussed earlier in this chapter, the amount of missing section seen in an electric log of a well is *not throw*, but rather vertical separation, unless, of course, the formations in the area have zero dip or the fault is vertical. Then the values for throw and vertical separation are the same. What is commonly measured on a seismic line for fault displacement is not vertical separation but throw, and more commonly, apparent throw, which is a measurement of fault displacement dependent upon the orientation of the seismic line in relation to both the bed and fault dip. The only case where we can actually measure the true throw of a fault, at a particular depth, is when the line is oriented ***perpendicular*** to the strike of the fault. This means that if several seismic lines, oriented at different angles to the strike of a fault, are used to estimate the throw of that fault, each line will yield a different value for the apparent throw for the same fault. Even if it were possible to shoot all seismic lines perpendicular to the strike of each fault being mapped, since the throw of the fault is **not** the fault component used in the construction of fault and integrated structure maps, the added work and expense would not be justified.

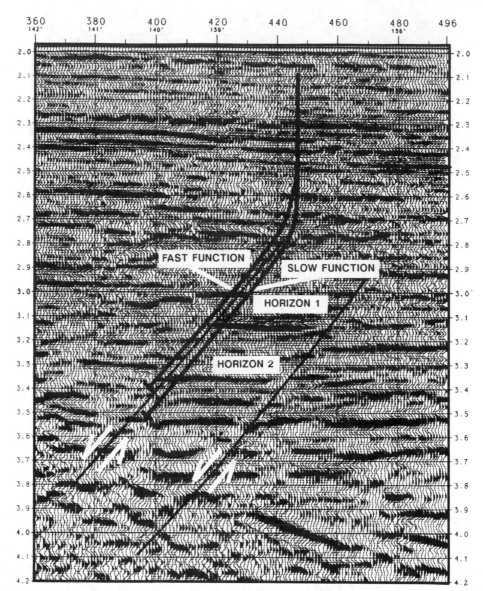

Figure 7-36   Seismic section shows the effect of velocity function uncertainty on the location of a directional well. (Prepared by C. Harmon. Modified from Tearpock and Harris 1987. Published by permission of Tenneco Oil Company.)

Fault displacement problems from seismic data can be easily reconciled by estimating the vertical separation of a fault, from seismic lines, instead of throw or apparent throw. If a seismic line has good correlations across a fault, it is very simple to measure the vertical separation of that fault. Remember, the size of a fault cut in a well log is the measurement of the fault's vertical separation. In order not to confuse components in our fault displacement measurements and constructed maps, the vertical separation of a fault must be measured on a seismic section. These depth converted vertical separation values can then be used to tie the well fault cut data and aid in the construction of the fault and structure maps.

Figure 7-37 illustrates the proper method for measuring the vertical separation from a seismic line. The line is an east-west line shot in the central Gulf of Mexico Basin by TGS Offshore Geophysical Company. It is an excellent line to use to define the procedure for measuring vertical separation, because the dip of the structure changes from east to west. At the eastern end of the line, the structure is essentially flat, with zero dip; to the west, the dip increases. Therefore, we can show the procedure for measuring the vertical separation in flat lying beds as well as those with significant dip. We also show the comparison of the values for vertical separation and apparent throw to reinforce the understanding that these fault components are not the same and that the use of throw in place of vertical separation can cause significant mapping errors.

We begin with the displacement for Fault A at **Horizon R**. The procedure is more easily followed by using the insert blowup in the upper left corner of the figure. The R Horizon is dipping at 16 deg to the west. To measure the displacement of the fault in terms of vertical separation, project the dip of the horizon from the upthrown fault block through the fault, as if the fault were not there, into the downthrown block until the projection is vertically over the intersection of the R Horizon with the fault in the down-thrown block. The vertical difference in depth from the projection, at − 10,900 ft, to the downthrown intersection of the horizon and fault, at − 11,760 ft, is 860 ft. This is the measurement of the vertical separation of the fault at this horizon. If a well were drilled through the fault at shot point 7865, so that the R Horizon is faulted out in the well, the missing section in the well log would be approximately 860 ft.

The measurement of the apparent throw of the fault at this horizon is the vertical difference in depth from the intersection of Horizon R with the fault in the upthrown

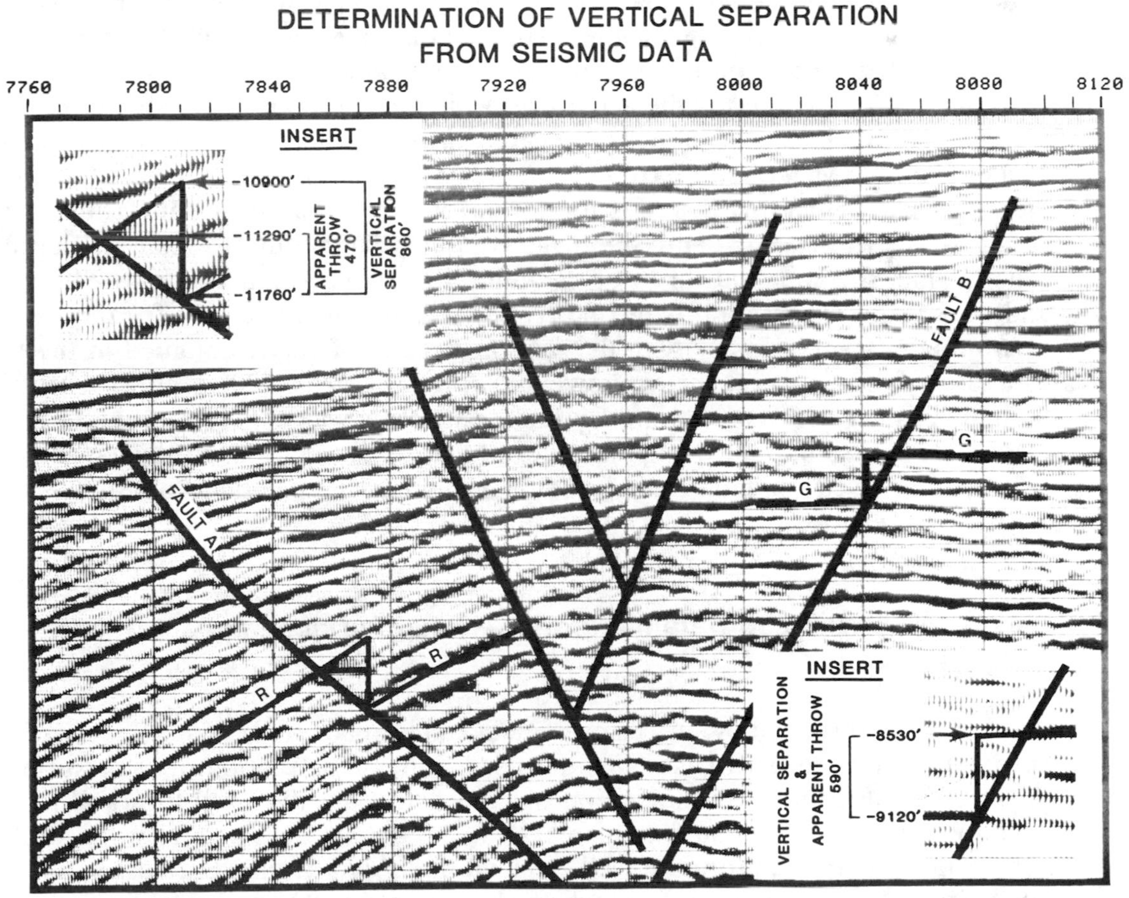

DETERMINATION OF VERTICAL SEPARATION FROM SEISMIC DATA

Figure 7-37 Seismic line Offshore Gulf of Mexico. The figure inserts illustrate the correct method for estimating the vertical separation of a fault from a seismic line when the beds are dipping (Bed R) and when the beds are horizontal (Bed G). In order to structure contour the correct displacement across a fault using geophysical data, the vertical separation must be estimated. Throw is not used in subsurface mapping. (Modified from Tearpock and Harris 1987. Published by permission of TGS Offshore Geophysical Company and Tenneco Oil Company.)

fault block ($-11,290$ ft) to the intersection of the horizon with the fault in the downthrown block ($-11,760$ ft), and this is 470 feet. ***The difference between the apparent throw and the vertical separation of the fault at the R Horizon is a significant 390 ft.***

Turning to the right side of the figure, we conduct the same procedure and measure the displacement of the **G Horizon** by Fault B, in an area in which the structural bed dip is essentially flat. Refer to the insert blowup in the lower right corner of the figure. First, measure the vertical separation by projecting the dip of the horizon in the upthrown block through the fault into the downthrown block until the projection is vertically directly over the intersection of the G Horizon with the fault in the downthrown block. The vertical distance between the depth of Horizon G at the fault intersection in the upthrown block at -8530 ft, and the depth of the horizon/fault intersection in the downthrown block, of -9120 ft, is 590 ft.

The apparent throw, measured as the vertical distance from the intersection of the horizon with the fault in the upthrown block to the intersection of the same horizon and fault in the downthrown block, is 590 ft. This is the same value obtained for the vertical separation. Since the formation dip at this location is zero degrees, we should expect these values to be the same, based on the definitions and discussion presented so far in this chapter.

If necessary, go back and review these procedures again until you thoroughly understand how to measure the vertical separation and clearly see how it differs from throw. It is essential to remember that whether the fault data are derived from well logs or seismic lines, the same fault component, for displacement, must be used in mapping faults and structures, and that component is *vertical separation*.

Figure 7-38 illustrates the accuracy of the technique outlined above in reconciling well log data with seismic data. The seismic line is a Tenneco Oil Company line in South Marsh Island (SMI), offshore Gulf of Mexico. The seismic line intersects the Signal Well No. 1 in SMI Block 67. There is a fault cut in the Signal Well with a missing section equal to 720 ft, at a log depth of 10,940 ft, which faults out Horizon A. Based on this seismic line, the apparent throw of the fault at Horizon A, marked on the line, is 505 ft, and the vertical separation is 695 ft. Notice that the measurement for the vertical separation agrees very closely with the missing section of 720 ft obtained in the well. The estimate of the apparent throw, however, differs by nearly 200 ft from the estimate of fault size by well log correlation. By this example, we again illustrate the importance of measuring the same fault component to obtain the value for fault displacement, which will later be used for fault and structure mapping, whether that value is derived from well log or seismic data.

Seismic and Well Log Data Integration—Fault Surface Map Construction

Since the measurement of fault displacement can be made from seismic and well log data, both can be posted on a base map and used in the construction of fault surface maps. Figure 7-39 illustrates the integration of fault data from two well logs and two seismic lines to construct a fault surface map. The fault cut data obtained from the well logs are posted next to the appropriate well, and the fault data from the two seismic lines are posted next to the shotpoint from which the data were obtained. Unlike fault data from a well log, which provides fault cut information from a single depth at a specific location, a seismic line presents continuous fault data over a series of shotpoints along the profile

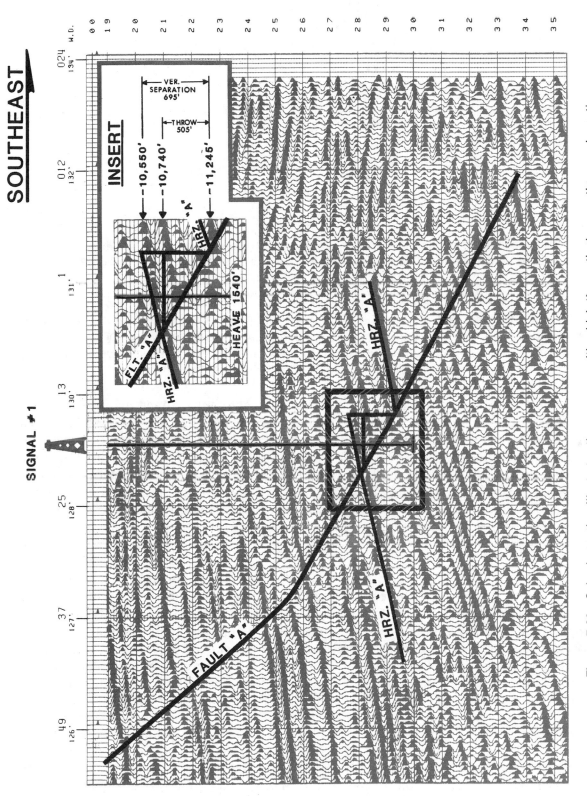

SOUTHEAST

SIGNAL #1

INSERT

VER. SEPARATION 695'

THROW 505'

−10,550'

−10,740'

HRZ. "A"

−11,245'

FLT. "A"

HRZ. "A"

HEAVE 1540'

FAULT "A"

HRZ. "A"

HRZ. "A"

Figure 7-38 Seismic section illustrates the accuracy with which the vertical separation can be estimated, as compared to fault cut data from a well log directly on the section line. Missing section in the signal No. 1 Well by correlation with surrounding wells is 720 ft. This compares very favorably with the vertical separation estimated from the seismic section of 695 ft. The apparent throw of the fault is only 505 ft. (From Tearpock and Harris 1987. Published by permission of Tenneco Oil Company.)

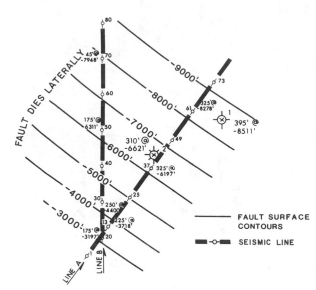

Figure 7-39 Fault surface map prepared using data from well logs and seismic sections. (Modified from Tearpock and Harris 1987. Published by permission of Tenneco Oil Company.)

of the line. This permits the fault surface to be mapped over a greater area than would be possible with well log data alone. The seismic also provides a tie to each well to confirm whether the fault cut observed in the well is the same fault surface seen on the seismic line.

Fault Displacement Mapping.   A review of each fault surface map presented in this text reveals that the size and subsea depth for each well log fault cut or seismic fault pick is posted on the base maps (i.e., 310 ft/−6621 ft). So far, in fault map construction, we have used only the subsea depth data to prepare fault contour maps. What about the posted vertical separation (fault size) data? Can these data be used for mapping; and if so, what is their importance? The vertical separation of the fault can and should be contoured when sufficient data are available. Such a map, called a *vertical separation map* or *throw map* by some, provides additional information on the growth history of the fault and additional data for the construction of integrated structure maps (see Chapter 8).

   In Fig. 7-40a, vertical separation contours for the fault are superimposed on the fault surface contours. This vertical separation is contoured in a dashed pattern with a contour interval of 200 ft. Notice, for the most part, that the vertical separation contours are *perpendicular* to the fault surface contours. This type pattern indicates that the fault is post-depositional and is therefore termed a **structural fault**. In other words, the fault is not a syn-depositional growth fault; it does not show an increase in displacement with depth. The deviation of the vertical separation contours, from their perpendicular trend, at a depth of about −3000 ft to −5000 ft indicates the time at which the fault was active.

   This information has application in evaluating a fault trap, because it graphically illustrates the depth of the geologic interval affected by post-depositional faulting. A late trap formed by a post-depositional fault such as the one in Fig. 7-40a can be evaluated as a prospect as long as migration of hydrocarbons occurred during or after the initiation of fault movement. Displacement mapping for growth faults is also very important and is covered later in this chapter.

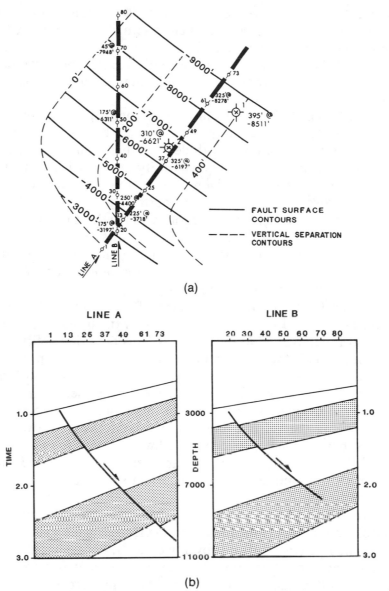

Figure 7-40 (a) Vertical separation contours added to fault surface contour map provide additional information which is helpful during fault and structure map integration (see Chapter 8). (Modified from Tearpock and Harris 1987. Published by permission of Tenneco Oil Company.) (b) Seismic lines A and B shown in map view in Fig. 7-40a. (From Tearpock and Harris 1987. Published by permission of Tenneco Oil Company.)

Seismic Pitfalls

As we have already mentioned, this text is not intended to instruct in the area of structural geology; nor is there any intent to discuss in detail the pitfalls of seismic interpretation. However, several seismic pitfalls can be avoided if fault surface and vertical separation mapping are undertaken. We address these subjects to emphasize the importance of constructing fault surface maps as part of any subsurface geologic study.

In extensional basins, do faults die with depth? The answer to this question is obviously yes; but, when a fault is interpreted as dying with depth, there has to be a geologic explanation for this occurrence (Chapter 9). Too often a fault is interpreted on a single seismic line as dying with depth, when it actually does not. A fault may appear to die with depth on a seismic line, but this does not mean that it actually dies with depth in the subsurface. What it does indicate is that a one-line interpretation can be very suspect, requiring additional subsurface work which may include evaluating additional seismic or well control, if it is available.

Seismic line B in Fig. 7-40b illustrates something commonly seen on seismic—a fault *"dying with depth."* Is it real or just an illusion? Line A shows the same fault; however, on this line the fault continues with depth rather than appearing to die at about 2.1 sec as it does on line B. The fault shown on these two seismic lines is the fault contoured in Fig. 7-40a. Referring to this figure, we see that the fault does not die with depth. The fault does, however, *die laterally* to the northwest. Looking at Fig. 7-40, notice that Line B is oriented north-south in the direction in which the fault is laterally dying. At shotpoint 50 on line B, the fault size is 175 ft, it is 45 ft at shotpoint 70, and 0 ft (it dies) at about shotpoint 77. Therefore, on line B the fault *appears* to die with depth. The key word is *appears*. This effect is simply a function of the orientation of the line in relation to the fault surface of this laterally dying fault.

In this particular case, the dilemma of incorrectly interpreting a fault dying with depth can be avoided if a detailed evaluation of the fault surface is conducted. First, Lines A and B can be tied at shotpoint 20. This fault tie confirms that the fault continues with depth. Second, the construction of a fault surface map, incorporating the well and seismic data, shows that the fault continues with depth. Third, as mentioned earlier in this section, whenever possible contour the vertical separation on the fault map as part of the geologic evaluation of a fault. In this case, the vertical separation contoured in Fig. 7-40a, using the seismic and well log data, clearly shows that the fault continues with depth, but dies laterally to the northwest. These three steps illustrate how correct and detailed fault surface mapping can present a better and more complete geologic interpretation of a fault and prevent incorrect subsurface interpretations that could later prove to be costly.

Figure 7-41 is a Gulf of Mexico offshore seismic line that illustrates an authentic example of a pitfall that could have been avoided with fault surface and vertical separation mapping. The line (Fig. 7-41a) shows two faults, both coincidentally lining up to look as if one fault trace could be drawn; however, there actually are two faults. This line was once used as a regional line prepared to illustrate a prospect. As shown in Fig. 7-41b, a fault correlation was forced where no fault exists (dashed fault line). This is significant, because by forcing a one fault interpretation, an upthrown trap is shown that really does not exist. Figures 7-42a and 7-42b illustrate a fault surface and resulting structure map from the forced correlation. Notice that the structure map identifies 500 ft of closure upthrown to the fault. The prospect, presented with the intent of bidding on the offshore lease sale block in which it occurred, was generated as a one line interpretation. At the time the prospect was presented, no fault map was prepared.

The prospect was reviewed using the seismic line in Fig. 7-41 and additional lines that were available in the surrounding area. First, based on the structure map in Fig. 7-42b and the one seismic line (Fig. 7-41b, forcing a one fault interpretation), the fault map shown in Fig. 7-42a was prepared. Second, using all the available lines, the fault and structural interpretation in Figs. 7-43a and 7-43b was made showing two faults. The shallower fault is interpreted as a radial fault that is dying laterally away from a salt dome to the west. It appears to die with depth on the seismic line (Fig. 7-41a), but actually the fault surface is simply migrating out of the plane of the line. The deeper fault is a buried growth fault that is also dying laterally, but in the opposite direction to that of the shal-

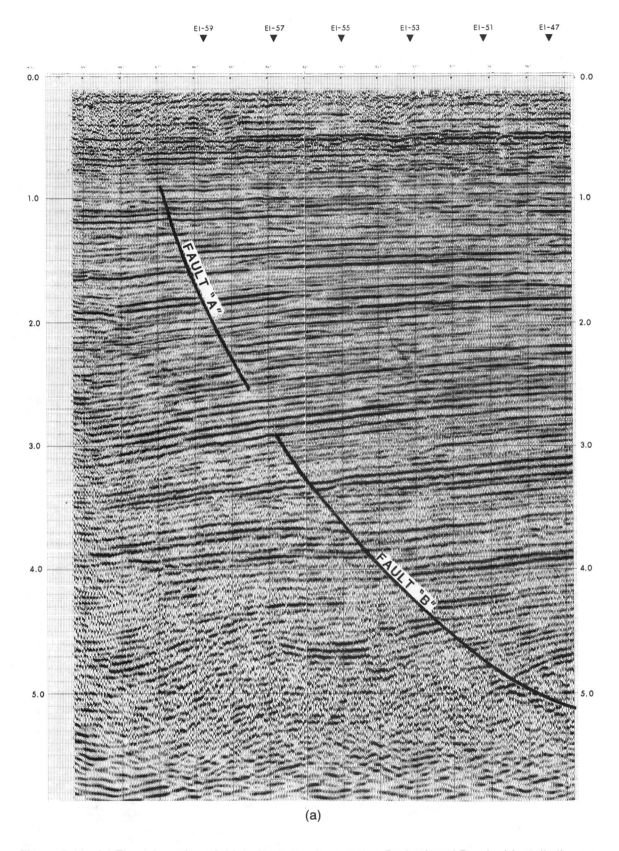

(a)

Figure 7-41 (a) The orientation of this seismic line is such that Faults A and B coincidentally line up and appear as if one fault trace could be drawn. (b) A one fault correlation forced through continuous reflectors (see dashed line). (From Tearpock and Harris 1987. Published by permission of Tenneco Oil Company.)

LINE A

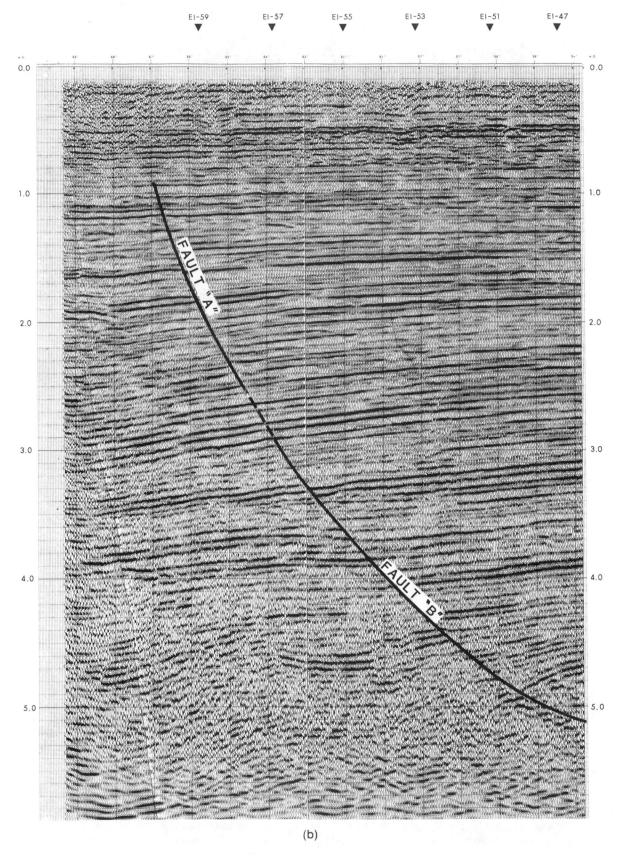

(b)

Figure 7-41 (*continued*)

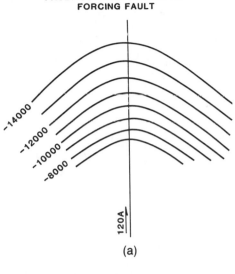

FAULT SURFACE IMPLIED BY
FORCING FAULT

—14000
—12000
—10000
—8000

120A

(a)

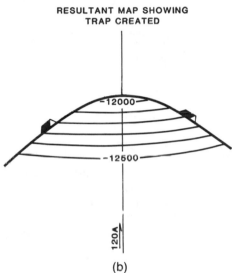

RESULTANT MAP SHOWING
TRAP CREATED

—12000—

—12500—

120A

(b)

Figure 7-42 (a) Fault surface map generated from the one fault interpretation (Fig. 7-41b). (b) Structure map of fault trap created by the forced interpretation. (From Tearpock and Harris 1987. Published by permission of Tenneco Oil Company.)

lower fault. The line just happens to be in the correct orientation to make the two fault traces appear to line up. Figure 7-43 illustrates the correct fault surface and structural interpretation based on all the available data. Notice that based on this new and correct interpretation, there is *no fault trap* and therefore *no prospect*.

The identification of this nonexistent prospect would not have occurred had the procedures outlined in this chapter been followed. First, no prospect should be based on a one line interpretation. There are too many **illusions** of geometry that can occur on any given line. Second, a fault map should always be prepared. Had a fault map been prepared based on the one line, one fault interpretation, the sharp outside bend taken by the fault surface (Fig. 7-42a) should have been suspect and further review of the fault undertaken. Finally, the construction of a vertical separation map would have shown a fault losing displacement with depth, then abruptly gaining displacement again, a very questionable and unlikely situation.

In Chapter 1, we stated that the only certainty in a subsurface interpretation, prepared from seismic and well log data, is that it is most likely incorrect. The best geologist is the one who comes up with the most accurate and geologically reasonable subsurface

CORRECT FAULT SURFACE MAP

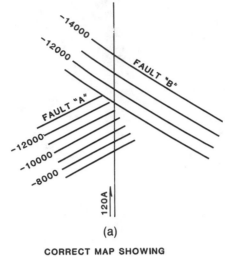

(a)

CORRECT MAP SHOWING
NON TRAPPING CONDITIONS

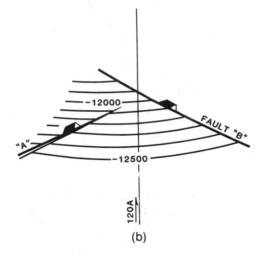

(b)

Figure 7-43   (a) Fault surface map for both Faults A and B. (b) Completed structure map constructed by integration of fault and structure maps shows no hydrocarbon trap. (From Tearpock and Harris 1987. Published by permission of Tenneco Oil Company.)

interpretation. Remember the two significant causes of errors in geological mapping: the failure to use all the data available and the incorrect application of the data. If we fail to make fault surface and vertical separation maps, when the data are available, how can we expect our interpretation and resultant maps to be accurate, geologically reasonable, and representative of the best interpretation of the subsurface?

GROWTH FAULTS

One of the principal petroleum-related settings for normal faults is that of major large-scale gravity slide structures. The normal faults in this category decrease in dip with depth to a low angle fault, often becoming bedding plane. These faults slide along a decollement and are called *detachment faults* (Chapter 9). Such fault systems are well known in areas such as the US Gulf of Mexico, Brunei, and the Niger Delta. These faults are called *listric growth normal faults* based on the Greek word listron, or shovel, because of their curved appearance in cross section. The major listric fault is commonly referred to as the master fault, if it has the largest related displacement. Numerous synthetic and antithetic (secondary) faults are often associated with the master fault.

These listric growth faults are syndepositional, often exhibiting a significant increase in stratigraphic thickening in the downthrown block or hanging wall and an increase in displacement with depth. This relationship indicates that there was movement along the fault surface while the sediments were being deposited. Such faults also are called contemporaneous because fault movement was occurring while the adjacent sediments were being deposited.

Growth faults are often associated with anticlines, called rollovers, which develop as a result of bending of the hanging wall fault block as it conforms to the curved fault surface (Suppe 1985). These rollover anticlines are one of the most important hydrocarbon traps associated with listric growth faults.

Estimating the Size of Growth Faults

The method for estimating the size of a post-depositional fault was outlined in Chapter 4. In this section, we present two methods for estimating the size (*vertical separation—displacement*) of a growth fault for any particular wellbore fault cut.

The Restored Top Method. Since the thickness of the stratigraphic section in the upthrown fault block of a growth fault is thinner than the equivalent stratigraphic section in the downthrown block, an estimate of the size of the fault in any particular well based on correlation with a well in the upthrown fault block will be smaller than an estimate of the fault size using a well in the downthrown fault block for correlation. Therefore, the procedure for estimating the size of a fault cut for a growth fault is not as straightforward as that for nongrowth faults. In most cases, at least two correlation well logs are needed: one located in the upthrown fault block and the other in the downthrown block. With the available electric well logs, the size of any particular wellbore fault cut for a growth fault can be estimated by a procedure called the *Restored Top Method*.

1. Choose a datum, faulted out of the well of interest for which the size of the fault cut is needed. This datum may be a marker, formation, or bed.
2. The log of a well in the upthrown fault block is then correlated with the deeper log section (the interval below the fault) in the faulted well in order to obtain an estimate of the upthrown restored top for the chosen datum.
3. The log of a well in the downthrown fault block is correlated with the upper log section (the interval above the fault) in the faulted well in order to estimate the downthrown restored top for the chosen datum.
4. The estimate for the size of the growth fault at the chosen datum is the difference in measured depths of the upthrown and downthrown restored tops for the chosen datum in the faulted well.

Using the cross section in Fig. 7-44, we illustrate the use of the restored top method. Fault 1 is a growth fault downthrown to the east. In Well No. 1, Beds A, B, and most of C are faulted out. Bed B is a horizon of interest to be mapped; therefore, there is a need to estimate the size of the growth fault affecting Bed B. Three wells are on the cross section: Well No. 2 in the upthrown fault block, Well No. 3 in the downthrown fault block, and Well No. 1, which has the horizon of interest faulted out. If necessary, go back to Chapter 4 and review the section on estimating restored tops before continuing this section.

First, estimate the size of the fault in Well No. 1 using the other two wells. By correlation with Well No. 2 in the upthrown fault block, the size of the fault in Well No. 1 is 190 ft; but by correlation with Well No. 3 in the downthrown block, the size of the fault is 315 ft. Since these two estimates for the size of the fault are different, which one should be used for mapping Bed B? The answer is neither estimate. The 190 ft is too small since it is based on correlation with a well in the stratigraphically thin upthrown block. The 315 ft is too large because this value was obtained by correlation with a well in the stratigraphically thick downthrown block.

How do we obtain an estimate for the size of the growth fault for structure mapping Bed B? The restored top method outlined earlier must be used. The upthrown restored top for Bed B in Well No. 1 by correlation with Well No. 2 in the upthrown fault block is at 7705 ft measured log depth (Fig. 7-44). The downthrown restored top in Well No. 1 for Bed B by correlation with Well No. 3 in the downthrown Block is at 7980 ft measured log depth. Using the restored top method, the size of the growth fault at Bed B is estimated as the difference in restored top depths for Bed B (7980 ft − 7705 ft = 275 ft). Therefore, in the preparation of a structure map on Bed B, the vertical separation value of 275 ft is used to contour across Growth Fault 1. Also, in using this method, two additional control points for mapping are estimated, these being the upthrown and downthrown restored tops in Well No. 1. Remember, these tops must be corrected to subsea depths before they are used for structure mapping.

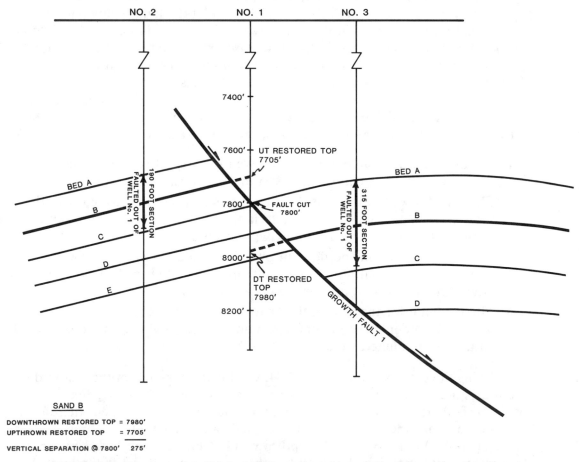

Figure 7-44 Cross section illustrates the restored top method for estimating the vertical separation for a growth fault. The restored tops are estimates using the procedures discussed in Chapter 4.

The estimated size of 275 ft for the fault cut for Bed B lies between the low value of 190 ft and the high value of 315 ft. This is the most accepted method for estimating the size of a growth fault for any particular bed, formation, or marker to be mapped.

The Single Well Method.   At times, it is possible to use only one well to closely approximate the size of a growth fault for the mapping of some specific marker or formation. This method is referred to as the *Single Well Method*. This method is illustrated in cross-sectional view in Fig. 7-45 and in map view in Fig. 7-46 for a growth fault with linear expansion.

In the cross section in Fig. 7-45 there are three wells: Well No. 3 lies in the upthrown fault block of the growth fault, Well No. 2 is in the downthrown fault block, and Well No. 5 is the well of interest. The growth fault has faulted out a significant stratigraphic section from the top of Horizon B through the very top of Horizon F in Well No. 5. For mapping purposes, we are interested in the size of the growth fault at Horizons B, D, and F.

First, we estimate the size of the fault in Well No. 5 with the other two wells. By correlation with Well No. 3, in the upthrown block, the estimated size of the growth fault is 200 ft, and with Well No. 2 in the expanded downthrown fault block the size is 400 ft, faulting out Horizons B through F. There is a two-to-one difference in the estimated size of the fault by correlation with these two wells. Which one is correct, or are they both correct; and which one, if any, should be used for structure mapping of the various horizons? We know that this is a growth fault, which means by definition that the size of the fault increases with depth. Therefore, it is reasonable to assume that the size of the fault increases from the B to the F Horizon; but, by how much and what is the size at each horizon?

The size of the fault at any horizon can be estimated by using both the *Restored Top Method* and the *One Well Method*. We will begin by estimating the size of the fault for Horizon B. The size can be estimated using the restored top method, the results of which are shown in the insert in Fig. 7-45. The upthrown restored top for the B Horizon is at 8050 ft, and the downthrown restored top is at 8250 ft. The difference between these two restored top values is 200 ft, which is an estimate of the size of the fault at Horizon B.

There is a simplified method for estimating the size of a growth fault that can be used under certain conditions. Notice in the cross section (Fig. 7-45) that Horizon B is at the top of the missing section in Well No. 5; therefore, we can say that most of the faulted interval in Well No. 5 can be represented by the thin upthrown section seen in Well No. 3. The size or vertical separation of the fault for Horizon B can therefore be estimated or closely estimated by only correlating Well No. 5 with the thin upthrown Well No. 3. By making this correlation, we obtain an estimate of 200 ft for the size of the fault. Observe the 200 ft of section faulted out of Well No. 5, illustrated on the well log stick for Well No. 3. We can check the accuracy of this simplified technique by comparing the result with that obtained from the restored top method. They are the same. Using the two methods independently, the size of the growth fault at the B Horizon is estimated to be 200 ft.

Now estimate the size of the fault for Horizon F. First, using the restored top method, the estimated size of the fault at this horizon is 400 ft. The results are again summarized in the insert in Fig. 7-45. Reviewing the log sticks in cross section, notice that the F Horizon is at the bottom of the missing section in Well No. 5. Therefore, the missing section is represented by the thick downthrown section seen in Well No. 2. So, we can use the simplified one well method and estimate or closely estimate the size of

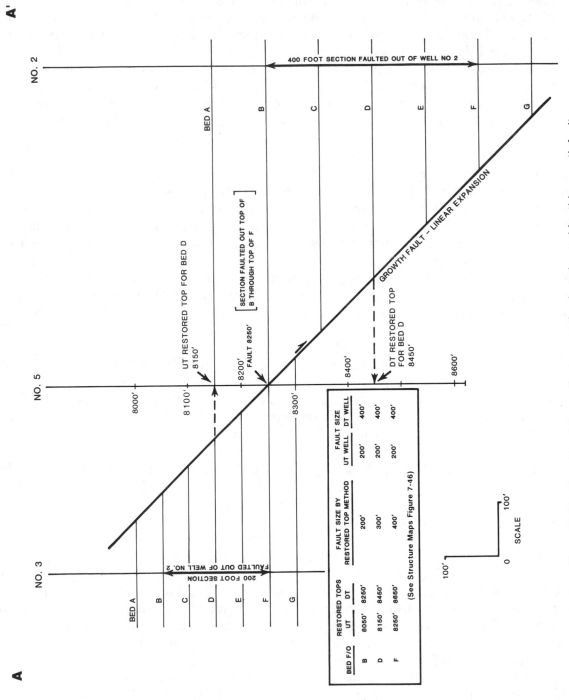

Figure 7-45 The size of the vertical separation is estimated for this growth fault using both the restored top and single well methods. The fault size is determined for mapping Beds B, D, and F.

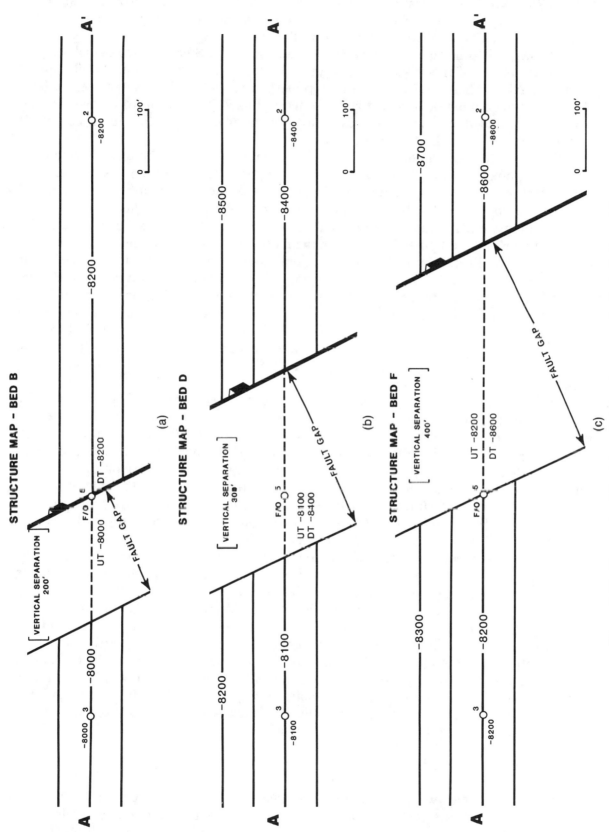

STRUCTURE MAP – BED B

VERTICAL SEPARATION 200'

F/O E

UT –8000
DT –8200

FAULT GAP

–8000
3 –8000

–8200
2 –8200

A

A'

0   100'

(a)

STRUCTURE MAP – BED D

VERTICAL SEPARATION 300'

F/O 5

UT –8100
DT –8400

FAULT GAP

–8200

–8100
3 –8100

–8500

–8400
2 –8400

A

A'

0   100'

(b)

STRUCTURE MAP – BED F

VERTICAL SEPARATION 400'

F/O 5

UT –8200
DT –8600

FAULT GAP

–8300

–8200
3 –8200

–8700

–8600
2 –8600

A

A'

0   100'

(c)

Figure 7-46 Segments of the integrated structure maps for Beds B, D, and F. Notice how the fault gap width changes with depth in response to the changes in vertical separation.

the fault at the F Horizon by correlating Well No. 5 with just one well, the thick down-thrown Well No. 2. With this correlation, the size of the fault at the F Horizon is estimated at 400 ft. Again, the estimate for the size of the fault is the same using the more detailed restored top method or the simplified one well method.

The last horizon requiring an estimated fault size is D. Since Horizon D is not near the top or bottom of the missing section in Well No. 5, the simplified one well method cannot be used. Therefore, the restored top method for estimating the size of the fault at the D Horizon must be used and is illustrated in Fig. 7-45. The upthrown restored top is at 8150 ft, and the downthrown restored top is at 8450 ft, resulting in an estimated fault size of 300 ft for the D Horizon.

Figure 7-46 shows a portion of the structure map for each of the three horizons B, D, and F. Looking at the cross section in Fig. 7-45 and the structure maps in Fig. 7-46, it is possible to formulate a rule of thumb for use of either the restored top or the simplified one well method. If the horizon being mapped is *at or near the top* of the missing section in the faulted well, the fault size (vertical separation) can be closely approximated by using the one well method and correlating the faulted well with a nearby well in the thin upthrown fault block. See the B Horizon in our example in Fig. 7-45. In map view this is represented in Fig. 7-46a, which is a portion of the structure map on Horizon B. The structure map also can be used to establish a rule of thumb. Notice that the faulted out Well No. 5 is near the downthrown trace of the growth fault. Therefore, we can state that if the faulted out well is *at or near the downthrown trace* of the growth fault on the structure map, the size of the fault for that horizon can be closely estimated by using the one well method and correlating the faulted well with a nearby well in the upthrown fault block.

If the horizon being mapped is *at or near the bottom* of the missing section in the faulted well, the fault size can be closely approximated by using the one well method and correlating the faulted well with a nearby well in the thick downthrown fault block. This was done for the F Horizon as shown in Fig. 7-45. In map view this is represented in Fig. 7-46c, which is a portion of the structure map on the F Horizon. The faulted out Well No. 5 is near the upthrown trace of the growth fault. Using the rule of thumb, we can state that if a well is *at or near the upthrown trace* of the growth fault, the size of the fault can be closely estimated by using the one well method and correlating the faulted well with a nearby well in the thick downthrown fault block.

If the horizon being mapped falls somewhere near the middle of the missing section in the well log (D Horizon, Fig. 7-45) or in map view, if the well is positioned near the middle of the fault gap (structure map D Horizon, Fig. 7-46b), then the restored top method must be used. It is important to point out that if time is not a factor, the restored top method should always be used because in addition to providing an estimate of the fault size, it also contributes two additional control points for contouring (the upthrown and downthrown restored tops). The faults used to illustrate the methods for estimating the size of a growth fault possess linear expansion and the beds are gently dipping in Fig. 7-44 and appear horizontal in cross-sectional view in Fig. 7-45. It must be pointed out, however, that these methods are applicable in areas with steeply dipping beds, as well as for a growth fault with nonlinear expansion.

Fault Surface Map Construction

Earlier in the chapter, we discussed the importance of preparing a fault surface map for all faults; this applies to growth faults as well. The procedures and guidelines for con-structing fault surface maps as outlined earlier are applicable to growth faults, with one possible addition. Recall that we stated it was good geological mapping practice to contour the fault displacement (vertical separation), as well as the fault surface (Fig. 7-40a). With a growth fault, however, it is absolutely necessary to contour both the fault surface and vertical separation whenever possible. The size of a growth fault changes with depth and can also change laterally very quickly; therefore, in order to prepare an accurate interpretation of the fault for future mapping and prospecting, both the fault surface and vertical separation should be mapped.

The vertical separation data for a fault can come from electric well logs or seismic

GROWTH FAULT

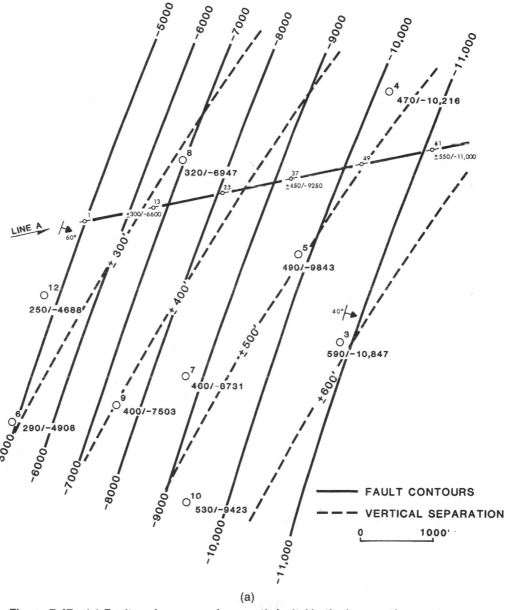

(a)

Figure 7-47 (a) Fault surface map of a growth fault. Vertical separation contours are placed on the fault map, adding valuable mapping information pertaining to the growth fault. (b) Vertical separation values are not contoured, but strategically placed on the fault surface map. A shortcut method. (From Tearpock and Harris 1987. Published by permission of Tenneco Oil Company.)

GROWTH FAULT

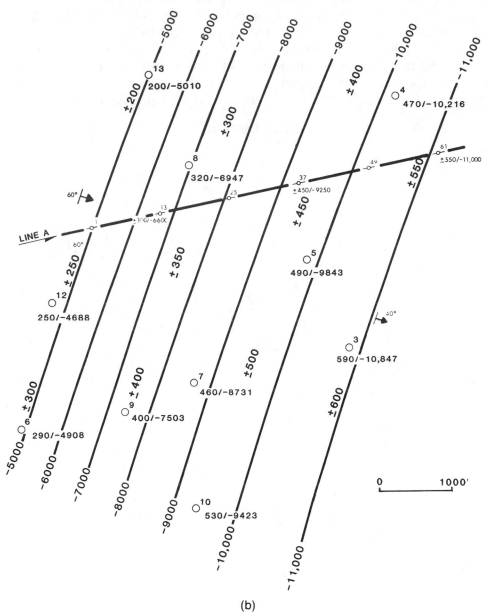

(b)

Figure 7-47 (*continued*)

sections. Once the vertical separation data are gathered, there are at least two ways to map the data. The vertical separation values can be contoured on the same base as the fault surface contours as shown in Fig. 7-47a, or they can be recorded at strategic locations on the fault surface map as shown in Fig. 7-47b. The actual contouring of the vertical separation is by far the more accurate method and the one we recommend.

Looking at the fault surface and vertical separation contour map in Fig. 7-47a (the map shown in the figure is only a segment of the entire fault), we can quickly observe a lot of valuable information about this growth fault. The sources of the fault surface and vertical separation data were both well logs and seismic sections. The fault surface contour spacing indicates that the fault has a listric shape with a maximum dip rate of 60 deg at around −5000 ft subsea, decreasing to 40 deg at around −11,000 ft. The vertical separation values next to each well and seismic shotpoint and the contours indicate that the size of the fault increases with depth from 250 ft at -4688 ft in Well No. 12 to 590 ft at $-10,847$ ft in Well No. 3. Based on the spacing of the vertical separation contours, the fault growth appears to be generally uniform over this area of the fault. Also, the contours show that the fault size increases laterally from the northeast to the southwest. Taking the -9000-ft fault contour as an example, in the southernmost portion of the map the size of the fault at this depth is 500 ft. Following along this -9000-ft contour to the northeast, the size of the fault decreases steadily to only 400 ft at the northern end of the map. Lastly, the contours point to the direction of maximum growth fault activity (major sedimentary depocenter for this fault) as being to the southwest. All of this information is important for future integration of this fault interpretation with structure and stratigraphic data to prepare detailed structure and isopach maps for use in the exploration and exploitation of localized hydrocarbon accumulations.

The pattern of fault surface and vertical separation contours is different for structural faults than it is for growth faults. Look once again at Fig. 7-40a, a fault map for a structural fault. Compare this map with that for the growth fault in Fig. 7-47a. Identify the differences observed and establish criteria that can be used to distinguish a growth fault from a post-depositional fault using the fault maps.

EXPANSION INDEX FOR GROWTH FAULTS

So far, we have discussed growth faults in terms of size (vertical separation) and depth/size relationships. We understand that movement on a growth fault is contemporaneous with local sedimentation. The simultaneous occurrence of faulting and deposition results in an increase in thickness of the stratigraphic section in the downthrown fault block. It is important to understand the timing related to fault growth and deposition because large accumulations of hydrocarbons are directly related to the timing of fault growth and hydrocarbon migration. To explore adequately for hydrocarbons in areas of growth faults, you need to know more about the fault than just its size and lateral or vertical extent.

Expansion Index refers to a technique used to study the timing and activity of a growth fault in terms of the expansion of the downthrown section. This study provides an analysis of the movement of a fault throughout its history. The Expansion Index technique was introduced to the oil and gas industry by Carl E. Thorsen (1963). Although Thorsen developed the technique by studying growth faults in Southeast Louisiana, this method of investigating and analyzing the history of growth faults can be applied to any area of the world where growth faults are present.

The procedure involves dividing the upthrown and downthrown stratigraphic sections into correlative units and then comparing the thickness of each unit in the upthrown and downthrown blocks. Since the correlative units can be positioned with reference to regional time-stratigraphic divisions, fault movement can be studied in terms of geologic time. This technique is similar to the method of study applied to structural uplift by Wadsworth (1953a). As all studied units are not of equal thickness, the analysis is done as a ratio rather than in absolute terms.

$$\text{Expansion Index} = \frac{\text{Thickness Downthrown}}{\text{Thickness Upthrown}}$$

For the purpose of illustration, the Expansion Index is usually presented in bar graph form. In this form, the fault movement with time is clearly displayed. Figure 7-48 is a simple example to show the determination and application of the Expansion Index for a growth fault. For simplicity of explanation, the thickness of the correlative units *a* through *k* in the upthrown fault block are all of equal thickness ($t = 1.0$). The left side of the figure contains a cross section of a growth fault with one well in the upthrown block and one in the expanded downthrown block. The thickness of each unit in the downthrown block is measured and the Expansion Index for each unit calculated and displayed as a bar graph on the right side of the figure. By analysis of the Expansion Indices of this growth fault, we conclude that movement on the fault began at time *j*, reached a maximum period of growth activity between times *f* and *g*, and the fault became inactive at time *b*. The maximum period of fault growth between times *f* and *g* involved a 2.1 to 1 expansion of the downthrown stratigraphic section.

Each growth fault appears to have its own unique *fingerprint* in reference to its Expansion Indices. The rate of movement on a growth fault increases from the initial movement to some maximum rate, and then decreases again until fault movement ceases. During its life, a growth fault may move in spurts or cycles, thereby having an irregular growth history. Over its life, each growth fault has its own unique growth history as a result of fault movement, the rate of subsidence, and the rate of sediment supply across the fault. Even over long distances, the fingerprint for a particular fault can be recognized, as shown in Fig. 7-49, although the Expansion Index may not be exactly the same across a single growth fault over a long distance. The Expansion Indices for this fault were calculated at separate locations two (2) miles apart. A review of the individual Expansion Indices at each location shows that they are similar enough to be identified as being from the same fault.

Because of the unique characteristics of the Expansion Index for each growth fault, in complexly faulted areas, the technique can serve as a lateral correlation tool for the identification of individual growth faults over long distances. The Expansion Index technique is applicable for use with either electric well logs or seismic sections. Because of

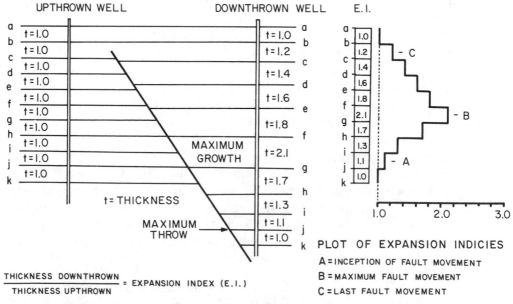

Figure 7-48 Expansion index can be used to measure the timing and rate of movement along a growth fault. Expansion index is commonly illustrated in bar graph form (see right side of figure). (From Thorsen 1963. Published by permission of the GCAGS.)

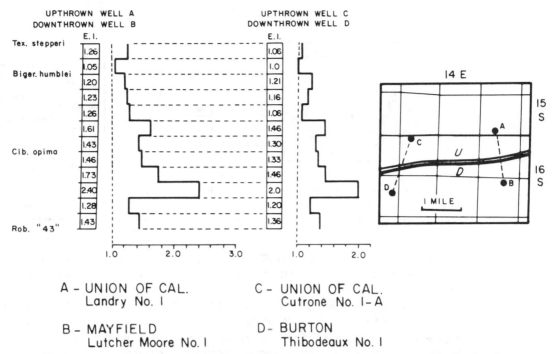

Figure 7-49 Comparison of expansion indices taken over 2 mi apart indicates that a growth fault's "fingerprint" can be recognized over long distances. (From Thorsen 1963. Published by permission of the GCAGS.)

the greater correlative stratigraphic detail found in well logs versus seismic sections, the Expansion Indices calculated using well logs are easier to prepare and more accurate. One important note needs to be made with regard to this technique and seismic sections. The correlative units identified for use on seismic sections must be **depth converted** before calculating the Expansion Index.

Since the thickness of the stratigraphic section in the upthrown block should nearly always be equal to or less than the thickness for the equivalent section in the downthrown block, the smallest Expansion Index that can be calculated for any specific unit is 1.0. Therefore, a calculated Expansion Index less than 1.0 using well logs or seismic data may indicate a bust in the correlation of the individual units across the fault. Therefore, the Expansion Index also serves as a vertical correlation aid for well log and seismic interpretations.

Figure 7-50 shows the analysis of three separate growth faults in South Louisiana. The Expansion Indices for the faults shown indicate that the age of each fault is different, and that they become progressively younger in a basinward direction (see map insert). The time of maximum fault growth, indicated by dark arrows, corresponds to the age of maximum sedimentation in the area across each individual fault. In certain geologic settings, such information is vital to the exploration for hydrocarbons. In the Miocene trend onshore and offshore Louisiana, for example, many of the major structural features are associated with the Louann Salt movement, which is directly tied to the timing of sediment loading across growth faults.

The analysis of the history of a growth fault by use of its Expansion Index provides information important to the exploration for hydrocarbons. There are at least eight benefits derived from analyzing the Expansion Index of a growth fault.

1. The index shows the time of fault inception.

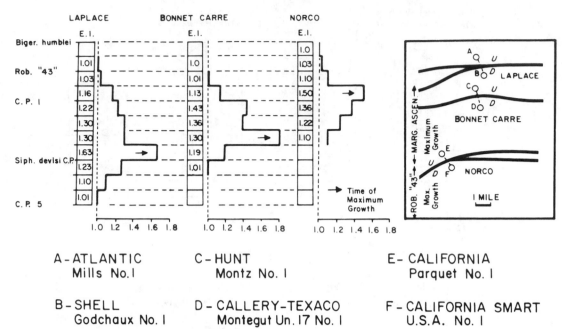

A - ATLANTIC
Mills No. 1

C - HUNT
Montz No. 1

E - CALIFORNIA
Parquet No. 1

B - SHELL
Godchaux No. 1

D - CALLERY-TEXACO
Montegut Un. 17 No. 1

F - CALIFORNIA SMART
U.S.A. No. 1

Figure 7-50 Expansion indices can be used to age-date growth faults with regard to their initial movement, time of maximum growth, and last movement. (From Thorsen 1963. Published by permission of the GCAGS.)

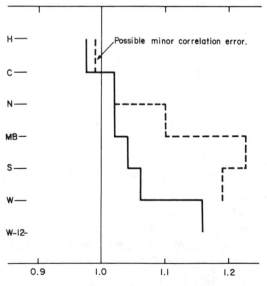

PLOT OF GROWTH INDICIES

$$\frac{\text{Thickness "low" well}}{\text{Thickness "high" well}} = \text{Growth Index}$$

—   STRUCTURAL GROWTH

Ⓞ   Sun No. 1 Long ÷ Sun No. 1 Andrews

---   FAULT GROWTH

⦿   Sun No. 1 Andrews ÷ Billups No. 1 Dreyfus

Figure 7-51 Plot of both the structural growth and fault growth histories from the Fordoche Field, Pointe Coupee Parish, Louisiana. (From *Typical Oil and Gas Fields of Southwestern Louisiana*, v. II, 1970. Published by permission of the Lafayette Geological Society.)

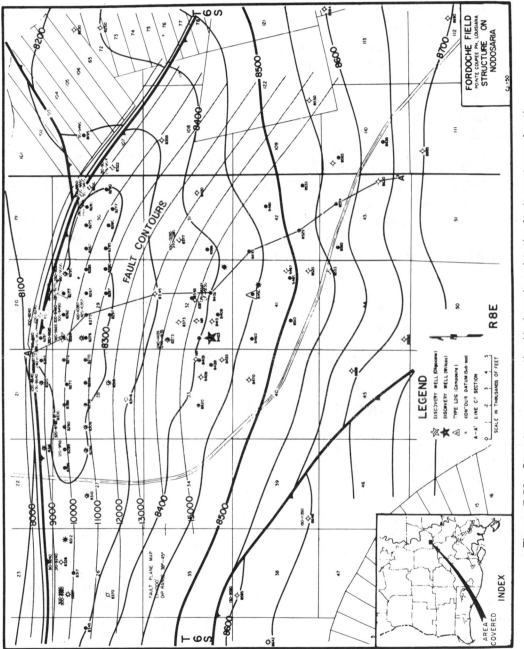

Figure 7-52 Structure map on the Nodosaria, with the fault contours for the down-to-the-south growth fault superimposed on the map. (From *Typical Oil and Gas Fields of Southwestern Louisiana*, v. II, 1970. Published by permission of the Lafayette Geological Society.)

2. It indicates the time of maximum fault movement (maximum expansion).

3. It shows the time of cessation of fault movement.

4. It provides a complete history of fault movement.

5. Regionally, it can be used to correlate stratigraphic units associated with maximum fault growth.

6. It can be used with structural growth indices to compare fault growth with structural growth of uplifted areas.

7. It can serve as a vertical correlation aid for seismic sections and well logs.

8. It can be used as a lateral correlation tool for recognizing specific faults over long distances using well logs or separate seismic lines.

Wadsworth (1953a) developed a technique for studying the history of structural uplifts similar to the expansion index. In his work, the assumption is made that structural uplifts result in the thinning of stratigraphic units deposited contemporaneously with structural growth. To prepare a *Structural Growth Index*, on-structure and off-structure stratigraphic thicknesses are compared instead of upthrown and downthrown thicknesses of a growth fault. A Structural Growth Index illustrated in plot form similar to the Expansion Index provides an analysis and timing of structural growth.

Figure 7-51 is an example of the combined use of a growth fault Expansion Index and a structural Growth Index to evaluate the interrelationship of structural growth and fault growth in the Fordoche Field in Pointe Coupee Parish, Louisiana. The plot indicates that the greatest structural growth predated the most active period of growth on the fault. The Fordoche Field is interpreted as a deep-seated anticline associated with a down-to-the-south growth fault (Fig. 7-52). The primary productive interval consists of a 6000-ft thick section of interbedded sands and shales ranging in age from Oligocene (Frio) to Eocene (Wilcox).

DIRECTIONAL SURVEYS AND FAULT MAPS

In Chapter 3, we discussed the importance of using the directional survey from a deviated well to calculate the position and depth of the borehole throughout its entire length. Recall that we illustrated the two basic ways in which directional well paths are presented on a map base, shown here again as Figs. 7-53a and 7-53b.

Using the simplified method of plotting a directional well shown in Fig. 7-53a, the only data required and plotted on the base map are the surface and bottomhole locations. At times the measured depth to TD is recorded next to the bottomhole location. Between the surface and bottomhole locations, a straight dashed line is usually drawn. This directional well plot provides absolutely no information about the position or depth of the wellbore in the subsurface between the surface and bottomhole locations. Such a plot is not helpful in the interpretation, construction, and evaluation of fault, structure, or isopach maps.

When directional survey data are actually plotted to provide detail regarding the position and subsea depth of the wellbore throughout its entire length, as shown in Fig. 7-53b, the plot has real value. Such a plot provides a visual guide (in map view) to the location and subsea depth of the wellbore anywhere along its path. Such a plot saves time in preparing subsurface maps and is extremely helpful in the interpretation, construction, and evaluation of fault, structure, and isopach maps. In this section, we look

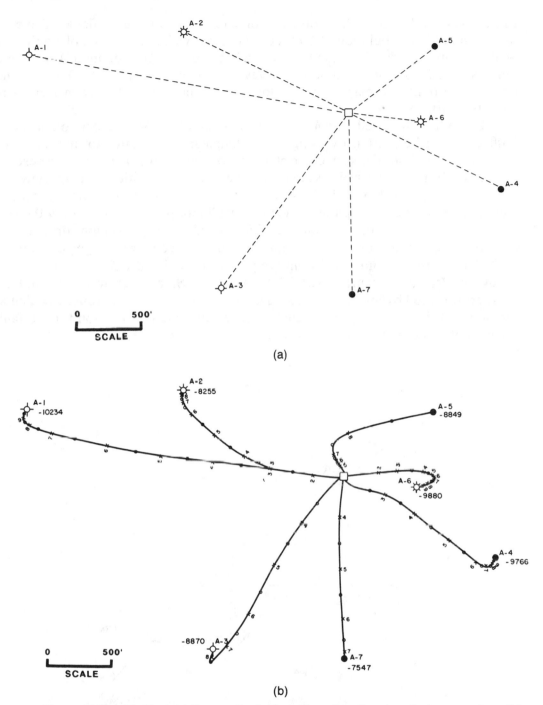

(a)

(b)

Figure 7-53 (a) Straight line method of plotting directional wells in map view. (b) Detailed plot of directional survey data indicating the location and subsea depth of each wellbore along its entire length.

at several important benefits of fault surface mapping derived from plotting the actual position and subsea depth, at evenly spaced increments (usually 500 ft or 1000 ft) of all directional wells on a base map. The benefits to structure and isopach mapping are covered in the appropriate chapters.

An example will best show the importance of detailed wellbore plots in the interpretation and construction of fault surface maps. The wellbore base map segments in

Figs. 7-54 and 7-55 are from an offshore oil and gas field. Table 7-1 contains a simplified tabulation of fault cut data from wells in the area of Fault C. For each well, the fault cut data provided are: (1) subsea depth, (2) size, and (3) rectangular coordinates (obtained from directional surveys and measured in feet). The rectangular coordinates provide, in map view, the location of the fault cut in each well as measured from the surface location.

Using the simplified base map construction in Fig. 7-54, the location of each wellbore fault cut must be plotted by using the rectangular coordinates to measure the position of the fault cut from the surface or platform location. An **X** is used to represent the fault cut for each individual well. Notice that the fault cut position in map view for several wells, such as Well No. C-1, does not coincide with the dashed directional path of the wellbore because this simplified directional wellbore path considers only the surface and bottomhole locations of the wellbore. Directional wells are seldom drilled in a straight line from surface to bottomhole, either because of wellbore design or natural wellbore drift. Therefore, the simplified wellbore plot for directional wells in many cases does not follow the true and accurate path of the wellbore. Since each wellbore fault cut in Fig. 7-54, represented by an X, was located using the rectangular coordinates calculated from the actual directional survey for each well, we assume the location for the fault cuts to be accurate (see plot for Well No. C-1).

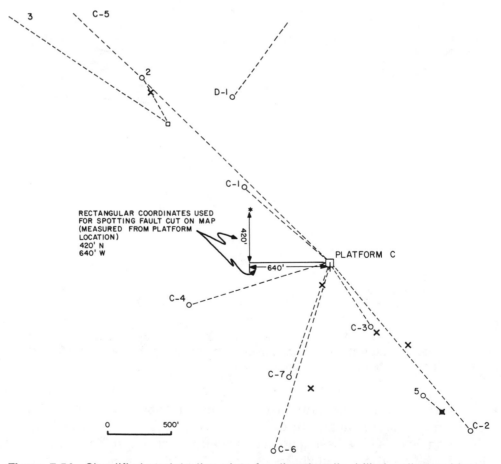

Figure 7-54 Simplified straight line plots for directionally drilled wells provide no information about the position and depth of the wellbore between the surface and bottomhole locations.

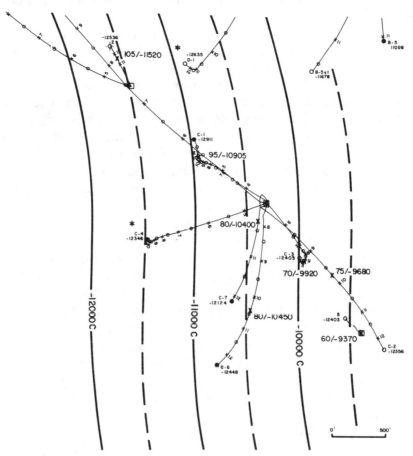

Figure 7-55 Directional surveys plotted accurately over the entire length of the well-bore are helpful in the interpretation, construction, and evaluation of fault, structure, and isopach maps.

In Fig. 7-55, we see the same base map segment depicting the actual path and subsea depth for each directional well plotted in 500-ft subsea increments. Such a plot is prepared once and placed on an original base map film from which paper or mylar copies can be made for all future subsurface fault, structure, and isopach mapping.

Take another look at the data for Fault C using the detailed wellbore base map in

TABLE 7-1

Fault Cut Data for Fault C

Well No.	Size	Subsea Depth	Rectangular Coordinates
C-1	95′	−10,905′	420′N, 640′W
C-2	75′	−9,680′	660′S, 620′E
C-3	70′	−9,920′	535′S, 350′E
C-4	?	?	
C-6	80′	−10,450′	1015′S, 140′W
C-7	80′	−10,400′	170′S, 65′W
#2	105′	−11,520′	225′N, 135′W
#5	60′	−9,370′	0′ , 0′
D-1	?	?	

Fig. 7-55. With this base, the only data needed to post the fault cut for each well are the subsea depth for the cut. For example, the subsea depth for the fault cut in Well No. C-1 is −10,905 ft. This datum point is posted on the map in Fig. 7-55 using the detailed directional well plot and is as accurate as the same point plotted in Fig. 7-54. By using the detailed base map, however, it was not necessary to calculate the rectangular coordinates from the directional survey for each fault cut nor to use these coordinates to measure the location of the cut from the platform location. The elimination of these steps saves mapping time. The example shown here is relatively simple, with only one fault cut per well and only about ten wells. But consider a field with one hundred wells and ten faults resulting in two or three fault cuts per well. This adds up to some 200–300 fault cut data points for which the rectangular coordinates must be calculated and then used to measure the location for each fault cut. These calculations and measurements, which can be time consuming, are unnecessary if the base map is prepared using the actual directional survey data to plot accurately the path and subsea depth for each wellbore along its entire length (Fig. 7-55).

If saving time were the only benefit derived from preparing a base map with the detailed directional wellbore plots, it would be worth doing; but, other benefits may be even more important. Look at Well No. C-4 in Fig. 7-55. No fault cut is assigned to this well for Fault C. Should there be a fault cut in this well? Was it missed during correlation? Is the well deviated in such a way that it does not intersect the fault, or is there no fault in the well? The answers to these questions are very important and play a vital role in the interpretation of Fault C. Where do we look for answers to these questions? If the base map has the detailed directional wellbore plots like that in Fig. 7-55, it might be the first place to start looking for answers, or at least it might provide a direction for further investigation.

Well No. C-4 in Fig. 7-55 does not have an assigned fault cut. Starting with the interpreted fault surface map for Fault C and the detailed wellbore plot for C-4, can we address any of the questions asked with regard to the absence of a fault cut in the well? The answer is yes. Well No. C-4 is drilled to a subsea depth of −12,346 ft and is an "S" shaped well. Look at the fault surface map along the wellbore path from −11,000 ft to −11,500 ft. At a fault surface depth of −11,000 ft, the wellbore is at a depth of −6300 ft or 4700 ft vertically above the fault surface. At a fault surface depth of about −11,500 ft, the wellbore is at a depth of −12,346 ft, or 846 ft below the fault surface. Therefore, if the fault surface interpretation is correct and the wellbore plot accurate, then somewhere between −11,000 ft and −11,500 ft the wellbore should have intersected Fault C (Fig. 7-56). Since the wellbore is vertical from −11,500 ft, the fault should have cut the wellbore at about −11,450 ft. Maybe the fault cut was overlooked or questionable during the initial correlation work.

With the new information obtained from the review of the fault map and detailed wellbore plot, the correlations in Well No. C-4 can be further reviewed. If there is a fault cut in the well, it will add another datum point to the fault surface interpretation; if the cut is not there, we may conclude that the present interpretation is incorrect and another fault interpretation is required, using the negative control from Well No. C-4. Negative control in this case means no fault cut in the well. If there is no fault cut, then the fault surface interpretation must be changed to honor the data. The possibility also exists that the fault passes below the well or the fault may die out before it reaches the well. All possibilities must be considered. In this case, however, further correlation of Well No. C-4 confirms a 90-ft fault cut at −11,480 ft, which was originally documented on the electric log as a questionable short section. This fault information can now be added to the fault map in Fig. 7-55.

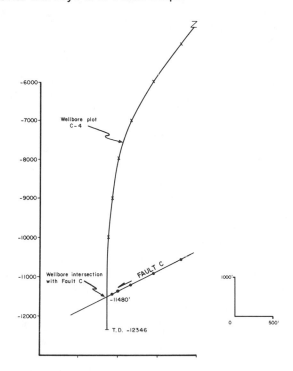

Figure 7-56 Schematic cross section along the path of Well No. C-4. The section shows that if the fault interpretation in Fig. 7-55 is correct, the well should intersect the fault at a depth of 11,480 ft.

The D-1 well, marked with an asterisk in Fig. 7-55, is similar to Well No. C-4 in that no fault cut is assigned to this well for Fault C. Assuming that the fault surface interpretation is correct and the wellbore plot is accurate, should Well No. D-1 intersect Fault C? And if so, at what subsea depth?

We have shown that the detailed plot of a directional well, including its position and subsea depth at regularly spaced increments on a base map, can save mapping time, be very beneficial in helping evaluate a fault interpretation, and provide another check for questionable fault cut correlations. Finally, supervisors, managers, and prospect evaluators can use the directional well data and techniques discussed here to quickly review or evaluate fault maps. When directional survey data are available, we strongly recommend that the deviated well paths be plotted on a base map to show the position and subsea depth of the borehole throughout its entire length.

Directional Well Pitfalls

When is a straight hole not a straight hole? When it is deviated. Many so-called vertical wells are not vertical at all, but have been drilled at a deviated angle. This deviation usually is not planned but is the result of natural wellbore drift. As we discussed in Chapter 3, the deviation angle of a drilling well is usually monitored in some way, such as with a Totco survey. Remember that a Totco records the deviation angle of the well but not its direction. Most states and the US Federal Offshore have regulations allowing a maximum deviation before an actual direction survey is required by law. In the state of Louisiana, for example, the maximum angle is now 5 deg.

The fault surface maps in Figs. 7-57a and 7-57b illustrate the kinds of problems that can arise when wells are assumed to be vertical or when no directional well data are available. The fault surface map in Fig. 7-57a was prepared using the available well data shown on the map. Each depth value represents the subsea depth at which this mapped fault was encountered in the adjacent well. The map does not look too bad, but it does

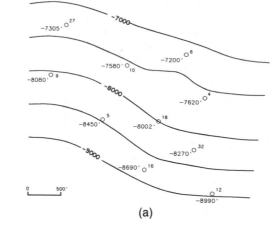

(a)

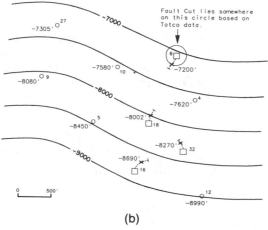

(b)

Figure 7-57 (a) Fault surface map prepared with the assumption that all the wells were vertical. Notice the unrealistic variations in dip in the northeast portion of the map. (b) Revised fault surface map based on new directional well data obtained for a number of wells. The new interpretation of the fault surface appears reasonable.

have some unusual variations in contour spacing (dip of fault) and indicates a significant problem in the northeast portion of the map. According to the available data on the base map and electric logs, all the wells are identified as being vertical. However, looking at the fault surface map, it is apparent that something is wrong. Either the fault cut correlations are incorrect, the wells are spotted at the wrong surface location, or some of these vertical wells are not vertical.

The answer to the problem is shown in Fig. 7-57b. Upon additional review of well files, five wells were found to be deviated as shown in the figure. By replotting the wellbores as deviated wells and respotting the fault cut data to correspond to the true wellbore deviation, the fault surface map shown in this figure was constructed.

Well No. 6 is of special interest. The only available directional data on this well was a Totco survey, which measures angle but not direction. In the case of Well No. 6, the assumption was made that the deviation was all in the same direction. Therefore, a circle whose radius is equal to the distance the well could have deviated from the vertical to a depth of −7200 ft was plotted. If the wellbore deviation was all in the same direction, the fault cut should fall somewhere on this circle. The direction of deviation that best fit the fault cut was used to spot the cut as shown in the figure. Notice that the contour spacing has evened out and the major contouring problem in the northeast was resolved.

Keep this example in mind, especially when mapping in areas with very old wells. A problem that might at first appear to be that of wrong correlations or an incorrect

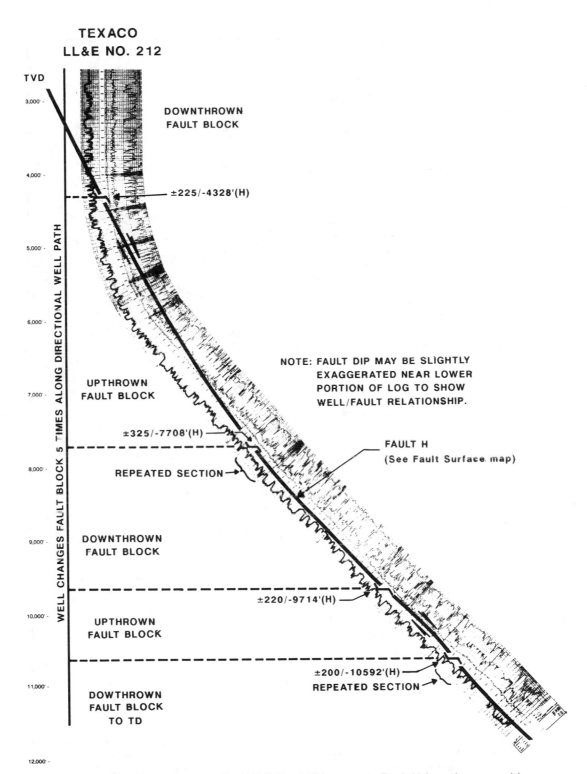

Figure 7-58 Directionally drilled Well No. 212 intersects Fault H four times resulting in two missing sections and two repeated sections. The well changes fault blocks five times. (Published by permission of Texaco, USA.)

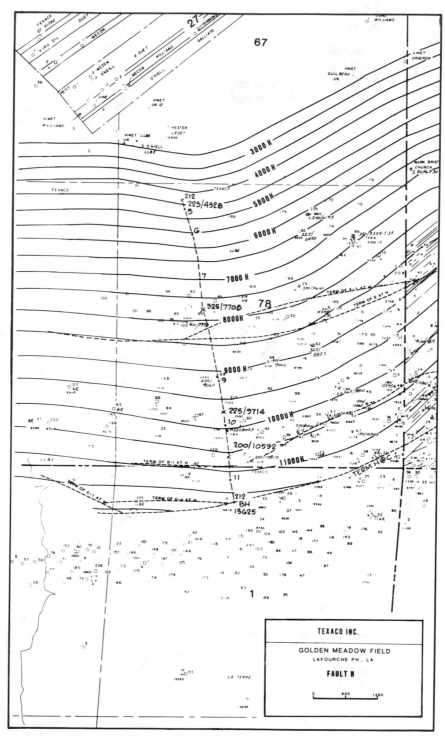

Figure 7-59 Portion of the fault map for Fault H. Note the four fault cut data points from Well No. 212. (Published by permission of Texaco, USA.)

interpretation might turn out to be nothing more than a problem of a vertical well not being vertical.

Fault Maps, Directional Wells, and Repeated Sections

In Chapter 4 we discussed the specific geometry between a normal fault and a directionally drilled well required for a repeated section (Fig. 4-27). We also looked at several examples of repeated sections, such as the one in the Texaco LL&E Well No. 212 (Fig. 4-29).

In the preparation of a fault surface map, a fault cut resulting in a repeated section is treated in the same manner as a fault cut resulting in a missing section. Figure 7-58 shows the Texaco LL&E Well No. 212 plotted in cross section along the wellbore deviation path. The well/fault geometry is quite unique in that the well crosses the same Fault H at least four times; twice in the normal sense, resulting in missing section, and twice in the reverse sense, causing two repeated sections. Therefore, the well actually penetrates the downthrown fault block three times and the upthrown block twice.

Figure 7-59 shows a portion of the fault surface map for Fault H in the vicinity of Well No. 212. The four fault cuts in the well and fault cut control from numerous straight holes and other deviated wells were used to prepare the fault surface interpretation shown. The fault surface has a generally listric shape, but displays little if any growth. Near the TD of the well, Fault H intersects and terminates against Fault C, which is a down to the-north fault. Faults E, E-1, and B-1 are interpreted as terminating against Fault H as shown in Fig. 7-59. These faults are not shown in the cross section in Fig. 7-58. This example shows the unusual geometries which can occur when directionally drilled wells encounter fault surfaces.

FAULT CUT SIZE—CORRECTION FACTOR AND DOCUMENTATION

The size of a fault cut as discussed in Chapter 4 is the measurement of the vertical thickness of the missing or repeated section. In vertical wells or straight holes, the thickness represented on a log is the *True Vertical Thickness*. Therefore, when correlating a fault in a well log with another log from a vertical well, the size of the fault cut is the measured amount of missing or repeated stratigraphic section found in the vertical well log.

For directionally drilled wells, the log thickness can be exaggerated or even too thin. Therefore, a true vertical thickness (TVT) correction must be applied to the log thickness to obtain its true vertical equivalent. The equations used for this conversion were reviewed in Chapter 4.

As discussed in Chapter 1, good quality work must be backed up by good documentation. This means recording all the data in some format that can be easily used or revised by the person conducting the work, supervisors or managers, other team members, or persons inheriting an area for future study. Fault cut data must be documented, particularly where deviated wells are used for correlation requiring correction factors.

CHAPTER 8

STRUCTURE MAPS

INTRODUCTION

The subsurface structure map is one of the primary tools used by a geologist to explore for hydrocarbons from the initial stage of exploration through the complete development of a field. Each subsurface structure map is a geologic or geophysical interpretation based upon limited data, technical proficiency, creative imagination, three-dimensional visualization, and experience. We consider the construction of a structure map to be an interpretive and creative process. No two geologists or geophysicists will construct a map exactly the same, even with the same data, because each uses the factors just mentioned to develop his or her interpretation. However, the interpretation must incorporate *valid structural principles*, and *correct and accurate mapping techniques*.

The importance and reliability of subsurface structure mapping increase with advancing stages of field development and depletion. Many management decisions are based on the interpretations presented on subsurface structure maps. These decisions involve investment capital to purchase leases, permit and drill wells, and to workover or recomplete wells, to name a few examples. A geologist must employ the best and most accurate methods to find and develop hydrocarbons at the lowest cost per net equivalent barrel.

Since faulted structures play such a significant role in the trapping of hydrocarbons, we devote a considerable portion of this chapter to the correct and accurate subsurface mapping techniques required to integrate a fault map interpretation into the construction of a completed structure map. A reasonable structural interpretation, in most faulted areas, begins with an accurate fault picture developed from fault surface maps using fault cut data from well logs and seismic sections (Chapter 7), followed by integration of these fault maps with the structural interpretation.

Many petroleum provinces involve multiple faults that result in extremely complicated structural relationships. The attempted reconstruction of a complex structure with isolated fault cut data from logs or seismic sections can result in erroneous geologic interpretations. Too often subsurface structure maps are prepared without giving much consideration to the three-dimensional geometric validity of the interpretation. The most accurate and sound structural interpretation in a faulted area requires: (1) construction of fault surface maps for all important trapping faults, (2) integration of the fault maps with the structural horizon maps, and (3) mapping of multiple horizons at various depths (shallow, intermediate, and deep) to justify and support the integrity of any structural interpretation (Tearpock and Harris 1987).

The exploration for and exploitation of hydrocarbons is creative work. Most of the time a geologist is dealing with geologic structures that are not visible on the surface. The formidable challenge of interpreting these unseen structures can only be accomplished with a clear understanding of basic geologic principles, familiarity with the geometry of formation and fault relationships, analysis of all available data, use of all technical capabilities, application of technical knowledge and skills, and imagination.

In this chapter, we concentrate on the technical knowledge and skills necessary to develop a geologically reasonable structural interpretation. Technical knowledge and skills fall into two categories: (1) a good understanding of the tectonic setting being worked, and (2) understanding and application of correct mapping techniques. The primary focus of this chapter is on the broad range of important structural mapping techniques; however, since the application of many techniques depends upon the tectonic style (type of structure and trap), we discuss and illustrate techniques as they apply to different tectonic settings and review a number of real-world examples.

Subsurface structure maps usually are constructed for specific stratigraphic horizons to show in plan view the three-dimensional geometric shapes of these horizons. These maps are constructed using correlation data from well logs and interpretations of seismic sections. Therefore, one of the primary factors that goes into developing a reasonable structural interpretation is correct correlations. Remember that *accurate correlations are paramount for reliable subsurface mapping*. Subsurface structure maps are no more reliable than the correlations used in their construction. Incorrect correlations will find their way, at some point, into the final interpretation. They may be incorporated into the fault, structure, or isopach maps and result in serious mapping problems. Therefore, it is essential that utmost care be taken in correlating logs and interpreting seismic sections.

Not every horizon within a stratigraphic sequence is suitable for structure mapping. A horizon that is not correlatable over a large area or one that is limited in areal extent may not be suitable. Maps on stratigraphic horizons of limited extent can be prepared after the overall structural interpretation has been developed from fieldwide or regional correlations.

A structure map is a form of contour map. As discussed in Chapter 3, marine shales exhibit distinctive characteristics over large areas. Therefore, they serve as good correlatable horizons for fieldwide or regional structure mapping. A structure contour map presents in plan view a two-dimensional interpretation of the three-dimensional shape of a specific stratigraphic horizon. Each contour connects points of equal elevation above or below sea level for a given stratigraphic horizon. A good structural interpretation requires three-dimensional thinking, as illustrated in the simplified block diagram in Fig. 8-1.

Broadly interrelated assemblages of geologic structures constitute the fundamental structural styles of petroleum provinces. These assemblages generally are repeated in regions of similar deformation, and the associated types of hydrocarbon traps can be

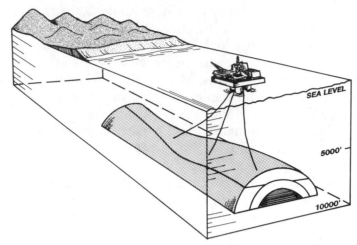

Figure 8-1   A three-dimensional view of an anticlinal structure 7000 ft below sea level.

anticipated (Harding and Lowell 1979). There are a number of petroleum-related tectonic habitats around the world; each results, to varying degrees, in different kinds of hydrocarbon traps that may require modified or different mapping techniques. In the first part of this chapter, we discuss numerous subsurface structure mapping techniques. These techniques are then reviewed as they apply to the following tectonic habitats and their associated hydrocarbon traps.

1. Extensional tectonics, including normal faulting and detached listric fault systems.
2. Compressional tectonics, including reverse faulting and fold and thrust belts.
3. Diapiric salt tectonics
4. Wrench fault tectonics

GUIDELINES TO CONTOURING

Review the five basic rules of contouring presented in Chapter 2. In addition to these basic rules, the following guidelines to contouring make a map easier to construct, read, and understand; they also help to ensure the technical accuracy and correctness of the completed map. Some guidelines covered in Chapter 2 are repeated here; many have been expanded, and additional guidelines are presented.

    1.   All contour maps must have a *chosen reference* to which the contour values are compared. A structure contour map usually uses sea level as the chosen reference. Therefore, the elevations on the map can be referenced above or below mean sea level. A negative sign in front of a depth value indicates that the elevation is below sea level.

    2.   The *contour interval* on a map should be constant. The use of a constant contour interval makes a map easier to read and visualize in three dimensions because the distance between successive contour lines has a direct relationship to the steepness of slope. Steep dips are represented by closely spaced contours, gentler dips by contours with a wider spacing. Figure 8-2 illustrates the confusion and difficulty involved in trying to visualize a contoured surface in three dimensions, when the contour interval is not constant over the mapped area. From fault block to fault block the contour interval changes from 100-

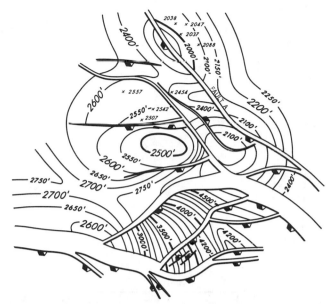

Figure 8-2 This structure map has an inconsistent contour interval randomly changing from a 50- to 100-ft contour interval from fault block to fault block. Such inconsistency in the contour interval makes a map difficult to visualize in three dimensions. Observe the change in contour interval upthrown and downthrown to Fault A. (From Tearpock and Harris 1987. Published by permission of Tenneco Oil Company.)

to 50-ft contours with no consistency. Notice upthrown to Fault A the contour interval is 50 ft; and downthrown, it is 100 ft and yet the contour spacing is about the same. This indicates that the area downthrown to Fault A has a much steeper dip than the area upthrown. Knowing the nature of the area and the minimal displacement on the fault, this is very unlikely.

The choice of a contour interval is an important decision. Factors to be considered include the density of data, the practical limits of data accuracy, the steepness of dip, the scale, and the purpose of the map.

3. *Contour spacing* depends upon the dip of the structure being mapped. For any given structure, the spacing of contours will vary at different locations unless the equal spacing method of contouring is used. Several nomograms can be conveniently used to compute contour spacing when the dip is known; likewise, the dip on a completed map can be determined by measuring the contour spacing. Figure 8-3 is a nomogram that relates the dip of beds to the horizontal distance between 100-ft contours. It can be used for determining dip or contour spacing for fault as well as structure maps.

4. All maps should include a *graphic scale* (Fig. 8-4). A graphic scale gives the viewer an idea of the areal extent of the map and the magnitude of the features shown. Maps are commonly reduced or enlarged for various reasons; without a graphic scale the values shown on the map become meaningless.

5. Every fifth contour is an *index contour*. It should be bolder or wider than the other contours and labeled with its value (Fig. 8-4).

6. *Hachured lines* should be used to indicate a closed depression (Fig. 8-5).

7. Contouring should be started in areas with the maximum number of control points (Fig. 2-8). The area or areas of maximum control often occur around structural highs or lows.

8. *Construct contours in groups of several lines* rather than one single contour at

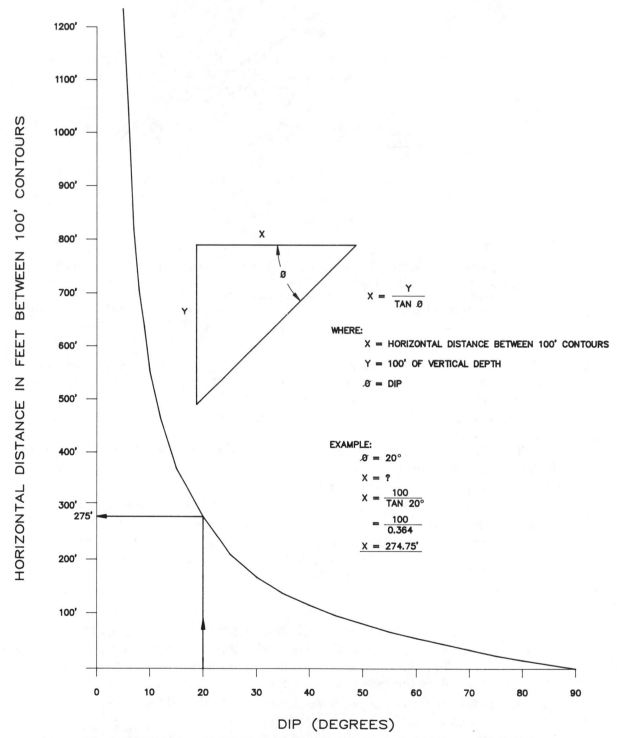

Figure 8-3 Nomogram of dip vs horizontal distance in feet between 100-ft contours. If greater accuracy is required for contour spacing or the determination of dip than that obtained from the graph, the equation shown in the figure can be used.

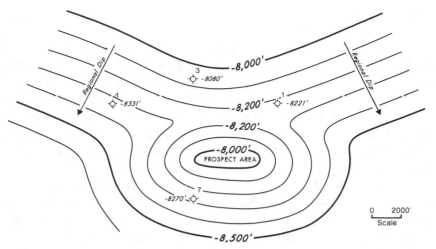

Figure 8-4 An example of a localized structural high indicated by a change in regional dip.

a time (Fig. 2-8). This method provides better visualization of the surface being contoured and results in more consistent contouring.

9. Initially, choose the simplest contour solution that honors the control and provides a realistic interpretation. The simplest solution may be the best, and it is usually easy to test. If problems arise with this solution, a more complex interpretation can be prepared.

10. Use a *smooth rather than undulating style* of contouring unless the data indicate otherwise. Some geologists argue that a smoothly contoured structure is not likely to occur in nature (Fig. 2-9). This may be true; however, it is better to keep the structure simple with smooth contours until the data indicate otherwise. It is possible to present a significant misinterpretation by placing unjustified wiggles in contours (Silver 1982).

11. Initially, a map should be *contoured in pencil* with lines lightly drawn so that they can be erased as the map requires revision.

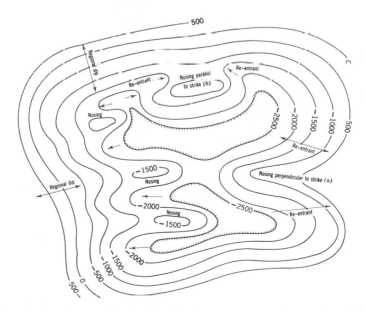

Figure 8-5 A diagrammatic structure contour map of a basin illustrating several important contouring guidelines. (From Bishop 1960. Published by permission of author.)

12. *Establish regional dip* whenever possible. Regional dip is the general direction of strike and dip for any given area. Regional dip may not be constant over a large area, but changes should be gradual. In areas of regional dip, contour lines have a certain degree of parallelism along regional strike. Any change in the dip rate may be an indication of local structures. In areas of minor or localized structures, contours extend away from regional dip. Such indications are important because in many petroleum provinces, such as areas along the US Gulf Coast, minor anomalous highs that break regional dip commonly are productive of hydrocarbons (Fig. 8-4).

If regional dip is interrupted by a localized structural high, re-entrants occur on each side of the minor uplift (Fig. 8-4 or Fig. 8-5). If the axis of the localized uplift parallels regional strike, the magnitude of the re-entrants may be small compared to re-entrants adjacent to a high that is perpendicular to regional strike (Fig. 8-5).

Any flattening or reversal of normal dip is a possible clue to local structures. Therefore, changes of this kind are extremely important. Local uplifts may have their axes perpendicular or parallel to the regional strike. When the axis of a local fold is perpen-

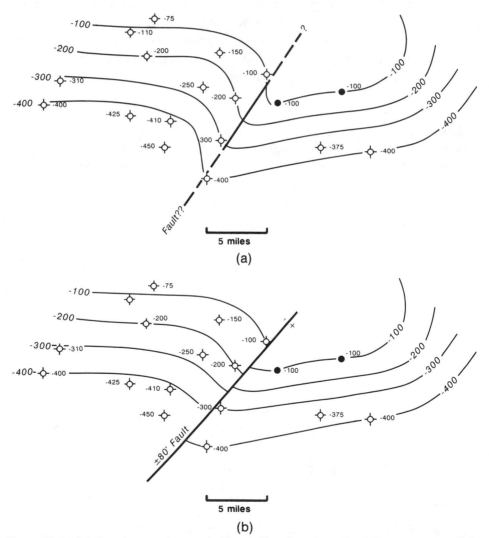

Figure 8-6 (a) An abrupt change in the strike direction of contours suggests the possibility of faulting. (b) Another interpretation of Fig. 8-6a that fits the geologic and hydrocarbon data includes a vertical fault not intersected by the wells. (Modified after Bishop 1960. Published by permission of author.)

dicular to strike, contours flare outward in a downdip direction and the distance between contour lines increases as the rate of dip decreases. A nosing or U-shaped projection results in the bottom of the U pointing basinward (Fig. 8-5). Nosings are flanked by re-entrants, the axes of which may be perpendicular or oblique to the regional strike. The re-entrants begin where the contours start to widen out and are less pronounced downdip until eventually they disappear.

If the axis of a structurally high area is parallel to regional strike, re-entrants are also parallel to strike. As the direction of regional dip reverses at the axis of the re-entrants, a high area results downdip, as shown in Fig. 8-5 (Bishop 1960).

13. *Contouring can be optimistic or pessimistic* depending upon your experience, corporate guidelines, and exploration philosophies. All contouring, however, must be governed by the general tectonic style, and optimism must be kept within geologically reasonable limits. At the same time, pessimistic contouring can condemn potentially prospective areas to the point that no exploratory drilling is undertaken. A good mapping philosophy to follow is to map neither optimistically nor pessimistically, but instead to use all of your technical expertise to map *realistically*.

14. In areas of either limited subsurface control or vertical faults, it is important to contour the limited data to reflect as simple a geologic interpretation as possible, rather than just to connect points of equal elevation. Therefore, any *radical change that occurs in the strike of the contours* may suggest faulting even though no fault has been recognized by well control. Figure 8-6a depicts such a situation. In these cases, all available data need to be reviewed, including production and pressure data to help resolve the geologic problem. In the example shown in Fig. 8-6, notice a significant change in contour strike in the area marked as a possible fault, although no fault is recognized in the wells. An interpretation that fits all the geologic and hydrocarbon data includes a vertical fault not intersected by the wells (Fig. 8-6b).

An abrupt increase in the rate of dip perpendicular to strike is a good indication of faulting. An increase in the rate of dip accompanied by an abrupt change in strike is very strong evidence of faulting (Bishop 1960). Increased dip might, alternatively, result from folding, but in most cases the increase is more abrupt where faulting is responsible.

15. *A change or reversal in the direction of dip* suggests the crossing of a fold axis (Fig. 8-5 or Fig. 8-7). Reversal of dip may occur over the crest of an anticline or the trough of a syncline. The amount of dip reversal is often a guide to whether the reversal is due to a regional change or local structure. An excessively steep dip may indicate the presence of a fault or steep fold, while a dip that flattens may be indicative of the crest of a fold or the bottom of a syncline.

16. *Structures may or may not have structural attitude compatibility (contour compatibility) across a fault.* The compatibility of structural attitude on opposite sides of a

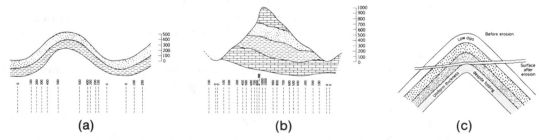

(a)   (b)   (c)

Figure 8-7 Structural highs and lows and unconformities can sometimes be recognized by their effect on contour spacing. (From Bishop 1960. Published by permission of author.)

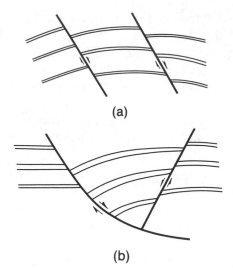

(a)

(b)

Figure 8-8 (a) Cross section shows structural compatibility across faults. (b) Cross section shows no structural compatibility across major growth fault.

fault depends primarily upon the size and type of fault. For example, many, if not most structures displaced by normal and reverse (structural) faults demonstrate structural compatibility across the faults (Fig. 8-8a). In contrast, many listric normal faults (such as growth faults) and thrust faults, with significant displacements, result in structures that are not compatible across the fault (Fig. 8-8b). The method of structure contouring across a fault depends upon whether there is a compatibility of structural attitude on both sides of the fault.

17. Structural highs, regardless of origin, usually tend to *flatten across the axis* with gentle dips across the top of the structure (Fig. 8-7a). Contour line spacing widens across the crest of the structure compared to spacing on the flanks. Synclines, like anticlines, also tend to flatten across their axes. The widening of contours is often even more pronounced in a syncline. If the data indicate a continuously steep slope up to the crestal high with little if any flattening of dip, this indicates that the surface is one that has been affected by erosion or indicates the presence of an unconformity (Fig. 8-7b and c).

18. *Closed structural lows* are not common. If possible, avoid closing lows with contours unless the data require it. The presence of closed lows often suggests an eroded surface or the presence of faulting. If the closed low is elongated, faulting is likely, and the greater the size of the closed low, the greater the probability of faulting.

19. *Contour license* refers to the geologist's right to contour a structure in a way that best fits the geologic and engineering data, and that best represents the types of structures present in the tectonic setting. The interpretive method of contouring as defined by Bishop (1960) best corresponds to what we call contour license.

20. Specific structural highs may be in the form of domes, anticlines, and noses. *Domal structures* are usually the result of local uplifting, such as a salt intrusion. With domes, the direction of dip is away from the central high and the dip rates are usually constant along strike, at least within each major fault block around the structure. Therefore, the contour spacing is often uniform along strike. In the dip direction, the highest areas along the uplift often have the steepest dips with a gradual decrease away from the uplift. Contour spacing is close near the crest of the structure, but widens with distance away from the crest.

Anticlinal structures generally appear as elongated domes. Their origin can be the result of compressional (fault-propagated folds associated with reverse faults) or extensional (rollover anticlines associated with listric growth normal faults) forces. In general, the direction of dip is away from the crestal area in two opposing directions. Since anticlines are often asymmetrical, with inclined axial surfaces, the dip rate and resulting contour spacing may vary around the anticline.

Structural noses which trend off local structures show dip away from the crest in three directions. Contour lines widen, indicating flatter dips in the area of a nose or an associated re-entrant. Contours become closer together immediately downdip of the local high until regional dip is attained again.

The best test of the three-dimensional geometric validity of any structure contour map is its predictability. How well does the interpretation hold together with additional data from the drilling of new wells or the shooting of new seismic lines. If the contours require major revision each time new data are obtained, there should be serious concern regarding the validity of the map. On the other hand, if only minor adjustments are required and hydrocarbon traps predicted by the mapping are successfully found, the interpretation may be considered valid. Remember that we always work with a limited amount of subsurface data which must be interpreted. Each geologist must use imagination, an understanding of local structures, the ability to visualize in three dimensions, technical knowledge and skills, and imagination to evaluate any number of possible (alternative) interpretations. Finally, a geologist must decide which interpretation, in his or her judgment, is the most reasonable. The 20 guidelines presented in this section should help you construct more accurate and reasonable structure contour maps.

SUMMARY OF THE METHODS OF CONTOURING

As discussed in Chapter 2, four distinct methods of contouring or combinations of methods commonly are used in the preparation of structure contour maps. These are (1) *mechanical*, (2) *equal spaced*, (3) *parallel*, and (4) *interpretive* (see Rettger 1929; Bishop 1960; and Dennison 1968).

1. *Mechanical Contouring.* In using this method of contouring, the assumption is made that the slope or angle of dip of the surface being contoured is uniform between points of control and that any change occurs at the control points. With this approach, the spacing of the contours is mathematically (mechanically) proportioned between adjacent control points. Mechanical contouring allows for little, if any, geologic interpretation and even though the map is mechanically correct, the result may be a map that is geologically unreasonable, especially in areas of sparse control (Fig. 8-9a).

2. *Parallel Contouring.* With this method, the contour lines are drawn parallel or nearly parallel to each other. This method does not assume uniformity of slope or angle of dip; therefore, the contour spacing can vary. Like the previous method, if honored exactly, parallel contouring yields an unrealistic geologic picture. It allows for some geologic license to draw a map a little closer to the real world, because there is no assumption of uniform dip (Fig. 8-9b).

3. *Equal-Spaced Contouring.* This method assumes uniform slope or angle of dip over an entire area or at least over an individual flank of a structure. Sometimes this method is referred to as a special version of parallel contouring. The advantage to this method, in the early stages of mapping, is that it can indicate the maximum number of structural highs and lows expected in an area of study (Fig. 8-9c).

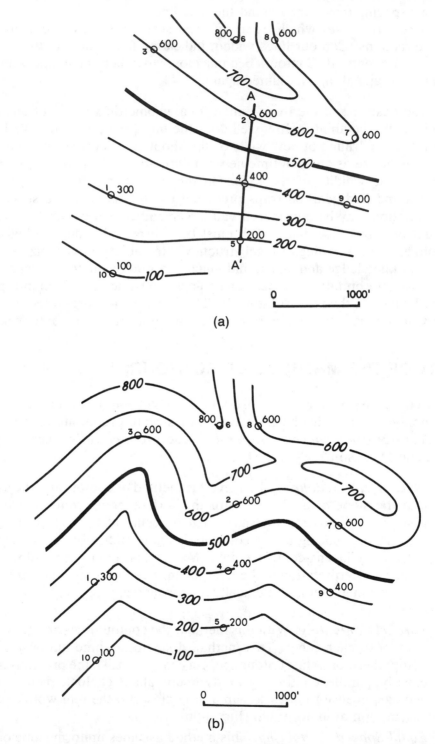

Figure 8-9 (a) Mechanical contouring method. (b) Equal-spaced contouring method. (c) Parallel contouring method. (d) Interpretive contouring method. (Modified after Bishop 1960. Published by permission of author.)

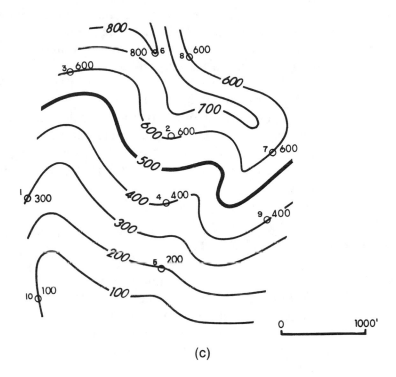

(c)

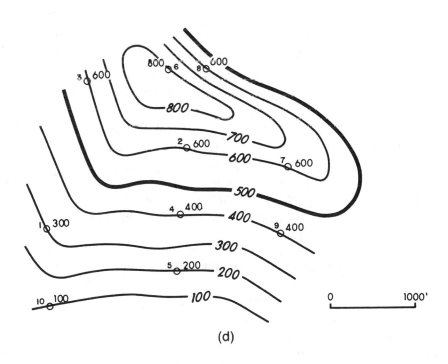

(d)

Figure 8-9 (*continued*)

4. *Interpretive Contouring*. With this method, the geologist has extreme geologic license to prepare a map to reflect the best interpretation of the area of study, while honoring the available control. No assumptions, such as constant bed dip or parallelism of contours, are made. Therefore, the geologist can use his or her experience, imagination, ability to think in three dimensions, and understanding and knowledge of the structural and depositional style of the geologic region to develop an accurate and realistic interpretation. This method is the most acceptable and the most commonly used method of contouring (Fig. 8-9d).

The specific method or combination of methods chosen for contouring may be dictated by such factors as the number of control points, the areal extent of these points, and the purpose of the map. No individual can develop an exact interpretation of the subsurface with the same accuracy as that displayed on a topographic map. What is important is to develop the most reasonable and realistic interpretation of the subsurface with the available data.

CONTOURING FAULTED SURFACES

The contouring of faulted surfaces adds complications in the contouring of both structural horizons and fault surfaces. A completed structure contour map of one or more horizons is usually the main objective in any mapping project. In order to construct a completed structure map, however, the faults themselves must be contoured and the fault maps integrated with the structure maps. This integration is required to support a reasonable geologic interpretation and to prepare accurate maps. In terms of map accuracy, this integration does the following (Fig. 8-10):

1. delineates the position of the upthrown and downthrown traces of the fault;
2. depicts the vertical separation of the fault for any particular mapped horizon;
3. defines the limits of fault bounded reservoirs; and
4. provides for the proper contouring of the mapped horizon across the fault.

Subsurface fault mapping was covered in detail in Chapter 7. In this section, we present the proper techniques for integrating a fault map with a structure map. Included in this section are: (1) techniques for positioning the upthrown and downthrown traces of a fault on a structure map; (2) construction of the fault gap or overlap; (3) the mapping of vertical separation versus throw; (4) the use of restored tops in structure mapping; (5) the application of contour compatibility across faults; and (6) the exceptions to contour compatibility.

Techniques for Contouring Across Faults

Normal Faults.   A **Fault trace** is defined as a line that represents the intersection of a fault surface and a structural horizon. Two fault traces are normally required to delineate a fault on a structure map. One line represents the upthrown trace and the heavier line represents the downthrown trace of the fault. One convention designed to indicate the direction of dip of the fault is to place a "*tent*" symbol on the downthrown fault trace. The structure map in Fig. 8-11 shows a fault displacing a contoured surface, using the conventional symbols described.

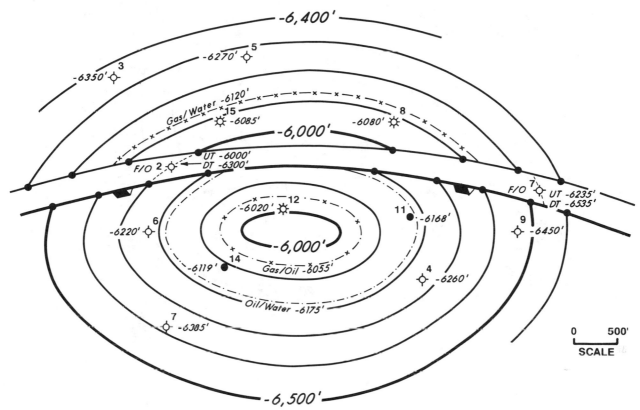

Figure 8-10 Integrated fault and structure map for the 6000-ft Sand. The darkened circles delineate the intersection of each structure contour with the fault contour of the same elevation.

Figure 8-11a shows the faulted structure map on the 8000-ft Sand. The fault (Fig. 8-11b) strikes east-west and dips to the south at an angle of 50 deg. The structure at the 8000-ft Sand level is a four-way (anticlinal) high. The fault has a cut or missing section equal to 400 ft based on electric log correlation. We saw in Chapter 7 that the missing section is a representation of the vertical separation of a fault.

The techniques presented in this section demonstrate the correct method for projecting established contours from one fault block across a fault into another fault block. Using the available data (Fig. 8-11a), contours are first established for the block with the best well control, which in this case is the upthrown block with four wells. These contours are extended to the upthrown trace of Fault 1. To contour across the fault, project the contours from the upthrown block through the fault into the downthrown block. This is shown in Fig. 8-11a by a set of dashed contour lines continued across the fault gap indicating what the structural attitude of the sand would have been if the fault were not there. *In other words, where would the contours have gone had the fault not been there?* Once the contours are projected through the fault gap to the downthrown fault trace, they are adjusted relative to the upthrown contour values using the amount of missing section, which in this case is 400 ft. The downthrown block is then contoured. For example, the −8500-ft contour in the upthrown block, when projected into the downthrown block, becomes the −8900-ft contour.

Contours may be projected for some distance within the fault gap. Notice how the −8300-ft contour is projected from the upthrown trace of the fault for some distance through the fault gap before it intersects the downthrown trace and enters the downthrown block as an −8700-ft contour. The mechanics of projecting contours, such as the −8300

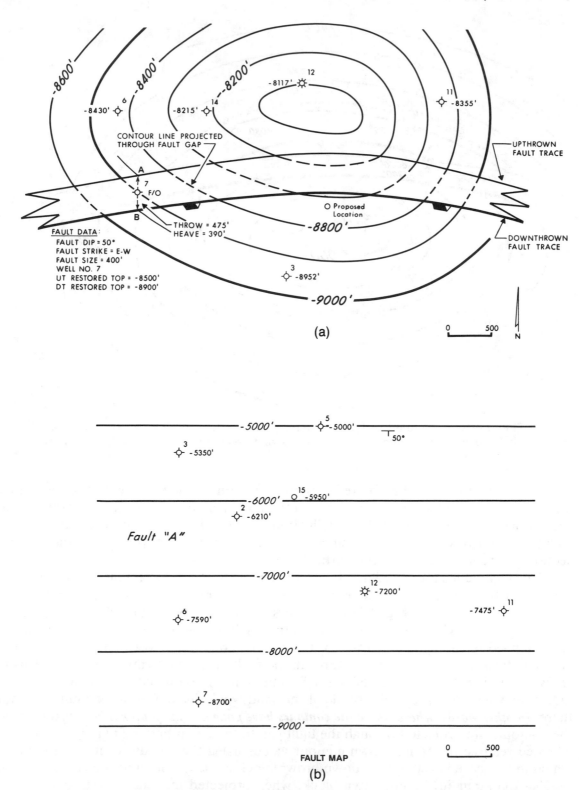

(a)

Fault "A"

FAULT MAP

(b)

Figure 8-11 (a) Faulted structure map on the 8000-ft Sand. The structure is cut by a 400-ft fault. The correct method for contouring vertical separation (missing section) is illustrated by the dashed contour lines. (b) Fault surface map for Fault 1 which strikes east-west and is dipping at 50 deg to the south. (Published by permission of Tearpock.)

-ft contour, through the fault as shown in this figure is the correct technique for contouring across a normal fault, using the vertical separation (missing section or fault cut) as obtained from well logs or seismic sections. The application of this technique correctly delineates the position of the upthrown and downthrown traces of the fault, thus establishing the fault gap. It also assures that the correct displacement has been mapped across the fault (Tearpock and Harris 1987; and Tearpock and Bischke 1990).

Some of you may still be asking yourselves why throw was not contoured across the fault. One reason is that no throw data are available for mapping, and throw is not the correct vertical displacement we want to map across the fault (refer again to Chapter 7). If, however, we want to know the fault throw and heave, their values can be determined by simple measurements once the structure contour map has been prepared as shown in Fig. 8-11 or by use of Eq. (7-1).

Throw is the difference in the vertical depth between where the fault intersects the formation in the upthrown block and where it intersects the formation in the downthrown block, *measured perpendicular to fault strike*. The fault shown in Fig. 8-11b strikes east-west; therefore, the throw can be determined by measuring across the fault in a north-south direction (see arrows in fault gap through Well No. 7). Using the points **A** and **B** on the map (Fig. 8-11a), the upthrown depth at point A is -8460 ft and the downthrown depth at point B is -8940 ft. The throw of the fault at this location is the difference between these two depths, or 475 ft. Mathematically, applying Eq. (7-1) the throw is estimated at 496 ft using an average bed dip of 13 deg and a fault dip of 50 deg. Considering the accuracy of graphical measurements on a contoured map, these two estimates for throw are in excellent agreement.

Heave, which is the horizontal distance across the fault gap from the upthrown to downthrown traces, measured perpendicular to the fault strike, is 390 ft. We can see for this particular example that the throw is about 80 ft greater than the vertical separation. For subsurface structure mapping, the measurements of throw or heave are usually measured for academic purposes and have no application in the actual contouring of a structure map. However, the throw and heave can be used to check a completed structure map using the graphical and mathematical methods shown in this chapter and Chapter 7. If the estimates for throw or heave calculated by both methods (graphical and mathematical) are reasonably close, you can conclude that the map construction is reasonable.

To further illustrate the proper construction of contours across a fault, we review Fig. 8-12, using the same data given in Fig. 8-11, with one exception. In this case, we map a horizon about 2000 ft shallower, placing the fault trace on the northern flank of the structure. The fault is now dipping in the opposite direction to the formation. The sand dipping generally to the north is displaced by the south dipping fault. The fault has a cut or missing section of 400 ft. From the data points available, the structure contours were first established in the upthrown block and extended to the upthrown trace of Fault 1. As shown in the figure, to contour across the fault it is necessary to project the contours from the upthrown block, through the fault gap, up to the downthrown trace of the fault, and then into the downthrown block. Ask yourself *where the contour would have gone had the structure not been faulted*. The construction from one block to the other is shown by a series of dashed contour lines placed from the upthrown trace, through the gap, to the downthrown trace. Once the contours have been projected into the downthrown fault block, they are adjusted in depth from the upthrown contour values by the size of the fault, which in this case is 400 ft. As an example, the -6000-ft contour in the upthrown block when projected into the downthrown block becomes a -6400-ft contour. As in Fig. 8-11, the same technique is followed for construction of all structure contour lines.

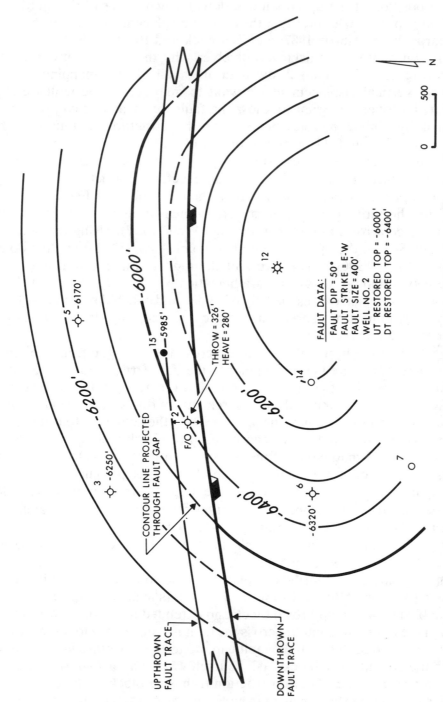

Figure 8-12 Portion of the structure map for the 6000-ft Sand showing the method for contouring vertical separation across a fault. (Published by permission of Tearpock.)

FAULT DATA:
FAULT DIP = 50°
FAULT STRIKE = E-W
FAULT SIZE = 400'
WELL NO. 2
UT RESTORED TOP = -6000'
DT RESTORED TOP = -6400'

THROW = 326'
HEAVE = 280'

CONTOUR LINE PROJECTED THROUGH FAULT GAP

UPTHROWN FAULT TRACE

DOWNTHROWN FAULT TRACE

-6000'
-6200'
-6400'

-6170'
-6250'
-6320'
-5985'

Now that the structure contours have been established in both blocks, the fault throw and heave can be measured. We graphically determine the throw of the fault by estimating the upthrown and downthrown structural depths using points **A** and **B** on the map. Notice that the values for throw and heave are different than those estimated in Fig. 8-11. The value for throw is now 335 ft, as compared with 480 ft in Fig. 8-11. The value for heave is 280 ft, as compared with 390 ft in Fig. 8-11. It is very important to note, however, that despite the changes in the values for throw and heave, the fault cut value in the wells in both cases remains 400 ft. The value for the fault cut has not been subjected to change, as it is representative of vertical separation and not throw. Remember, throw and heave are dependent fault slip variables which change with variation in the apparent attitude of the fault or formation. In this case (Fig. 8-12), the relative direction of the fault has changed from dipping in the same general direction as the formation (Fig. 8-11) to dipping in the opposite direction (Fig. 8-12) to the formation. It is this change in fault dip direction (fault strike is essentially unchanged) that causes the different values for throw and heave shown in the two figures. Using Eq. (7-1) and the data on the map near Well No. 2 (average bed dip is -14.5 deg), estimate the throw across the fault at points A and B and compare this estimate with that obtained graphically from the map.

Mapping Throw Across a Fault. Previously, we mentioned that subsurface fault cut data is vertical separation. Let us assume for a moment, however, that a geologist *incorrectly considers the fault cut as throw* and contours across the fault as if the missing section were throw. In Fig. 8-13, the fault cut and structural data are exactly the same

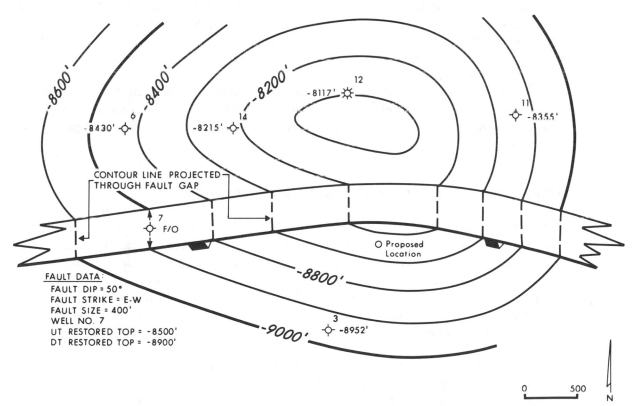

Figure 8-13 Fault and structure data are the same as shown in Fig. 8-11a. For this interpretation, the missing section is contoured across the fault incorrectly as if it were throw. Compare the downthrown fault block with that shown in Fig. 8-11a. (Published by permission of Tearpock.)

as that in Fig. 8-11; therefore, we can compare the results of this (throw contoured) map with the map in Fig. 8-11.

Using the available data, the contours are first established in the upthrown fault block and extended to the upthrown trace of Fault F-1. When contouring throw across a fault, the strike direction of the contours changes at this point and becomes perpendicular to the strike of the fault (Fig. 8-13). The contour is then projected through the fault gap to the downthrown trace of the fault. The strike direction of the contours is again changed to conform to what it was in the upthrown block.

Follow the −8500-ft contour through its construction in order to gain a good understanding of the technique. The strike direction of the −8500-ft contour in the upthrown block is established by the surrounding well control. At the intersection with the upthrown trace of the fault, the strike direction of the contour line changes abruptly to a strike direction that is perpendicular to fault strike. In this example, as in the one in Fig. 8-11, the strike direction of the fault surface is east-west. Therefore, treating the vertical separation as if it were throw, the contour is projected through the fault gap in a north-south direction, perpendicular to the fault strike. At the intersection of the contour and the downthrown trace of the fault, the contour strike direction changes again to conform to the strike direction in the upthrown fault block. Once the contour is projected into the downthrown fault block, its depth is adjusted from the upthrown contour value by the size of the fault (400 ft) to −8900 ft.

In the downthrown fault block in Figs. 8-11 and 8-13, there is a proposed well location. Considering the correctly contoured map (Fig. 8-11), the depth at which the proposed well is estimated to penetrate this horizon is −8720 ft, while at the same location based on the incorrectly contoured map (Fig. 8-13), the well is estimated to penetrate the horizon at −8640 ft. The depth to the horizon is mapped 80 ft shallower based on the incorrect map. Contoured depth differences of 80 ft can often make the difference between a successful well and a dry hole, or result in a well that is not drilled in the optimum position on the structure.

Based upon the nature of the contouring technique (mapping vertical separation incorrectly as throw), *the magnitude of error becomes greater when mapping near or on the crest of a structure*. This is very critical since hydrocarbons are often trapped near the crest of structures.

Error Analysis—Procedure for Checking Structure Maps

From normal mapping parameters, the values for the vertical separation, fault dip, and bed dip can be determined. With these data, Eq. (7-1) (Tearpock and Bischke 1990) can be used to calculate the throw at any point along a fault. If the calculated value for throw agrees fairly well with the value determined graphically from the map, we can conclude that the interpretation is reasonable. If throw was mapped as the missing or repeated section in place of vertical separation, the value for throw determined mathematically will not compare favorably with the determined graphically and the map would be in error.

The nomogram presented in Fig. 8-14, and derived from Eq. (7-1), can also be used to check a completed structure map. This nomogram can be used during daily operations to check the consistency of structure contour maps, and to ensure that a structure map has been contoured correctly across any existing faults. **In Fig. 8-14, the fault and bed dips are taken to be clockwise.** If the bed dips exceed the fault dip and if the beds and fault dip are in the same direction, the AE/AC must be added to 1.0 in the figure. This

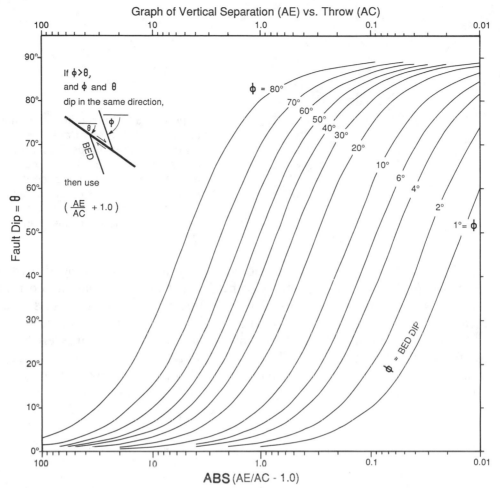

Figure 8-14 Nomogram of vertical separation vs throw. (Published by permission of Tearpock and Bischke.)

is the case of a repeated section due to a normal fault. Use the data in Fig. 8-11a and the nomogram (Fig. 8-14) to verify the dip of Fault 1 shown in Fig. 8-11b.

The relationship shown in Fig. 8-15 is used to conduct error analyses on incorrectly contoured maps (i.e., how much error is introduced into a map which assumes that vertical separation is throw). By this time it should be readily apparent that if vertical separation is mapped as if it were throw, the errors involved could result in a well that totally misses its target. The nomogram in Fig. 8-16, derived from the relationship in Fig. 8-15, can be used to analyze the magnitude of error (Tearpock and Bischke 1990). From Fig. 8-15:

$$AE = A'C$$

Therefore from the Law of Sines

$$BA \sin(\theta - \phi) = A'C \sin(\pi/2 - \theta)$$

As

$$BA = TA/\sin \phi$$

$$\frac{TA}{A'C} = \frac{\sin \phi \cos\theta}{\sin(\theta - \phi)} \tag{8-1}$$

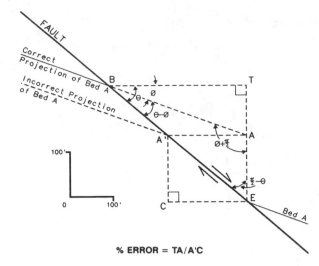

150' = Missing Section by Log Correlation
T   = Correctly Projected Depth Level
A'   = Incorrectly Projected Depth Level
A'C = Vertical Separation Incorrectly
       Mapped as Throw
AE   = Vertical Separation
TE   = True Throw

Figure 8-15 The relationship shown here can be used to conduct error analysis on incorrectly contoured structure maps. Based on the bed dip and intersection of Bed A and the fault in the downthrown block (footwall), observe the difference between the correct vs incorrect depth of intersection of Bed A and the fault in the upthrown block (hanging wall). (Published by permission of Tearpock and Bischke.)

The error is taken relative to a horizon that is incorrectly mapped as throw, or relative to length A'C in Fig. 8-15, which is the percent distance that the correctly contoured horizon is away from the incorrectly contoured horizon. Thus, Fig. 8-16 can be used to measure the error that is introduced by improper contouring techniques.

An examination of Fig. 8-16 shows that unacceptably large errors are introduced for faults dipping at angles of less than 30 deg (or greater than 150 deg) and for beds dipping at less than 5 deg (or greater than 175 deg, recognizing that 180 deg is a lower dip than 175 deg). For faults dipping at 45 deg, bed dips of 10 deg or more can introduce

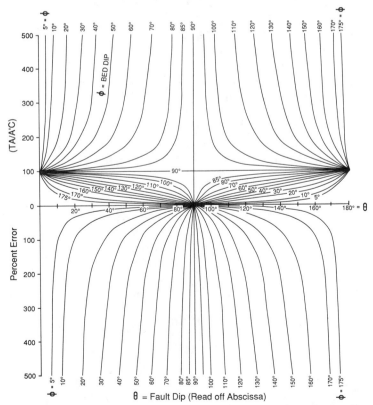

Figure 8-16 Nomogram used to measure the percent error on an incorrectly contoured structure map. Fault and bed dips are taken to be clockwise from 0 deg to 180 deg. (Published by permission of Tearpock and Bischke.)

unacceptable errors. Notice that if the beds are dipping in the same direction as the fault, then the errors encountered are correspondingly larger (see lower half of the figure) than a situation in which the beds are dipping in the opposite direction to the fault. If the bed dip is about equal to the fault dip, as is often the case along the flanks of salt domes, then the errors involved in mapping vertical separation as throw can readily exceed several hundred percent.

In Fig. 8-16, the error is relative to the depth of an incorrectly contoured horizon or a horizon mapped as throw. Thus, a bed that dips at 135 deg (or 45 deg to the west) into a fault dipping at 45 deg (to the east) will have an error of 50%, which means that the correct depth to the bed under consideration is 50%, or one-half the vertical distance away from the incorrect level. Thus, relative to the correctly contoured bed, the error relative to the depth that should have been correctly contoured as vertical separation, is

$$\text{Error} = \frac{(\text{incorrect depth} - \text{correct depth})}{\text{correct depth}}$$

or

$$\frac{(1 - 0.5)}{0.5} = \text{a } 100\% \text{ error}$$

From this discussion, we can conclude that the errors involved in substituting throw for vertical separation are larger than most interpreters may realize or be willing to accept. **Equations (7-1) and (8-1) and the nomograms in Figs. 8-14 and 8-16 are powerful tools that can be used on a daily basis to help generate correctly contoured maps and to evaluate the maps of others.** Such analysis should be routinely conducted when evaluating prospect maps. Use Fig. 8-16 to analyze the magnitude of error in the incorrectly contoured map in Fig. 8-13.

The Legitimate Contouring of Throw

The contouring of throw is a legitimate contouring technique. Where all fault data are in terms of throw, such as in mining or outcrop geology, the construction of contours across the fault by mapping throw is an accepted technique. With regard to petroleum subsurface mapping, however, the technique is, for most cases, not valid. First, well log fault cut data are not throw. Second, most seismic fault data given as throw are actually apparent throw. Third, if throw is measured from a seismic section, it cannot be tied to fault cut data from well logs for mapping, since fault cut measurements are of vertical separation. Finally, throw often varies significantly along the strike of a fault; therefore, if throw is to be mapped, an almost infinite number of fault throw values are required to integrate a fault with a structure contour map.

When throw is the desired fault component to map, as in mining or outcrop mapping, the technique is sometimes incorrectly used. We shall look at one of the main errors made in mapping throw. Remember from previous discussions that *true throw is measured perpendicular to the strike of the fault surface and not necessarily perpendicular to the fault trace.* Therefore, to project the structure contours through a fault gap, the fault surface map must be available to determine the strike of the fault. Since the strike of a fault can change across the mapped area, it is good practice when mapping throw to place the fault map under the structure map to obtain the strike direction so that all the

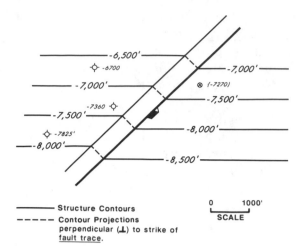

Figure 8-17 Structure contours illustrate the incorrect method for mapping throw perpendicular to the strike of the fault trace (see dashed contours in fault gap).

contours can be projected through the fault gap perpendicular to the strike of the fault at any location along the fault.

At times, contours are projected through a fault *perpendicular to the fault trace rather than the strike of the actual fault surface.* This can result in serious mapping errors. If the strike directions of the fault itself and the fault trace are extremely close, the trace may be used to project contours with minimal error. But this is not always the case. *The line of intersection between two inclined planes with different strikes (such as a formation and fault) is not parallel to the strike of the formation or the fault.* Therefore, fault traces are not normally parallel to the strike of either the formation or fault. Figure 8-17 shows such an example. The map in Fig. 8-17 is contoured incorrectly by projecting the structure contours from the upthrown fault block through the fault gap *perpendicular to the trace* of the fault. In Fig. 8-18, the contours are projected correctly through the fault gap *perpendicular to the strike of the fault surface* (see line indicating fault strike). Notice that the contour value of the point labeled **X** in the downthrown fault block is mapped 95 feet deeper using the incorrect technique. Depending upon the size of the fault and the difference in attitudes between the fault surface and formation, the magnitude of contour errors on completed maps can vary from being insignificant to being very significant. When mapping throw, take the time to use the technique correctly.

You might consider this discussion on the correct technique for mapping throw as more of an academic exercise than one having some practical importance; but remember

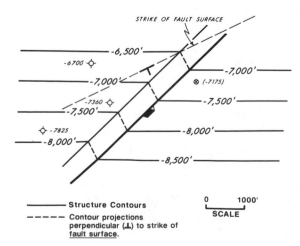

Figure 8-18 Structure contours illustrate the correct method for mapping throw perpendicular to the strike of the fault surface (see dashed contours in fault gap).

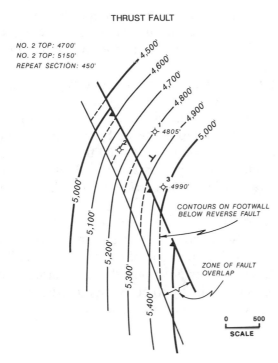

THRUST FAULT

NO. 2 TOP: 4700'
NO. 2 TOP: 5150'
REPEAT SECTION: 450'

CONTOURS ON FOOTWALL
BELOW REVERSE FAULT

ZONE OF FAULT
OVERLAP

0   500
SCALE

Figure 8-19 The correct method for contouring repeated section (vertical separation) across a reverse fault is illustrated by the solid and dashed contours in the fault overlap.

that in mining geology and some outcrop mapping, where the actual fault surface can be touched and throw values physically measured, this mapping technique is valid. You should, however, consider one important point. *If the fault data consist of both actual throw measurements from a mine or an outcrop and well log fault cut data, they cannot be used together in structure mapping because the well log fault cut data represent vertical separation.* Equation 7-1, however, can be used to convert the throw data to vertical separation, which can then be used for mapping.

Only if the dip of the formation being mapped is zero or nearly so, or if the fault is vertical, are the values for vertical separation and throw essentially the same. In these cases, they can be used together in fault and integrated structure mapping.

Later in this chapter, we will examine a petroleum-related generic case study which further illustrates the importance of mapping the missing section correctly as vertical separation rather than as throw in petroleum subsurface mapping. It can make the difference between drilling a successful well and a dry hole.

Reverse Faults. The technique presented for contouring across a normal fault is also applicable for contouring across a reverse fault. Reverse faults and overthrusts, however, produce a fault overlap rather than a fault gap. Figure 8-19 shows the technique for contouring across a reverse fault. The 4500-ft Sand, which dips generally to the west-northwest, is displaced by a west-southwest dipping reverse fault. The fault size or vertical separation determined from well log correlation is 450 ft.

The technique of contouring across a reverse fault is easier than that for a normal fault because there is no projection of contours through a fault gap. With a reverse fault, the hanging wall is thrust up and over the footwall, resulting in an overlap of structural horizons. Therefore, the hanging wall is contoured right up to the hanging wall cutoff at the upthrown fault trace. Likewise, the footwall is contoured up to the footwall cutoff at the downthrown trace of the fault. Since the fault blocks overlap, the strike direction of the contours established in the block that is contoured first (the block with the most control) serves as a guide to the contouring of the other fault block. As with the normal fault example, to guide the strike direction of the contours, consider how the structure would be contoured if the fault were not there. Wells positioned in the fault overlap serve as a guide to the contouring of both fault blocks.

Referring back to Fig. 7-29 in Chapter 7, the thickness of the repeated section or vertical separation resulting from a reverse fault can be calculated by measuring the vertical distance from the top of the mapped horizon in the hanging wall to the top of the same horizon in the footwall. Notice that Well No. 2 in Fig. 8-19, in the fault overlap, has penetrated the top of this horizon twice; the first time at 4700 ft and the second time at 5150 ft. The vertical difference between these two tops is equivalent to the repeated section (vertical separation), which is equal to 450 ft. *The vertical separation can therefore be seen directly in the fault overlap.*

There is one problem with the construction of a reverse fault overlap; it results in significant clutter, which can be confusing. Some confusion is eliminated by dashing the contours on the footwall beneath the fault, within the zone of fault overlap, as illustrated in Fig. 8-19. Another method of eliminating clutter is to pull the fault blocks apart and present each fault block separately. This is a good way to construct a structure map with a reverse fault, especially if hydrocarbons are present and isopach maps are required. The method is shown later in this chapter.

FAULT AND STRUCTURE MAP INTEGRATION

In this section, we present the technique for integrating a fault surface map with a structure map. The correct application of this technique is essential for accurate structure mapping in faulted areas. The technique provides the following important contributions to the structure contour map interpretation:

1. an accurate delineation of the fault location for any mapped horizon;
2. the precise construction of the upthrown and downthrown traces of the fault;
3. the correct width of the fault gap or overlap; and
4. the proper projection of structure contours across the fault based on fault cut data from well logs or seismic (vertical separation).

Normal Faults

The step-by-step method of integrating a fault map with a structure map for a normal fault is illustrated in Figs. 8-20 and 8-21. Figure 8-20 is a fault surface map constructed from fault data from the seven wells shown on the figure. The fault strikes generally north-south, is slightly concave to the west with a dip of 45 deg, and has a vertical separation of 400 ft. The horizon being contoured is called the *7000-ft Sand*. The subsea tops for the sand for each well are shown on the partly completed structure map in Fig. 8-21a. In Well No. 3, for example, the top of the 7000-ft Sand is at a subsea depth of 7045 ft.

The 7000-ft Sand is cut by Fault A. A review of the depth of the fault cuts and the structure tops for each well indicates that five of the seven wells are in the upthrown block, and that only Wells No. 4 and 5 are in the downthrown fault block. Considering one of the general contouring guidelines of beginning structure contouring in the area or fault block with the most control, the contouring of the 7000-ft Sand begins in the upthrown fault block. The initial structure contouring indicates an anticlinal structure with

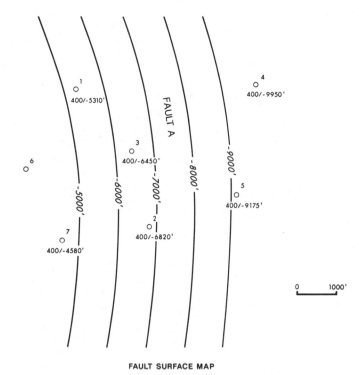

FAULT SURFACE MAP

Figure 8-20 Fault surface map for Fault A constructed from well log fault cut data from seven wells. The fault map has a 1000-ft contour interval.

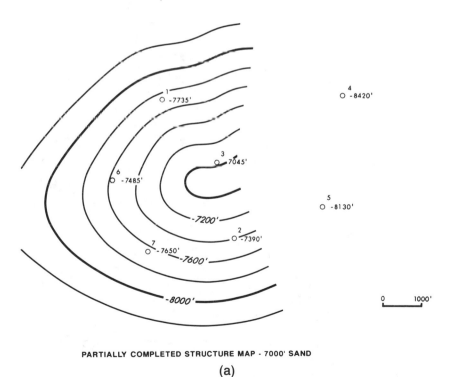

PARTIALLY COMPLETED STRUCTURE MAP - 7000' SAND

(a)

Figure 8-21 (a) Partially completed structure map on the 7000-ft Sand. (b) Integration of the fault and structure maps to identify the intersection of the fault surface with the upthrown fault block of the 7000-ft Sand. (c) Structure map shows the delineation of the upthrown trace of Fault A and the projection of form contours into the downthrown fault block. (d) Integration of the fault and structure maps to identify the intersection of the fault surface with the downthrown fault block of the 7000-ft Sand. (e) The final, integrated structure map for the 7000-ft Sand.

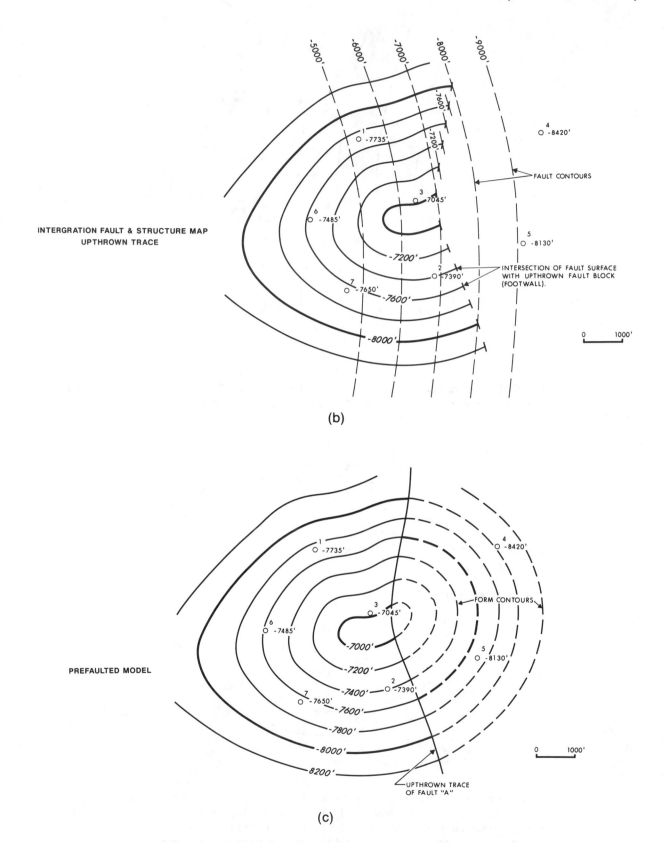

(b)

(c)

Figure 8-21 *(continued)*

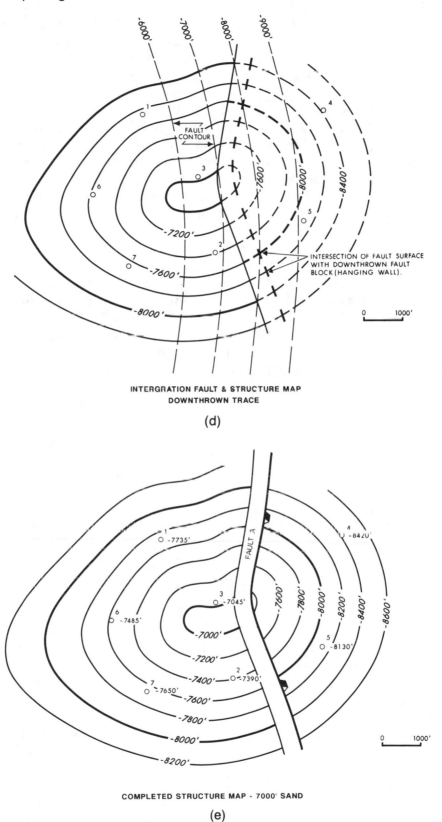

INTERGRATION FAULT & STRUCTURE MAP
DOWNTHROWN TRACE

(d)

COMPLETED STRUCTURE MAP - 7000' SAND

(e)

Figure 8-21 *(continued)*

a slightly elongated east-west axis (Fig. 8-21a). The contours are not continued across the entire map area because at some point the contours in this upthrown fault block intersect Fault A. By underlaying the fault map on the structure map, the approximate location of this intersection can be determined and the contouring stopped in the vicinity of the intersection.

The next step is to overlay the structure map (Fig. 8-21a) onto the fault map (Fig. 8-20) as shown in Fig. 8-21b to determine the upthrown trace of the fault. The upthrown trace occurs where structure contours in the upthrown block intersect fault contours of the same elevation. These intersections are highlighted on the map by a small mark placed at the end of each contour line as shown in the figure. Because the fault map has a contour interval of 1000 ft and the structure map has a 200-ft contour interval, the precise position of the intersections for each contour line should be located by using 10-point proportional dividers. Dividers allow subdivision of the 1000-ft fault contours into 200-ft intervals anywhere on the map without the clutter of actually drawing the extra contours (see Chapter 2, Fig. 2-11). For a small portion of the fault map, 200-ft contours are shown between the -7000-ft and -8000-ft depths. These contours serve two purposes: (1) to illustrate the accuracy of the intersection of the fault and structure contours for the -7000-ft, -7200-ft, -7400-ft, -7600-ft, 7800-ft and 8000-ft elevations; and (2) to show that the construction of 200-ft fault contours over the entire map would result in unnecessary clutter. Two hundred-foot (200-ft) contours can, however, be sketched in localized areas, as shown in the figure, to determine precisely the intersections. Once used for this purpose, the additional contours are then erased. After all the intersections have been identified, the upthrown trace of the fault is accurately constructed simply by connecting all the marks with a smooth line, as shown in Fig. 8-21c.

The next step is to project the structure contours through the upthrown trace of the fault into the downthrown fault block. There are two points of control in the downthrown fault block that must be honored: the structural top in Well No. 4 is -8420 ft, and in Well No. 5 it is -8130 ft. Earlier in this chapter we presented the technique for projecting contours from one fault block across a fault into the adjacent fault block. This technique must be used to complete the construction of the anticline in the downthrown fault block and delineate the downthrown trace of the fault. An easy way to remember the technique to *guide your contouring through the fault is to consider how the structure would be contoured if the fault were not there.* Often it is best to project the contours into the opposing fault block as form contours (contours without values) in order to complete the structural picture, as shown in Fig. 8-21c.

Once contours are projected through the fault into the downthrown block, their values are adjusted from those in the upthrown block by the size of the fault, which in this case is 400 ft. Thus, the projection of the -7400-ft contour from the upthrown block becomes a -7800-ft contour in the downthrown block. Continue this procedure for all contours projected through the fault. In Fig. 8-21d, the downthrown contours have been assigned structural elevation values and now the downthrown trace of the fault can be determined. Once again, the fault map is placed under the structure map and the intersections of the structure contours in the downthrown block with the fault contours of the same elevation are identified and indicated by small marks.

Finally, connect all the marks with a heavy smooth line to accurately delineate the downthrown trace of the fault. Place a symbol on the downthrown trace to show the direction of fault dip and the integrated structure contour map for the faulted anticlinal structure is complete (Fig. 8-21e). By correctly integrating the fault and structure maps, we have: (1) accurately delineated the position of the fault on the structure; (2) precisely constructed the upthrown and downthrown traces of the fault; (3) established the actual

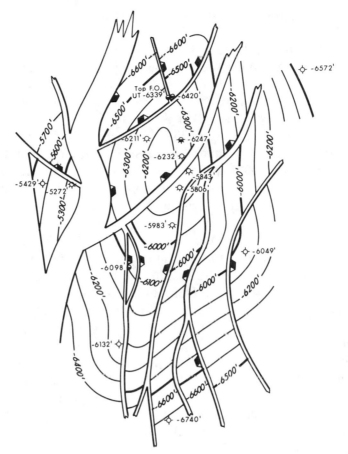

Figure 8-22 An integrated structure map of a very complexly faulted anticlinal structure. Each fault was integrated with the structural interpretation as shown in Fig. 8-21.

width of the fault gap; and (4) projected the structure contours correctly across the fault. **An understanding of this technique is paramount to the correct and precise integration of a fault and structure map**; therefore, take the time to review the process again.

For the example shown in Figs. 8-21, the process was relatively easy because the structural pattern is a simple anticline cut by only one fault. But, regardless of the complexity of the structure and fault pattern, the procedure is basically the same. Figure 8-22 shows the complexly faulted anticlinal structure of an oil and gas field. A fault surface map was prepared for each of the 13 individual normal faults, and the integration technique was used to prepare the completed structure map. With a complex structure such as the one shown here, the logistics are more involved and it takes more time to prepare the fault maps and integrated structural interpretation, but the techniques are the same.

Restored Tops—An Aid to Structural Integration.   In Chapter 3 we mentioned that in dealing with normal faults, a formation being mapped may be faulted out from one or more wells in a field. It is often possible, however, to estimate an upthrown and downthrown restored top for the formation missing from the faulted well(s). A restored top is an estimated top for a specific marker or formation that is faulted out of a well. In other words, *a restored top is an estimate of the depth the top of a formation or marker would have had in that fault block if it were not faulted out.* The procedure for estimating restored tops for both straight and deviated wells was discussed in detail in

Chapter 3. In Chapter 7, we also showed the importance of restored tops in estimating the size of growth faults.

In this section, we discuss how these restored tops are used to provide additional control points for fault and structure map integration. In Fig. 8-10, notice that Wells No. 2 and 7 have the formation faulted out and that estimated restored tops are indicated next to each well. For Well No. 2, the upthrown restored top (UT) is −6000 ft and the downthrown restored top (DT) is −6300 ft; for Well No. 7 the UT is −6235 ft, and the DT, −6535 ft. The vertical difference between the DT and UT restored tops should be equal or nearly equal to the size (vertical separation) of the fault. In this case, the difference is 300 ft, which is equal to the size of the fault based on well control; consequently, the restored tops appear consistent with the available data.

Procedures for projecting contours through a fault and the integration of a fault map with a structure map have been discussed, so we can review the structure map in Fig. 8-10 and determine the size of the fault. A darkened circle marks the intersection of each structure contour with the fault surface contour at the same elevation. By connecting these marks, the upthrown and downthrown fault traces are delineated as shown on the map. Any contour may be taken from its intersection with the upthrown fault trace, and projected through the fault to the downthrown trace to estimate the size of the fault. For example, the −6100-ft contour in the upthrown block projected through the fault gap intersects the −6400-ft contour in the downthrown block, indicating a fault displacement (vertical separation) of 300 ft. The vertical difference in depths between each of the UT restored tops and its associated DT restored top is also 300 ft (for example, 6535 ft − 6235 ft = 300 ft). Therefore, the estimates for the restored tops in Wells No. 2 and 7 are reasonable and provide additional control points for contouring the structure map.

Restored tops are located in the well itself, not at the upthrown and downthrown traces of a fault (see Figs. 3-32 and 3-33). For Wells No. 2 and 7 (Fig. 8-10), the restored tops are located right at the well location. For example, using the restored tops in Well No. 7, project the −6235-ft contour line from the well into the upthrown fault block parallel to the other contour projections. The contour line intersects the upthrown fault trace and enters the upthrown fault block at a depth equal to −6235 ft. Likewise, project the DT restored top of −6535 ft into the downthrown fault block. This contour enters the downthrown block at a structural elevation equal to the depth of the restored top.

How are these restored tops used as an aid in structure mapping? Depths for restored tops are honored in the same way as any other well control point during structure mapping. For example, the UT restored top for Well No. 2 is −6000 ft; therefore, the −6000-ft structure contour in the upthrown block honors this control point at Well No. 2. Likewise, the −6300-ft contour in the downthrown fault block is projected through the fault gap to intersect with Well No. 2, whose DT restored top is −6300 ft. Each UT and DT restored top provides two additional points of control to aid in structure contouring in and around a fault. The contouring of the structure map in Fig. 8-10 confirms that the four restored tops were used as control points in the construction of the final structure map.

In areas of limited well control, restored tops in faulted wells provide significant structural information which is often necessary for the preparation of a realistic and accurate structure map. Considering the well control in Fig. 8-23a, how would you contour these data points? The data can be contoured in a number of ways. Figure 8-23b shows one interpretation which appears to be reasonable. The map does honor all the established well control and was used to propose the two drilling locations shown on the map. The first location is upthrown to Fault C in Reservoir C-3. An oil show in Well No. 14 establishes the downdip limit of oil at −9245 ft. The proposed well is designed to penetrate

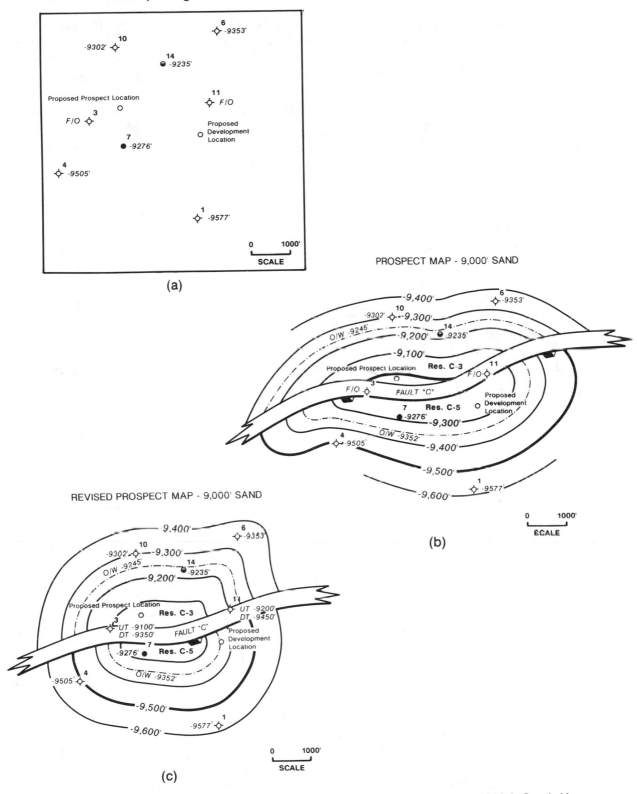

Figure 8-23 (a) Base map with posted data for the top of the 9000-ft Sand. How would you contour the data? (b) Structural interpretation of the 9000-ft Sand using all the well data except for the restored tops in Wells No. 3 and 11. Two development wells are proposed based on this interpretation. (c) Revised structural interpretation of the 9000-ft Sand using all the available well data including the restored tops for Wells No. 3 and 11. Compare this interpretation with that shown in Fig. 8-23b.

the reservoir in the optimum position for maximizing the well's drainage efficiency. Based on this interpretation, the volume of anticipated recoverable oil updip of the oil/water contact is 6480 acre-feet. Considering a reasonable recovery factor for this area of 450 barrels per acre-foot, this prospect represents 2,916,000 barrels of potentially recoverable oil. Downthrown to Fault C, a development location is proposed to maximize the drainage efficiency for the east-west elongated Reservoir C-5. Based on this structural interpretation, the volume of estimated reserves is 1,107,000 barrels of recoverable oil.

The interpretation appears reasonable except for the two wells within the fault gap in which the 9000-ft Sand has been faulted out. For whatever reason, no restored tops were estimated for these wells to incorporate into the structural interpretation. Figure 8-23c is a structure contour map for the same two reservoirs using the UT and DT restored tops for the 9000-ft Sand in Wells No. 3 and 11 in the interpretation. It is obvious that the four restored tops are very important data needed to develop a more accurate interpretation of this structure. The addition of the restored top data has a significant impact on the interpretation of the overall geometry of the structure and the proposed well locations. For the upthrown Reservoir C-3, the volume of recoverable hydrocarbons has been reduced to 3480 acre-feet or 1,566,000 barrels of recoverable oil, a reduction in volume of **46%**. The map for Reservoir C-5 in the downthrown block, using the restored tops, indicates that the volume of Reservoir C-5 is smaller by **42%** compared to the map in Fig. 8-23b. Potentially recoverable hydrocarbons decrease from 1,107,000 barrels to 648,000 barrels. More importantly, the proposed development location for Reservoir C-5 is actually downdip of the oil/water contact, and if drilled would result in a dry hole.

Information provided by the UT and DT restored tops was very critical in this example. The tops are valuable mapping data that should not be ignored, particularly in areas of limited well control. These extra data points guide the structure contours into and through the fault gap. Figure 8-23 makes this point very clear. Upthrown and downthrown restored tops improve the accuracy of a structure contour map, and in this case reduced the size of the two prospective reservoirs.

Restored tops are also helpful in generating prospects where none previously existed. Figure 8-24a shows a portion of a structure contour map in a South Louisiana oil field. Notice Well No. 22 in the western portion of the map, upthrown to Fault A, has an oil show with an oil/water contact at 6458 ft TVD (true vertical depth). Well No. 25 to the east is wet and Well No. 65 is faulted out of this sand by Fault A. Based on the structural interpretation shown in this figure, the volume of potentially recoverable oil is insufficient to justify the drilling of an updip development well into Reservoir A-1. The structural closure is too small, and the target area between the oil/water contact and the upthrown trace of the fault too close to risk the drilling of a well. During the preparation of this structure map, the trace of Fault A was constructed based on isolated fault cut data without the use of a fault map and without estimated restored tops for the faulted out Well No. 65.

The structure map in Fig. 8-24b was constructed using a fault map for Fault A to contour across the fault, the UT and DT restored tops for this sand in Well No. 65, and maps on shallower and deeper horizons in which the tops in Well No. 65 were present. These additional mapping data change both the configuration of Reservoir A-1 and the volume of potentially recoverable hydrocarbons. This accurately *integrated structure contour map* delineates a reservoir of sufficient size (369,000 barrels of oil) and closure to be re-evaluated as a potentially drillable prospect. The UT and DT restored tops are used in Fig. 8-24b to guide the strike direction of the contours into the fault and to honor the vertical separation, thus changing the structural configuration of the reservoir.

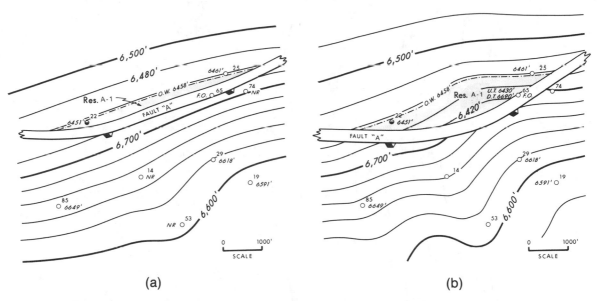

(a)   (b)

Figure 8-24   (a) Structure map on the 6450-ft Sand shows Reservoir A-1 of insufficient size to justify a development well. This map was constructed without the integration of a fault map, and the restored tops for Well No. 65 were not used. (b) Revised structure map on the 6450-ft Sand, Reservoir A-1. The revised interpretation was prepared using the restored top data from Well No. 65 and the integration of the fault map for Fault A with the structural interpretation. Also, shallower and deeper structure maps were used to guide the structure contouring. Reservoir A-1 is now of sufficient size to propose a development well location.

We cannot sufficiently emphasize the importance of using all available data when preparing subsurface maps of any type. We always work with limited data, so to ignore or not use valuable data, for any reason, is unthinkable. Our job is to develop the most reasonable and accurate interpretation possible from available data. The use of upthrown and downthrown restored tops is a necessary part of preparing a structure contour map. The time required to estimate these tops and incorporate them into the integrated structural picture is minimal; but it can have a significant impact on the final interpretation, as shown in the examples presented.

Reverse Faults

The method of integrating a reverse fault surface map with a structure map is illustrated in Figs. 8-25 and 8-26. We do not describe the procedure in as much detail as we did for the integration of normal faults, since the procedure is basically the same. There are seven wells from which fault and formation data were obtained. Figure 8-25 shows the reverse fault surface map constructed using the data from the seven wells. The fault strikes generally north-south with a dip rate of 35 deg to the west and has a displacement (vertical separation) of 300 ft.

The technique for integrating a reverse fault with a structure map is often easier than that for a normal fault because there is no projection of contours through a fault gap. In the case of a reverse fault, the hanging wall is thrust up and over the footwall, resulting in an overlap of structural horizons. Therefore, the hanging wall is contoured up to the hanging wall cutoff at the upthrown trace of the fault, and the footwall is

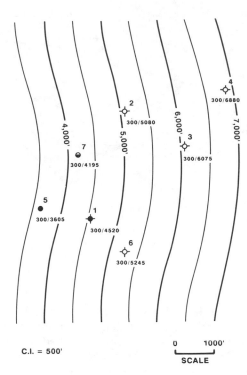

C.I. = 500'

0   1000'
SCALE

Figure 8-25 Fault surface map for reverse Fault 1 based on well control from seven wells. Depth values are positive, indicating they are above sea level.

contoured up to the footwall cutoff at the downthrown trace of the fault. Since the fault blocks overlap, the strike direction of the contours established in one fault block is used to guide the direction of the contours in the opposing block.

The faulted structure for this example (Fig. 8-26) is the nose of a plunging anticline that has been penetrated by seven wells. Wells No. 2, 3, and 6 penetrated the footwall, Wells No. 5 and 7, the hanging wall, and Well No. 1 is in the fault overlap, thereby penetrating the horizon twice—first at 4760 ft in the hanging wall, and second at 4460 ft in the footwall. Since more wells penetrate the northern flank of the structure, this is a good place to start the structure contouring. Notice that the contour values are positive, indicating this structure is above sea level.

We begin contouring this structure in the footwall fault block. Place the fault surface map under the structure map as shown in Fig. 8-26a. The fault trace on the hanging wall occurs where the structure contours in the hanging wall intersect fault surface contours of the same elevation. These intersections are highlighted on the map as small marks placed at the end of each contour. Likewise, the fault trace on the footwall is located where the structure contours in the footwall intersect fault contours of the same elevation. Again, these intersections are highlighted by small marks placed at the end of each contour (Fig. 8-26a). For accuracy, 10-point dividers may be used to locate the exact points of intersection between structure and fault contours.

The difference in elevation between contours in the footwall and those in the hanging wall equals 300 ft, which is the value of the vertical separation determined from well log correlation. For example, the 4000-ft contour in the footwall within the zone of overlap is positioned directly under the 4300-ft contour in the hanging wall. This same 300-ft difference in contour elevation is maintained for each of the structure contours in the mapped area.

Finally, connect all the marks representing the intersections of the fault surface with the footwall and hanging wall structure contours with smooth lines to accurately delineate

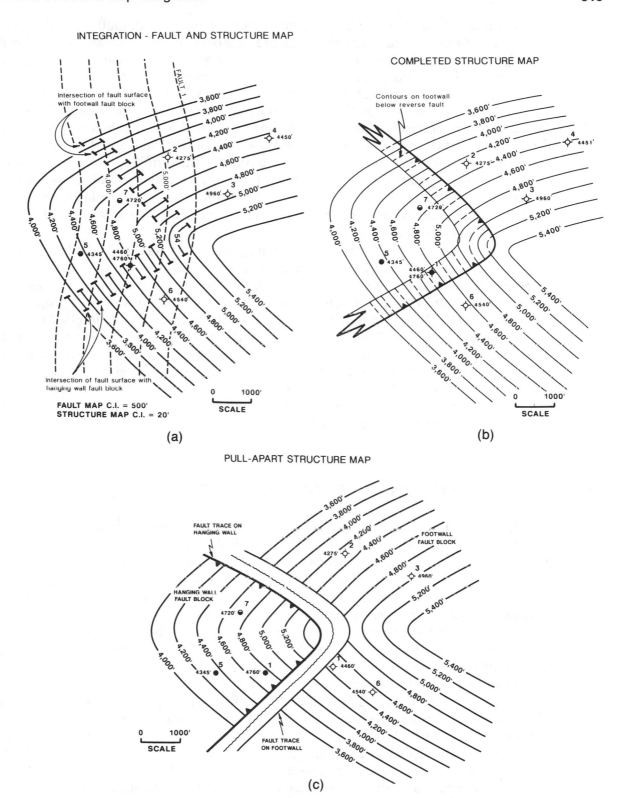

Figure 8-26 (a) The fault map is placed under the structural interpretation to obtain the intersection of the fault surface with both the footwall and hanging wall fault blocks. (b) The final integrated structure map for the 5000-ft Sand. (c) A pull apart map of the final integrated structural interpretation. This type map separates the footwall and hanging wall fault block to show them separately without the clutter of overlapping contours.

the fault traces (Fig. 8-26b). Place a symbol on the hanging wall fault trace to show the direction of fault dip and dash the structure contours in the footwall, under the fault overlap for clarity. Having completed these various steps, the integrated structure contour map for the faulted plunging anticlinal nose is complete.

By correctly integrating the fault and structure maps, the following is accomplished: (1) the position of the fault on the mapped horizon is accurately delineated; (2) the traces for the fault are precisely constructed; (3) the actual width of the fault overlap is established; and (4) the contours are projected correctly in the fault overlap. Although the actual fault surface map is striking north-south, the trace of the fault on this horizon strikes northwest-southeast on the northern flank of the structure and southwest-northeast on the southern flank of the structure. **The curved shape and position of the trace for Fault 1 on this structure map could not be intuitively constructed without the integration of the fault and structure maps.**

We mentioned earlier that a structure map with a reverse fault overlap is cluttered and can be confusing because of the double set of contours in the overlap. One way to minimize the clutter and confusion is to dash the contours in the footwall in the zone of overlap (Fig. 8-26b). Another solution to the clutter and confusion is to prepare a *pull apart map*, as shown in Fig. 8-26c. With this map, the two separate fault blocks are pulled apart enough to show each block separately without any overlap. Such a map should be constructed only after the fault and structural integration is complete in order to eliminate any possibility of contouring errors. In this example, the hanging wall is productive of hydrocarbons. Eventually, net sand and net oil isopach maps will be constructed for the hanging wall reservoir. These maps can be prepared more easily if the reservoirs are pulled apart as shown in Fig. 8-26c.

FAULT TRACES AND GAPS—SHORTCUTS AND PITFALLS

The following discussion refers primarily to fault traces and gaps formed by normal faults; however, all the geometric relations are also applicable to fault overlaps formed by reverse faults. *Fault traces* are the two lines on a structure map representing the intersection of a fault surface with the structure contoured surface in the upthrown and downthrown fault blocks. Together, the fault traces form the overall trace for a fault on any particular mapped horizon. *Fault gap* is defined as the horizontal distance between the upthrown and downthrown fault traces *measured perpendicular to the fault trace* as depicted on a completed structure contour map. When the integration technique presented earlier is used, the fault gap is *automatically* constructed, since it is the horizontal distance between upthrown and downthrown traces.

With regard to the trace of a fault and the resultant gap or overlap, there are some common misunderstandings that we intend to clarify. Also, we present some shortcuts for delineating fault traces and gaps, and the pitfalls associated with these shortcuts.

There is widespread belief that the horizontal width of a fault zone or fault gap equals the size of the fault, as obtained from well logs. It is thought that the fault gap, may therefore be scaled off with an engineer's scale, using the horizontal reference scale on the map. Thus, for a fault of 500 ft, the width of the fault gap can be mechanically constructed by scaling off a 500-ft horizontal gap between the upthrown and downthrown traces of the fault using an engineer's scale or dividers. This **technique is often *incorrect*, however, and should be employed only under special conditions.** The technique assumes that the positions of the upthrown and downthrown traces of the fault are known. Fault trace positions, however, are very difficult to define without the use of the fault/structure map integration technique. Also, the width of a fault gap is a function of the angle and direction of dip of the fault, as well as the dip of the formation. Therefore, on a single structure contour map, variations in the attitude of either a fault or the formation may result in changes in the width of the fault gap from place to place, although the size of the fault itself has not changed.

There is a general rule that should be kept in mind regarding fault gap. *A fault dipping updip (in the opposite direction of bed dip) will have a thinner fault gap than a fault of similar size dipping downdip (in the same direction as the bed dip).* Look again at Figs. 8-11 and 8-12 to see the application of this general rule. In Fig. 8-11, the fault is dipping in the same general direction as the formation, so the fault gap is wider than that shown in Fig. 8-12 for the same size fault in which the fault is dipping in the opposite direction to bed dip. In Fig. 8-27, we see the same effect. The fault cuts this horizon north of the crest and is dipping in the opposite direction to the mapped horizon; therefore, the fault gap is thinner in this area than on the flank of the structure. On the flanks, where the fault cuts this horizon south of the structural axis, the fault is dipping in the same general

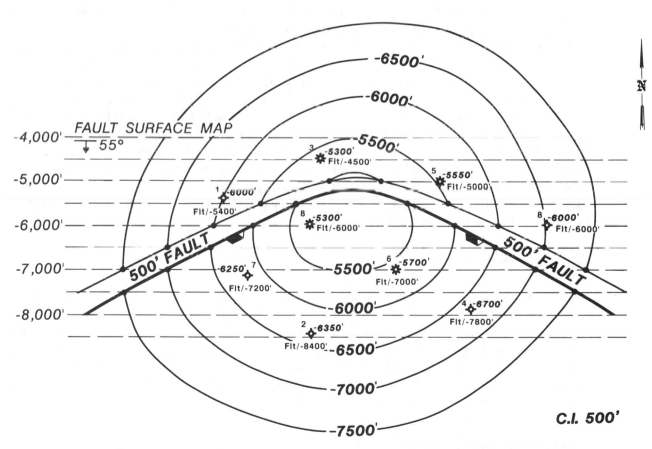

Figure 8-27 A fault surface map superimposed onto a completed structure map to show that the fault traces have been determined by the integration of the fault and structure maps. Observe the change in the width of the fault gap across the structure. The change gives the bold appearance that the fault is smaller near the crest of the structure. (Prepared by J. Bollick. Published by permission of Tenneco Oil Company.)

direction as the mapped horizon, resulting in a wider fault gap. If a deeper horizon were mapped, one in which the fault were entirely to the south of the crest of the structure, the width of the fault gap near the crest would be greater.

As a simple exercise, project the well control on the structure map to a horizon 1000 ft deeper and use tracing paper to integrate the fault with this deeper horizon (i.e., change the depth in Well No. 3 from 5300 ft to 6300 ft). What observation can you make about the change in width of the fault gap and strike direction of the fault trace?

In Fig. 8-10, the curvature of the fault trace and width of the fault gap change across the structure from 550 ft on the flanks, to 350 ft near the crest. Although the fault has a constant size of 300 ft, nowhere does the gap width measure 300 ft. Figures 8-11 and 8-12 show a significant difference in the width of the fault gap, which ranges from an average 420 ft in Fig. 8-11 to 290 ft in Fig. 8-12, despite the fact that the fault on each map is the same fault with a vertical separation of 400 ft. Finally, examine Fig. 8-27, which shows a map of a fault surface striking east-west and dipping to the south at 55 deg, superimposed on the structure map of the 5500-ft Sand. The upthrown and downthrown fault traces and thus the fault gap were constructed using the integration technique. Notice how the fault gap thins near the crest of the structure, causing the size of the fault to **appear smaller** on the crest than on the flanks. The size of the fault only **appears** to change across the structure because the width of the fault gap changes.

The width of a fault gap depends upon the angle and direction of the fault and formation dip and the intersecting strike relationship between the two. Therefore, the **width of a fault gap for any given fault on a structure map cannot be *intuitively* determined.** It is extremely difficult at best, regardless of your ability to see in three dimensions, to predict the position of fault traces and the width of a fault gap by looking at fault cut data, or even a fault surface map and a structure map independently. Examine Fig. 8-27 again and ask yourself whether you could have predicted the curvature of the overall fault trace and the variation in width of the fault gap on this final structure map simply by reviewing the fault map and structure map data separately without integration. Most likely, your answer would be "No." We emphasize that if the construction of the upthrown and downthrown fault traces is done properly by integrating the fault and structure maps, the width of the fault gap is automatically constructed without any need for guesses or estimations (Tearpock and Harris 1987).

Rule of "45"

Figure 8-28 shows a special condition in which the fault traces can be delineated with reasonable accuracy, without integrating a fault and structure map and where the fault gap is equal in width to the size of the fault. The technique used for this situation is known as the *Rule of 45*. If a fault is inclined at 45 deg and the beds are approximately horizontal (zero dip), the position of the traces can be identified with reasonable accuracy without integration, and the width of the fault gap is equal to the size of the fault, measured off using the map scale as reference (see map insert in Fig. 8-28).

Tangent or Circle Method

Another technique, referred to as the *Tangent or Circle Method* (Lyle 1951; Bishop 1960), can be used to quickly delineate the approximate position of the fault traces in areas where the beds are nearly horizontal. Lyle introduced this technique in 1951 for mapping faults in southwest Texas and Bishop (1960) diagrammatically illustrated its use.

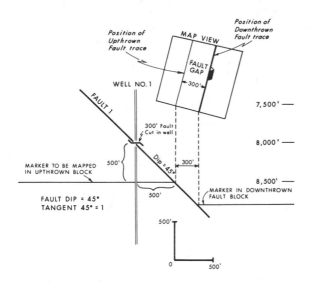

Figure 8-28 The "Rule of 45." On any given mapped horizon, the horizontal distance from a well to the nearest fault trace is equal to the vertical difference in depth from the fault cut to the horizon in the well, if the fault is dipping at 45 deg and the mapped horizon is horizontal.

Figure 8-29a is a base map showing the depth of each fault cut and the depth to the top of a key horizon in each well. The Tangent or Circle technique is used here to quickly prepare a structure map including the fault traces. This technique is good for rapidly checking a completed structure map for accuracy of the positioning of the fault traces.

1. The horizontal distance between the fault and the well equals the **cotangent** of the dip angle of the fault multiplied by the vertical distance between the cut point and the formation datum in the well (Fig. 8-29b). Cotangent of 45 degrees is equal to 1.

2. The well serves as the center point of a circle, the radius of which equals the distance between the fault and the well (radius = AC cotan θ; see Eq. 8-2) as determined in step 1 (Fig. 8-29c).

3. A tangent common to two or more circles determines the fault trace. The exterior tangent is the fault trace for wells on the same side of the fault; the interior tangent is the fault trace for wells on the opposite sides of the fault (Fig. 8-29c). Refer again to Chapter 7, Fig. 7-9 and the accompanying text for determining on which side of a fault a well is located for any mapped horizon.

4. Finally, contour the formation tops assuming continuity of structural attitude across the faults (Fig. 8-29d).

If the beds are horizontal, the fault gap width can be determined as shown in Fig. 8-29b. Then the fault gap can be drawn on the structure map rather then representing the fault as one line, as shown in Fig. 8-29d. This method of estimating the fault trace is useful if the dip of the fault surface can be determined or at least assumed with some degree of certainty and if the formation being mapped is approximately horizontal. Compare the mapping results in Fig. 8-29d with those shown in Figs. 8-29e and 8-29f. Figure 8-29e is a fault map for the two faults and Fig. 8-29f shows the fault traces constructed by the integration of the fault and structure maps. The position of the fault in Fig. 8-29d compares very favorably with the position of the fault in Fig. 8-29f, showing that in an area of relatively flat beds, the Tangent method can provide a good approximation for positioning the fault on a structure map. However, we caution that this technique does not work in areas of dipping beds.

In Fig. 8-29c or 8-29d, the fault traces are constructed as straight line segments with sharp angular changes in strike direction. In order to prepare a more accurate map, the

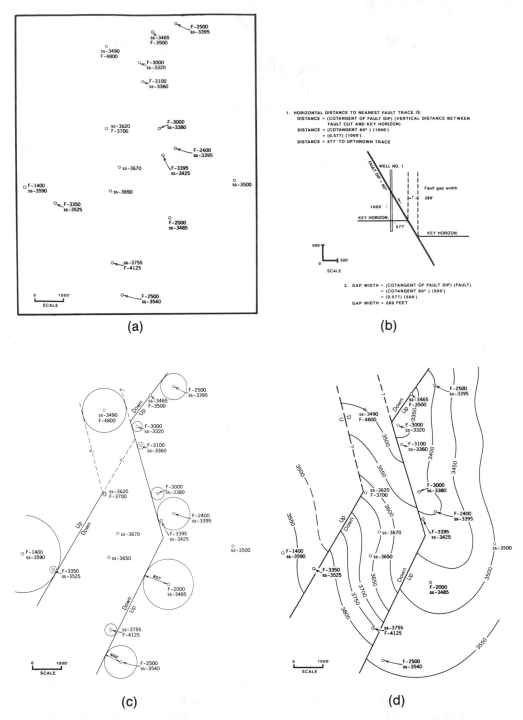

Figure 8-29 (a) Base map with the depth to fault cuts (F-3100 ft) and sand tops (55 ft–3320 ft) posted. (Modified after Bishop 1960. Published by permission of author.) (b) Determination of the horizontal distance from a well to the nearest fault trace using the "Cotangent Rule" and the width of the fault gap. (c) Assumed angle of fault = 60 deg. The position of the fault trace is determined by the "Circle Method." (Modified after Bishop 1960. Published by permission of author.) (d) Structure contour map. Position of the fault determined by the "Circle Method." (e) Integration of mapped horizon and fault surface maps to determine the position and width of the fault gaps. (f) Final integrated structure map. Compare position of fault traces with those shown in Fig. 8-29d. (From Bishop 1960. Published by permission of author.)

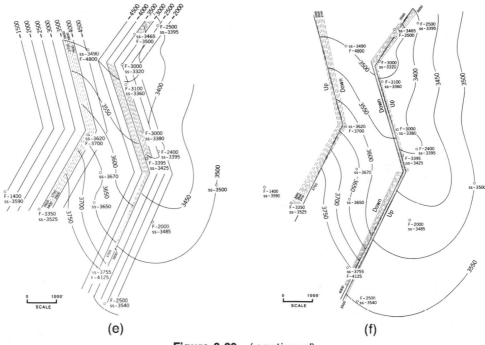

(e)   (f)

Figure 8-29   (*continued*)

change in strike direction can be smoothed to best fit the overall pattern of the faults in the area being mapped. The fault surface map for Fig. 8-29e and the integrated structure map in Fig. 8-29f were also constructed as straight line segments to make the results comparable with those in Fig. 8-29d.

The Tangent method is not accurate if the formation is steeply dipping because the wellbore is then not perpendicular to the bedding plane. The correct construction of the fault traces and fault gap for faults on a steeply dipping structure, such as a piercement salt dome, is especially critical because the fault position has a significant impact on any proposed location of an exploration or development well. Figure 8-30a shows a portion of a structure map on the flank of a piercement salt dome. The depth to the mapped horizon and the size and depth of the fault cuts are shown next to each well. Since the fault is a perfect plane with a dip of 45 deg, the structure map in Fig. 8-30a was prepared (incorrectly) using the rule of 45 to determine the position of the fault traces and the width of the fault gap. As an example, look at Well No. 6 shown in the figure insert. The fault cuts the well at −9510 ft and the top of the horizon intersects the well at −9395 ft. From the figure insert we see that Fault A cuts the well 115 ft below the mapped horizon; therefore, based on the 45-degree trigonometric relationship of 1 to 1, the downthrown trace of the fault was located 115 ft west of the location of Well No. 6 and the horizon in the well is in the downthrown block. Also, since the fault has a dip of 45 deg, the width of the fault gap was scaled to equal the size of the fault (400 ft). Using this method, the fault cut in Well No. 3 does not fit as part of the same fault identified in Wells No. 2, 5, 6, and 10; therefore, a second fault "B" was postulated as shown on the figure, resulting in an untested fault trap updip of Well No. 5.

The fault in this example has a constant dip of 45 deg and in this respect, the above methods were appropriate. However, the beds are **not** horizontal; they dip quite steeply (about 45 deg in the upstructure position). Therefore, neither the rule of 45 nor the Tangent method is appropriate for the preparation of the structure contour map in Fig. 8-30a. Figure 8-30b shows the correctly contoured structure map for this horizon, and the fault surface map. The fault is a peripheral fault paralleling the face of the salt; the beds

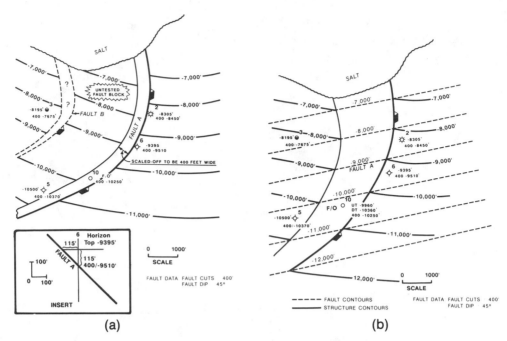

Figure 8-30 (a) Portion of a structure map on a steeply dipping piercement salt dome. The map was incorrectly prepared using the "Rule of 45" to position the fault traces. Although the fault is dipping at 45 deg, the "Rule of 45" is not applicable since the beds are not horizontal. (b) Correctly contoured structure map based on the integration of the structural interpretation with the fault surface map. Compare this map to that shown in Fig. 8-30a.

are dipping at about 45 deg at −7000 ft and flatten to about 32 deg between −11,000 ft and −12,000 ft. Since the beds are steeply dipping, the fault map was required for integration of the fault and structure maps to delineate accurately the fault traces on the structure map. The figure shows the fault surface map superimposed upon the completed structure map. The upthrown and downthrown traces of the fault and therefore the resulting fault gap were constructed using the integration technique discussed earlier in the chapter.

Observe that although the fault has a constant vertical separation size of 400 ft, as determined from well log correlation, and the fault is a perfect plane dipping at 45 deg, the trace of the fault is curved and nowhere along its trace is the gap 400 ft wide. Measuring across the fault gap at the location where the −8000-ft structure contour in the upthrown block intersects the upthrown trace of the fault, the width of the fault is approximately 1700 ft. At the intersection of the −10,000-ft structure contour and the upthrown trace of the fault, the width of the fault gap has decreased to only 1200 ft. Study this figure very carefully, for it shows how important it is to integrate the fault and structure maps on steeply dipping structures to accurately position the fault traces and delineate the fault gap.

As a general rule of thumb, we consider any structure with a dip greater than 10 deg as steeply dipping for purposes of fault and structure map integration. However, to prepare a structure map as accurately as possible regardless of the structural dip, the integration technique should always be used to delineate the position of the fault traces and the width of the fault gap.

Figure 8-31 shows three different sets of integrated fault and structure contour maps that result in significantly different fault traces and gaps. For example, the set on the left

CONSTRUCTION OF SUBSURFACE CONTOURS
FAULT TRACES

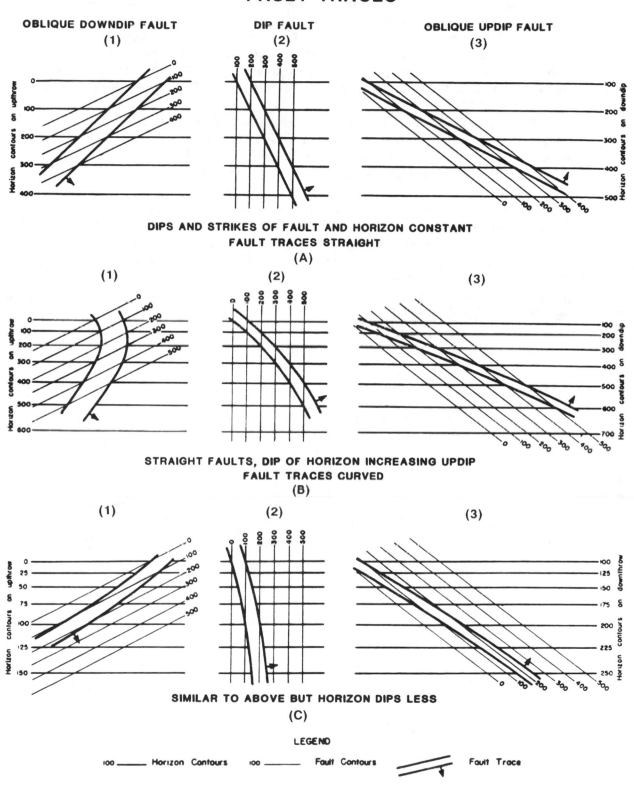

OBLIQUE DOWNDIP FAULT
(1)

DIP FAULT
(2)

OBLIQUE UPDIP FAULT
(3)

DIPS AND STRIKES OF FAULT AND HORIZON CONSTANT
FAULT TRACES STRAIGHT
(A)

(1)   (2)   (3)

STRAIGHT FAULTS, DIP OF HORIZON INCREASING UPDIP
FAULT TRACES CURVED
(B)

(1)   (2)   (3)

SIMILAR TO ABOVE BUT HORIZON DIPS LESS
(C)

LEGEND

100 ———— Horizon Contours   100 ——— Fault Contours   ⟍⟍ Fault Trace

(All Faults have 100' Vertical Separation)

Figure 8-31   Three different sets of integrated fault and structure maps illustrating significantly different fault traces. (Published by permission of Tenneco Oil Company.)

side of the figure consists of completed structure maps A(1), B(1), and C(1), all of which use the same oblique downdip fault with a size of 100 ft to integrate with the structure. A(1) shows a mapping horizon with a dip that is constant, B(1) shows a steep dipping horizon which increases in dip in the upstructure direction, and C(1) is similar to B(1) except that the horizon dip is less. Despite the constant fault surface in all three cases, each integrated structure map results in a strikingly different fault trace pattern and fault gap width. Take a few minutes to review the different fault traces and gaps for sets 2 and 3 in Fig. 8-31.

The nine separate patterns shown in Fig. 8-31 and those in Figs. 8-10, 8-26, 8-27, and 8-30 reinforce our statements that the final fault traces and fault gap for any particular set of fault and structural conditions are not intuitively predictable and, in fact, are often impossible to predict without actually integrating the two surfaces. The precise positioning of the fault traces and delineation of the width of the gap or overlap (in the case of a reverse fault) can mean the difference between generating and not generating a prospect, the difference between drilling a successful well versus a dry hole, or the difference between a reservoir whose calculated volume matches its performance versus a reservoir that appears to either have over- or under-produced the volumetric estimate of reserves. Taking shortcuts to identify the positions of fault traces and to determine the widths of fault gaps is a risky business unless the beds are relatively flat lying and the fault dips are very close to 45 deg.

New Circle Method

We have emphasized the importance of integrating a fault and structure map in areas of dipping beds to accurately delineate a fault trace. However, we have developed a mathematical relationship that can be used in areas of dipping beds to estimate the approximate position of the fault traces, if there is sufficient well control. These equations can be used as a quick check when reviewing structure maps.

Two equations are required for this method. The first equation estimates the distance from a well (radius of a circle) to the nearest fault trace, and the second estimates the heave of the fault, which is used to locate the position of the second fault trace. The strike direction of the fault *must be known* in order to determine the direction in which to measure heave. The equations are shown here and illustrated in Figs. 8-32 and 8-33.

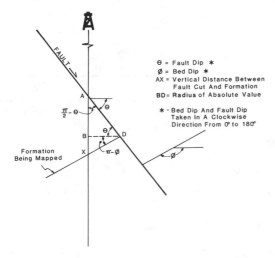

Figure 8-32 Diagrammatic illustration of the fault and formation parameters used to derive Eq. (8-2).

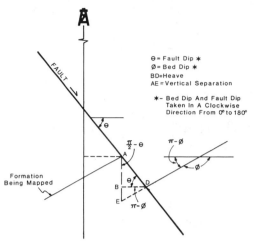

Figure 8-33 Diagrammatic illustration of the fault and formation parameters used to derive Eq. (8-3).

Equation to Determine Radius

From Law of Sincs (Fig. 8-32)

$$\frac{\sin(\pi/2 - \theta)}{DE} = \frac{\sin(\theta + \pi - \phi)}{AE}$$

and

$$\cos(\pi - \phi) = BD/DE$$

Therefore:

$$\frac{\sin(\pi/2 - \theta)}{BD/\cos(\pi - \phi)} = \frac{\sin(\theta + \pi - \phi)}{AE}$$

or

$$BD = \frac{AE \cos(\pi - \phi) \sin(\pi/2 - \theta)}{\sin(\theta + \pi - \phi)}$$

Therefore:

$$\text{Radius} = \left| \frac{- AE \cos \phi \cos \theta}{\sin(\theta - \phi)} \right| \qquad (8-2)$$

where

$$\phi = \text{bed dip}$$

$$\theta = \text{fault dip}$$

θ and ϕ are taken clockwise from 0 deg to 180 deg

If $\phi = \pi$, then eq. 8-2 becomes

$$\text{Radius} = \frac{-AE \cos(\pi) \cos \theta}{\sin \theta}$$

$$= \frac{AE \cos \theta}{\sin \theta}$$

Radius* $= AE \cot \theta$

Example:

$$\theta = 30 \text{ deg east}$$

$$\phi = 20 \text{ deg west or } 180 \text{ deg} - 20 \text{ deg} = 160 \text{ deg}$$

$$AE = 1000 \text{ ft}$$

Using Eq. (8-2):

$$\text{Radius} = \left| \frac{AE \cos 160° \cos 30°}{\sin(30° - 160°)} \right|$$

$$= \frac{1000(0.94)(0.866)}{0.766}$$

$$\text{Radius} = 1063 \text{ ft}$$

Equation to Determine Heave

Again utilizing the Law of Sines (Fig. 8-33)

$$\frac{\sin(\pi/2 - \theta)}{DE} = \frac{\sin(\theta + \pi - \phi)}{AE}$$

and

$$\cos(\pi - \phi) = BD/DE$$

$$DE = \frac{BD}{\cos(\pi - \phi)}$$

Therefore:

$$BD \text{ (Heave)} = \frac{AE \sin(\pi/2 - \theta) \cos(\pi - \phi)}{\sin(\theta - \phi)}$$

$$BD = \left| \frac{AE \cos \theta(-\cos \phi)}{\sin(\theta - \phi)} \right|$$

(8-3)

which is the same form as the radius equation 8-2.

* This is the equation used for the tangent method when the beds are horizontal.

Example:

$$\theta = 60 \text{ deg}$$

$$\phi = 45 \text{ deg}$$

$$AE = 1000 \text{ ft}$$

$$BD \text{ (Heave)} = \left| \frac{1000 \cos 60° \cos 45°}{\sin(60° - 45°)} \right|$$

$$BD \text{ (Heave)} = \left| \frac{1000(0.5)(0.707)}{0.2588} \right|$$

$$BD = 353.5/0.2588$$

$$\mathbf{BD = 1366 \text{ ft}}$$

In order to use these equations, *the strike direction of the fault must be known.* These equations should be used only as a quick look or check, and do not take the place of the fault/structure integration technique.

Fault Gap Vs Fault Heave

Before leaving the subject of fault gap, we shall briefly discuss the fault component **heave** which is often misused or mistaken for fault gap. **Fault gap** is the horizontal distance between the upthrown trace and downthrown trace of a fault *measured perpendicular to the strike of the fault traces.* **Fault heave** is defined as the horizontal distance between the upthrown and downthrown traces of a fault *measured perpendicular to the strike of the fault surface itself.* The width of a fault gap is equal to fault heave *only* where the beds are horizontal or nearly so, and the strike of the fault is parallel to the strike of the formation. Considering most other situations, the width of the fault gap and the fault heave are not the same. Therefore, the *term fault gap or overlap (in the case of a reverse fault) is not synonymous with fault heave.* One final note: heave is a slip-related fault component that only has limited application in applied subsurface mapping.

The completed structure map in Fig. 8-30b clearly illustrates the difference between fault gap and fault heave. Using the completed structure map, measure the heave for Fault A at the location where the −9000-ft contour in the upthrown block intersects the upthrown trace of the fault. First, draw a line perpendicular to the strike of the fault surface at the intersection of the −9000-ft structure contour with the −9000-ft fault contour. This perpendicular line intersects the downthrown trace of the fault at some point. Using the horizontal reference scale on the map, measure the length of this perpendicular line, which is the measurement of fault heave at this location along the fault. Now, take a minute to measure the throw of the fault along the same line. If done correctly, the perpendicular line intersects the downthrown trace of the fault at a depth of −10,825 ft so the horizontal measurement for the fault heave is 1825 ft. Since the fault has a dip of 45 deg, the throw of the fault must equal the heave, or 1825 ft (10,825 ft − 9000 ft = 1825 ft). We can easily check the measured result for fault throw by using Eq. (7-1). To use the equation, the bed dip must be averaged for the interval between the

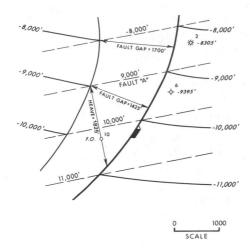

Figure 8-34 An enlargement of Fig. 8-30b illustrating the measurements of fault gap and fault heave.

−9000-ft contour in the upthrown block to the −10,825-ft contour in the downthrown block because the gap is too wide and bed dip changes very rapidly in this area. For the equation, use 38 deg as an average bed dip. Also, use Eq. (8-3) to estimate the fault heave mathematically and compare the results with the measurement shown in Fig. 8-34. As a final measurement, measure the fault gap at the same intersection of the −9000-ft structure contour and upthrown fault trace.

These measurements are shown in Fig. 8-34. Notice for this example that there is a significant difference between the strike direction of the fault gap and that for the fault heave. Fault heave and fault gap are different fault components that cannot be referred to interchangeably. As another exercise, measure the fault overlap and heave for the reverse fault shown in Fig. 8-26 and compare the results.

Subsurface geologists should, as a general rule, avoid using the terms "throw" and "heave" and should always refer to measurements derived from log correlations and map construction by their correct-names, "vertical separation" and "fault gap width" (Tearpock and Harris 1987).

STRUCTURE MAP—GENERIC CASE STUDY

In the section, **Techniques for Contouring Across Faults**, we presented the technique for projecting established contours from one fault block across a fault into the opposing fault block to contour correctly the displacement resulting from a fault. The fault size or displacement to which we are referring, as determined from well log correlations or seismic data, is vertical separation. In the same section, the incorrect technique of contouring displacement as throw was also presented. Now we look at a generic case study which illustrates the impact of preparing a map with the incorrect, as opposed to correct, contouring technique.

This field example shows an anticlinal high cut by a large 600-ft down-to-the-south normal Fault A and a second smaller 100-ft normal compensating down-to-the-north Fault B (Fig. 8-35). The structure map in Fig. 8-36a was prepared with the incorrect assumption that the missing section in the wellbores is equal to throw (note the dashed contours in the fault gap running perpendicular to the strike of the fault). Based on this interpretation, two development wells were proposed. Location, ''X,'' upthrown to Fault A in Reservoir A, is estimated to penetrate the reservoir updip and to the east of Well No. 12 at a subsea

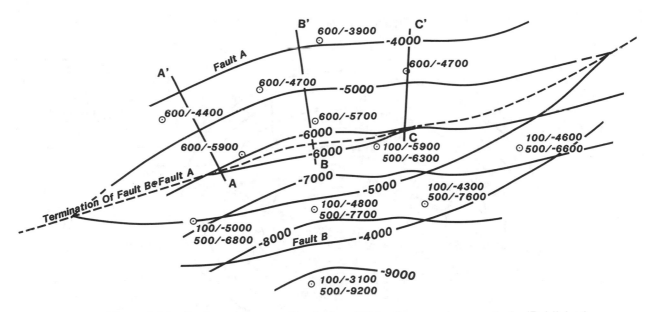

Figure 8-35 Fault surface map (Faults A and B) for the generic case study. (Published by permission of Tenneco Oil Company.)

depth of 4960 ft. The purpose of the well is to improve the drainage efficiency of this reservoir. Location "Y," upthrown to Fault B, is estimated to penetrate Reservoir B at a subsea depth of 5360 ft and is designed to be drilled in the optimum structural position to drain the attic reserves in this reservoir.

The structure map in Fig. 8-36b was constructed correctly using the missing section in the wellbores as vertical separation. Observe the dashed contours in the fault gap showing the projection of the contours from one fault block to the next. This correct interpretation shows a different structural picture. First, Well X is estimated to penetrate the formation at a subsea depth of 5030 ft, which is the depth to the oil/water contact in Reservoir A. Therefore, Well X, if drilled, would be a dry hole. The penetration point for Well Y, in Reservoir B, is estimated at −5415 ft, or 55 ft deeper than that shown in Fig. 8-36a. Based on the correctly contoured map, the proposed well is about 800 ft west of the optimum position to efficiently drain the remaining reserves in Reservoir B.

The correctly contoured map has the following impact on these two reservoirs and proposed wells.

Reservoir A (upthrown to Fault A)

1. It eliminates the drilling of a dry hole.
2. It improves the volumetric reserve estimate. In this case, a 36% reduction in reserves.

Reservoir B (upthrown to Fault B)

1. It improves the volumetric reserves estimate for the reservoir (a reduction of 11%).
2. It improves the configuration of the reservoir, impacting the location of the proposed development well.

INCORRECT INTERPRETATION

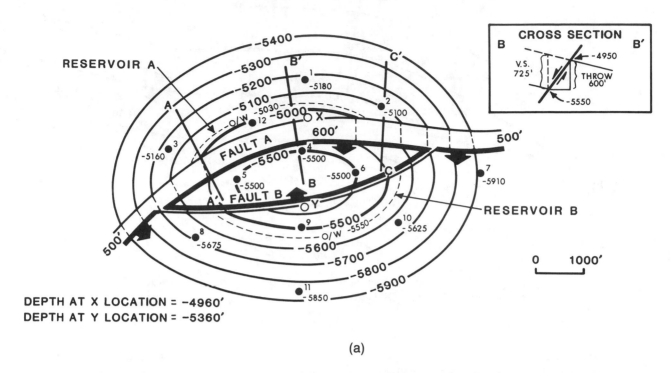

DEPTH AT X LOCATION = –4960'
DEPTH AT Y LOCATION = –5360'

(a)

CORRECT INTERPRETATION

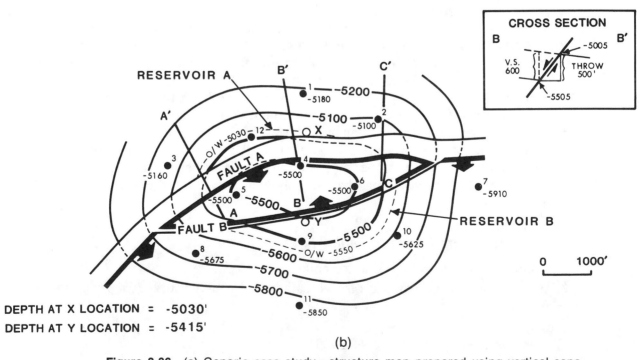

DEPTH AT X LOCATION = –5030'
DEPTH AT Y LOCATION = –5415'

(b)

Figure 8-36 (a) Generic case study—structure map prepared using vertical separation incorrectly as if it were throw. (b) Generic case study—structure map prepared using vertical separation to correctly contour across Faults A and B. (Published by permission of Tenneco Oil Company.)

The magnitude of error created by incorrectly mapping vertical separation as if it was throw is greatest near or on the crest of a structure as illustrated in this generic case. On both the east and west flanks of the structure, where the strike direction of the contours is nearly perpendicular to the strike of the fault, the projection of contours through the fault is almost identical for both structure map interpretations. When the strike direction of the structure contours is perpendicular to the strike of the fault, lines parallel to the contour strike give the appearance that the formation has an apparent dip of zero degrees or is horizontal. Therefore, whether the contours are projected through the fault as if the fault were not there, mapping the vertical separation; or the strike direction of the contours is projected through the fault perpendicular to the strike of the fault to map throw, the direction of contour projection through the fault is similar. However, as the crest of the structure is approached, the projection of the contours is radically different for each structure map. Review the −5000-ft contour starting in the upthrown block of Fault A and follow the projection across the fault into the downthrown fault block on both maps; notice the significant difference in contour projections. Here again we emphasize that it is very important to contour correctly over the entire map, but it is especially critical in and around the crest of a structure where hydrocarbon accumulations are expected to occur.

There are three cross-section lines shown on each structure map. The B-B′ cross section for each map is shown in the upper right corner of Figs. 8-36a and 8-36b. The B-B′ section in Fig. 8-36a shows that if the throw of the fault were 600 ft as mapped, then the vertical separation in cross section B-B′ would have to be 725 ft; therefore, the missing section in the wells would also have to be 725 ft. The B-B′ cross section in Fig. 8-36B shows a vertical separation based on a fault size of 600 ft, which agrees with the missing section in each well, and a throw across the fault at this location of 500 ft. This cross section geometry is supported by the well log data and the completed structure map. As a cross section exercise, complete sections A-A′ and C-C′ for both maps and evaluate the results. If you are interested, use Eq. 7-1 again to test the validity of the structure contour map in Fig. 8-36b.

THE ADDITIVE PROPERTY OF FAULTS

In the preparation of detailed structure maps, the size (vertical separation) of most intersecting faults is very close to additive. If two faults join into one, the size of the "surviving" fault is equal to the sum of the sizes of the two initial faults. This property is referred to by several names, including **Conservation of Fault Size, Conservation of Vertical Separation, Conservation of Fault Throw, and the Additive Property of Faults.**

The additive property of faults, simply defined, states that, *in areas where two or more faults have an intersecting relationship, the sum of the fault size must be conserved at the intersection.* As shown in Fig. 8-37a, if two faults are downthrown in the same general direction as in this example of a bifurcating fault system, the size of the resultant Fault 1 is equal to the sum of the size of the two initial Faults 1 and 1-A. If the two initial faults are downthrown in opposite directions toward one another, as in the compensating fault example shown in Fig. 8-37b, the size of the resultant Fault 1 is the difference in the size of the two initial Faults 1 and 2. Finally, if the two initial faults are upthrown in opposite directions as shown in Fig. 8-37c, the size of the resultant Fault 1 is the difference in the sizes between the two initial Faults 1 and 2.

The sum of the size (vertical separation) of intersecting faults should equal zero at

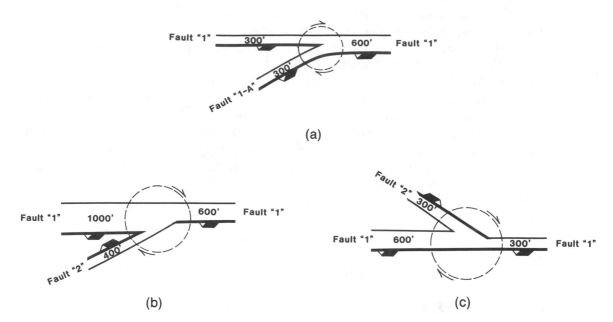

Figure 8-37   Conservation of vertical separation at fault intersections.

the point of intersection (see dashed circles in Fig. 8-37). In other words, the size of the fault(s) on either side of the intersection must be in balance.

This additive property is very important to keep in mind when preparing structure maps involving intersecting faults, and the property is also quite helpful in quickly evaluating or reviewing already completed structure maps. A simple method of checking an intersecting fault pattern is to imagine a circle around the fault intersection, as shown in Fig. 8-37, and, starting at the 12 o'clock position, move clockwise adding up the size of the faults dipping in the same direction and subtracting the size of the faults dipping in the opposite direction. Upon completion of a traverse around the circle, the additions and subtractions should equal or nearly equal zero (Fig. 8-37).

A failure in the test of the additive property of faults can indicate a structure mapping problem. Often it is easy to identify the mapping bust, but difficult to determine why there is a problem. Sometimes the problem is the result of incorrect contouring, a misunderstanding of the property, or a correlation bust. The example shown in Fig. 8-38a is probably the result of incorrect contouring and failure to integrate the fault and structure maps. This completed map shows an intersecting compensating fault pattern with Fault A dipping to the north with a size of 400 ft at the intersection (12,400 ft − 12,000 ft = 400 ft) and Fault B dipping to the southwest and intersecting Fault A. Fault B is getting smaller to the south so that at the intersection, the size is only 100 ft. Therefore, the size of the resultant Fault A east of the intersection must be 300 ft (400 ft − 100 ft = 300 ft); however, it is contoured as 450 ft on the map.

There appear to be three separate problems with this structure map: (1) Although this is a compensating fault pattern, the map is prepared incorrectly, depicting a bifurcating pattern where the individual fault sizes for Faults A and B were added. The 400 ft for Fault A added to the 100 ft for Fault B equal 500 ft. Notice that the fault sizes do not even add correctly, which is another problem related to the completed map. (2) It is obvious that this map was prepared without being integrated with fault maps. When fault and structure maps are integrated, the fault traces and gaps are properly constructed. Thus, an error of constructing fault traces that depict the addition of fault sizes rather

COMPENSATING FAULT SYSTEM

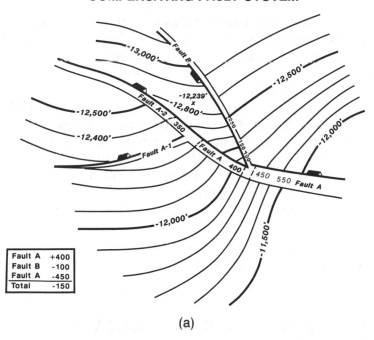

Fault A	+400
Fault B	-100
Fault A	-450
Total	**-150**

(a)

BIFURCATING FAULT SYSTEM

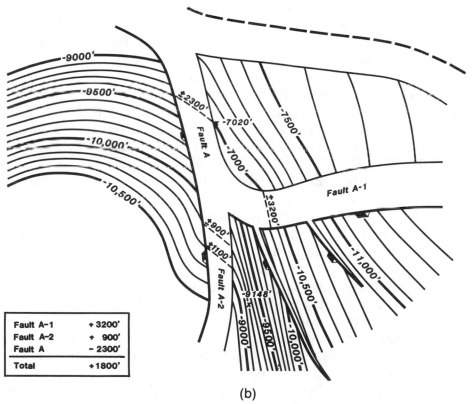

Fault A-1	+ 3200'
Fault A-2	+ 900'
Fault A	- 2300'
Total	**+1800'**

(b)

Figure 8-38 (a) Vertical separation is not conserved around the intersection of Faults A and B. This mapping bust appears to be the result of incorrect contouring and the failure to integrate the fault and structure maps. (b) The failure to conserve the vertical separation around the intersection of Faults A-1 and A-2 in this example is significant. This mapping bust is probably due to a seismic correlation mis-tie across one or both of the intersecting faults. (Modified from Tearpock and Harris 1987. Published by permission of Tenneco Oil Company.)

then their subtraction, such as in the case of this compensating fault pattern shown in Fig. 8-38a would not occur. (3) There is a problem with the contouring across Fault A at the intersection. The contours on opposite sides of Fault A east of the intersection should have an elevation difference of 350 ft rather than the 450 ft contoured on the map, since this is a compensating fault pattern.

Using tracing paper, recontour this map (Fig. 8-38a) to reflect an acceptable structural interpretation around the fault intersection. Finally, examine the structure map further for mapping busts in contouring across Fault A-2 and the intersection of Fault A-2 with A-1. See if you can identify and correct these problems.

Figure 8-38b shows another example of a completed structure map that fails to honor the conservation of vertical separation. Unlike the error in Fig. 8-38a, which is minor as errors go, the error in Fig. 8-38b is quite significant. A test of the additive property of faults around the intersection of Faults A, A-1 and A-2 indicates an error on the order of 1900 ft. Such a significant error is not due to incorrect contouring, but is most likely the result of a seismic correlation mis-tie across one or more of the intersecting faults.

INTEGRATION OF SEISMIC AND WELL DATA FOR STRUCTURE MAPPING

In Chapters 5 and 7, we covered the basic techniques for integration of fault data determined from seismic sections. In this section, we discuss the additional data provided by seismic sections to assist in the construction of subsurface structure maps. The lateral resolution of seismic data contributes information away from well control, and helps define a larger, more continuous surface than is mappable from well data alone.

This discussion assumes that several conditions have been met before the seismic data are used to make a structure map. *The first condition* is that the seismic lines have been approximately tied into the event that is correlative with the actual horizon being mapped. In some areas this may be simply a matter of using the correct velocity function to locate the proper time event that corresponds to the horizon. In more complex areas, the tie to seismic information may require the use of a synthetic seismogram or VSP to locate the proper event to map.

The second condition is that there should be no significant velocity changes over the area being mapped. If a lateral velocity gradient appears to be present, a gradient map may be required to deal with this problem. If the seismically derived horizon tops do not agree with the well tops across the area being mapped, there are several potential reasons for this disagreement. First, a velocity gradient may be present. These are common in some areas and rare in others. Second, the well log or seismic correlations may be wrong. Look hard at both sets of data before modifying or discarding one of them. Third, the wrong event may be interpreted as corresponding to the horizon being mapped. This can cause problems, especially in areas of extreme thinning and thickening of the stratigraphic intervals.

If the velocity field is uniform over the area, then seismic data can be extremely valuable in subsurface mapping. Seismic data provides: (1) estimates of structural elevations for the horizon being mapped; (2) continuous dip information along the seismic line; (3) accurate structural data points at the intersection of faults with the horizon along the seismic line; and (4) an estimate for the vertical separation across any fault.

Figure 8-39 shows an example of seismically derived data that can be posted to aid in the construction of a subsurface structure map. The first and most obvious data points

SEISMIC INTEGRATION–STRUCTURE MAPPING

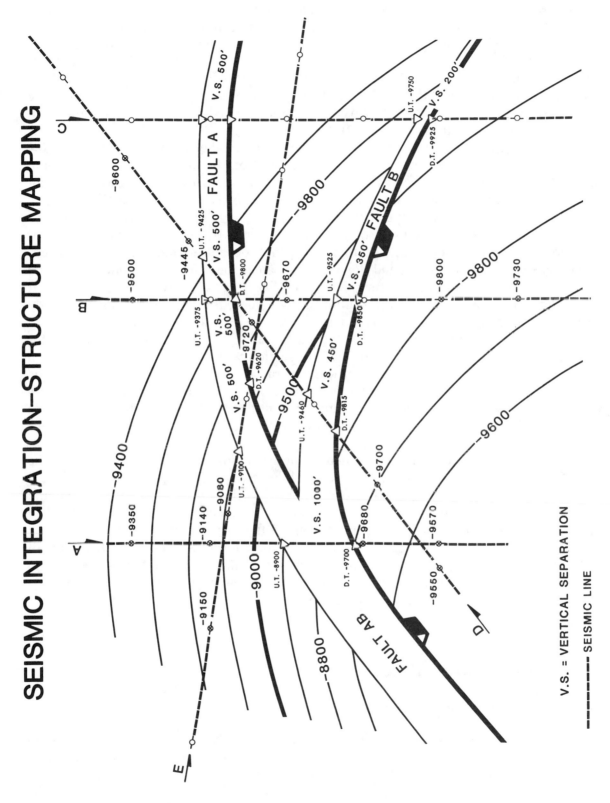

V.S. = VERTICAL SEPARATION

----- SEISMIC LINE

Figure 8-39 Completed structure map illustrating seismically derived subsurface data used to prepare the map. (Prepared by C. Harmon. Modified from Tearpock and Harris 1987. Published by permission of Tenneco Oil Company.)

are the horizon depths posted at the shotpoint locations. These values are seismic time picks converted to depth with the appropriate velocity function. Second, fault data is posted along the intersection of the faults with the horizon. For example, line D illustrates how to post the upthrown and downthrown elevations for the mapping horizon at the intersection of the horizon with the fault along the line of section. For Fault B, the upthrown elevation of the intersection of the horizon with the fault is −9460 ft, and the downthrown elevation at the fault/horizon intersection is −9815 ft. Finally, the vertical separation data estimated across the fault are posted and used to aid in the mapping (i.e., 450 ft for Fault B on Line D).

When using seismic data, be aware of the difference between the apparent throw values and the vertical separation and label the latter on the map. The estimate for apparent throw is the difference in elevation between the upthrown and downthrown intersections of a horizon with the fault along the line of section. Notice the changing values for apparent throw along fault A; these changing values are a function of the orientation of the line of section crossing the fault (lines B, D, and E). The values for the vertical separation along this portion of the fault do not change, however. For example, based on the upthrown and downthrown values measured along lines E, B, and D, the apparent throw of Fault A is 520 ft across line E, 425 ft across line B, and 375 ft across line D. Considering these apparent throw values, the fault could be incorrectly interpreted as decreasing in size or dying to the east. But the estimate of the vertical separation across each line shows that the size of the Fault A is a constant 500 ft, for mapping purposes in the area east of the intersection with Fault B. Contrast this with Fault B, which shows a decreasing vertical separation as the fault dies laterally to the east; however, the apparent throw, as measured from line D, is 355 ft, and 325 ft from line B, suggesting a fault of almost constant apparent throw. It is the vertical separations that are used for mapping displacement across faults whether the data are from well logs or seismic sections.

The contouring techniques presented at the beginning of this chapter apply to the use of seismically derived data, as well as log data. Seismic data offer some additional advantages that are lacking from well control, such as the ability to post upthrown and downthrown mapping control points at fault/horizon intersections. These additional data give two more control points to help in contouring the final structure map.

In summary, seismic data allow us to extend the interpretation beyond the limits of well control. It helps in the construction of a continuous subsurface picture of the horizon being mapped and provides additional, more complete information about the characteristics of the faults that intersect these horizons.

OTHER MAPPING TECHNIQUES

The Mapping of Unconformities

An *unconformity* is a surface of erosion or nondeposition that separates strata of different ages. The development of an unconformity involves several stages of activity. The first is the deposition of the initial sediments; second, the area is subject to uplift and subaerial erosion or nondeposition, developing an erosional (unconformity) surface; and third, the deposition of younger sediments above the unconformity. Unconformities are present in all geologic settings, but are especially common on steeply dipping structures.

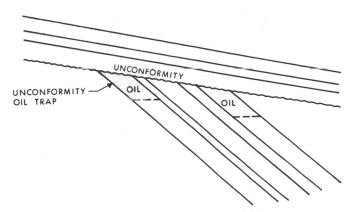

Figure 8-40 Typical hydrocarbon trap formed by an angular unconformity.

Unconformities alone or in combination with faults or stratigraphic anomalies, such as sand pinchouts, serve as excellent hydrocarbon traps (Fig. 8-40); it is therefore important to recognize and map unconformities in the subsurface. In Chapter 4 we presented the guidelines to recognizing unconformities by log correlation and dipmeters, and distinguishing faults from unconformities. In this section, we discuss general techniques for mapping an unconformity in the subsurface and integrating that unconformity map with a structure map.

An unconformity map is a type of contour map. It differs from a structure map in that the contours are on an unconformity surface rather than some stratigraphic marker or formation. Unconformity maps are not normally made to stand alone; instead, they are integrated with a structure map to delineate the intersection of the unconformity with the horizon being mapped. This intersection marks the termination of the horizon at the unconformity. If the horizon is productive of hydrocarbons, this termination marks the limit of the hydrocarbons.

Although there are various kinds of unconformities, for petroleum exploration and exploitation we are primarily interested in two types: (1) angular unconformity, and (2) disconformity. An *angular unconformity* is one in which the rocks above and below the unconformity are not parallel with one another, with the rocks below the unconformable surface typically dipping at a steeper angle then those above. Figure 8-41 is a north-south electric log cross section through eastern Beauregard Parish, Louisiana, showing a good example of an angular unconformity determined by log correlation (Lock and Voorhies 1988). Good, easily traceable resistivity markers in the section below the unconformity are seen to be progressively cut out at the unconformity surface. Notice that correlating down the logs the section is interrupted in each well at the same stratigraphic marker. From the Tribal Oil No. 1 to the Sinclair No. 5, about 150 ft of section is missing as a result of the unconformity.

A disconformity is defined as an unconformity in which the rocks on opposite sides of the unconformity are parallel. In the subsurface, disconformities are particularly difficult to recognize. In some cases, paleontological data may be helpful, in the absence of physical evidence from well logs or seismic data.

Unconformities occur on a variety of scales ranging from very local in extent, such as those associated with meandering stream channels across a flood plain, to those covering many miles. A major unconformity in the area around the western flank of the Sabine Uplift straddling the Louisiana-Texas border serves as the trap for the giant (5-billion barrel) East Texas Oil Field (Fig. 8-42). Actually, this giant field results from the

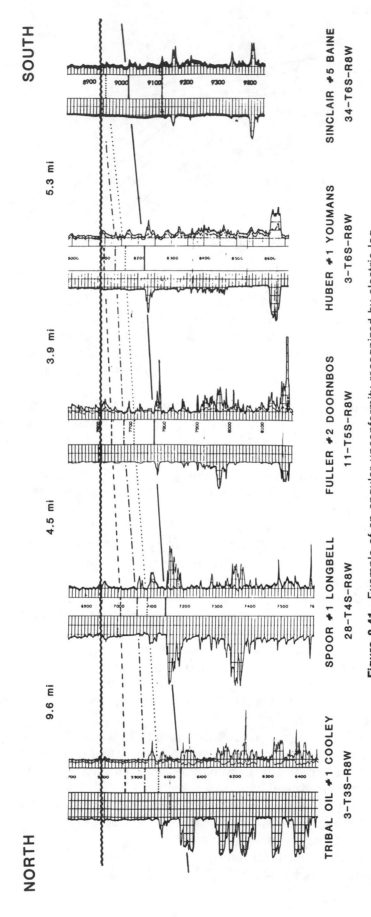

Figure 8-41 Example of an angular unconformity recognized by electric log correlation. (From Lock and Voorhies 1988. Published by permission of GCAGS.)

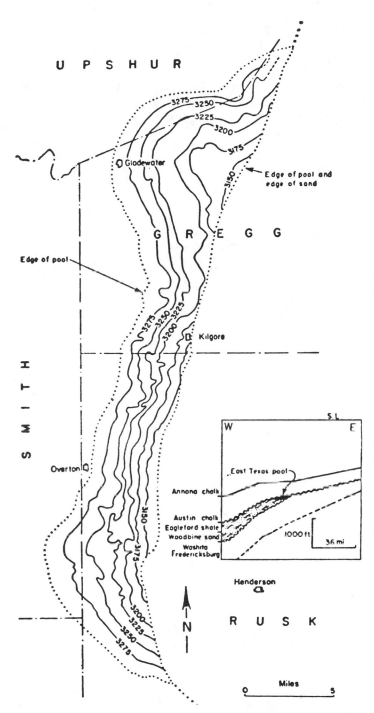

Figure 8-42 Structure map on top of the Woodbine Sand in the East Texas pool. As shown in the cross section insert, the intersection of two unconformity surfaces marks the eastern boundary of this unconformity trap. (From *Geology of Petroleum 1/E.* By Levorsen, Copyright © 1954 by W. H. Freeman and Company. Reprinted by permission.)

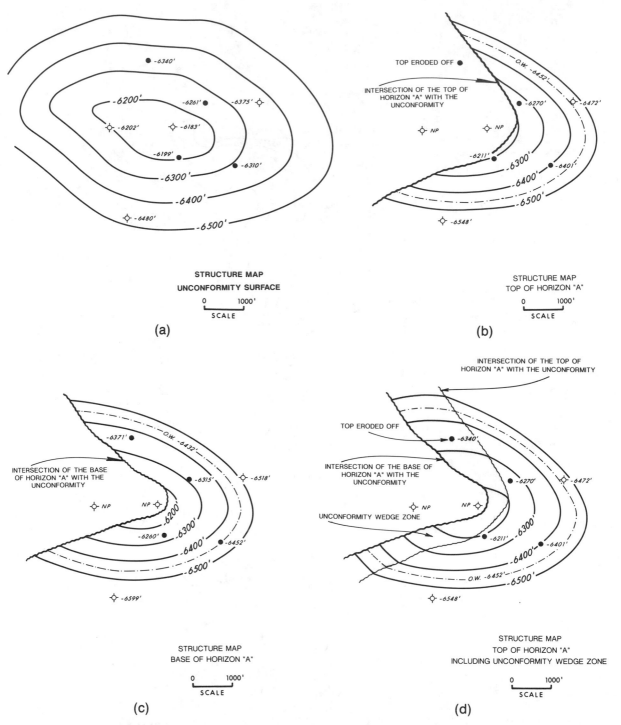

Figure 8-43 (a) Structure map on an unconformity surface based on both electric log and seismic correlation data. (b) The western limit of the oil reservoir on the top of Horizon A is defined by the intersection of the unconformity with top of structure for Horizon A. (c) The intersection of the base of Horizon A with the unconformity surface defines the western limit of this oil reservoir on the base of Horizon A. (d) An unconformity wedge zone is defined as the area between the unconformity traces on the top and base of the sand unit. The wedge zone is shown as the shaded area on the structure map on top of Horizon A. If the wedge zone is of significant size, it must be mapped because large quantities of hydrocarbons may exist within the wedge zone.

intersection of two unconformities creating the updip oil trap in the Woodbine Sandstones. Unconformities are very common around steeply dipping structures, especially salt domes. An excellent example of a major angular unconformity associated with a salt dome was shown in Fig. 4-35, Chapter 4, in East Cameron Block 293 Offshore, Louisiana.

Mapping Techniques. Figure 8-43a is a map of a deformed unconformity surface based on log correlations and seismic data. Figure 8-43b shows the integration of the top of Horizon A, which is an oil-bearing sand, with the unconformity delineating the western limit of the sand against the unconformity. The termination of the top of the sand against the unconformity is determined by overlaying the structure map on the A Sand onto the unconformity map. The intersection of each structure contour for the top of the sand with each unconformity contour at the same elevation delineates the position of the unconformity on the top of the A Sand (Fig. 8-48b). This integration is basically the same as that for a fault and structure map.

Depending upon the scale of the map, thickness of the sand, dip of the beds, and other considerations, a map on the top of a particular formation may not be sufficient to accurately show the effect of the unconformity on that formation. At times, a structure map on the base of the formation or sand is required, in addition to a map showing the unconformity wedge between the top and base of the sand. Figure 8-43c is a structure map on the base of Sand A. This map delineates the position of the intersection of the unconformity on the base of the sand. The procedure for determining the position of the unconformity on the base of the sand is the same as that just shown for the top of the sand.

Notice that the unconformity traces on the top and base of Sand A are not in the same position, indicating that there is an *unconformity wedge zone* between the two traces. This wedge zone can be mapped on the structure map on the top of the sand as shown in Fig. 8-43d. This map is prepared by first overlaying the structure map on the top of the sand onto the structure map on the base sand and transferring the trace of the unconformity on the base to the structure map on the top of the sand. There are now two unconformity traces on the top of structure map. Between these two traces, the sand goes from full thickness at the position of the unconformity on the top of structure map, to zero thickness at the position of the unconformity trace on the base of sand structure map. Since the structure in the wedge zone is actually that of the unconformity itself, next overlay the top of the structure map onto the unconformity map and draw the structure contours in the wedge zone parallel to the contours on the unconformity map. For example, the -6200-ft contour on the top of Sand A, at the trace of the unconformity, follows the -6200-ft contour on the unconformity in the wedge zone until it intersects the -6200-ft contour on the base, which is at the unconformity trace on the base of the sand. Notice that the contour changes strike direction in the wedge in order to parallel the unconformity contours. This procedure is followed for all the contours on the map. The result is a structure map on the top of Sand A showing the intersection of the unconformity on the top and base of sand and the wedge zone between them. Observe that the unconformity wedge zone for this sand comprises a significant area of the oil reservoir; in such situations the wedge zone must therefore be contoured.

Mapping Across Vertical Faults

High angle and vertical faults, particularly those with small displacements relative to the overall structure, tend to offset the structure without any change in the structural attitude across the fault. *With vertical faults, the fault size represents both the vertical separation*

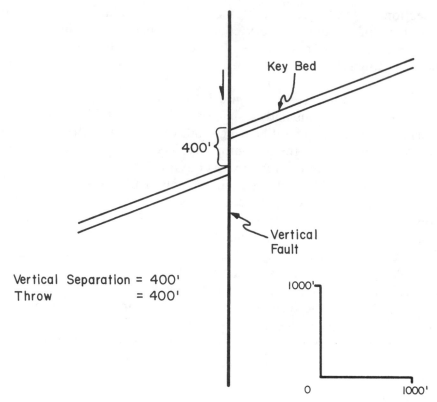

Figure 8-44 Cross section shows that the values for both vertical separation and throw are the same across a vertical fault.

and throw. Figure 8-44 shows that the fault size in terms of throw is 400 ft and the vertical separation is also 400 ft. They are the same because there is no projection of bed dip across a fault gap or overlap. Therefore, we need only concern ourselves with one value for fault size when contouring across a vertical fault; and since, for many cases, there is continuity of structural attitude, the strike direction of the contours is maintained when mapping across the fault. In very complexly faulted areas, however, the strike direction of the contours may be affected from fault block to fault block, particularly if any rotation of the fault block occurs.

Figure 8-45 depicts a structure cut by two vertical faults. The contours from one block are projected across the fault into the next fault block. The contours are projected across the fault as if the fault were not there, representing a continuity of structure across the fault, although each contour is adjusted for the size of the fault. Since the fault is vertical, there is no fault gap or overlap, so the fault is depicted as a single line on the finished structure map.

One of the biggest problems with a vertical fault, besides recognizing the presence of the fault, is estimating the size of the fault for contouring. The fault cannot be intersected by a vertical well. An estimate of the size of the fault must come from its intersection with deviated wells, seismic data, or sufficient well control within each fault block to contour both the upthrown and downthrown blocks separately and use the offset contours to determine the size of the fault.

Structure Top Vs Porosity Top Mapping

Subsurface structure maps are drawn on specific stratigraphic units to depict the three-dimensional geometric shape of the geologic structures being mapped. Once the geometry

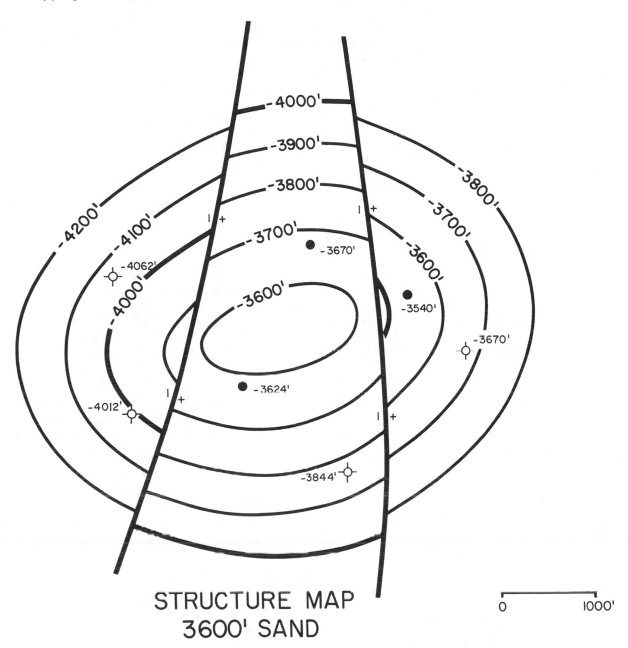

STRUCTURE MAP
3600' SAND

0   1000'

Figure 8-45 Completed structure contour map on the 3600-ft Sand cut by two vertical faults. The structure contours are drawn from one fault block to the next assuming structural continuity across the faults.

of the structure has been determined, the primary effort is focused on the mapping of all hydrocarbon-bearing formations.

At times, for various reasons, a structure map is prepared on the top of a good marker bed or resistivity character that is correlatable in all or most of the wells in a region or field instead of on an actual hydrocarbon-bearing formation or sand. In some cases this may be done because the hydrocarbon-bearing sand is discontinuous or has great vertical variation not reflected in the true shape of the structure. Therefore, it is necessary to prepare a structure contour map first on a stratigraphically equivalent top or marker in order to construct a map that conforms to the true structure of the field or region. This marker may be a few feet or several hundred feet above the actual hydro-

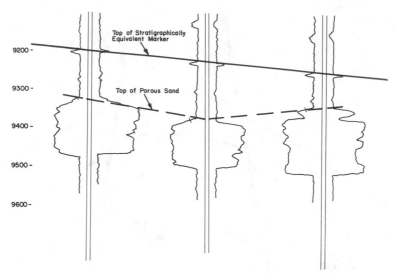

Figure 8-46 Electric logs from three wells. The upper stratigraphic marker conforms to true structure and is used to construct a map representing the true structural framework of the area. The top of the thick productive sand member does not conform to structure, but represents a porosity top. It must be mapped separately to delineate the actual reservoir configuration.

carbon-bearing sand(s). Once the structural framework is prepared by contouring the data from the stratigraphically equivalent marker, a second map, called a *Sand*, or more correctly, a **Porosity Top Map**, is required on the top of any hydrocarbon-bearing reservoir rock for the purpose of delineating the actual configuration and limits of the productive formation(s) (Fig. 8-46).

It is quite common for the upper portion of a particular formation or sand to be composed of nonreservoir quality rock. This nonreservoir quality rock is often referred to as a *tight zone or tight streak*. Although the top of sand may represent the actual stratigraphically equivalent top, it does not constitute reservoir quality rock. Therefore, the structure maps prepared to interpret the true structure often cannot be used to evaluate the reservoir itself.

Once a structure map is completed, the next step is to prepare a Top of Porosity Map for accurate delineation of the reservoir, and for later use in the construction of net hydrocarbon isopach maps. Two parameters are considered in evaluating the importance of separately mapping the top of porosity: (1) the thickness of the tight zone, and (2) the relief of the structure. A thick tight zone has a greater effect than one that is thin. Low relief structures introduce greater error in delineating the limits of a reservoir than steeply dipping structures, particularly if the low relief structure contains a reservoir with bottom water.

Figure 8-47a shows a structure map and cross section for the 6000-ft Sand. This sand consists of nonreservoir quality rock in the upper 75 ft of the sand. The same reservoir is mapped on the top of the porous sand or porosity top in Fig. 8-47b. Notice that by mapping on the top of the sand, in which the upper 75 ft consists of nonreservoir quality rock, the limit of the reservoir (gas/water contact) is extended beyond the true gas/water contact as mapped on the top of porosity (Fig. 8-47c). Even though no net pay is assigned to the tight zone, the productive area of the reservoir mapped on the top of the sand unit is larger. In turn, the volume of the reservoir is also larger than that mapped on the porosity top. In this case, the volume, based on net gas isopach maps, is larger by **32%**. This added reservoir area (Fig. 8-47c) created by mapping on the top of the sand does not contain hydrocarbons and therefore is not productive; consequently, the volume of recoverable hydrocarbons based on this map is overestimated.

The decision to prepare a separate map on the top of porosity, when the upper portion of a sand unit is not productive, needs to be made on a reservoir by reservoir basis. Depending upon the geometry of the reservoir and thickness of the tight zone, the difference in volume between a map on the top of the unit and a map on the top of porosity may be too insignificant to warrant additional mapping. In Chapter 10, we discuss in greater detail the impact of nonproductive zones in isopach mapping.

Contour Compatibility—Closely Spaced Horizons

Structure maps on vertically close horizons should exhibit, in most cases, compatible contour configurations. In order to maintain compatibility during structure contour mapping of two or more vertically consecutive horizons, a completed map should be used as a guide for the next horizon to be mapped. The simplest way to do this is to lay the completed map under the base map for the next horizon to be mapped and use it as a

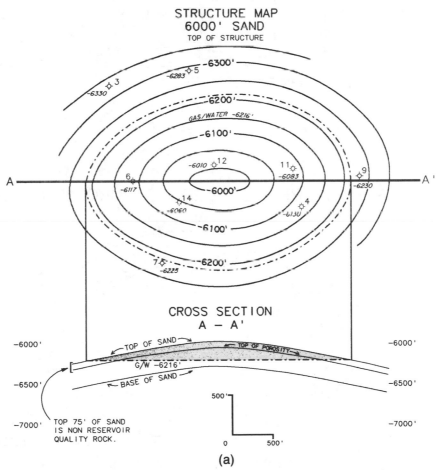

(a)

Figure 8-47 (a) Structure map on top of the 6000-ft Sand with a gas/water contact at a depth of −6216 ft and cross section A-A' illustrating (1) the top of the sand unit, and (2) top of porous sand, and base of sand. (b) Structure map on the top of porosity for the 6000-ft Sand, with the gas/water contact at a depth of 6216 ft, and cross section A-A'. (c) Mapping on top of porosity versus top of structure results in a 32% reduction in volume.

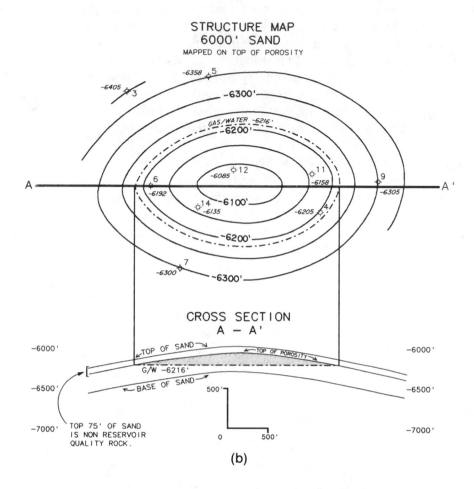

STRUCTURE MAP
6000' SAND
MAPPED ON TOP OF POROSITY

CROSS SECTION
A – A'

(b)

STRUCTURE TOP VERSUS POROSITY TOP

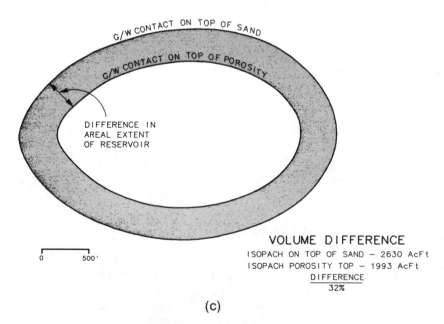

VOLUME DIFFERENCE
ISOPACH ON TOP OF SAND – 2630 AcFt
ISOPACH POROSITY TOP – 1993 AcFt
DIFFERENCE
32%

(c)

Figure 8-47   (*continued*)

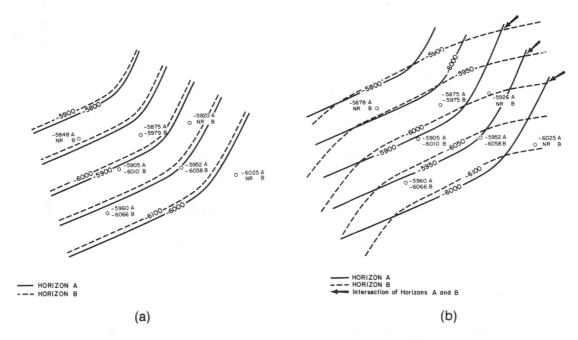

Figure 8-48 (a) Structure contour maps on Horizons A and B showing structural compatibility. The configuration of Horizon A was used to guide the contouring of Horizon B, especially in areas of limited or no well control. (b) An alternate interpretation of the structure contour map on Horizon B drawn independently and not taking advantage of the configuration of Horizon A, which has additional subsurface well control. A geologically impossible situation is shown; a contour of a given value (i.e., −5950 ft) on Horizon A crosses a contour of the same value (i.e., −5950 ft) on Horizon B.

guide for contouring the new horizon. This technique usually ensures that consecutively mapped horizons possess structure or contour compatibility.

Figure 8-48 illustrates the use of this technique and also shows the problems that can arise if vertically consecutive horizon maps are not overlaid during structure contouring. In Fig. 8-48a, the completed structure map for Horizon A was used to guide the contouring of Horizon B. Notice the compatibility in the structural configuration of the two horizons. Figure 8-48b illustrates the kind of contouring errors that can arise when a completed map is not used to guide the structure contouring of another vertically close horizon. Notice that in the area of good well control, the contours are similar to those in Fig. 8-48a. The major problem occurs in the areas of limited control on Horizon B. Observe in Fig. 8-48b that several contours on Horizon A actually cross contours of the same value on Horizon B (see for example the −5900-ft contour) which incorrectly indicates a stratigraphic thickness between Horizons A and B of zero feet. We know, however, from the available well data that there is about 100 ft of gross stratigraphic section between these two horizons.

There is a contouring rule which states that contours of the same value cannot cross on separately mapped horizons. If contours of the same value, mapped on different horizons, intersect or cross, then the interval between the two horizons would have zero or negative thickness, which is a physical impossibility. This is a serious mapping error that occurs because completed structure maps are not used to guide the contouring of other vertically close horizons.

A major mapping bust was shown in Fig. 6-11 (a and b) in Chapter 6. In the situation shown, the two horizons were not vertically close (separated by 800 ft of section in the off-structure position) and yet the structure contour maps for the P5 and P7 Horizons crossed in the upstructure position. To avoid such errors, we recommend that completed structure maps always be used to guide the contouring of other horizons, whether they are above or below the completed map.

APPLICATION OF CONTOUR COMPATIBILITY ACROSS FAULTS

The method of contouring across faults using the vertical separation (missing or repeated section in a wellbore) requires the structure, individual beds, or markers being mapped to have the same or similar structural attitude on opposite sides of the fault, as shown in Figs. 6-5, 6-8, 7-15, 8-10, and 8-22. This means that the use of the technique requires *contour compatibility* across the fault. For many tectonic settings, the compatibility of structural attitude across faults is more the rule than the exception; for others it is not. The use of the missing or repeated section (vertical separation) for contouring across faults, shown earlier in this chapter, does have application on a worldwide basis in nearly every tectonic habitat, including such areas as those shown below.

> Thin-skinned Extensional Areas
> > Gulf of Mexico—onshore and offshore
> > West coast of Africa—Gabon and Niger Delta areas
>
> Crystalline Basement—Involved Extensional Areas
> > East African Rift Area
> > North Sea
> > Red Sea
> > South America
>
> Thin-skinned Compressional Areas
> > Southern Mexican Ridges
> > Western United States and Canada
>
> Diapiric Salt Tectonic Areas
> > Gulf of Mexico
> > North Sea
> > Northern German Salt Basin
>
> Wrench Fault Tectonic Areas
> > Western United States

The use of seismic data is very helpful in determining whether there is structural (contour) compatibility across faults in any area being mapped (Figs. 7-37 and 7-38). To be a good mapper, you must have a good understanding of the tectonic setting being

---

Figure 8-49 Observe the differing bed attitude upthrown and downthrown to the major growth Fault A. The rollover anticline, containing all the hydrocarbon accumulations, displays a compatibility of bed attitude across Faults B, C, and D. (Modified from Tearpock and Harris 1987. Published by permission of Tenneco Oil Company.)

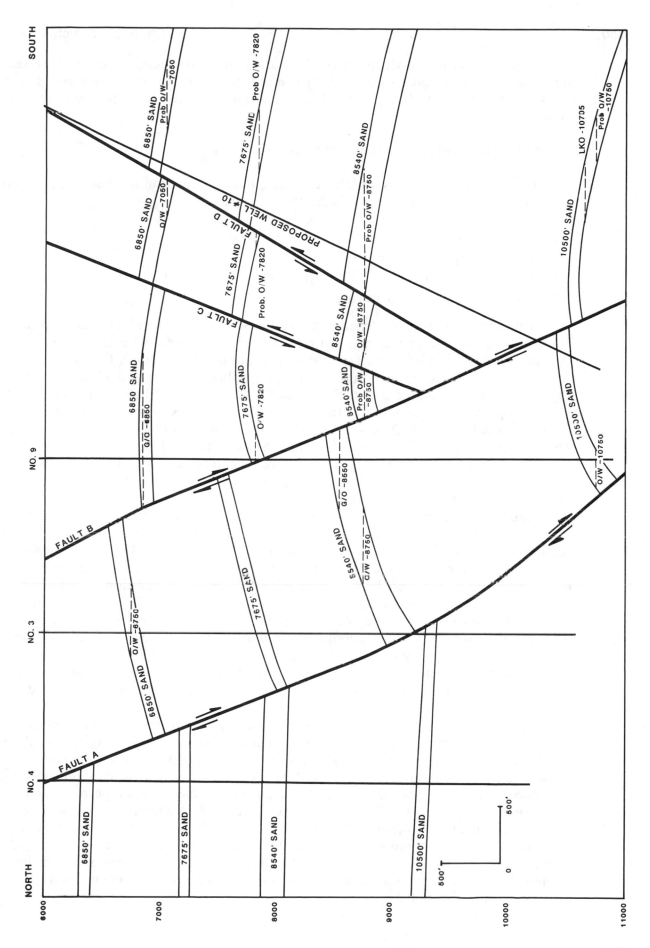

worked and detailed knowledge of the individual structures being mapped. It is very important to know where to apply a particular technique, such as mapping across faults, and when the structure being mapped is an exception to the rule requiring a different technique.

Exceptions to Contour Compatibility Across Faults

Four key geologic situations usually result in differing structural attitudes and display structural incompatibility across a fault. These include:

1. normal growth faults exhibiting large vertical separation;
2. thrust or large reverse faults;
3. intermediate/late-stage wrench faults; and
4. ramps relating to rapidly dying faults.

A normal growth fault, depending upon its size and the local amount of contemporaneous deposition, may or may not possess structural (contour) compatibility across the fault. Usually, the larger the fault the less chance of compatibility. For example, Fig. 7-44 in Chapter 7 shows a growth fault with compatible beds across the fault. However, Fig. 8-49 illustrates a large growth fault (Fault A) that does not have bed compatibility from the upthrown to downthrown fault blocks. Figure 8-50 is a cross section from West Cameron Block 17 Field, Offshore, Gulf of Mexico, which exhibits both characteristics. The structural attitude is compatible across the fault in the shallow section, gradually becoming noncompatible with depth. There are several reasons for this change in structural attitude, including the activity of the large Miocene Embayment Growth Fault shown to the far left in the cross section.

Figure 8-25, shown earlier, is an example of an area cut by a reverse fault that exhibits structural (contour) compatibility across the fault, so the vertical separation (the repeated section in this case) is used to map across the fault. In Figure 6-25 (Chapter 6), however, there are significantly different structural attitudes on opposite sides of the blind thrust fault; in this case, the technique of contouring the vertical separation across the fault (the repeated section) cannot be used.

Vertical separation (missing or repeated section) cannot be used to contour across a fault where there is no compatibility of structural attitude across the fault. For such structures, it is necessary to map each fault block **separately and independently using the available well or seismic control. Each contoured fault block must be individually integrated with the fault surface map for the displacing fault to determine the upthrown and downthrown fault traces** (Tearpock and Harris 1987).

Observe in Fig. 8-49 that although the vertical separation mapping technique cannot be used across the major growth Fault A, the faulted rollover anticline containing all the hydrocarbon accumulations does exhibit structural compatibility across all faults. The missing section (vertical separation) must be used to map across all the faults that cut the rollover anticline downthrown to the major growth fault. This type of situation is quite common in both extensional and compressional areas. Major normal growth, thrust, or wrench faults, may downdrop, thrust, or slide one fault block with respect to the other to such a degree that there is no compatibility of structural attitude across these major faults. The associated structures formed by these growth, thrust, or wrench faults, if faulted, however, often possess structural compatibility. This is shown clearly in the cross section in Fig. 8-49.

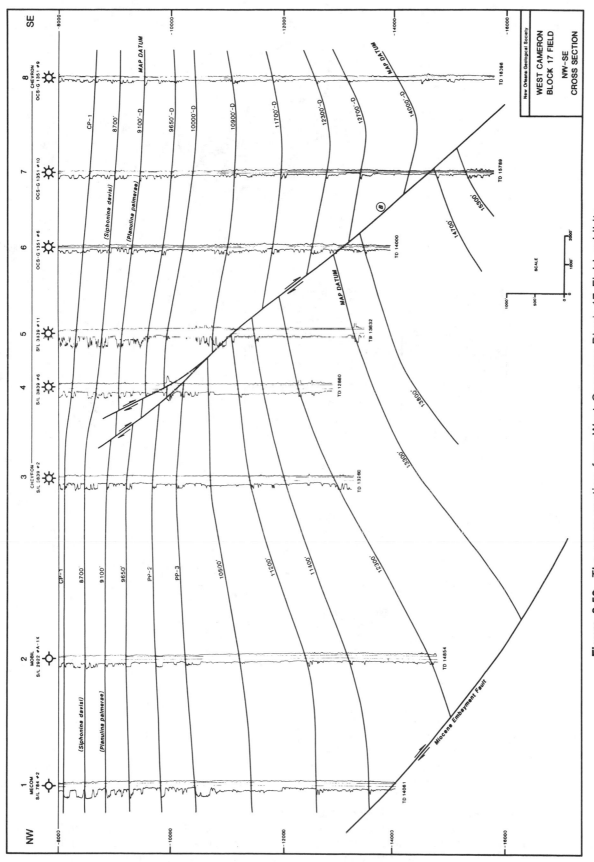

Figure 8-50 The cross section from West Cameron Block 17 Field exhibits structural compatibility in the shallow section and structural incompatibility in the deeper section across the same normal fault. (Published by permission of the New Orleans Geological Society.)

351

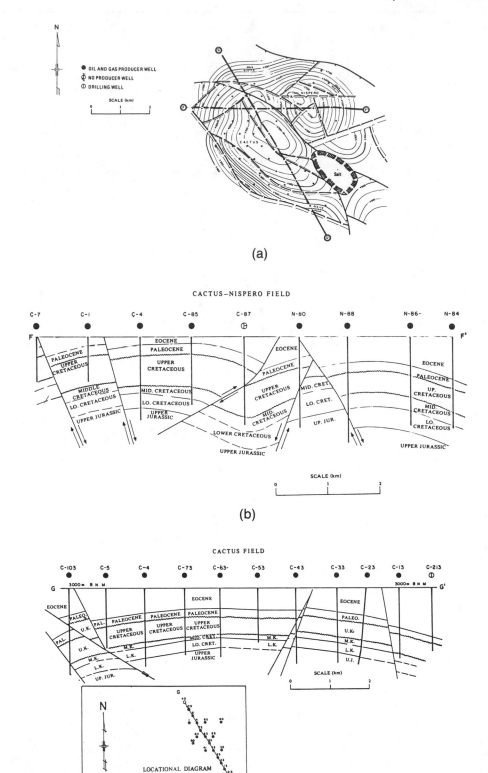

Figure 8-51 (a) Structure map of the Middle Cretaceous, Cactus-Nispero Field, Southern Zone, Mexico. The structure is a piercement salt structure cut by normal and reverse faults. (b) Geologic cross section F-F′ through the Cactus-Nispero Field. (c) Geologic cross section G-G′ through the Cactus Field. (From Santiago 1980. Published by permission of the American Association of Petroleum Geologists.)

A good example of a field exhibiting both compatible and noncompatible structural attitudes across faults, within the same field, is seen in the Cactus-Nispero Field in the Southern Zone, Mexico (Fig. 8-51). Figure 8-51a is a structure map on the middle Cretaceous showing this piercement salt structure cut by both normal and reverse faults. Figures 8-51b and 8-51c are two cross sections through the field, both of which are shown in plan view on the structure map. A careful review of the structure map and cross sections shows that the structure exhibits contour compatibility across most of the faults and therefore the vertical separation should be used to map across these faults. Two areas provide major exceptions. In cross section F-F', a loss of compatibility is present across the reverse fault cutting Well No. C-87. On the structure map, the large normal fault just north of the salt intrusion does not exhibit contour compatibility from the upthrown to downthrown blocks. To contour these two areas accurately, each block must be contoured separately and then integrated with the fault map for each associated fault.

All these examples stress the importance of understanding the three-dimensional geometry of any structure being mapped. The accuracy of the mapping depends upon the use of the correct mapping techniques, which often depends upon the geometry of the structure itself. Different techniques may be used within the same field.

MAPPING TECHNIQUES FOR VARIOUS TECTONIC HABITATS

In this section, we review the mapping techniques applicable to four major tectonic habitats and their associated hydrocarbon traps. These habitats include: (1) extensional, (2) compressional, (3) wrench fault, and (4) diapiric settings. In addition to discussing the various types of hydrocarbon traps in each tectonic setting, we illustrate the fault trace construction and formation contour construction for each of the fault patterns observed in these settings and present a field example where possible.

Extensional Tectonics

For extensional tectonics, we include nongrowth normal faulting and detached listric growth fault systems, since both often occur together within this setting. In many areas of the world, hydrocarbons are trapped upthrown to normal faults. An important exception to this is found in the Gulf of Mexico, where hydrocarbons are commonly trapped downthrown, as well as upthrown to normal faults. Often, these downthrown traps are rollover anticlines associated with normal listric growth faults.

Faults are the primary trapping mechanism for hydrocarbons in extensional habitats, although four-way closures, unconformities, and stratigraphic traps are also important. Our primary discussion centers around five different types of fault patterns: (1) single nongrowth (structural) faults, (2) compensating, (3) bifurcating, (4) intersecting, and (5) growth fault systems.

The construction techniques for all fault patterns that exhibit structural compatibility across a fault are the same as those used for the working model (Fig. 8-21) to arrive at an integrated structure map. For the examples of each fault pattern, we show at least two mapped horizons which have a stratigraphic interval of constant thickness. For the purpose of detailing the correct mapping techniques, the data for the examples are somewhat idealized. However, the techniques are applicable in both simple and complex petroleum geological mapping. Where possible, actual examples for each of the fault patterns are presented.

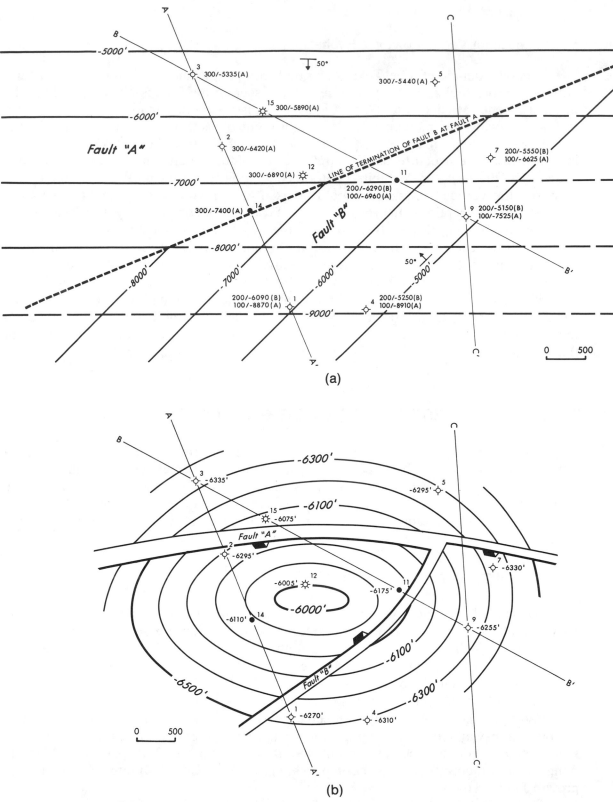

Figure 8-52 (a) Fault surface map for a compensating fault system including Faults A and B. (b) Integrated structure map on the 6000-ft Sand. (c) Fault and structure maps (6000-ft Sand) superimposed to illustrate the details of the integration technique. (d) Integrated structure map on the 7000-ft Sand. Upthrown and downthrown restored tops in Wells No. 9 and 14 were used to aid in the structural interpretation. (e) Fault and structure maps (7000-ft Sand) superimposed to show the accurate construction of the faulted 7000-ft Sand map using the integration technique. (f) Overlay of fault traces for both the 6000-ft and 7000-ft Sands. The figure shows that all fault intersections fall on the line of termination. The location of the fault intersections on any structural horizon can be predicted from the strike of the line of termination.

OVERLAY OF FAULT AND STRUCTURE MAP - 6000' SAND

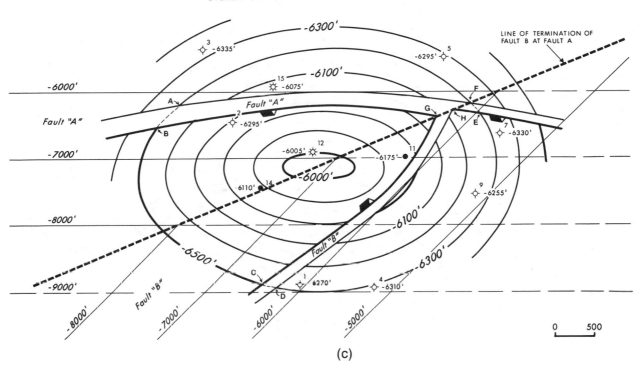

(c)

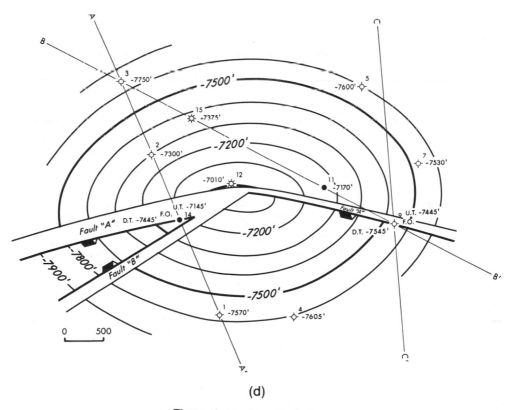

(d)

Figure 8-52 (continued)

OVERLAY OF FAULT AND STRUCTURE MAP - 7000' SAND

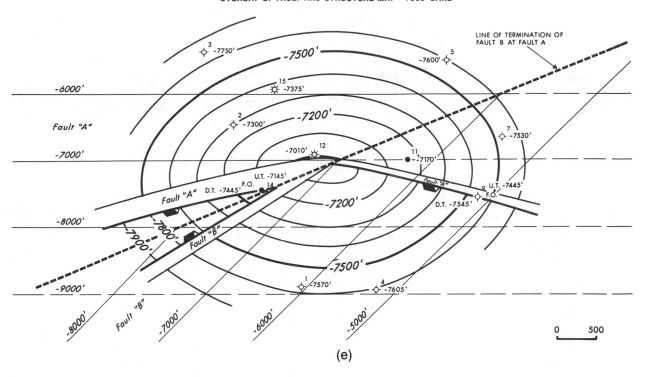

(e)

OVERLAY OF FAULT MAP AND STRUCTURE MAPS

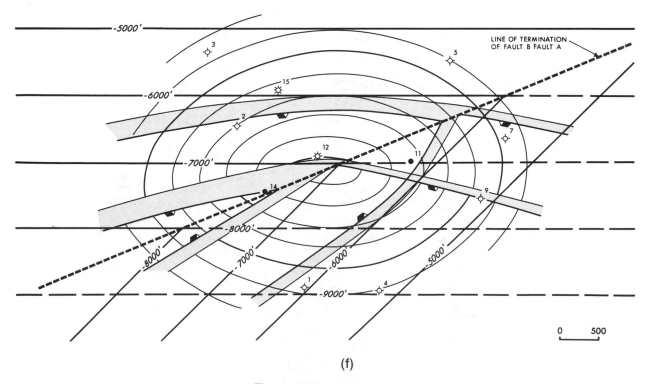

(f)

Figure 8-52 (continued)

Compensating Fault Pattern. Figure 8-52a shows the fault contour map for Faults A and B, which together represent a typical compensating fault pattern. If necessary, refer again to Chapter 7 and review the major aspects of this fault pattern. Several key points must be re-emphasized here. First, the line of termination must be shown on the fault map; its importance will become clear as we review the structure maps on the 6000-ft and 7000-ft Sands. Second, vertical separation is conserved across fault intersections. Observe that throughout the area where two faults are present (i.e., the area south of the line of termination of Fault B at Fault A in our example), Fault A has 100 ft of vertical separation and Fault B has a vertical separation of 200 ft. Only Fault A, with a vertical separation of 300 ft, is present north of the termination line.

Figure 8-52b is a completed structure contour map on the 6000-ft Sand. The structure is a faulted anticline. Figure 8-52c shows the fault map for Faults A and B superimposed onto the completed structure map for the 6000-ft Sand. Using this figure, *we review the construction of the upthrown and downthrown traces for each fault, the construction of the intersection of the two faults, and the proper bed contour construction across each fault using the vertical separation (missing section from each wellbore).*

Begin with the northern upthrown fault block and the −6200-ft contour and follow its complete construction over the entire structure. The intersection of this structure contour with the fault contour of the same value defines the upthrown trace of Fault A at point A. The contour is projected through the fault, as if the fault were not there and adjusted in the downthrown block by the size of the fault (300 ft) to become the −6500-ft contour. The intersection of this −6500-ft structure contour with the fault contour of the same value defines the downthrown trace of Fault A at point B. The −6500-ft contour downthrown to Fault A crosses the axis of the structure and intersects with the −6500-ft contour for Fault B at point C defining the downthrown trace of Fault B at this location. The contour is now projected through Fault B as if the fault were not there and then adjusted by 200 ft (representing the vertical separation for Fault B) in the upthrown block of Fault B to become the −6300-ft contour. This −6300-ft structure contour intersects the −6300-ft fault contour at point D, defining the upthrown trace of Fault B at this location. This contour once again crosses the axis of the structure, this time on the eastern side of the anticline, and intersects with the downthrown trace of Fault A at point E. Project the contour through the fault gap of Fault A and adjust its value by 100 ft for the vertical separation at point F. Finally, continue this contour in the upthrown block of Fault A until it intersects with point A, which was our starting point. This procedure is continued for each structure contour on the map to arrive at the integrated structure map shown.

The heavy dashed line (Fig. 8-52c) striking southwest-northeast is the line of termination of Fault B at Fault A. *The intersection of the fault traces for both faults* falls on this line of termination. For this map, there are two trace intersections: (1) the intersection of the downthrown traces of Faults A and B shown at point G, and (2) the intersection of the upthrown trace of Fault B with the downthrown trace of Fault A shown at point H. Since all trace intersections must fall on this line, it is very important to place this termination line on the fault map. By following the strike of the line, we can predict the location of the fault intersections for other horizons either shallower or deeper in the subsurface because all intersections must fall on this line. Observe that north of the line of termination, only Fault A is present, with a size of 300 ft, while south of the line, both Faults A and B, with vertical separations of 100 ft and 200 ft, respectively, are present. Take a minute and test the additive property of faults around the fault intersection on the completed structure map.

This construction technique of integrating the fault and structure maps removes all the guesswork with regard to: (1) fault trace construction; (2) the location of all faults on the structure map; (3) the location of all fault intersections; and (4) the width of the fault gaps. The integration of the two maps constructs all these features automatically without the need for guesses or estimates. Take one or two additional structure contour lines and go through the same procedure just described to make sure you fully understand the technique for integrating a fault map with a structure map.

Figure 8-52d is a structure map on the deeper 7000-ft Sand, which shows that the faults have migrated with depth. The technique of construction is exactly the same as presented for the 6000-ft Sand. In Fig. 8-52e, we again superimpose the fault map onto the completed structure map. Choose several structure contours and follow each across the mapped area to review the construction techniques as discussed for the preparation of the completed structure map for 6000-ft Sand.

Finally, the completed fault traces for both the 6000-ft and 7000-ft Sands, and the general trend of the structure contours is superimposed onto the fault map for Faults A and B in Fig. 8-52f. Looking at each set of fault traces (those for the 6000-ft and 7000-ft Sands), can you detect any difference between them? Use 10-point proportional dividers or an engineer's scale to measure the width of the fault gap for Fault A on both horizons. Notice that the gap is wider for Fault A on both sides of the line of termination on the structure map for the 7000-ft Sand, when compared with the fault gap on the structure map for the 6000-ft Sand. Remember the general rule of fault gap width presented earlier. **A fault dipping updip (in the opposite direction of bed dip) will have a thinner fault gap than a fault of the same or similar size dipping downdip (in the same direction as the bed dip).** Notice that Fault A on the 6000-ft Sand is on the northern side of the structural axis and is dipping generally in the opposite direction to the formation. On the deeper

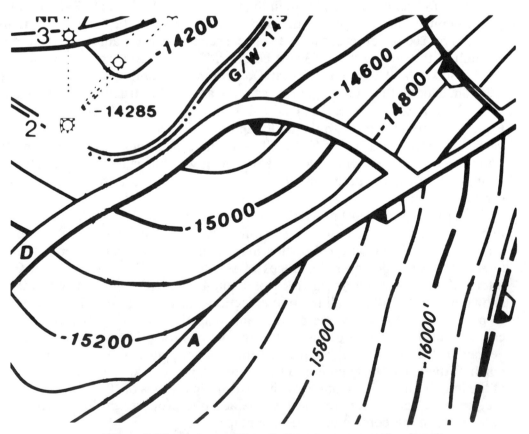

Figure 8-53 Compensating fault system or mapping bust?

structure map for the 7000-ft Sand, Fault A has crossed the axis of the structure and is now dipping more or less in the same general direction as the formation. Therefore, the fault gap is wider when mapped on the structure map for the 7000-ft Sand.

With the technique of integrating the fault and structure maps, **this change in fault gap width is automatically built into the integration.** It is unlikely that such a change in gap width could have been intuitively incorporated into the completed structure contour maps had the fault and structure map integration technique not been used. For this example, the change in gap width is significant; but, depending upon the configuration of the structure and the geometry of the fault, the change in width can be minor. Let the integration technique be the tool used to establish fault gap widths.

The completed structure map in Fig. 8-22 illustrates two good compensating faults to the large down-to-the-east normal fault. Fault surface maps were prepared for each fault and they were integrated with the structure map to arrive at the completed map.

Look at the completed structure map in Fig. 8-53. What type of fault pattern is represented by the intersection of Faults A and D? Is it a compensating fault pattern or is it a mapping error?

Bifurcating Fault Pattern.   The fault map in Fig. 8-54a shows a fault contour map for Faults A and A-1, which together form a bifurcating fault pattern. Refer again to

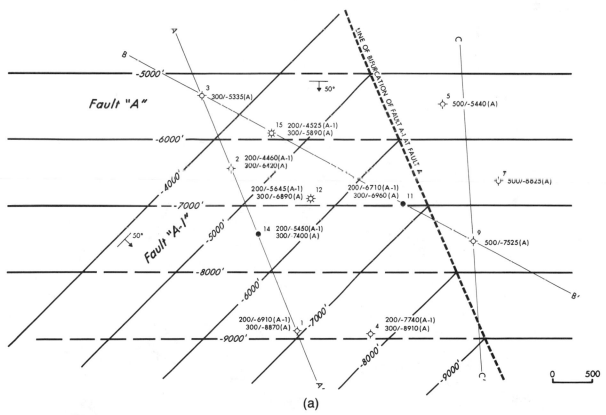

(a)

Figure 8-54   (a) Fault surface map for a bifurcating fault system including Faults A and A-1. (b) Integrated structure map on the 6000-ft Sand. (c) Fault and structure maps (6000-ft Sand) superimposed to show the accuracy of the integration technique to position the fault traces and intersections, and determine the width of the fault gap. (d) Integrated structure map for the 7000-ft Sand. (e) The fault map superimposed onto the 7000-ft structure map. (f) Overlay of fault traces on both the 6000-ft and 7000-ft Sands. As with the compensating fault pattern, all intersections fall on the line of bifurcation.

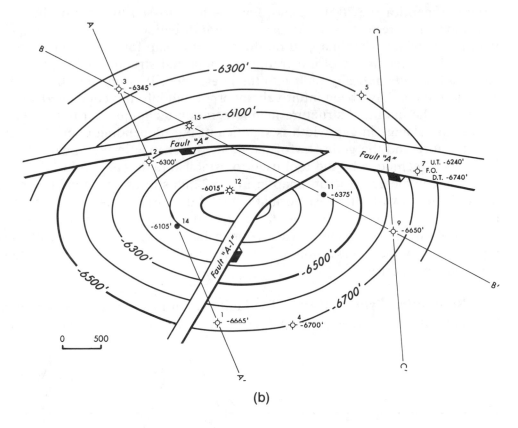

(b)

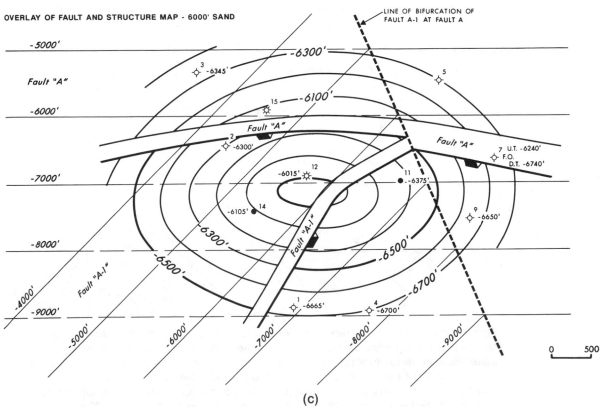

OVERLAY OF FAULT AND STRUCTURE MAP - 6000' SAND

(c)

Figure 8-54 (*continued*)

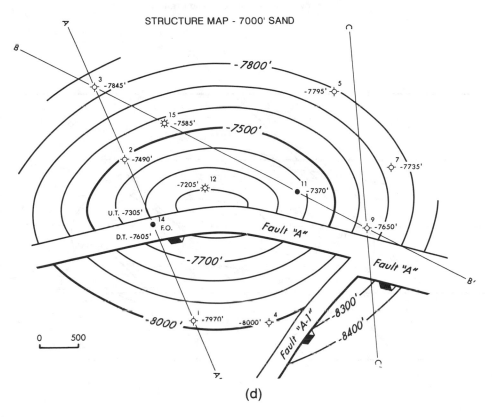

(d)

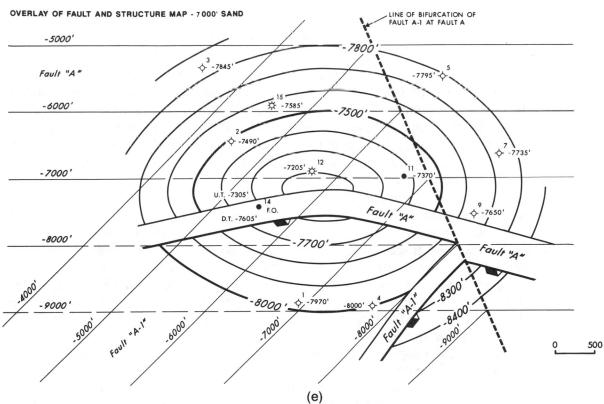

(e)

Figure 8-54   *(continued)*

OVERLAY OF FAULT MAP AND STRUCTURE MAPS

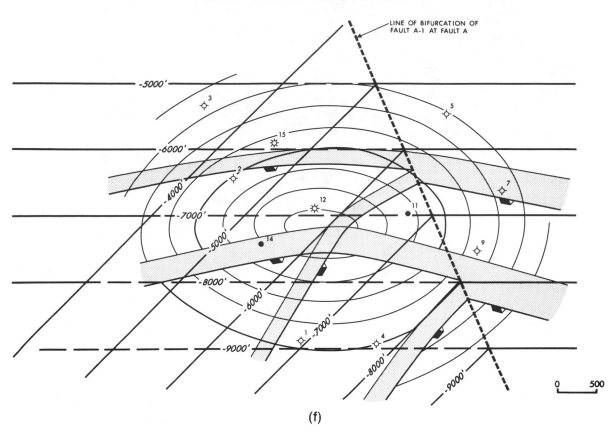

(f)

Figure 8-54   *(continued)*

Chapter 7, if necessary, to refresh your memory on the characteristics of a bifurcating fault pattern. Figure 8-54b is the completed structure map on the 6000-ft Sand. Figure 8-54c shows the fault map for Faults A and A-1 superimposed onto the structure map for the 6000-ft Sand. Choose several structure contours upthrown to Fault A and go through the integration technique to check the position of the fault traces and the fault intersections, the width of the fault gaps, and the contour construction for the 6000-ft Sand across the two faults.

Fault traces are relatively easy to delineate when there are a number of structure contours intersecting fault contours, as in the case shown in Fig. 8-54; however, in areas where fault contours are semi-parallel to the structure contours, it is more difficult to delineate the fault traces. In these situations, you must interpolate depth values to precisely delineate the position of the fault traces.

Figure 8-54d is the completed structure map on the 7000-ft Sand. Figure 8-54e shows the fault map superimposed onto the structure map. Observe that the trace intersections for Faults A and A-1 fall directly on the line of bifurcation. Figure 8-54f shows the fault traces, as seen on the completed structure maps for both the 6000-ft and the 7000-ft Sands with the fault map superimposed. Notice how the fault intersection migrates from north to south along the path of the line of bifurcation. Also observe that the width of the fault gap for Fault A is wider on the structure map for the 7000-ft Sand than it is on the 6000-ft Sand structure map. This occurs here for the same reason discussed for the compensating fault example.

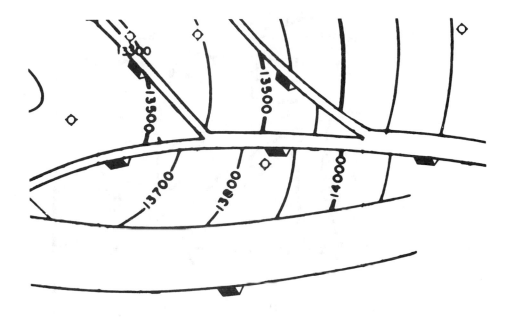

Figure 8-55 Portion of the structure map on the Nodosaria Sand in South Lake Charles Field illustrating a bifurcating fault system. (Published by permission of the Lafayette Geological Society.)

Figure 8-55 is a portion of the structure map on the Nodosaria Sand in South Lake Charles Field in Louisiana, and Fig. 8-56 is a portion of the U Sand structure map from West Cameron Block 192 Field, Offshore Gulf of Mexico. Both figures show a good example of the mapping of a bifurcating fault system. The fault system in South Lake Charles actually has two bifurcations. Take a minute to check the construction of these completed maps in the area around the bifurcations, by applying the additive property of faults.

Intersecting Fault Pattern. Figure 8-57a illustrates a typical intersecting fault pattern composed of Faults A and B. Cross section A-A' is shown in Fig. 8-57b. In Chapter 7 in the section dealing with the intersecting fault pattern, the fault maps were contoured with the assumption that neither fault surface was offset (Fig. 8-57b). In other words, the two faults were considered contemporaneous rather than of two different ages. The fault traces for the structure maps shown in this figure were constructed with this assumption. Notice, however, in Figs. 8-57c–8-57f **that all the fault traces are offset.** Figure 8-57d shows, for example, that although the fault surfaces themselves are not offset, the fault traces on the completed structure maps show **significant offsets for all the fault traces** at the fault intersections. The positions of the offset traces are accurate for the constructed

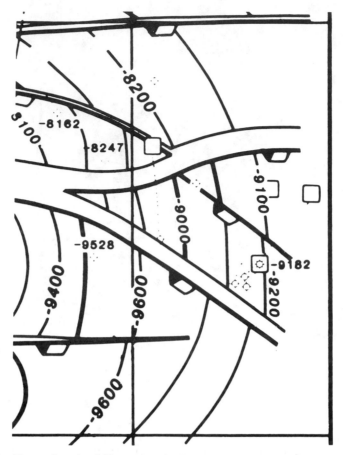

Figure 8-56 Example of a bifurcating fault system from West Cameron Block 192 Field. (Published by permission of the New Orleans Geological Society.)

fault surface maps. If the faults are of two different ages, however, then the completed maps are subject to error around the intersection. If the faulting is contemporaneous, then the traces are accurate, as shown on the completed structure maps.

Compare the completed structure maps in Figs. 8-57d and 8-57f. There is a pronounced change in the configuration of the fault traces shown on these two maps. The change is the result of the intersecting geometries of the two sands with Faults A and B at different depths. Without the integration of the fault and structure maps, the accurate delineation of these fault traces, particularly for the 7000-ft Sand, would be difficult, if not impossible.

Finally, observe that all the fault trace intersections fall on the line of intersection. In this case of intersecting faults, there are four fault intersections (Figs. 8-57e and 8-57g). This construction again emphasizes the necessity of placing the line of intersection on the completed fault maps.

Seismic data may be necessary to determine the relative ages of faults that did not form contemporaneously, for construction of a more accurate fault surface map showing the offset of the earlier fault by the later one. Also, a graphical solution to the construction of the fault surfaces and their effects on stratigraphic horizons can be undertaken. A graphical technique was presented by George Dickinson in the Bulletin of the American Association of Petroleum Geologists in May, 1954. The method is not detailed here; however, if it is absolutely necessary to prepare precise fault and structure maps reflecting the accurate geometry of intersecting faults of different ages, then the Dickinson method should be considered.

Figure 8-58, from Dickinson's paper, shows a completed structure map for a horizon above a deep-seated salt dome in Texas. Such a complex intersecting fault pattern is not uncommon in areas of salt tectonics. In this example, Fault 3 is indicated as a through fault. A test was planned for the upthrown block to the east. In order to test the block at an optimum structural position, it was necessary to accurately locate the position of the intersected Fault 5, so the graphical solution outlined by Dickinson was used to prepare the fault and structure maps.

In addition to the intersection of normal faults, there are areas around the world such as western Texas, California, southern Oklahoma, and the southern zone of Mexico, for example, that exhibit intersecting normal and reverse fault relationships. The Dickinson method is also very helpful in developing a structural picture of these intersecting faults. For many situations, however, particularly those involving small faults, the short-cut method of assuming contemporaneous faulting, shown earlier in this section, is often adequate for the construction of the fault and completed structure maps.

We cannot leave this subject of intersecting faults without covering an important misunderstanding with regard to intersecting fault traces on completed structure maps. There seems to be a popular belief that when two faults of different ages intersect, the

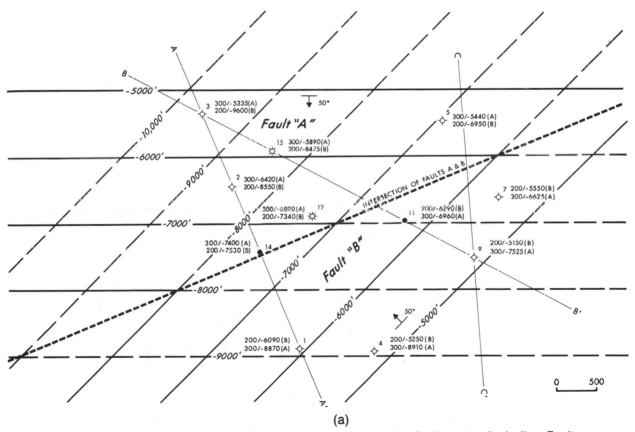

(a)

Figure 8-57   (a) Fault surface map for an intersecting fault system including Faults A and B. Fault B does not terminate at the line of intersection. (b) Cross section A-A′ illustrates the shortcut method of constructing the fault map as if the fault surfaces are not offset. (c) Integrated structure map on the 6000-ft Sand. (d) Completed structure map on the 6600-ft Sand. Observe that both fault traces are offset at the intersection. (e) Fault surface map and the structure map on the 6600-ft Sand superimposed to show the accuracy of fault trace construction. (f) Completed structure map on the 7000-ft Sand. (g) Overlay of the fault surface map and the 7000-ft structure map. Notice that all fault trace intersections fall directly on the line of intersection.

CROSS SECTION A - A'

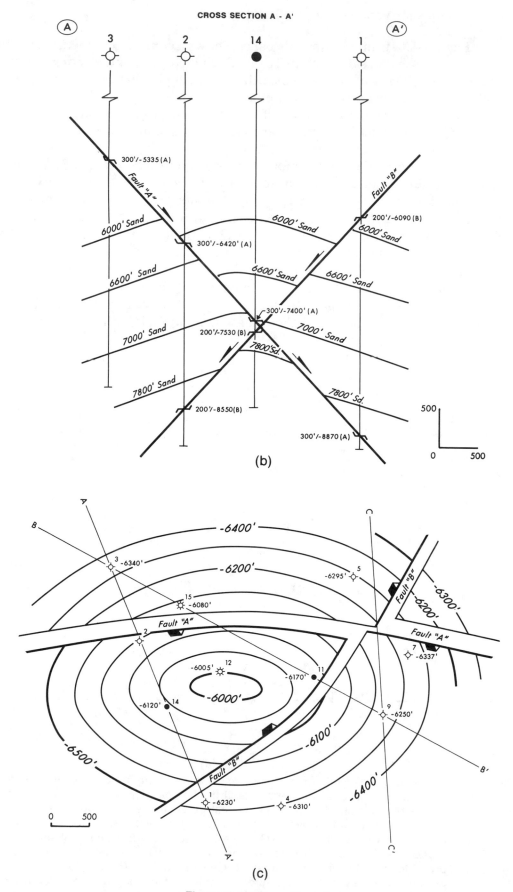

(b)

(c)

Figure 8-57 (*continued*)

366

STRUCTURE MAP - 6600' SAND

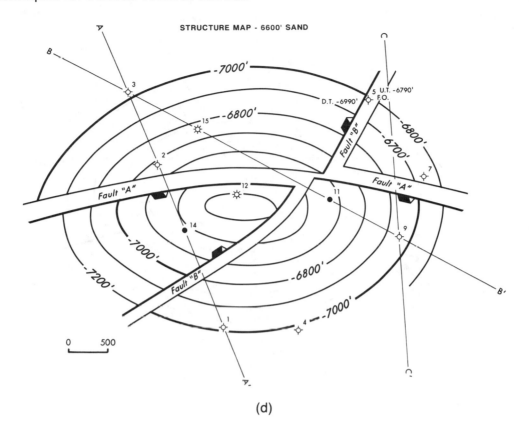

(d)

OVERLAY OF FAULT MAP & STRUCTURE MAP - 6600' SAND

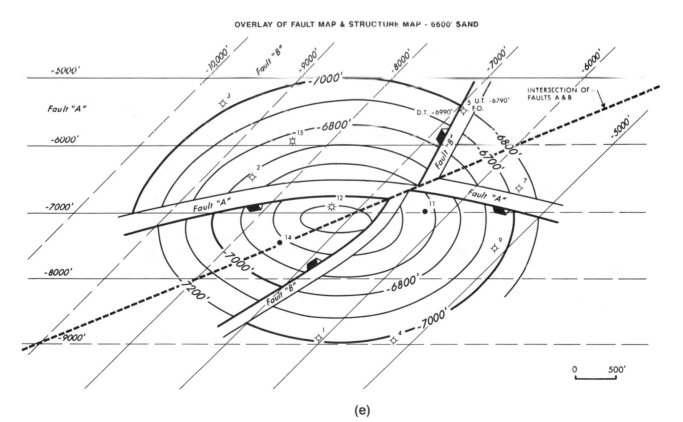

(e)

Figure 8-57   (*continued*)

STRUCTURE MAP - 7000' SAND

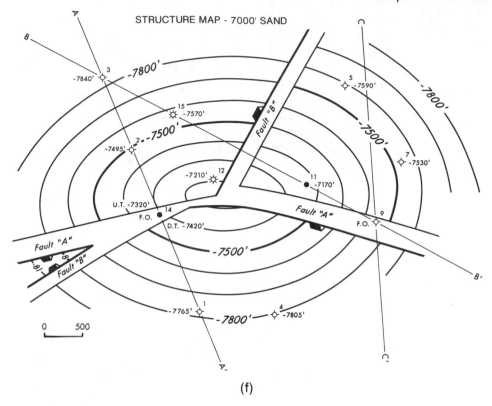

(f)

OVERLAY OF FAULT AND STRUCTURE MAP - 7000' SAND

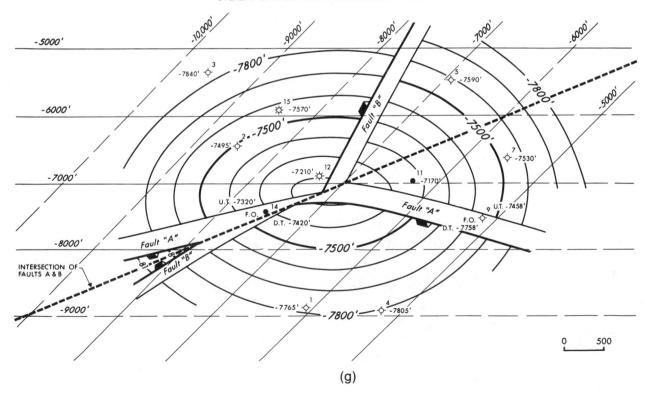

(g)

Figure 8-57 (*continued*)

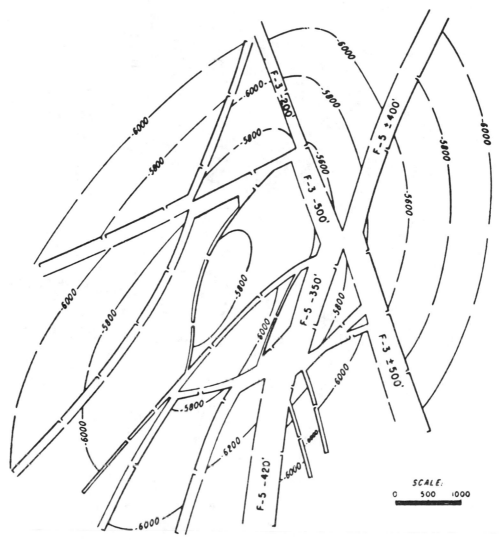

Figure 8-58 Structure contour map over a deep-seated salt dome in Texas illustrating several intersecting faults. (From Dickinson 1954. Published by permission of the American Association of Petroleum Geologists.)

fault trace of the intersected or older fault is not offset on a completed structure map, as shown in Fig. 8-59. We strongly emphasize that this idea is **not** correct for most cases. *The intersections of the intersected or older fault on any mapped horizon are always offset both horizontally and vertically, except in two special cases, where the offset is entirely vertical.* These exceptions occur where (1) either the beds are horizontal and the faults intersect at right angles (Fig. 8-59), or (2) the intersection of the (older) intersected fault and the mapped horizon are perpendicular to the strike of the intersecting (younger) fault (Dickinson 1954).

A structure map with intersecting faults showing the intersected (older) fault traces unoffset must meet one of the two special cases or it is constructed incorrectly. Usually this type error occurs because: (1) the mapper does not understand the relationships of intersecting fault construction, or (2) no fault maps were prepared and used to integrate with the structure map to arrive at an accurate solution for the intersecting faults. Figure 8-60 is a portion of a completed structure map with an intersecting fault pattern. A review of the map quickly shows that the pattern does not meet the requirements for either of

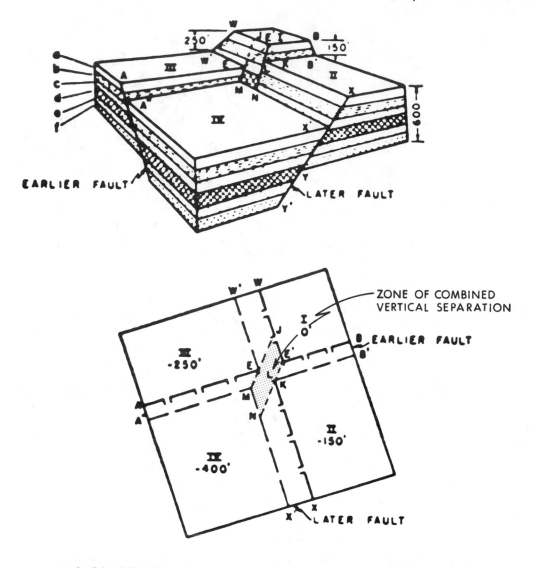

SUBSURFACE MAP ON TOP OF UPPER LAYER OF BLOCK.

Figure 8-59 Block model and map view of intersecting normal faults in horizontal strata. Fault dip is 60 deg and the movement is all dip slip. (Modified from Dickinson 1954. Published by permission of the American Association of Petroleum Geologists.)

the two special cases; however, the fault traces were prepared showing no offset for the assumed older Fault A. We therefore conclude that this map was prepared without the aid of a fault map and is incorrect.

If a pair of interesting fault traces is shown on a completed structure map with neither trace offset and the traces resemble a "*cross*" (Fig. 8-61), the construction is incorrect. Such errors as the ones shown in Figs. 8-60 and 8-61 can place suspicion on the rest of the structural interpretation.

If the area you are working is complexly faulted with intersecting faults and extreme accuracy is required, it is advisable to become familiar with the Dickinson method for resolving the geometry of fault intersections.

Combined Vertical Separation: In Chapter 7, we discussed the combined vertical separation that results when two faults of different ages intersect. In Fig. 8-59, the zone of combined vertical separation is shown in map view for this intersecting fault example

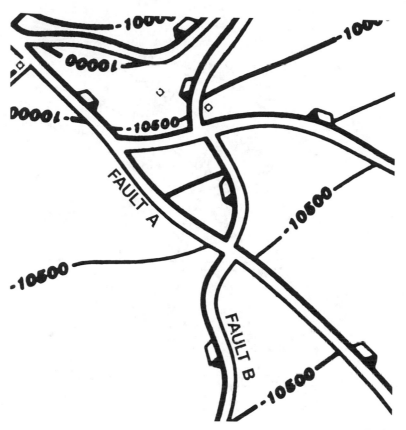

Figure 8-60 Portion of a structure map showing intersecting faults.

and is the area between J, E', N, and M. A well drilled within this zone will cross only one of the two intersecting faults, but the interval shortening or missing section is equal to the algebraic sum of the fault cuts for both faults. For this example, the earlier fault has a size of 150 ft and the later fault is 250 ft; therefore, the size of a fault cut in the area of combined vertical separation is 400 ft.

Growth Faults. A growth fault is a special type of normal fault. A growth fault is a normal fault which is contemporaneous with sedimentation. There are several primary characteristics of growth faults:

1. They are usually arcuate in shape and concave toward the basin.
2. Fault dip decreases with depth, often becoming a bedding plane fault (Chapter 9). We refer to this shape as *listric*, based on the Greek word *listron*, or shovel, because of its curved shape in cross section.
3. The size of the fault normally increases with depth. Displacements of several thousand feet are not uncommon.
4. The time equivalent strata in the downthrown block are thicker than those in the upthrown block (see Chapter 7, Estimating the Size of Growth Faults and Expansion Index for Growth Faults).
5. A rollover anticline often develops in the downthrown block of a growth fault as a result of collapse of the hanging wall block toward the fault.
6. Secondary faulting is normally associated with growth faults. These faults can be synthetic (dipping parallel to the master fault) or antithetic (dipping toward the master fault).

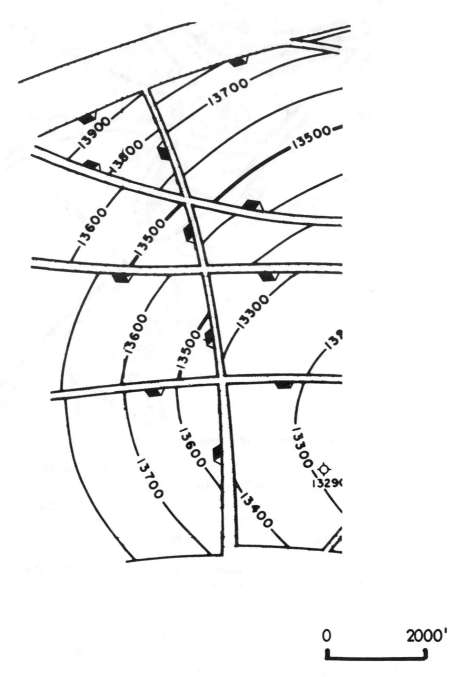

0   2000'

Figure 8-61   Completed structure map showing several intersecting faults with each of the trace intersections forming a "cross." (The final construction is incorrect. No fault traces are offset.)

A majority of growth faults dip toward the basin (south, in the Gulf of Mexico) and strike parallel or semi-parallel to the coastline. However, growth faults can dip in any direction, including up-to-the-coast. Along strike, most growth faults are generally concave basinward.

Faults are not perfect planar surfaces of slip; they generally have undulations of some type. As two fault blocks slip past one another, there must be deformation in at least one fault block, because rocks are not strong enough to support large voids. This

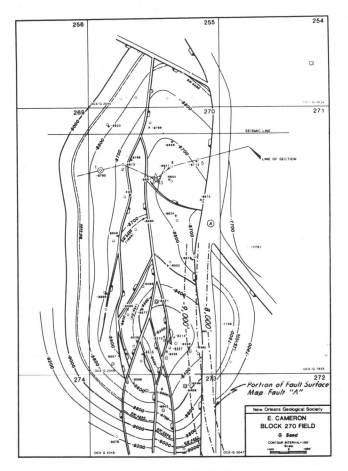

Figure 8-62 Structure map on the G Sand Unit, East Cameron Block 270 Field. A portion of the fault surface map for Fault A is superimposed on the structure map. The primary hydrocarbon trap is a faulted rollover anticline formed in response to Growth Fault A. (Published by permission of the New Orleans Geological Society.)

is the main reason why many major rollover folds exist within the hanging wall fault blocks, formed by bending of the fault blocks as they slip over nonplanar (listric) fault surfaces. This mechanism of folding is called **collapse folding** (Hamblin 1965). The rollover anticline, associated with listric growth faults, is one of the most widely recognized folds, as seen in such areas as the Gulf of Mexico (United States), Brunei, and Niger Delta.

For contouring purposes, a growth fault may or may not show structural compatibility across the fault. The primary factors are the size of the fault and the amount of deposition in the downthrown block. Usually the larger the fault, the less chance of compatibility. If structural compatibility does exist, the missing section (vertical separation) can be contoured across the fault; but, if compatibility is absent, then each fault block must be contoured separately. At times, only the upthrown or downthrown block of a growth fault is mapped, usually because: (1) of a lack of good control or correlation across the fault, or (2) only one block is productive and that is the only block mapped.

Figure 8-62 is a structure map on the G Sand in East Cameron Block 270 Field Offshore, Louisiana. This field is productive from 19 sands trapped both upthrown and downthrown to the west-dipping growth Fault A. Rollover into Fault A has formed a north-south trending anticline which is the primary hydrocarbon trap. This field has produced in excess of 38 million barrels of oil and condensate and nearly 1 trillion cu ft of gas.

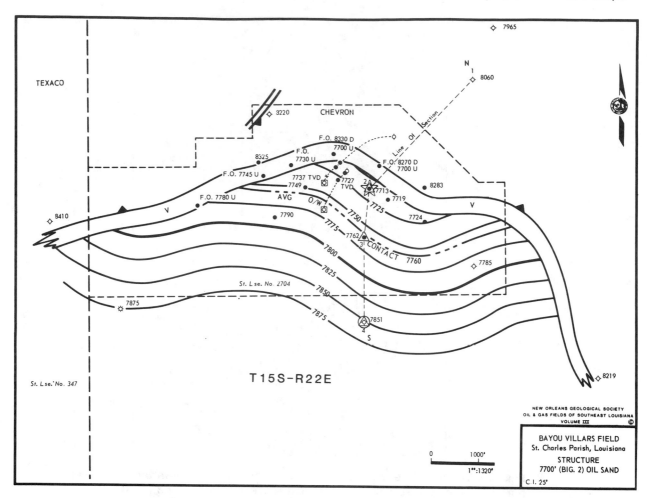

Figure 8-63 Structure map on the 7700-ft (BIG.2) Oil Sand upthrown to Fault V in Bayou Villars Field, St. Charles Parish, Louisiana. Fault V is a growth fault. (Published by permission of the New Orleans Geological Society.)

Superimposed on the southern half of the structure map is a portion of the fault surface map for Fault A from −8000 ft to −9000 ft. The structure contours drawn up-thrown and downthrown to the fault, in this area, indicate reasonable compatibility in structure across this growth fault, which has a size of about 800 ft at this level. The existence of this structural compatibility does not, however, ensure such compatibility along fault strike nor with depth (recall the cross section in Fig. 8-50 for West Cameron 17 Field). Notice that the rollover anticline is complexly faulted with secondary synthetic and antithetic faults. Although the fault maps for these faults are not published, it is easy to recognize that there is good structural compatibility across all the secondary faults; therefore, the vertical separation is used to contour across these faults.

We mentioned earlier that in most areas of the world, hydrocarbons are trapped upthrown to normal faults and that the US Gulf of Mexico was an exception in that downthrown blocks with rollover anticlines, as well as upthrown fault blocks, are often productive. There are growth faults, however, that do not develop rollover anticlines and are not productive in the downthrown fault block. Figure 7-13, in Chapter 7, is a fault surface map of growth Fault V in the Bayou Villars Field, St. Charles Parish, Louisiana. Fault V is an up-to-the-coast fault dipping to the north. For the most part, north-dipping faults are not considered very prospective. A number of such traps are produc-

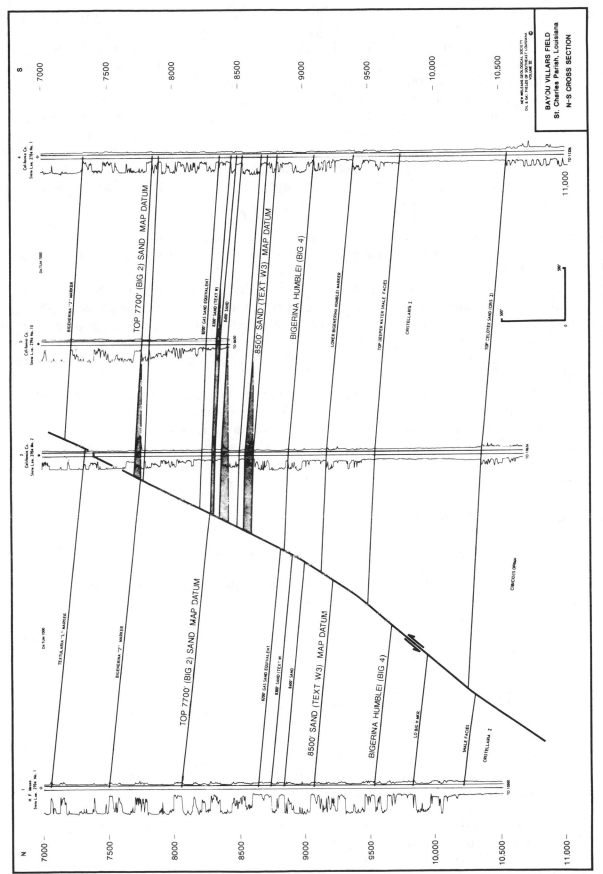

Figure 8-64 North-south cross section through Bayou Villars Field. The structure exhibits good continuity across the growth fault with little, if any, associated rollover downthrown to the fault. (Published by permission of the New Orleans Geological Society.)

375

tive, however, and the Bayou Villars Field is an excellent example. The field has produced in excess of 13 million barrels of oil and condensate from five Middle Miocene Sands.

Fault V is a growth fault with a vertical separation of 130 ft at a depth of −2400 ft. The deepest well control indicates that the vertical separation increases to 600 ft at a depth of −9000 ft. The dip of the fault measures 63 deg shallow and decreases to about 50 deg at a depth of −9000 ft. Figure 8-63 is the structure map on the 7700-ft (Big. 2) Sand. Notice on the fault map (Fig. 7-13) and on the structure map that the fault surface has a **double bulge** instead of exhibiting a simple, smooth arcuate shape. This bulge provides room for enough additional structure, upthrown to the fault, to become a good trap for hydrocarbons.

Notice that there are five wells faulted out in this sand. The restored tops have been estimated in each well and used to aid in the construction of the structure map. As mentioned earlier, the use of restored tops is often vital to development of a reasonable interpretation.

Figure 8-64 is a cross section through the field (see structure map for the location of the line of section). Although there is growth in the downthrown fault block, evidence suggests little, if any, rollover in the downthrown block. There is also very good structural compatibility across the fault, which implies that were there enough well or seismic control in the downthrown block, the vertical separation could be used to contour across this growth fault.

One final note on growth faults. It is often observed that the most productive interval in a growth fault complex is the stratigraphic section deposited during the most active fault movement. Therefore, when prospecting around growth faults, it is important to understand the history of the fault movement. The technique of plotting a growth fault's *Expansion Index* (see Chapter 7) is most helpful in evaluating the fault history and its potential for hydrocarbon accumulations.

Diapiric Salt Tectonics

A *diapir* is defined as a body of rock that moves upward by piercing and displacing the overlying strata. The word comes from the Greek verb diaperein, meaning "to pierce," and was first used in the Carpathian Mountains of Romania. The most common use of the word is in association with salt and shale intrusions. Diapiric salt intrusions are associated with large accumulations of hydrocarbons in many areas around the world including the US Gulf Coast, the southern zone of Mexico, Gabon, Senegal, the Canadian Arctic, the North Sea, northern Germany, Romania, the Zagros Mountains of Iran, and the Ukraine of the Soviet Union. These salt intrusions, commonly called *salt domes*, are often structurally and stratigraphically very complex. They come in all shapes and sizes from small, needle-like spines less than 2000 ft across, such as the Rabbit Island Salt Spine (Fig. 8-65), to very large salt massifs such as the Marchand-Timbalier-Caillou Island Salt Massif, which is more than 27 mi long and 13 mi wide at a depth of 20,000 ft (Fig. 8-66). Regardless of the actual size of the salt mass, most are associated with significant quantities of hydrocarbons. For example, the Rabbit Island Field has estimated reserves of 1.5 trillion cu ft of gas and 55 million barrels of condensate and oil. The fields associated with the Marchand-Timbalier-Caillou Island Salt Massif have combined ultimate reserves of over 500 million barrels of oil. Figure 8-67 shows the extent of salt structures in the northwestern Gulf of Mexico and adjacent interior basins.

Several types of normal faulting are commonly associated with most piercement, as well as many deep-seated, salt structures. If the regional stress field is nearly isotropic, the normal faults develop in an outward radiating pattern from the salt structure, called

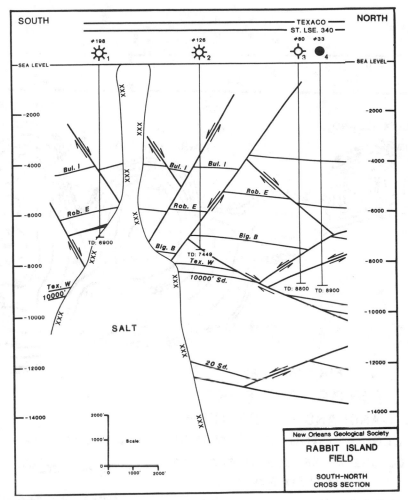

Figure 8-65 Cross section showing the needle-like Rabbit Island Salt Spine which is less than 2000 ft across to a depth of about 7000 ft. (Published by permission of the New Orleans Geological Society.)

radial faults. If the regional stress field is more anisotropic, or the salt exhibits a variable growth history in different areas around the dome, the fault pattern may have a more preferred orientation, resulting in *peripheral faults* that often are parallel to the face of the salt or ride down the flank of the salt.

Growth of a salt structure causes the salt to pierce progressively shallower strata. The overlying strata are first pushed upward, stretched, and finally pierced by the salt. Graben blocks often form in the sediments over the salt intrusion because of extension. The dip of associated faults can be toward the dome, as well as away from it. Although not directly associated with the salt uplift, regional growth faults are often found on or near salt structures. In certain areas of the world, salt intrusions are associated with reverse faults, as well as normal faults such as the Cactus-Nispero Field in the Southern Zone, Mexico (Fig. 8-51).

Hydrocarbon Traps. Intrusive salt structures have an extremely varied and complex geometry resulting in numerous types of hydrocarbon traps. Any single salt dome may have more than one type of trap depending upon the history of the salt structure and surrounding sediments. Figure 8-68 (Halbouty 1979) shows an idealized section

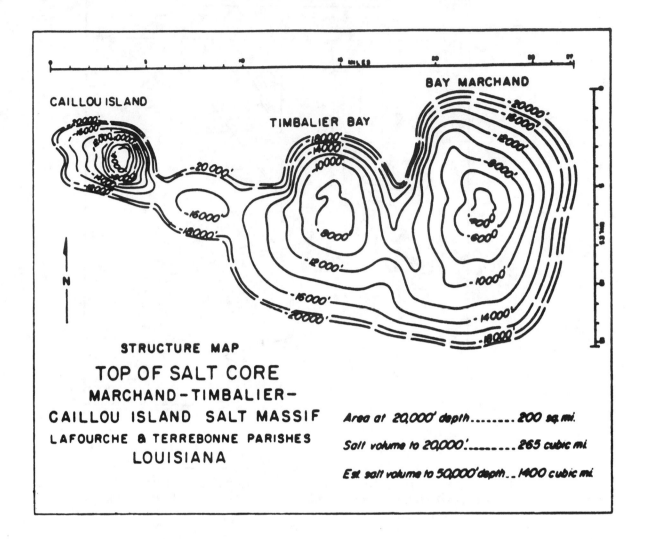

Figure 8-66 Structure map on the top of the Marchard-Timbalier-Calliou Island Salt Massif. (From Atwater and Forman 1959. Published by permission of the American Association of Petroleum Geologists.)

through a salt dome illustrating the more common types of hydrocarbon traps, including: (1) a simple domal anticline, (2) graben fault traps over the dome, (3) porous cap rock (limestone or dolomite), (4) flank sand pinchouts, (5) traps beneath an overhang, (6) traps against the salt itself, (7) unconformities, (8) fault traps downthrown away from the dome, and (9) fault traps downthrown toward the dome. Fault traps like 8 and 9 can also be considered first fault out of the basin type traps. In addition to the traps shown in Fig. 8-68, radial and peripheral faults also serve as excellent hydrocarbon traps.

Considering their complexity, salt structures require very precise and detailed mapping in order to exploit all the hydrocarbon potential. When working with salt-related

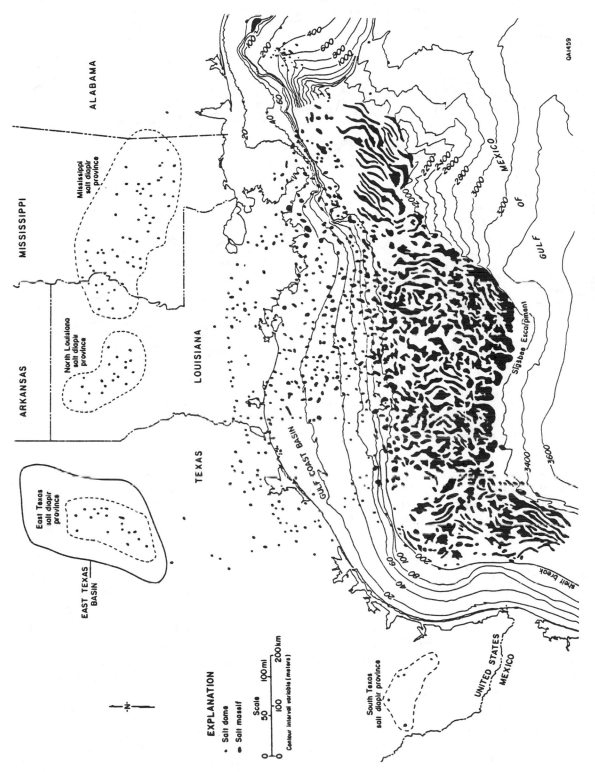

Figure 8-67 Salt structures in the Gulf of Mexico and adjacent interior basins. (From Seni and Jackson 1983. Published by permission of the American Association of Petroleum Geologists.)

379

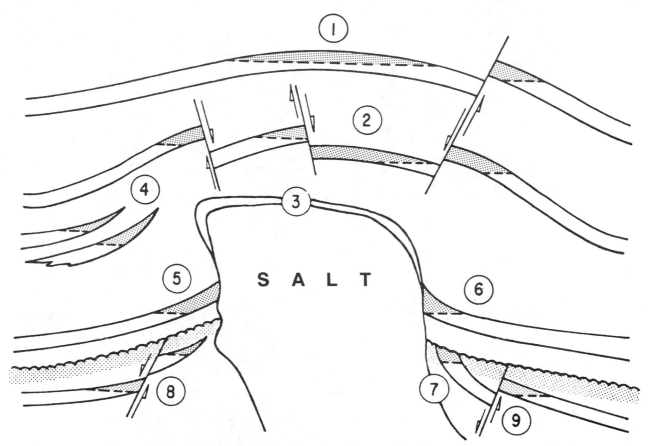

Figure 8-68 An idealized diapiric salt structure showing common types of hydro-carbon traps. (From *Salt Domes,* by Michel T. Halbouty. Copyright 1979 by Gulf Publishing Company, Houston, Texas. Used with permission. All rights reserved.)

structures, maps are needed for the salt itself, and for salt/sediment, salt/fault, and fault/sediment interfaces, in addition to associated unconformities and top of diapiric shale, if present.

Contouring the Top of Salt. No special techniques are required to contour the top of salt. The key to making a good salt map is having sufficient data. Often only a limited number of wells penetrate salt, and commonly the quality of seismic data near salt is poor. Figures 8-69a–8-69c show three different contour maps on the top of salt illustrating the variety in the shapes of salt intrusions. The top of the salt contour map for Main Pass Block 299 Field (Fig. 8-69c) is of special interest because the salt dome has an overhang on the south and southwest flanks. Notice how the contours under the overhang are dashed, making the map easier to use when preparing structure contour maps, as well as making the salt map easier to read.

Salt/Fault Intersection. Figure 8-70a illustrates the method for contouring the intersection of a salt structure and a fault. The technique is basically the same as that presented earlier for contouring the upthrown or downthrown traces of a fault. The salt/fault intersection occurs where the structure contours on the salt intersect the fault contours of the same elevation. Like all intersecting surfaces presented in map view, the salt/fault intersection should be delineated as shown in the figure.

Salt/Sediment Intersection. The salt/sediment intersection represents the termination of a sedimentary surface against the salt mass. This interface is important for two reasons: (1) it often serves as the seal for the trapping of hydrocarbons; and (2) it delineates the limit of the potentially productive hydrocarbon-bearing formation. Figure 8-70b shows a portion of a salt contour map (same as in Fig. 8-70a) and the line of termination of the 8000-ft Sand against the salt. The termination of the sediments occurs where the structure contours intersect salt contours of the same elevation.

Completed Structure Picture. The mapping of all intersections and the integration of the fault, salt, and structure maps results in a completed structural interpretation, such as that shown in Fig. 8-70c for the 8000-ft Sand in the southeast portion of this field. This example is relatively simple, but it does illustrate the use of the mapping techniques. Salt domes are normally associated with highly faulted, highly complex structures such as the one shown in Fig. 8-71, which is a structure map on the Grand Isle Ash at Grand Isle Block 16 Field, Offshore Louisiana. This structure map is unique in that it contains

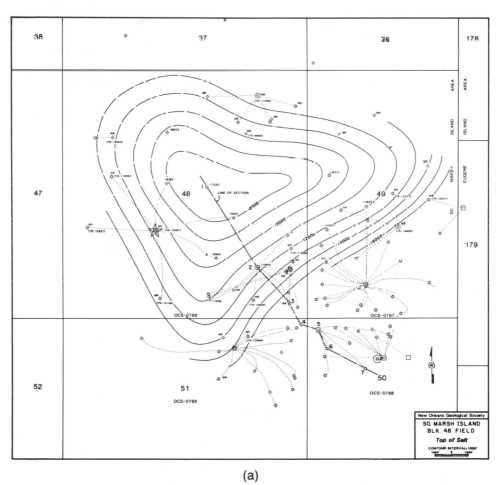

(a)

Figure 8-69 (a) Structure map top of salt, South Marsh Island Block 48 Field, Offshore, Gulf of Mexico. (Published by permission of the New Orleans Geological Society.) (b) Structure map top of salt, Cameron Meadows Field, Louisiana. (Published by permission of the Lafayette Geological Society.) (c) Top of salt, Main Pass Block 299 Field, Offshore, Gulf of Mexico. Observe the dashed contours to clearly illustrate the salt overhang. (Published by permission of the New Orleans Geological Society.)

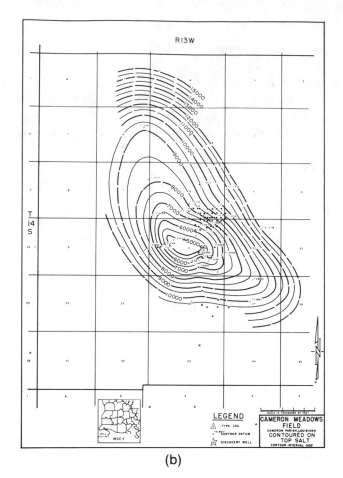

(b)

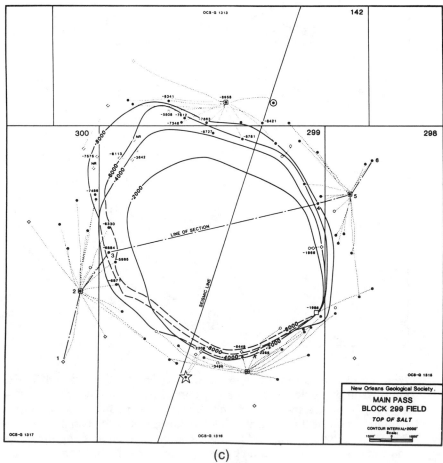

(c)

Figure 8-69 (*continued*)

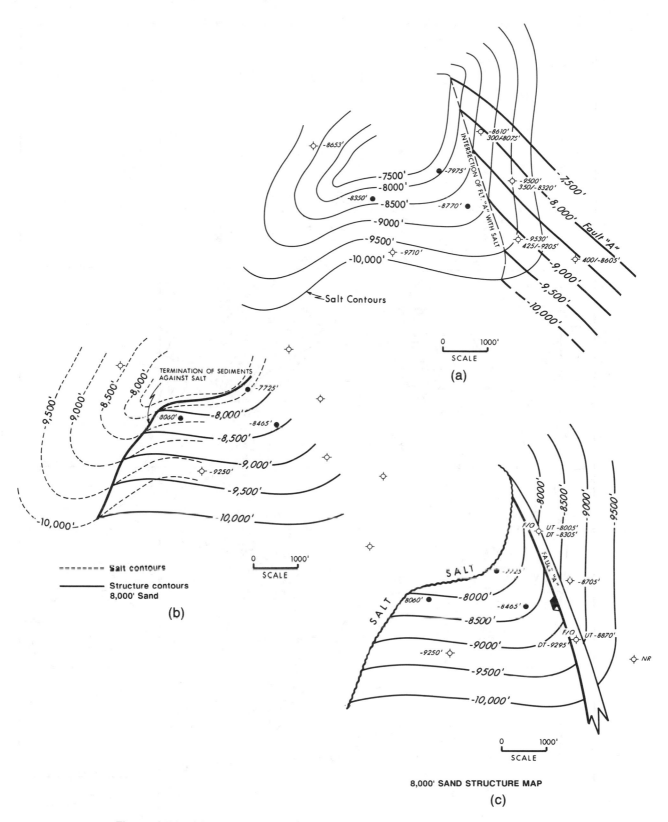

Figure 8-70 (a) Integration of the contour map on top of salt and Fault A delineates the intersection of the fault with salt. (b) Integration of the contour map on top of salt and the structure contours for the 8000-ft Sand. The salt/sediment boundary is located where the salt contours intersect the fault contours of the same elevation. (c) Completely integrated structure map on the 8000-ft Sand, Reservoir A, in the southeast portion of this piercement salt structure. This oil reservoir is bounded to the north and west by salt, to the east by Fault A, and to the south by an oil/water contact at a depth of −8605 ft.

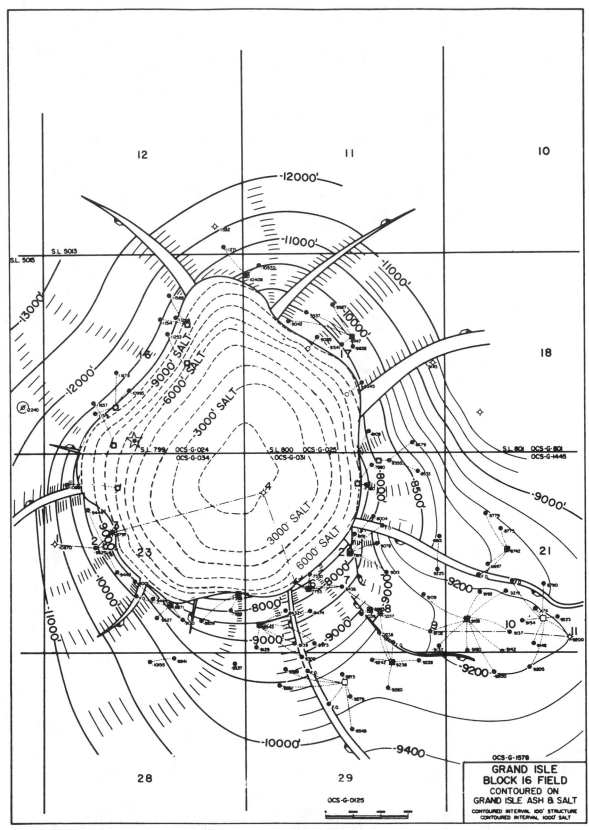

Figure 8-71 Structure contour map on the Grand Isle Ash and salt at Grand Isle Block 16 Field, Offshore Louisiana. The outline of the salt at this mapping level is defined by the intersection of the salt and structure contours at the same elevation. (Published by permission of the Lafayette Geological Society.)

384

both the contours on the salt and structure. It is the intersection of these salt and structure contours that delineates the salt/sediment boundary.

As mentioned in Chapter 6, steeply dipping structures, such as those formed due to salt uplift, require the layout of a number of cross sections to help develop an accurate geologic interpretation and aid in the structure map construction. Typical cross sections for steeply dipping structures such as a piercement salt dome are designed to incorporate both straight and deviated wells. Initially, problem-solving type cross sections are laid out to help resolve correlation problems and aid in developing the structural interpretation. Later, during the advanced stages of mapping, these sections can be converted to finished illustration sections for display and presentation. For further information on the use of cross sections as an aid to structure mapping of salt structures, refer again to Chapter 6.

Considering the tectonics of diapiric salt intrusions and the uplifting effect on the surrounding sediments, the general structural attitude of these sediments is usually one of parallelism or semi-parallelism to the salt intrusion. Therefore, more often than not, the structure contours for any given horizon around a salt dome tend to parallel or semi-parallel the salt contours. This is illustrated in Fig. 8-71 for Grand Isle Block 16 Field.

Around certain salt structures, subsurface data show structure contours intersecting salt at a sharp angle. When this happens, it may be an indication of several possible situations, including: (1) the possibility of a salt overhang, such as the one shown in Fig. 8-72 for the Bethel Dome, Anderson County, Texas, or (2) an unrecognized peripheral fault sliding down or near the face of the salt, causing the strike direction of the contours to *appear* to turn into the salt. Actually, the contours are striking into the downthrown side of the peripheral fault. Figure 8-73 is an example of such a situation. Notice on the southwestern portion of the salt dome that the contours appear to strike directly into the

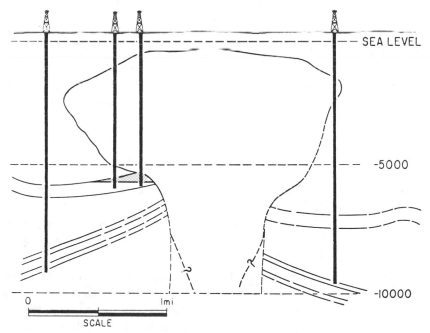

Figure 8-72 Generalized cross section of Bethel Dome, Anderson County, Texas, showing a hydrocarbon accumulation below the salt overhang. The significant amount of overhang might indicate that the dome is detached from the salt source bed at depth. (From *Salt Domes,* by Michel T. Halbouty. Copyright © 1979 by Gulf Publishing Company, Houston, Texas. Used with permission. All rights reserved.)

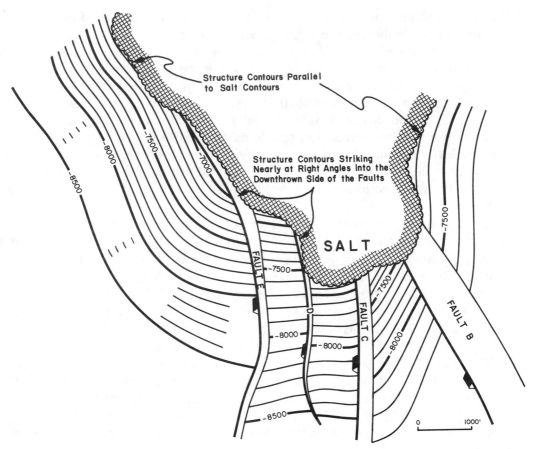

Figure 8-73 Structure contours striking into the downthrown side of Faults E and D give the appearance that the contours are striking into the salt.

salt. Due to the complex nature of the faulting pattern, Faults C, D, and E, which appear as radial faults shallow in the section, become peripheral faults with depth, paralleling the southwest flank of this salt structure. As a result, the structure contours tend to strike at various angles into the fault. In contrast, observe the general conformity of the structure contours with the salt on the western and eastern flanks of the structure not affected by peripheral faulting.

Wrench Fault Tectonics

Wrench faults are defined as high angle or vertical strike-slip faults that form under horizontal compression. A wrench fault system may have great linear extent, such as the San Andreas fault complex, where major crustal plates are involved, or it may occur within a local or subregional area as a limited system of finite length. These local wrench systems are sometimes referred to as *compartmental* because of the independent deformation on either side of the fault (Fig. 8-74a). In this type of deformation, a structure which seems to be cut by a strike-slip fault need not have a *severed* portion offset on the other side of the fault, since the faulting and deformation are contemporaneous and independent on either side of the fault (Fig. 8-74b). Vertical movement, of varying degrees, is commonly associated with wrench fault systems, as well as fault block rotation.

The primary hydrocarbon traps associated with wrench faults are anticlines that straddle the wrench system. These anticlines which may be faulted by either normal or reverse faults are good traps because they form early and commonly develop large closures sufficient to trap economic quantities of hydrocarbons. Our interest is in the mapping techniques that are applicable to these wrench fault systems. Figure 8-75 shows an example of a wrench fault system with offset faulted anticlines on each side of the fault. In many cases, the faults cutting the anticlines are small and simply offset the structures with little, if any, change in structural attitude across the fault.

Two different types of mapping techniques are required to map these wrench fault structures. Normally, the structures on either side of the main wrench fault must be mapped independently (Fig. 8-75) because structures tend to terminate against the wrench fault (Bischke, Suppe, and Pilar 1990). As for the individual anticlines that form on either side of the main wrench fault, the vertical separation technique is often applicable for mapping across the faults that cut these anticlines. In other words, the faulted anticlines very commonly exhibit contour or structural compatibility across the small normal or reverse faults that cut the anticlines, unless some type of rotation occurs along the fault.

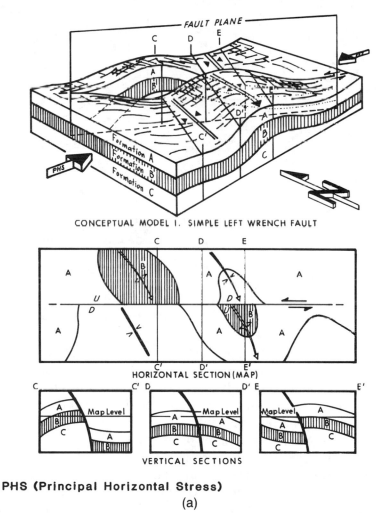

PHS (Principal Horizontal Stress)

(a)

Figure 8-74 (a) Conceptual model of simple left wrench fault: Top—block diagram; middle—plan view; bottom—cross-sectional views. (Froim Stone 1969. Reprinted by permission of the Rocky Mountain Association of Geologists.) (b) Block diagram illustrating different deformational patterns on opposite sides of a finite wrench fault. (From Bell 1956. Published by permission of the American Association of Petroleum Geologists.) (c) The change in direction of asymmetry or fold frequency across the fault is due to a different response to the compressional forces. (From Brown 1982. Published by permission of the American Association of Petroleum Geologists.)

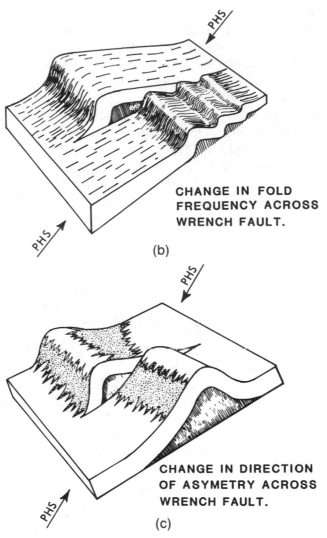

CHANGE IN FOLD
FREQUENCY ACROSS
WRENCH FAULT.

(b)

CHANGE IN DIRECTION
OF ASYMETRY ACROSS
WRENCH FAULT.

(c)

Figure 8-74 (*continued*)

Figure 8-76 is a structure map from the Rosecrans Oil Field, California. Although there is currently some debate as to whether the Inglewood fault system is a true wrench system, or a transpressional system, it can be used to illustrate the mapping techniques applicable for a wrench fault system. Notice that the anticlines adjacent to the Inglewood Fault are cut by small reverse faults. The cross section insert and the structure map show that for the most part there appears to be structural compatibility across these small faults. With the use of well control and seismic data, the vertical separation for each fault can be determined and used to contour across the faults.

Notice the two faults labeled C and D on the structure map. Based on this structure map and available cross sections, these faults appear to have imparted some rotation, in addition to dip-slip motion, resulting in a loss of structural compatibility across the faults. Therefore, it may be difficult, if not impossible, to contour across these faults using the vertical separation. The structures on either side of Faults C (Compton Thrust Fault) and D may require independent contouring. However, north of Fault D the structure again has good contour compatibility across the last two faults shown on the map.

Figure 8-77 is a structure map on the Ranger Zone in the Wilmington Field in the Los Angeles Basin, California. The anticlinal structure that forms the Wilmington Field

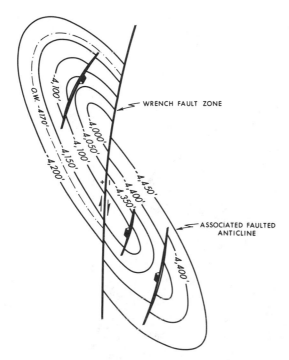

Figure 8-75 Typical wrench fault system with associated faulted anticlines.

is 11 mi long and 3 mi wide. This giant field has produced over 1.2 billion barrels of oil. During the Middle Miocene, compressive stresses formed a north-south couple folding the strata and establishing the present northwest-southeast Wilmington structure. During its complex structural history, the anticline developed a series of normal faults as a result of tensional forces acting along the structural axis. These normal faults are small, ranging in size from 100 ft to 400 ft and dip at angles between 45 deg and 65 deg. Many of the faults are sealing and therefore divide the field into at least seven major fault blocks (Mayuga 1970).

A review of the structure map in Fig. 8-77a and the cross section in Fig. 8-77b suggests very good structural (contour) compatibility across all the faults in the field. Therefore, all structure and fault map integration, in this field, would use the vertical separation technique for contouring across the faults. The Wilmington Field is another example of a wrench fault associated structure that exhibits good internal structural compatibility across the faulted anticline.

Compressional Tectonics

For our discussion on **compressional tectonic settings**, we include both high angle reverse and thrust faulting since they often occur together in this setting. Also, this section centers on fold-and-thrust belts, which include forearc, backarc, and collisional belts, since they make up the most prolific compressional habitat. The most common hydrocarbon trap is the hanging wall anticline, which includes such structures as fault propagated folds (snake heads), fault bend folds, and duplex structures (see Chapter 9). For example, in the Wyoming-Utah backarc fold-and-thrust belt fields (Fig. 8-78), nearly all the hydrocarbons are trapped in the hanging wall of the Absaroka Thrust. They include such fields as Painter Reservoir, Whitney Canyon, Ryckman Creek, and Anschutz Ranch Fields (Lamerson 1982). Nearly all these fields are found in asymmetric anticlinal folds with the steep limb to the east. Collisional zones, such as the Zagros collisional belt of Iran,

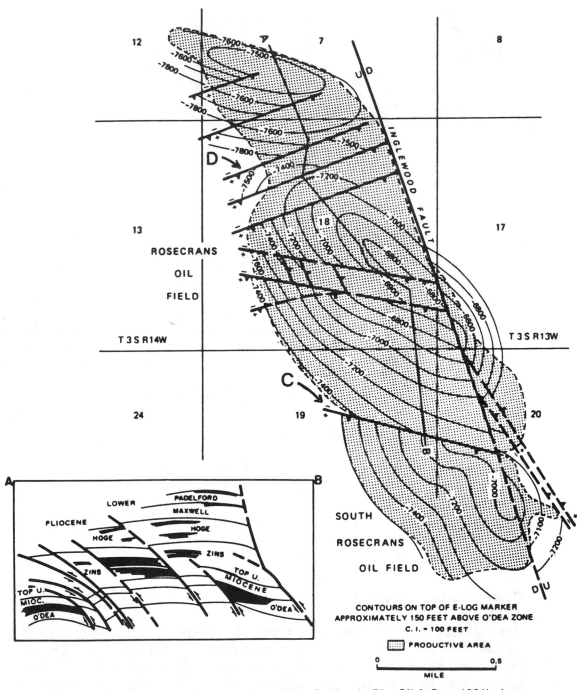

Figure 8-76 Rosecrans field structure (After California Div. Oil & Gas, 1961) shows a distinct pattern of reverse faulted anticlinal folds oriented obliquely to the Inglewood Fault. (Modified from Harding 1973. Published by permission of the American Association of Petroleum Geologists.)

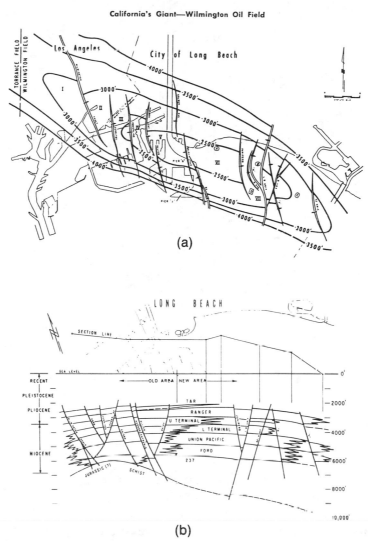

California's Giant—Wilmington Oil Field

(a)

(b)

Figure 8-77 (a) Structure contour map on the top of the Ranger Zone, Wilmington Oil Field, Southern California. The normal faults that cut the field exhibit several different patterns: (1) single, (2) compensating, (3) bifurcating, and (4) intersecting faults. Lower Pliocene Repetto sections show thicker sediments in the downthrown blocks, suggesting that for some faults, displacement was contemporaneous with sediment deposition (growth faults). (b) East-west geologic cross section along the axis of the Wilmington structure. The section shows that there is excellent structural continuity across the faults. Length of section about 8 miles. (From Mayuga, 1970. Published by permission of the American Association of Petroleum Geologists.)

are some of the most prolific of all fold-and-thrust belt types. At one time, the Zagros belt accounted for 75% of the world's fold-and-thrust belt production.

An extensive literature search yields very few examples of contoured fault maps and integrated structure maps for compressional structures. There are volumes of published fault and integrated structure maps for extensional petroleum areas, but they appear to be scarce for compressional areas. The inverse is true for balanced cross sections. A significant number of balanced cross sections have been published for compressional structures, but very few are published in extensional areas.

Fault surface mapping is not commonly done in compressional areas, although there is *no* reason why it should not be done; and indeed the construction of fault surface maps

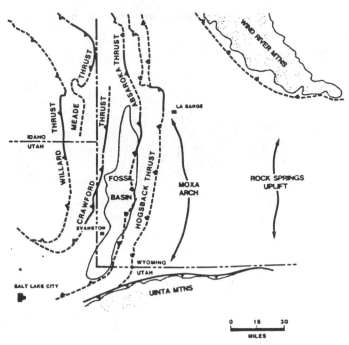

Figure 8-78 Maps showing the position of the fossil basin on the hanging wall of the Absaroka Thrust Plate. (From Lamerson 1982. Reprinted by permission of the Rocky Mountain Association of Geologists.)

would aid in the understanding of the geology. One possible reason for not making fault maps and integrating them with structure maps is that, in many cases, hydrocarbons are trapped in the hanging wall anticline, only slightly disrupted or controlled by reverse or thrust faulting; therefore, fault maps are not prepared. Another reason is that balanced sections are often constructed; therefore, the construction of fault surface maps is assumed to be unnecessary. We believe, however, that the mapping of all related faults is an *essential* part of any structural interpretation. In previous sections of this text, we showed that cross sections and seismic sections can misrepresent the actual structure because of the orientation of the line of section. We also mentioned in Chapter 7, with regard to the tying of seismic lines, that any nonvertical line can tie; the test of a valid fault interpretation is the preparation of a fault surface map that is geologically reasonable based on the available data.

We have shown that fault trace construction for any mapped horizon, particularly in areas of steeply dipping beds, **cannot be done intuitively**. The accurate delineation of a fault trace for any mapped horizon requires the integration of a fault map with the structure map for that horizon. Remember, a fault trace on a specific horizon may look nothing like the fault surface map itself. The construction of a fault trace on a structure map based on a few well cuts and seismic lines is guesswork, and is, more often than not, incorrect. We therefore encourage the construction of fault surface maps in compressional areas.

Reverse Faults.   Earlier in this chapter, we showed the techniques for construction of a reverse fault map and the integration of this map with a structure map (Figs. 8-25 and 8-26). In this section, we look at intersecting reverse faults. Unlike the case involving normal faults where the fault appears as a formation discontinuity on the map (a gap), no plan view formation discontinuity is present on a structure map with a reverse fault.

The contours on the upthrown block (hanging wall) overlie the contours on the down-thrown block (footwall).

The example in Fig. 8-79 is drawn in a form similar to the normal fault examples except that fault transport is reversed to show what happens under compression. The pattern of intersecting faults illustrates the mapping techniques, but the geometry may not be an accurate predictive model. In other words, reverse faults on a plunging anticlinal structure may result from a volume problem created as planar beds are forced into the compound curvature of the nose. Such faults may not necessarily intersect, because the compression may be accommodated on different stratigraphic levels by separate faults. Intersecting faults do occur in compressional areas, however, and therefore the understanding of the mapping techniques for these faults is important.

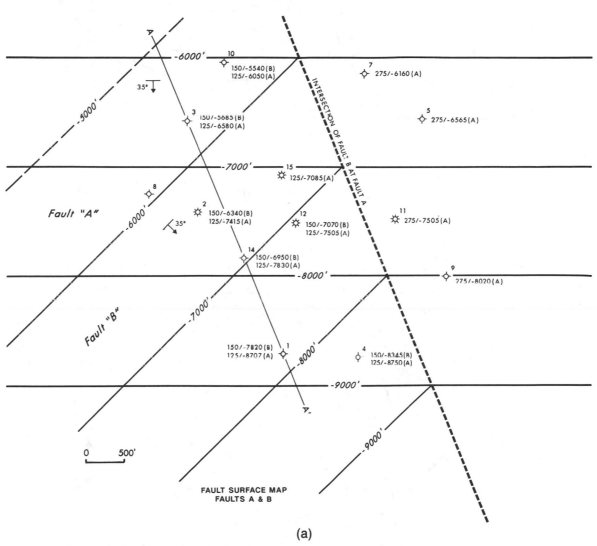

(a)

Figure 8-79   (a) Fault surface map for reverse Faults A and B. (b) Integrated structure map on the 6000-ft Sand. Contours are dashed on the footwall block in the area of fault overlap for clarity. (c) Fault map superimposed onto the structure map to show the accuracy that is achieved by the integration of the two maps regarding: (1) fault trace construction, (2) position of faults, (3) fault intersections, and (4) proper bed contour construction across each fault. (d) Completed structure map on the 7500-ft Sand. Contours dashed on footwall in area of fault overlap.

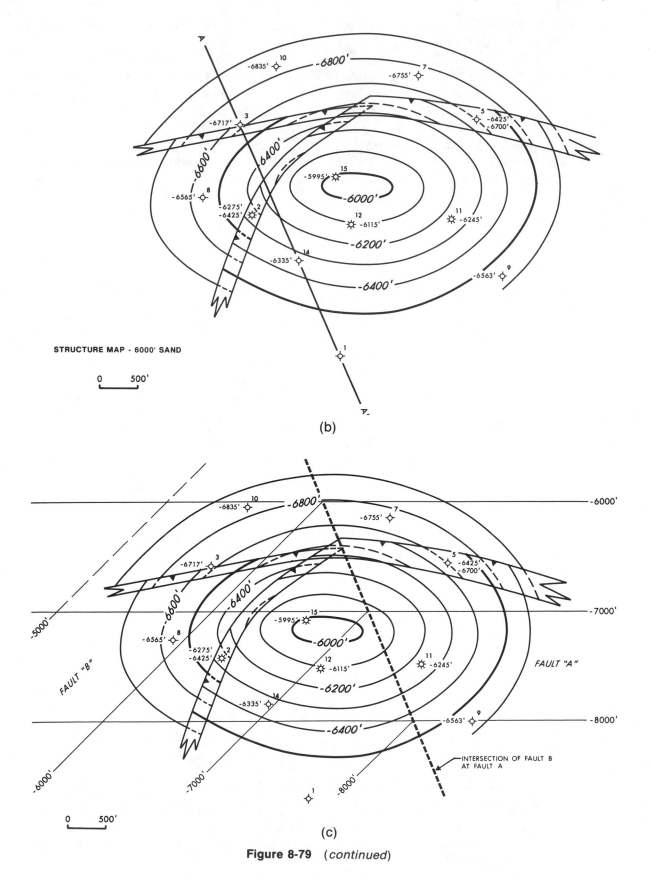

STRUCTURE MAP - 6000' SAND

0   500'

(b)

FAULT "B"

FAULT "A"

INTERSECTION OF FAULT B
AT FAULT A

0   500'

(c)

Figure 8-79 (*continued*)

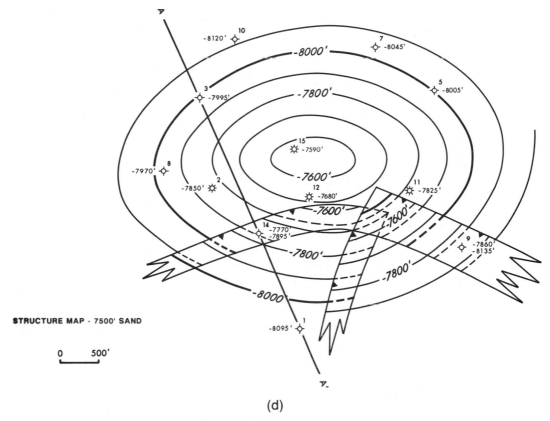

STRUCTURE MAP - 7500' SAND

0   500'

(d)

Figure 8-79 (*continued*)

Figure 8-79a shows the fault contour map for reverse Faults A and B. Since we are dealing with intersecting faults, the line of intersection must be shown on the fault map. Also, the additive property of faults must again be considered; throughout the area where two faults are present (west of the termination line in our example), Fault A has 125 ft of vertical separation and Fault B has 150 ft of vertical separation. Only Fault A with a vertical separation of 275 ft (150 ft + 125 ft = 275 ft) is present east of the line of intersection. Therefore, the vertical separation is conserved across the line of intersection.

Figure 8-79b is a completed structure map on the 6000-ft Sand faulted by Faults A and B. Figure 8-79c shows the fault map superimposed on the completed structure map. This figure demonstrates the construction of the upthrown trace (trace on the hanging wall) and downthrown trace (trace on the footwall) for each fault, the construction of the intersection of the two faults, and the proper bed contour construction across each fault using the vertical separation (repeated section from well log data). We have discussed in detail this type of construction with a number of other examples; therefore, we do not detail the construction here, but instead recommend that you pick at least two contours and review the construction techniques used to complete this structure map. Notice that the fault intersections fall on the line of termination. Figure 8-79d is the completed structure map on the deeper 7500-ft Sand; take a moment and review its construction.

As with normal faults, the construction technique of integrating a fault with a structure map removes all the guesswork with regard to: (1) the location of the fault on the mapped horizon, (2) fault trace construction, (3) the location of all fault intersections, and (4) the width of the fault overlaps. The integration of the two maps constructs these features automatically without the need for guesses or estimates.

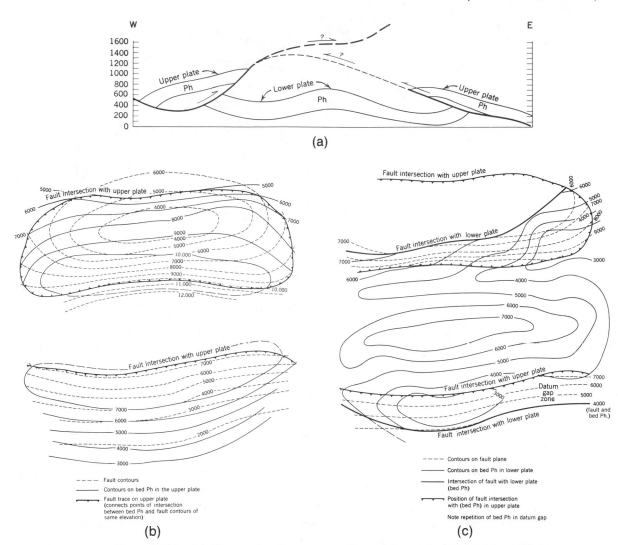

Figure 8-80   (a) Diagrammatic cross section of thrust fault (see Mount Tobin, Nevada, USGS map series GQ7). (Modified from Bishop 1960. Published by permission of author.) (b) Part of upper plate and fault surface map. Delineates the intersection of the fault with the upper plate. (c) Overlay of fault map and lower plate to show the intersection of the fault with the lower plate. (From Bishop 1960. Published by permission of author.)

Thrust Faults.   Most thrust faults have displacements large enough to offset the two fault blocks so that there is no continuity of structure (contour continuity) across the fault. In such cases, each fault block must be contoured independently and then integrated with the fault map or maps.

Figure 8-80a is a diagrammatic cross section of the Mount Tobin Thrust Fault, Nevada. Notice that there has been significant movement of the upper plate over the thrust fault. Therefore, the upper and lower plates of the Ph Bed require independent contouring and fault integration. Figure 8-80b shows the Mount Tobin Thrust Fault map superimposed on the upper plate structure map. Observe in the upper portion of the figure that the shape of the fault trace on the upper Ph Bed is shown as a closed oval (in plan view). This type of configuration is not arrived at intuitively; it requires the integration of the fault and structure maps for accurate delineation.

Figure 8-80c shows the integration of the fault map with the lower plate (the trace on the upper plate is also shown on the map). This example demonstrates the detail required to contour the upper and lower plates and integrate these plates with the fault map to accurately determine the position of all the fault traces. Take a few minutes and study the map to be sure you understand how the two maps were integrated to arrive at the finished structure map.

Earlier, we discussed the technique of preparing a *pull apart map* for compressional areas, or the use of a *flap map*. Such a map allows the contouring of a given horizon all the way to its footwall cutoff, which may be a considerable distance from the leading edge of the hanging wall block. With this technique, the mylar or paper used for mapping is cut along the leading edge of the hanging wall block, and new material is spliced in underneath to allow mapping of the subthrust surface to its fault cutoff. In complex areas with substantial thrust transport, this eliminates the confusion of dashed versus non-dashed contour lines. The maps can be separated to show the whole surface of the mapped horizon in both the upper block and subthrust. They are also very helpful during isopach mapping.

REQUIREMENTS FOR A REASONABLE STRUCTURAL INTERPRETATION

At the end of any mapping project, the accuracy of the reconstructed subsurface geology depends upon the completeness of the work undertaken. We cite 11 requirements that must be met to ensure the best and most accurate subsurface interpretation.

1. Correct correlations (log and seismic).
2. A good understanding of the tectonic setting being worked.
3. A clear understanding of basic geologic principles including the intersecting geometry of formations and faults.
4. Three-dimensional validity of the interpretation.
5. The use of **all** the available data.
6. An understanding of the accuracy of the data.
7. The use of correct and accurate mapping techniques.
8. The construction and integration of all required maps.
9. Construction of cross sections (balanced, if possible).
10. Multiple horizon mapping.
11. Good documentation of all completed work.

If all these requirements are met, there should be a high degree of confidence that the geologic interpretation is reasonable and accurate with respect to the available data.

Multiple Horizon Mapping

Almost any set of fault and structural data can be forced to fit on one particular horizon. The test of the fault and structural framework is whether the interpretation fits on a series of horizons at various depths. Therefore, one of the most important of the 11 requirements is the preparation of integrated multiple horizon structure maps. Multiple horizon mapping means the preparation of structure maps on several horizons, to ensure that the inter-

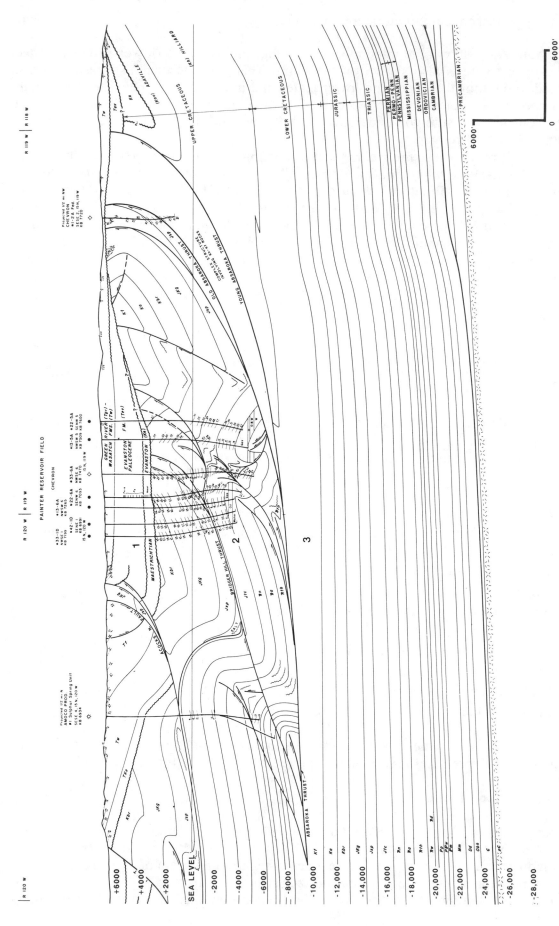

Figure 8-81 Cross section through Painter Reservoir Field. The Sub-Evanston Unconformity and the Absaroka Thrust Fault are interrupting events that result in different structural geometry above and below these events. (From Lamerson 1982. Reprinted by permission of the Rocky Mountain Association of Geologists.)

preted structural framework maintains continuity at all levels. Normally, at least three horizons are required to establish confidence in the interpretation; however, depending upon the size of the area being mapped and the number and vertical distribution of pay horizons, additional integrated structure maps at various depths may be required.

Once the structural framework (fault pattern and structure) is established by constructing several fieldwide or regional structure maps at various depths, the mapping of all pay horizons between these initially mapped horizons becomes a much easier task.

Discontinuity of Structure with Depth.   If the structural geology of an area maintains continuity with depth, the mapping of at least three horizons, at various depths, establishes and supports the structural interpretation. However, there are many instances where the structural continuity is interrupted with depth by one or more of three primary geologic events: (1) unconformities, (2) thrust faults, and (3) listric growth faults. In these cases, the structural geology above the interrupting event most likely will not conform to the structure below. Therefore, separate interpretations are required above and below the interrupting event.

Figure 8-81 is a cross section through the Painter Reservoir Field in Wyoming. Observe that the structural continuity is interrupted by two different events, first by the Sub-Evanston Unconformity at about 3500 ft, and second at the Absaroka Thrust Fault. Notice how the structure of the area changes above and below each interrupting event. Such major events must be recognized to develop a good structural interpretation and are very important in the search for hydrocarbons.

CHAPTER 9

STRUCTURAL GEOMETRY AND BALANCING

INTRODUCTION

Structural balancing is based on the intuitively satisfying concept that the interpreter must not create nor destroy volume during the interpretation process (Goguel 1962). Thus, an interpreted map or cross section, whether it be a geologic or seismic section, should volumetrically *restore* without overlaps or voids in the stratigraphic section. An analogy might be a child who removes a new block puzzle from a box and places it on the floor. Once all of the pieces of the puzzle have been removed from its container, the puzzle can be restored to its initial state by placing each block back into its proper position. The first attempt by the child at restoring the puzzle may result in most of the pieces being placed into the box with one or two pieces remaining on the floor. A second attempt could result in all of the pieces being placed in the box, but with some of the pieces being tilted at various angles or forced to fit.

The geophysicist or geologist experiences similar problems when attempting to retrodeform (restore) geologic and/or geophysical data. Of course, the correct solution to a puzzle is one that has been perfectly restored to its initial position.

The benefits of balancing are fundamental to correct geologic interpretations. Nature contains no holes or mass overlaps; thus, a section which does not balance cannot be geologically reasonable on simple geometric grounds. Unfortunately, a balanced section, although physically reasonable, need not necessarily result in the correct geologic interpretation. Balancing is *not unique* and two geologists can produce two balanced sections that are not alike. Obviously, the more complete the data set and the better the interpretive techniques, the more likely that the balanced section will reflect reality.

Balancing is still a relatively new discipline, and new techniques and interpretations are being developed yearly. Thus, in the final analysis, we tend to believe that correct interpretation tempered by a mass conservation concept is the key to valid geologic constructions. If the structural interpretation is correct, then balancing techniques can be used to *quantify* the interpretation.

This chapter is primarily designed to present most of the known structural mapping *techniques* and is not intended primarily as a text for interpretation or structural geology. The reader should already have a good understanding of structural geology. Several excellent texts exist on structural geology and balancing (Billings 1972; Suppe 1985; Woodward, Boyer, and Suppe 1985; Marshak and Mitra 1988).

The ultimate goals of balancing are to restore complexly deformed rock to its initial state or to its correct palinspastic restoration and to determine the sequence of events. Such information can be very useful to the geologist or geophysicist. Not only is the geometry of the structure better understood, resulting in better and more accurate reservoir maps, but geologic trends such as sand lines can be more accurately located. An understanding of the timing of the structural events should aid in oil migration studies, and define how and where fluids may have entered the structure. If the geometry of the structure is understood, then this knowledge can be utilized to more accurately process seismic data, which in turn results in an even better understanding of the geometries. Balancing can also be effectively utilized to *check* your assumptions and interpretations. Lastly, balancing tends to keep the interpreter more focused. If the section does not balance, then perhaps it is time to reconsider the interpretation.

Balancing can be subdivided into two disciplines: classical balancing, which was primarily developed by Goguel (1962), and Dahlstrom (1969) and his co-workers, and nonclassical balancing, which was primarily developed by Suppe (1983, 1985) and his students and co-workers. Most of the concepts that were developed in the introduction can be attributed to Goguel and Dahlstrom.

MECHANICAL STRATIGRAPHY

For many years, structural geologists have argued about the mechanical properties of the upper crust—does it exhibit elastic and/or frictional behavior as indicated by earthquakes, or is it viscoelastic or viscoplastic as indicated by the bent strata in the hinge areas of folds? Could time be a factor? Do the sedimentary strata buckle out (Biot 1961) or do the strata follow faults within the sedimentary section (Rich 1934)? Although all of these mechanisms are possible, the evidence now strongly suggests that the deformation that occurs in petroleum basins is primarily controlled by brittle (low temperature) deformation processes, and that the viscous deformation expressed by fold trains (Fig. 9-1) is confined to metamorphic belts (Tearpock and Bischke 1980). The fold style depicted in Fig. 9-1 with its near constant wavelength is not commonly observed in petroleum basins, and thus another deformation mechanism is required to explain the folds which trap hydrocarbons. This mechanism appears to be frictional deformation. Davis et al. (1983), Dahlen et al. (1984), and Dahlen and Suppe (1988), have formulated a frictional or brittle theory of crustal deformation that applies to both compressional and

FOLD TRAIN OBSERVED IN
METAMORPHIC BELTS

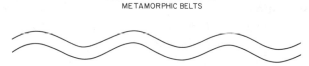

Figure 9-1 Example of a fold train commonly observed in metamorphic belts.

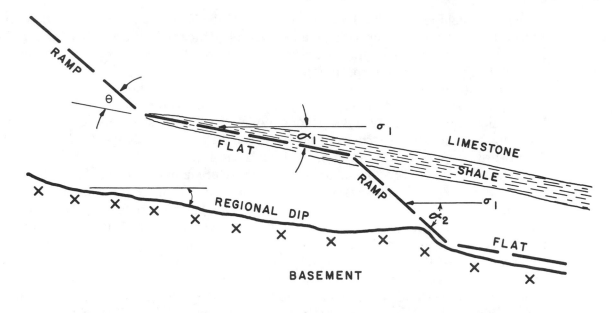

Figure 9-2 Cross section of ramp geometry. For explanation, see text. (Modified after Rich 1934. Published by permission of the American Association of Petroleum Geologists.)

extensional regimes. The theory resolves the overthrust paradox (Smoluchowski 1909; Hubbert and Rubey 1959) and is consistent with the geologic and seismic information collected from petroleum basins. Our intention here is to apply this theory and its observations to our areas of interest. Those readers who maintain an interest in mechanics can consult the references listed at the end of the textbook.

The frictional theory of crustal deformation states that when folds form, the maximum principal stress (σ_1) is inclined slightly to the bedding surfaces (Fig. 9-2). The rock will then fracture along angles which are dependent on the pore pressure and the intrinsic strength of the rock. The weaker the rock, the lower the angle between σ_1 and the fracture.

For example, consider an alternating sequence of limestone and shale layers (Fig. 9-2). Intuitively, the shale layers seem to be weaker than the better consolidated limestone layers, and it is well known that shales can contain abnormally high fluid pressures that drastically weaken these rock types. The theory states that as shales are weaker than limestone, the angle (α_1) between σ_1 and the fractures in shales must be smaller than the angle (α_2) between σ_1 and the fractures in limestones (Fig. 9-2). As σ_1 is slightly inclined to the bedding, the fractures in the shales are more subhorizontal than the fractures in the limestones. This leads to the primary conclusion of this section: In more competent or stronger rocks, the fractures will form at a high angle to bedding, and in the weak overpressured shales, the fractures tend to form parallel or subparallel to bedding.

If motion along these fractures causes them to coalesce, then a decollement will form along that *flat lying* area that may follow shale (or evaporite) horizons for tens of kilometers (Davis and Engelder 1985). In areas where the weaker layers gain strength or are pinched or faulted out, the decollement may *ramp* to a higher structural level (Fig. 9-2). As these ramps must pass through rocks which are stronger and have lower pore pressures than shales, the angle (α_2) between σ_1 and the fractures will be larger. Thus, ramps have higher angles with respect to the bedding than do the flat lying portion of thrust faults (Fig. 9-2).

When the ramp connects to a weaker layer on a higher structural level, the ramp transforms into a flat. Once a network of ramps and flats is formed and a large force is applied to the back of the wedge-shaped region in Fig. 9-2, the strata above the flats and ramps will begin to move along the fault. Material will begin to slide along the flats and up the ramps. Eventually, folds will begin to form in a manner that was initially described by Rich (1934), but this process is the subject of a later section.

The angle at which the ramp steps up from the bedding is called the cutoff or step-up angle (θ in Fig. 9-2). This angle is often characteristic (**fundamental**) to a particular fold-thrust belt and depends on both the pore pressure in the rock and the rock type. Similar relationships may exist in extensional terrains. The characteristic cutoff angle from several fold-and-thrust belts is generally less than 20 deg and tends to vary within several degrees of its mean value. In Taiwan the characteristic step-up angle is 13.3 deg +/− 2.4 deg (Suppe and Namson 1979; Dahlen et al. 1984). An attempt must be made to determine this angle prior to a balancing study. This step-up angle will be utilized to balance your structures.

There appear to be at least three methods which give insight into estimating the characteristic step-up angle. Field studies or a literature search can be conducted in the area of interest. As the step-up angle is the angle between the flat and the ramp (Fig. 9-2), field measurements or a description of this relationship will provide the required answer. A second, less direct measurement technique is to observe a well-imaged ramp and flat on a seismic section, but remember to first depth convert the seismic section. Lastly, the strata riding up the ramp will have the same dip as the ramp and therefore the same dip as the step-up angle (Fig. 9-2). Therefore, a study of the dips across an area may give insight into the characteristic step-up angle. For this method, it is first necessary to know the regional or undeformed dip of the area. For example, suppose that an area has no regional dip. It therefore follows that the undeformed beds will have zero dip. Strata which have moved up ramps and are deformed may dip at 12 deg. The characteristic step-up angle is therefore 12 deg. We might, however, be faced with a situation in which 20% of the dips are near zero, 30% are 3 deg, and 50% are about 9 deg or greater. The problem here is attempting to decide whether the regional dip is zero or 3 deg and whether the step-up angle is 9 deg or 12 deg, or greater? Often this matter is resolved by finding that one of these choices simply works better than the other during the restoration process.

CLASSICAL BALANCING TECHNIQUES

In previous sections we introduced the concepts of *volume conservation and brittle deformation*, which we intend to apply to petroleum basins and not to metamorphic belts which often lie adjacent to our areas of interest. Here we will develop these concepts in a manner that can lead to the mapping of structures that better define our prospect areas.

The volume conservation concepts which are developed in this section, although rigorous in their general application, do not precisely specify how this volume is to be conserved. The interpreter is to conserve volume, but the techniques do not define exactly how this volume is to be conserved. Thus, a significant degree of artistic license is left to the interpreter. For this reason the classical techniques developed by Goguel (1962) and Dahlstrom (1969) are ultimately qualitative in their approach. No graph or formula constrains the interpretation.

Volume Accountability Rule

The basic principle behind all balancing techniques is that *nature and not the interpreter* can create or destroy rock units, and that the interpreter should account for all of the present or pre-existing volume. Thus, the actual concept is one of volume accountability. Most geologists will be quick to point out that geologic compaction, particularly in growth structures, changes volume with time. In addition, fluid flow through rock can remove volume by pressure solution, and the like, and this volume reduction can be significant (Groshong 1975; Engelder and Engelder 1977). Arguments of this type, although correct, should not be substituted for lazy thinking. We have discovered that even thinking about growth structures in terms of strict volume conservation has forced the development of new balancing and interpretation techniques. If the structure does not balance volumetrically, then what process is causing the imbalance? The conservation of volume principle at least brackets the error or helps define the amount of compaction, etc. In the case of widespread volume removal, regional balancing and structural analysis would indicate that another process is occurring and to what extent. We normally find, however, that these volume reduction processes are not a major concern, and that the interpreter normally can think in terms of volume conservation, while being prepared for alternatives.

The economic issue that needs to be addressed here is much more practical and is much more likely to confront the interpreter on a daily basis than is pressure solution. Interpreters often unknowingly have a tendency to introduce mass overlaps and gaps into their interpretations. Often these gaps or overlaps are confined to a particular region of their cross sections or to a particular structure. For example, a given cross section upon retrodeformation has twice as much volume between sp (shotpoint) 320 to sp 420 (at about 1.5 sec to 2.2 sec) and no volume between sp 285 to sp 400 (at about 2.8 sec to 3.1 sec). An obvious question thus arises: does this volume incompatibility affect the viability of the prospect, and would a better interpretation enhance or detract from the prospectivity of the area? Therefore, balancing literally attempts to take the "holes" out of our interpretations.

Area Accountability

In the section on Mechanical Stratigraphy, we described the petroleum basin as a low temperature regime subject to brittle (i.e., frictional) deformation. In such an environment flow, elongation and flattening are not of primary importance, and thus the three-dimensional volume problem can be reduced to two dimensions. In other words, we shall assume that material is not entering or leaving the geologic cross section, and therefore the problem can be reduced to two dimensions. Notable exceptions to this rule would be shale and salt diapirs, which are often three-dimensional phenomena. These structures, which are associated with withdrawal and rim synclines surrounding the diapir (Trusheim 1960), contain a wealth of information which defines the salt flowage. Time dependent deformation is presently poorly understood but will some day be utilized to balance salt diapirs in three dimensions. Another exception is the bifurcating normal fault structure, which moves material out of the plane of cross section. Techniques for studying this type of deformation are briefly addressed in the section on Extensional Structures. In the meantime, however, and as long as the deformation is brittle and the transport direction is subperpendicular to the fault trace, the three-dimensional problem can be reduced to a two-dimensional cross section that is subperpendicular to the strike of the fault.

Bed Length Consistency

If we accept the premise that petroleum-bearing rocks are brittle and form at low temperature, then the two-dimensional problem can be linearized (Goguel 1962). In other words, if there is no large scale material flow within or without the plane of the two-dimensional cross section, then the seismic reflection or bed length before deformation will remain the same after deformation (Fig. 9-3). This logic will hold true for the thickness of each bed involved in the deformation, which means that the folding will be of the concentric type. Thus, bed length can be utilized to balance cross sections. If a sedimentary sequence is 2 km long before deformation, it must remain 2 km long after the deformation. The bed may be bent and it may be broken, but it will still be 2 km long.

Although the logic inherent in the above statement may seem self-evident, it appears to be one of the primary causes of the "*so called balanced*" cross section which is prevalent throughout the literature. The above logic implies that if one measures the bed lengths across a prospect, and the bed lengths are equal on all levels, then the cross section will balance. In practice, however, small changes in the lengths of lines can result in significant volume changes that result from inaccuracies in, or a lack of, dip measurements. This follows from the trigonometric relationship that at low angles the length of the adjacent line is about equal to the hypotenuse (Fig. 9-4). Consequently, one can see that the line segment AB is about equal to AC, even though the thickness AX is not equal to the thickness CZ. Therefore, one can often check existing cross sections by simply observing whether beds or formations are subject to unexplained thickness variations. If these thickness variations are not due to logical variations in stratigraphic thickness, then the interpretation should be subjected to further analysis.

DEFORMATION MAP

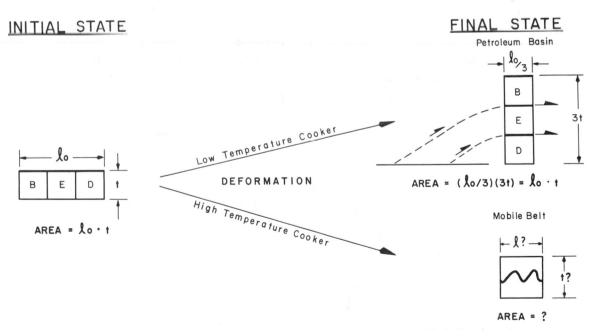

Figure 9-3 Deformation map of petroleum basins vs metamorphic belts. Low temperatures tend to preserve cross-sectional volume, whereas in metamorphic belts, material will flow in and out of the plane of cross section.

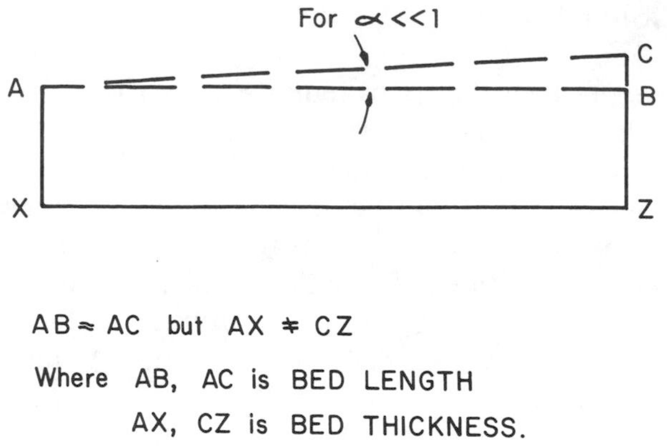

For $\alpha \ll 1$

A ⌐‑‑‑‑‑‑‑‑‑‑‑‑‑‑‑‑‑‑‑‑‑‑‑‑‑‑‑⌐ C
  └‑‑‑‑‑‑‑‑‑‑‑‑‑‑‑‑‑‑‑‑‑‑‑‑‑‑‑⌐ B

X └‑‑‑‑‑‑‑‑‑‑‑‑‑‑‑‑‑‑‑‑‑‑‑‑‑‑‑‑‑┘ Z

AB ≈ AC but AX ≠ CZ

Where AB, AC is BED LENGTH

AX, CZ is BED THICKNESS.

Figure 9-4 Noticeable changes in bed thickness result in small changes in bed length.

Pin Lines

You can check the validity of any cross section by measuring bed lengths, while keeping an eye out for variations in the thickness of units. This is accomplished through the use of pin lines (Dahlstrom 1969). In this procedure, one tries to locate regions that are *not* subject to deformation (such as shear or bedding plane slip, etc.) and then affix these regions to the basement by driving a pin vertically through the cross section. Bed length consistency is then measured relative to these pin lines (Fig. 9-5). Dahlstrom realized that bed length consistency must be preserved in both two and three dimensions and that if the bed length consistency does not hold from one section to another, then the interpretation is likely to be in error. Figure 9-5 is modified from Dahlstrom with Fig. 9-5a signifying the undeformed pin state. If the unit is concentrically folded and displaced a distance S, then the bed length (l_0) of the concentric fold after deformation should be the same length as it was before deformation (Figs. 9-5a and 9-5b).

In Fig. 9-5b, the length (l_0) of the folded unit is not the same as the length of the underlying undeformed unit (i.e., l is not equal to l_0), and the unit has been shorted a distance S (compared Figs. 9-5b and 9-5c). In Fig. 9-5c dipping beds overlie flat beds, which is the classic indication of a geometric discontinuity or decollement.

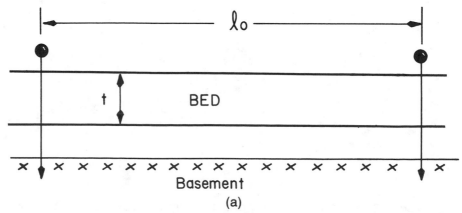

Figure 9-5 Pin lines and bed length consistency. (a) Undeformed bed state. (b) and (c) Deformed bed. (Modified after Dahlstrom 1969. Published by permission of the National Research Council of Canada.)

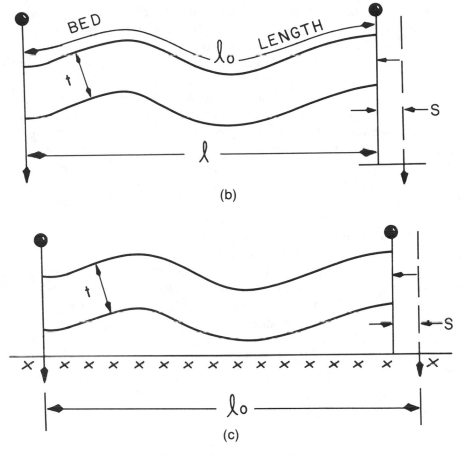

(b)

(c)

Figure 9-5   (*continued*)

STRUCTURAL OUTLINE MAPPING (LONG WAVELENGTH DOMAIN MAPPING)

Detailed mapping tends to concentrate on the shortest wavelength features that are resolvable in a data set. At the other end of the spectrum is *structural form or outline mapping*, which tends to concentrate on the *longest wavelength features* that are present within the data set. Thinking on this scale forces the interpreter to focus on the more *general or regional* aspects of the problem, rather than on specifics. Thus, structural outline mapping is another concept which can be applied to prospects in areas of complex deformation.

Structural outline mapping offers several advantages. This general type of *shape* mapping can be done very rapidly and areas as large as a basin can be finished in a day or two. The mapping should be conducted on the same scale as the other longer wavelength data sets (such as gravity or magnetics, etc.) so that these maps can be directly compared to one another. Therefore, a general understanding of the shape of the structures in a basin can be obtained in a very short period of time. Furthermore, the major features are the only features that are mapped, so they are highlighted on the map. The structural outline map is an accurate but generalized representation of the geometry of the major structures within the basin. This long wavelength mapping may not and perhaps may never precisely agree with the more detailed time horizon maps, particularly in areas of more complex deformation. For this reason, it is a good idea to conduct the long and the short wavelength studies independent of each other to avoid preconceived ideas, circular reasoning, and seeing only what we are looking for. In the end, the two studies conducted at widely divergent scales should be compared and satisfactorily rationalized.

An example of how structural outline mapping is conducted for thrust structures is as follows. One should think only in terms of the longest wavelength structural features and not in terms of time or stratigraphy. Map only the major changes in dip (the major dip domains), the major changes in structural shape, the top and the base of the structures. **Shape is of prime importance.** The time horizons are largely neglected, although when mapping the base or the termination of a structure, some attempt may be made to stay near the same time horizons. Focus on the shape and the location of major structural changes or discontinuities. These features are to a large extent independent of geologic or stratigraphic correlation, and thus discrepancies between different types of maps may indicate correlation errors.

The result will be a map which shows the location of structural changes with respect to their general shape and not strictly with respect to time. What will be represented is the width and the crests of anticlines, the bottom of basins, the locations and trends of major faults, cross faults and cross cutting features, the locations where the front, back, and sides of folds begin and end. The line which defines where the side of the fold ends marks the trend of the lateral (side) ramps which formed the anticlines (Wilson and Stearns 1958). This trend may also be the location of good gas plays (Wheeler 1980). Lastly, structural outline mapping by its very nature is more general in its approach, and the probability of making an error is consequently reduced. Thus, although shape mapping need not strictly follow time or stratigraphic horizons, major deviations between the general and detailed maps could indicate mistakes or correlation errors, and so on.

CROSS SECTION CONSISTENCY

So far, we have generalized the concept of brittle or frictional deformation to a two-dimensional cross section. As deformation is three-dimensional, the brittle deformation imposes constraints on the adjoining cross sections; therefore, folds or faults once formed must not terminate abruptly. The deformation must be taken up in another form. In other words, *slip must be consistent although not necessarily conserved*, from cross section to cross section. However, the slip can die as the result of deformation in the cores of folds.

For example, if a complex structure exhibits three thrust faults with 3 mi of slip, then it is very likely that a nearby cross section will also contain three thrust faults of similar shape and form that also contain about 3 mi of slip. If these three thrust faults radically change position and/or shape, then some *intervening transverse structure* must exist to accommodate the deformation. Such intervening structures are called **transfer zones**, and these structures exist in compressional (Dahlstrom 1969) as well as tensile environments (Gibbs 1984). Transfer zones often occur as tear or cross faults which form at high angles to the major structural trend. Furthermore, these transverse structures are often responsible for changes in the trends and shapes of structures from cross section to cross section. Figure 9-6 illustrates a transfer by lateral shear from one fault bend fold to another (see section on Fault Bend Folds). In Fig. 9-6a, the displacements on fault 1 are compensated for by displacements on fault 2 (see left-hand side of diagram Fig. 9-6a). The sum of the displacements on fault 1 and fault 2 remain constant; thus, as the slip on fault 1 decreases, the slip on fault 2 increases. The resulting structures caused by the lateral shear are shown in Fig. 9-6b.

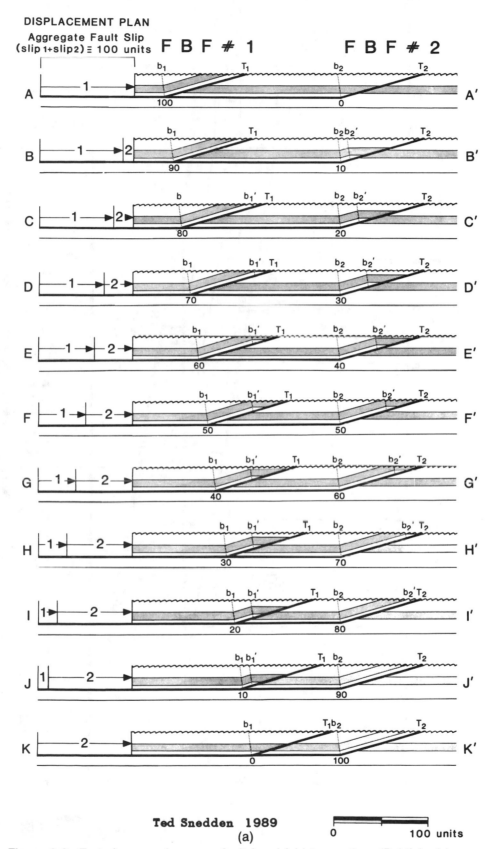

DISPLACEMENT PLAN
Aggregate Fault Slip
(slip1+slip2) ≡ 100 units

F B F # 1   F B F # 2

Ted Snedden 1989
(a)

0   100 units

Figure 9-6 Transfer zone from one fault bend fold to another. (Published by permission of Ted Snedden.)

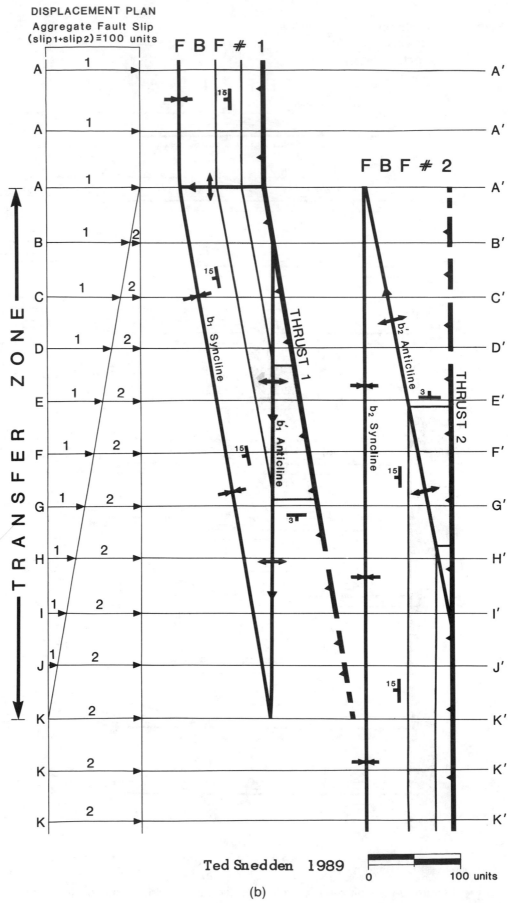

Ted Snedden 1989

(b)

Figure 9-6 (*continued*)

BOW and ARROW RULE

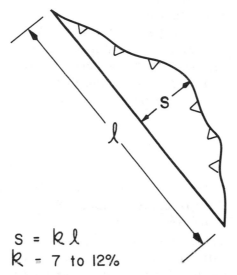

$$s = k\ell$$
$$k = 7 \text{ to } 12\%$$

Figure 9-7 Bow and Arrow Rule. Slip perpendicular to fault strike is approximately 10% of the fault length. (Modified after Elliott 1976. Published by permission of the Royal Society of London.)

Therefore, we see that small changes are apparently permissible from cross section to cross section, but how much change is permissible? Elliott (1976) has to some extent answered this question with the **Bow and Arrow Rule** (Fig. 9-7). This rule states that the amount of displacement can vary along a fault zone, but at an amount equal to 7% to 12% of its strike length. For example, suppose you mapped deformation along a large fault zone that has a total length of 10 mi. From the Bow and Arrow Rule, one would predict that the *maximum dip-slip* motion on the fault would be on the order of 0.7 mi to 1.2 mi. Now let us also assume that the amount of displacement along another fault is known to *increase* along a 10-mi portion of the fault zone. We can now predict that not only is there at least 1.4 mi to 2.4 mi of dip-slip motion, but also that the fault is at least 20 mi long. Elliott developed the Bow and Arrow Rule for thrust faults, but a similar rule may also exist for normal faulting. In general, the larger the magnitude of the normal fault, the greater the probability that the fault extends laterally.

Dahlstom also pointed out that in a given geologic environment, there is likely to be a consistency of structural forms or styles, and that only a limited number of structural shapes occur within a basin or orogenic belt. Furthermore, these structural styles tend to occur on cross section after cross section. This consistency in structural form also constrains the interpretation and puts limits on the balancing procedures.

CROSS SECTION CONSTRUCTION

There are presently two methods available for extrapolating dip data to depth—the Busk method of segmented circular arcs (Bush 1929), and the Kink method, which stresses the long planar limbs exhibited by most folds (Faill 1969, 1973; Laubscher 1977; Suppe 1985; Boyer 1986). Both methods assume that the folding is concentric or that bedding or formation thickness (in the absence of more detailed information) remains constant.

The Busk or the Kink method can be utilized to extrapolate any type of dip data. The data can be in the form of outcrop dips, formation tops or bottoms taken from outcrop or well log information, or dipmeter data obtained from well logs. *It is important, however, to be consistent in the use of the data.* For example, a formation top is projected to an adjacent formation top only if the formations being mapped do not change thickness, which is often the case over short distances. A dipmeter recording coming from within a formation is *not* projected to a dipmeter reading in an adjacent well unless these recordings are on the same stratigraphic level. In other words, it is important to understand that you are projecting the angle or curvature of a bed, and you should not project from one dip recording to another based solely on the level at which the measurements are taken.

Busking

Normally, dip data measured from surface outcrops, well logs, or seismic sections will not lie along the plane of cross section. Thus, the data must be projected to the plane using the methods discussed in Chapter 6. Let us assume for simplicity that the data, measured from outcrop, is shown in Fig. 9-8a. Normals (lines perpendicular to dip) are drawn from the position of the dip measurement data. These normals intersect at a point that represents a radius of curvature for an arc (point O in Fig. 9-8b) which is utilized to project the beds between the two data points A and B. A compass centered at point O is extended so that it has a radius OA, and then an arc is constructed from point A to line D. This procedure is then repeated for point B using radius OB (Fig. 9-8c). The result of this exercise is two arc segments AE and BF which define a curved layer AE-BF which has a constant thickness AF or BE. If another data point G is introduced, the normal to this adjoining data point will intersect at a different location, point O', and now several different radii (O'B, O'G, OI) are utilized to complete the depth to layer extrapolation (Fig. 9-8d).

The method can be visualized as consisting of several adjoining regions or domains in which *the curvature of the beds is constant,* and at the intersection of these domains the curvature of the beds changes abruptly. The Busk method is therefore a *curved dip*

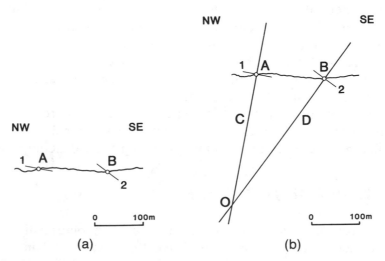

Figure 9-8 Busk Method Approximation. The sedimentary beds are projected to depth along circular arcs. (Modified from Marshak and Mitra 1988.)

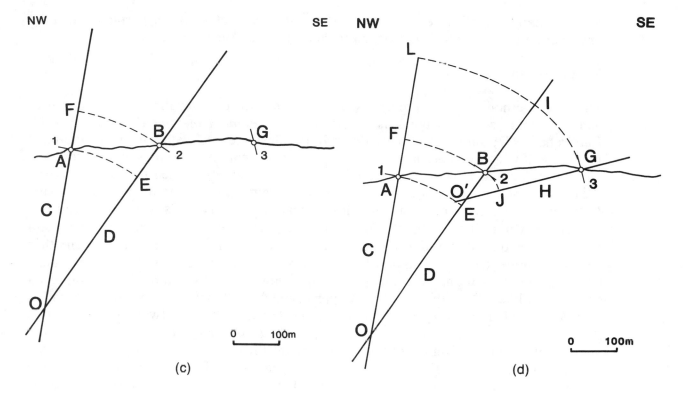

(c)

(d)

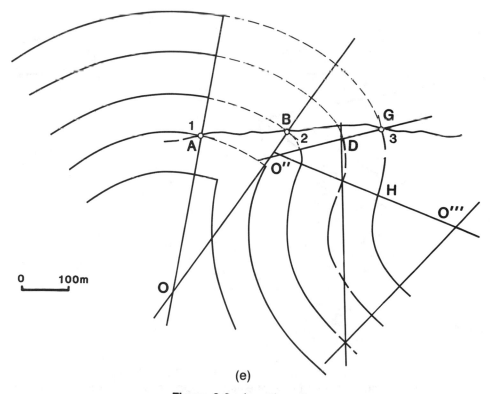

(e)

Figure 9-8 (*continued*)

domain method that is best applied to the hinge areas of folds, in particular to fault propagation folds, but suffers from an inability to retrodeform easily.

Kink Method

The next method that has proven extremely useful for extrapolating data to depth or along a cross section is the **kink** or *constant dip domain method* (Faill 1969, 1973; Laubscher 1977; Suppe and Chang 1983). In the Busk method, data which was mutually related, and which was assumed to exhibit a common curvature, represented a common curvature domain. However, we could have just as readily bisected the angle between the dip data points and created two regions of constant dip related to the two data points. In the limit, or where those data are closely spaced, both methods would be identical.

As shown in Figure 9-9a, the first task in kinking layers is to project the dip at data point B in the direction of data point A. Next, place two triangles adjacent to each other so that the upper triangle (X) is *parallel* to dip A and can be moved over the lower triangle (Y) (Fig. 9-9a). Now move the upper triangle upward past the dip B data point and construct a line CD so that point D is approximately halfway between points A and B (Fig. 9-9b). When working with real data, point D need not be halfway between points A and B, and its position will depend on where the beds change dip. After *bisecting* the angle between lines CD and DB with a protractor, the dip data A is projected up to the domain boundary line with the triangle (line AE, Fig. 9-9c). The triangles are then moved to a new position so that one of them is parallel to the dip B data and this triangle is moved down to continue the dip A line into the domain B data zone (line EF, Fig. 9-9c). The result is two dip domains with each domain containing a constant dip and a

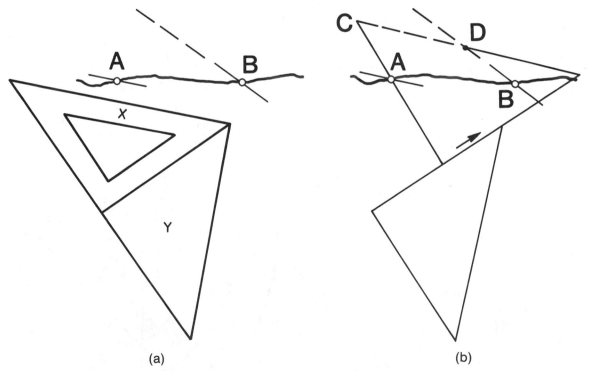

(a)   (b)

Figure 9-9 Kink Method Approximation. The sedimentary beds are projected to depth along planar surfaces. The method applies to a majority of folds which possess subplanar limbs.

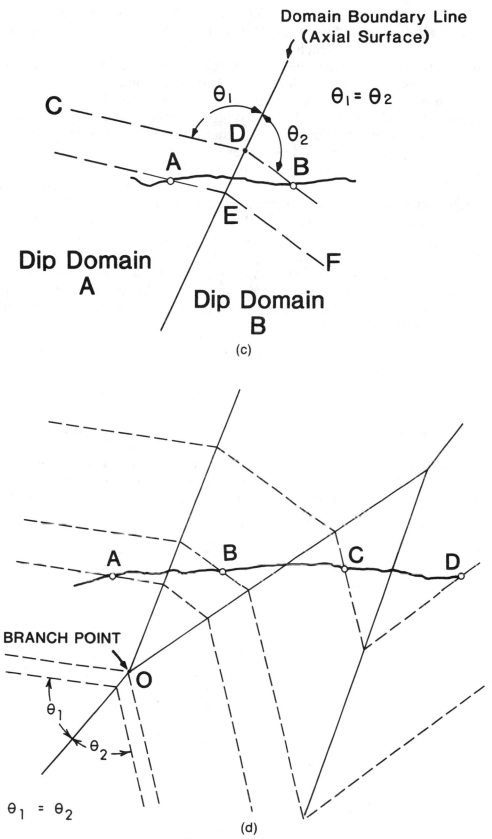

Figure 9-9 (*continued*)

theoretical bed of constant thickness (DE). The process can be repeated as additional data are introduced (Fig. 9-9d). Notice that in Fig. 9-9d, dip domain B converges and terminates at point O, which is called a *branch point*. Only two dip domains exist beneath the branch point, where three domains exist above the branch point. Notice that the axial surfaces bisect the beds both above and below the branch points.

In many folded areas, extensive regions of relatively constant dip adjoin smaller regions of rapidly changing dip. This is often seen on seismic sections. These relationships suggest that many folds possess limbs that have a uniform or near constant dip, but have hinge zones that are curved. As a result of this uniformity in dip, the kink method is readily adapted to work in low temperature fold belts.

When applying the constant dip domain method, always remember to bisect the angle between the data sets, thereby creating two adjoining and individual dip domains. Usually the data are generalized. This can be accomplished by taking two triangles and aligning them so that the top triangle can be passed across the data. In this manner, the triangle can be used as a filter to generalize or average the data. Areas of different generalized dip are defined as individual or separate dip domains, and the dip is then assumed constant within each domain. The method also works very well with seismic or well data. This procedure is in fact an application of Snell's Law as applied to balancing (Suppe 1988) (Fig. 9-10).

$$t_1/t_2 = \sin(\alpha_1)/\sin(\alpha_2)$$

If this procedure is judiciously applied, the cross section is more likely to line length and area balance.

When mapping using the kink method, one will find that, as the formations change thickness, the theoretical level of the formation as predicted by the kink method will deviate from the observed level. Thus, periodic adjustments in bed thickness must be made, usually at the position of the axial surfaces or at the domain boundary line (Fig. 9-9c). Our preference is to follow the observed formation or sequence boundary in regions

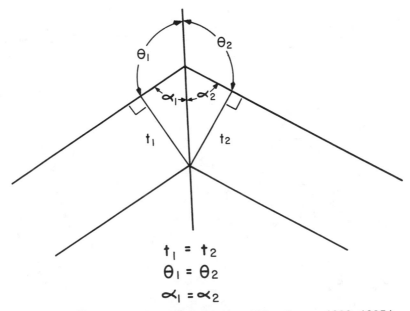

Figure 9-10 Kink Method Geometry. (After Suppe 1980, 1985.)

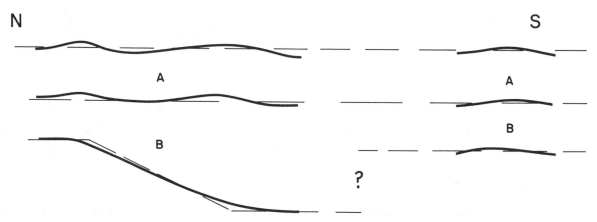

Figure 9-11 Example utilizing the uniform thickness approximation. Major change in the thickness of Unit B, but not in Unit A, implies that a structure or stratigraphic change is present in the region which lacks data.

of onlap, etc., even though this results in a divergence of once parallel lines. If units above the unconformity do not change thickness dramatically, little harm is done by accurately representing what happened to the strata.

In areas of good data, the bisected dip domain data will ensure proper line length and area balancing. In regions where the data are poor or nonexistent, the kink method can be used to extrapolate the theoretical level of the units being mapped. Even under these conditions, the uniform thickness assumption can be a very powerful tool. Assume, for example, that you are mapping units A and B in Fig. 9-11 from the north but that you encounter a region where no data exist. Mapping toward the no-data area from the south results in a good match on unit A but a poor match on unit B. What would you conclude in this case? The mismatch could result from either a dramatic change in facies or from a fault that stopped growing prior to the deposition of unit A.

Before proceeding to the next section, let us examine Figs. 9-8e and 9-9d in light of the two methods. If you apply the Busk method to the steep limb data (GH arc, Fig. 9-8e), you are inclined by intuition to terminate the plunging arc, but where? Yet, given adjoining surface data, the kink method provides an estimate of the maximum depth of the stratigraphic units (Fig. 9-9d).

DEPTH TO DETACHMENT CALCULATIONS

A method of determining the depth at which folding terminates can be attributed to Chamberlin (1910) and to Bucher (1933), who applied the method to determine the depth to detachment in the Jura Mountains. If the sequence that you are studying consists of a train of folds, then each fold must be isolated and studied separately. In this method, one measures the length (l_0) of a marker or reference bed, the present length (l), and the *average* amount that it has been uplifted ($\bar{u}$) above the undeformed level of the marker bed (Fig. 9-12). The amount of shortening that the unit has experienced is defined as

$$S = l_0 - l$$

The area of uplift times the present length ($\bar{u} \times l$) is then equated to the amount of

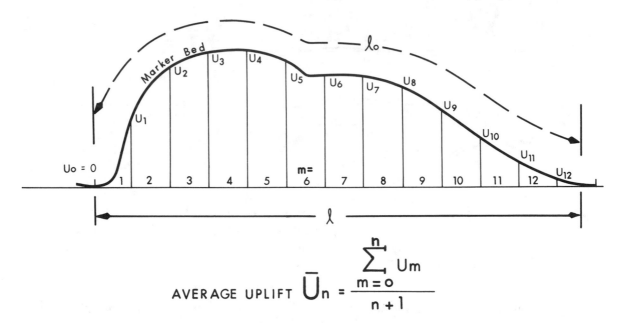

$$\text{AVERAGE UPLIFT} \quad \bar{U}_n = \frac{\sum_{m=0}^{n} U_m}{n+1}$$

Figure 9-12 The average amount that a marker bed has been uplifted ($\bar{U}n$) can be determined by measuring equally spaced line segments that are drawn between a base level and the marker bed, and then averaging the line lengths.

material that enters the structure from the sides ($S \times d$), where d is the depth to detachment (Fig. 9-13).

It therefore follows that (Bucher 1933)

$$d = l \times \bar{u}/S$$

Alternatively, if the depth to detachment is known, then the method can be utilized to check fold shape.

A closely related method developed by Goguel (1962) and employed by Laubscher (1961) has been commonly used in the petroleum industry, as its application generally yields the most accurate prediction of depth to detachment. This method also assumes that no material is entering the structure from below as in a duplex (see the section on Duplexes) and that all of the material in the core of the structure is derived from the sides of the structure. Mitra and Namson (1989) have pointed out that these assumptions are invalid if there is interbed shear (i.e., distortion of the vertical pin line) or if material is transferred out of the area of the cross section as occurs in fault bend folds.

If the material enters the structure from the sides, then the area within the core of a structure (A_u) at a given reference level is measured, as is the final length (l) and the initial length (l_0) of a reference or marker bed (Fig. 9-13). The shortening as before at the reference level is

$$S = l_0 - l$$

The area within the core of the structure beneath the reference bed (A_u) is assumed to be equal to an equivalent volume that comes in from the side (A_S) (Fig. 9-13). The area A_u can be obtained by planimetry.

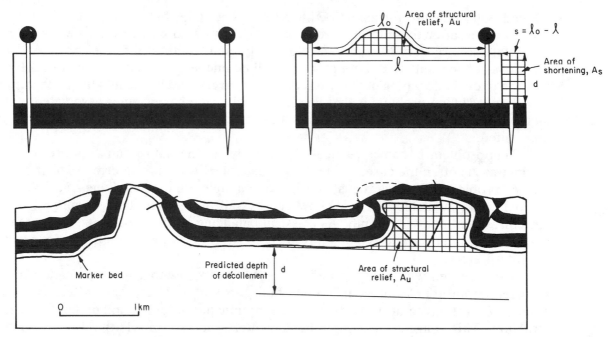

Figure 9-13 Depth to detachment calculation. The amount of material entering the cross section from the sides is equal to the material that has been uplifted above base level. (Modified after Laubscher 1961; Suppe 1985. Published by permission of the Swiss Geological Society.)

Therefore:

$$A_s = S \times d$$

where d = depth to detachment, and as

$$A_u = A_s$$

it follows that:

$$A_u = (l_0 - l)d$$

and

$$d = A_u/(l_0 - l)$$

NONCLASSICAL METHODS

Introduction

In the section on Classical Balancing Techniques we described a number of powerful techniques and concepts, developed by Goguel and Dahlstrom. The application of these techniques should not only enable you to more critically assess other people's cross sections, but also aid you in constructing better cross sections on your own.

Dahlstrom emphasized that within a given area, only a limited number of geologic structures are likely to exist. He also realized that these *structures must area and line length balance, but exactly how does the interpreter accomplish these tasks?* One obvious method is to measure formation bed lengths to check for balance, but this can only be accomplished after the interpretation is *finished*. In addition, two interpreters given the same data set are very likely to place lines of equal length at different positions within the cross section, although both products will be line length balanced sections! How are we then to evaluate which of the two sections is **correct**, and how could the interpretations be *improved*? One problem is that line length balancing has no rules associated with the method, other than that the bed lengths must be consistent and that the structural styles are limited.

This problem becomes particularly acute when the data in an area are undercon-strained, as is often the case, and leads to what John Suppe has referred to as the *"blank paper"* problem (Woodward et al. 1985). For example, you are studying an area in which the only data available are at shallow depths and these data strongly suggest that the structures continue with depth. As classical balancing lacks constraints, any attempt to continue the interpretation to depth is likely to result in as many interpretations as there are interpreters.

Those of you who have worked with structures know that several relationships are recurrent from area to area. In normal fault terrains, the faults are often listric, they contain rollover structures, antithetic and synthetic minor faults, and perhaps a keystone structure. This suggests that some fundamental process controls the development and formation of normal faults. In the compressional regime, folds often tend to be either symmetric as described by Gwinn (1964) in his work on the Appalachians, or they are asymmetric as in the Rockies (Link 1949). Geologists have noticed that when folds are present, faults also seem to exist in association with the folding (Woodward et al. 1985; Jones 1971). As many different regions around the world possess thrust belts that contain within them symmetric and asymmetric folds, some fundamental process seems to control the orogenic process! If we could develop realistic models of these fold-and-thrust belts, then the petroleum industry would have powerful tools in which to aid interpretation.

Hence, we enter the world of formulas, graphs, and models. Perhaps a word of caution is required at this time for the new geology student. Although models can be very powerful tools (e.g., the revolutionary plate tectonic model), the improper appli-cation of a correct model to the wrong situation will, of course, only result in error. To make matters worse, even model balancing is nonunique. Different interpreters, applying the same model to a given structure, are likely to generate similar results, as we shall see during the exercises, but the skeptic will point out that this is merely an artifact of being schooled in the same interpretations techniques.

Before we enter the exciting and demanding world of **kinematic modeling**, we need to restate that this book is designed primarily to present subsurface mapping techniques and is not a complete reference on interpretation per se. In the balancing sections of this book, the mapping techniques are often difficult to separate from the interpretation, as you must choose which technique to apply to a given structure, and this choice involves interpretation. Let us caution you that other interpretation techniques exist which do not involve any particular mapping technique, such as growth sedimentary patterns and struc-tures (Medwedeff and Suppe 1986). These growth patterns are often extremely helpful in determining which model or technique to apply to the structure, and thus we also recommend that the student enroll in one of the appropriate interpretation courses.

Suppe's Postulates

When presented with the problem of a poor or nonexisting data set, several approaches are open to the interpreter. Solutions to this problem seem to involve the following.

1. Collect more and/or better data.

2. Make more assumptions in order to solve the structural problem. If data are lacking or are unobtainable, it is still possible to solve the structural problem providing that you can extrapolate known data, using known geologic principles, into the area of interest. For example, in Figs. 9-8e and 9-9d, we were confronted with the problem of extrapolating the units within the limb of a fold to greater depths, even though no data existed in this region. We solved this problem by assuming that the Busk or Kink methods were appropriate. Certainly, other assumptions could be made to arrive at a solution. In this sense, assumptions can replace data.

3. Invent more powerful interpretation methods and techniques so that you can extrapolate existing data into the "blank paper" areas. Solutions to the above problems may involve:

   Solution 1: This area is left to the data contractors.

   Solution 2: Assumptions
   a. "Thrust faults step up abruptly from a decollement and do not have continuously curved listric shapes."
   b. "All thrust faults (producing a given structural style) in a given area step up at approximately the same angle."
   c. "Layer parallel slip in a thrust sheet is limited to that caused by changes in dip" (Suppe 1988). This is another way of stating that the kink method applies at all times.

   Solution 3: This is the subject of the remaining sections in the balancing chapter.

Fault Bend Folds

Our examination of seismic sections from various portions of the world (e.g., Australia and through the Pacific rim to Alaska, western and eastern United States and Venezuela, etc.) indicate that there are two commonly recurring fold styles within the low temperature portions of thrust belts: the **symmetric or fault bend fold type** (Figs. 9-14 and 9-15) (Rich 1934; Suppe 1983), and the **asymmetric or fault propagation fold type** (Fig. 9-24) (Link 1949; Suppe 1985). We stress here that complications in these structures often exist, such as multiple and back thrusts, and other thrust-related geometries are also present (Fig. 9-16). We wish to emphasize, however, that these two structural styles are the simplest types of folds that are commonly present in petroleum basins.

Fault bend folds were described by Rich (1934) in the Pine Mountain thrust region of the Appalachians, where he recognized that this fold style consisted of *symmetric* anticlines (Fig. 9-17). Rich also recognized that these folds were associated with thrust faults and he postulated that the folds were the result of the **"thin skinned"** deformation. Notice that if motion were to occur along the decollement in Fig. 9-2, material would ride up the ramp and onto the flat. Rich recognized that if this occurs, anticlines and synclines would form (Fig. 9-18). This example was eventually modeled, utilizing a volume conservation concept (Suppe and Namson 1979; Suppe 1980; and Suppe 1983), and the kinematics of the process are as follows.

A decollement on a lower structural level (Y level, Fig. 9-19a) ramps to a higher stratigraphic level (X level). Motion along the fault and the conservation of volume principle cause the beds to ride up the ramp and roll through axial surface BY. This causes the back dip panel (or flap, BYY'B' to form (Fig. 9-19a). The two axial surfaces (BY and B'Y') and the back dip panel terminate at the fault surface, as they are produced by the bend in the decollement as the beds move up the ramp. Similarly, the beds moving

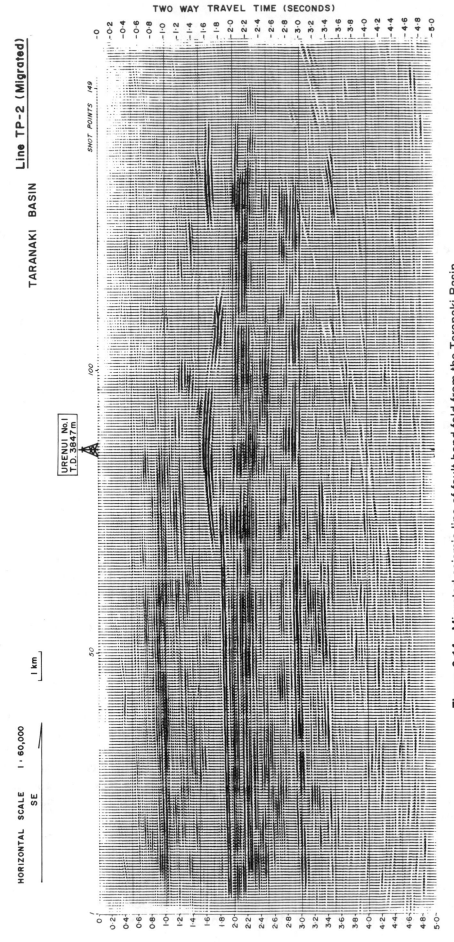

Figure 9-14 Migrated seismic line of fault bend fold from the Taranaki Basin, New Zealand. A symmetric fold is imaged in the vicinity of the well and sp 100 between the two way travel times at 1.5 sec to 1.9 sec. In the vicinity of sp 100, dipping beds overlie flat beds, indicating a decollement. (From Seismic Atlas of Australian and New Zealand Basins, by Skilbeck and Lennox 1984. Published by permission of Earth Resources Foundation, University of Sydney.)

Figure 9-15 Fault bend fold in Hudson Valley, New York, located on Route 23 about 300 meters west of New York Thruway. (Compliments of Jon Mosar.)

up the ramp and onto the flat must roll through axial surface AX, which forms the frontal dip panel AXX'A' (Fig. 9-19a). As the beds roll through axial surface AX, they are assumed to experience *bedding plane slip* within the frontal dip panel. This slip produces shear in the frontal limb of the fold [Fig. 9-19(b)] and causes the frontal limb to dip at a higher angle than the back dip panel. This point is emphasized here as it will be applied to the solution of more complicated problems in the section on Duplexes. As the beds

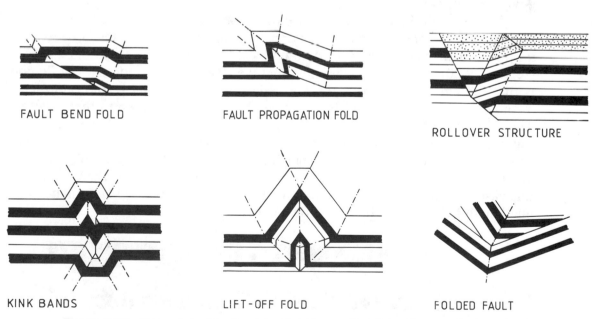

FAULT BEND FOLD   FAULT PROPAGATION FOLD

ROLLOVER STRUCTURE

KINK BANDS   LIFT-OFF FOLD   FOLDED FAULT

Figure 9-16 Examples of fault related fold types. (Published by permission of John Suppe.)

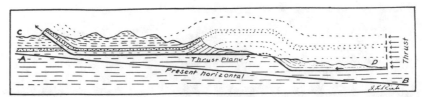

Figure 9-17 Fault bend fold forming over a step-up on a thrust fault. (From Rich 1934. Published by permission of the American Association of Petroleum Geologists.)

roll through the active axial surface BY and AX, a fracture porosity is likely to form in the deformed beds.

As the fold grows, fault slip increases, the dip panels extend in width, and point Y′ migrates toward point X [Fig. 9-19(b)] until the fold attains its maximum amplitude. When this occurs, axial surface B′Y′ has migrated to the top of the ramp and point Y′ reaches the upper footwall cutoff [point X in Fig. 9-19(b)]. The fold now extends by the lateral motion of axial surface AY′ away from axial surface B′X (Fig. 9-19c). As the two dip panels move away from each other, and as the fold has reached its maximum amplitude, no material is currently rolling through the AY′ axial surface. This surface has become inactive. However, material continues to roll through the B′X and BY surfaces, probably fracturing the rock.

The resulting *idealized* fold shape, caused by simple step-up of material along a ramp and onto a flat, has a frontal dip panel that contains slightly higher dips than the back dip panel (β is usually slightly greater than θ; Fig. 9-19c). Thus, the fold geometry, particularly at cutoff angles of less than about 20 deg, is roughly symmetric.

The mathematics of this model can be summarized in the form of a graph (Fig. 9-20). This model vigorously utilizes the *kink method*, and as this method conserves volume, line length, and bed thickness, it is **not** necessary to retrodeform a solution which is derived from Fig. 9-20. If the data conform to the angles presented in Fig. 9-20, the interpretation will **automatically retrodeform.** Thus, the graphical methods presented in this section possess several advantages which are amenable to working in a production environment.

Let us apply Fig. 9-20 to the case of a fault that steps off a decollement at a 20-deg (initial) cutoff angle and ramps to an upper flat that parallels the lower decollement (Fig. 9-21a). This means that $\phi = \theta$. Also notice that when $\phi = \theta$, θ can not exceed 30 deg (see Fig. 9-20). The other assumption that we shall make for purposes of demonstration is that the amount of slip on the lower decollement is equal to the ramp length. This means that the axial surface (B′Y′ in Fig. 9-19b) has moved up to the top of the ramp. The initial cutoff angle θ can now be read off of the abscissa and projected upward on Fig. 9-20 until this line intersects the $\phi = \theta$ line. Next, the dip of the frontal flap (β) can be read off of the more steeply dipping lines on Fig. 9-20, which in this case is about 23 deg (also see Table 9-1); and the axial angle (γ) can be read off the ordinate, which in this case is 78.5 deg. The final solution, shown in Fig. 9-21b, will automatically area and line length balance, but there is a final check that should be made.

Figure 9-18 Model of fault bend fold constructed from paper sheets. (From Rich 1934. Published by permission of the American Association of Petroleum Geologists.)

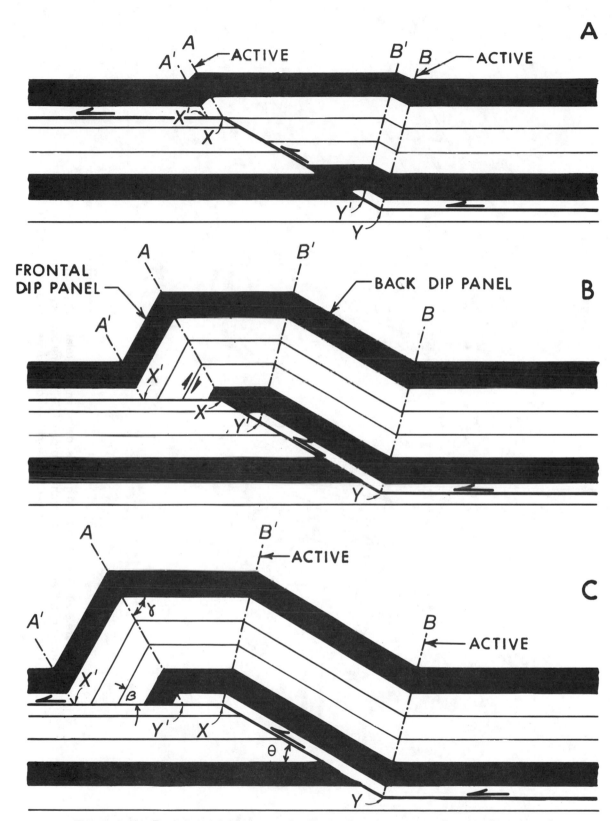

Figure 9-19 Fault bend fold kinematics illustrating the progressive development of beds riding up a thrust ramp. The beds are deformed by the active axial surfaces. (Modified after Suppe 1983, 1985. Published by permission of the American Journal of Science.)

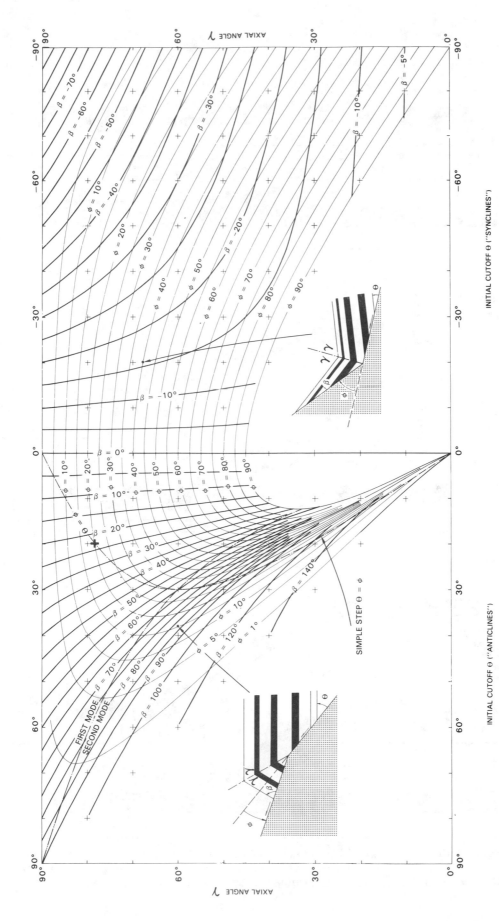

Figure 9-20 Fault bend fold graph showing angular relationships between the initial cutoff angle (θ), the frontal dip panel (β), and the axial surface angle (γ). (From Suppe 1983. Published by permission of the American Journal of Sciences.)

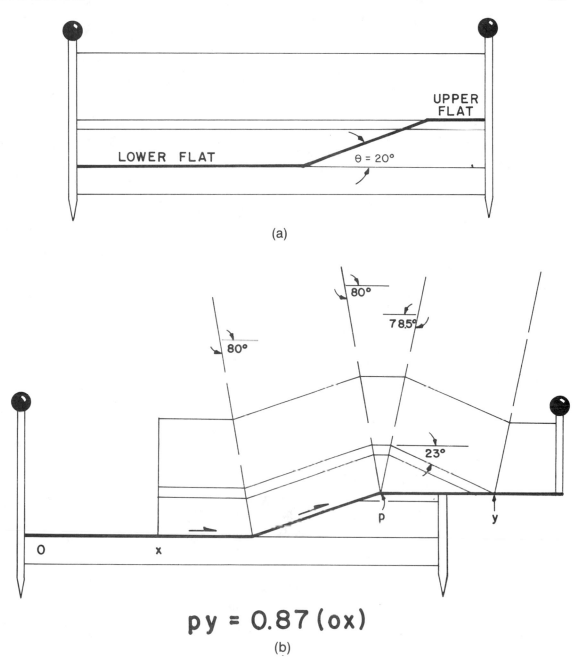

$$py = 0.87(ox)$$

(b)

Figure 9-21 Fault bend fold exercise for beds ramping up a 20-deg cutoff angle.

Previously we stated that as the beds rolled through axial surface AX (Fig. 9-19a), the deformation was accommodated by bedding plane slip within the frontal dip panel. This is required to conserve both volume and bed thickness, and causes angle β to be larger than angle θ. Thus, some of the slip which rode up the ramp is *consumed within* the beds of the frontal dip panel, and causes the amount of slip along the upper flat to be less than the amount of slip along the lower flat. The amount of slip to be expected along the upper flat can be determined from Fig. 9-22. Again a line is projected from the 20-deg cutoff angle upward off the abscissa to the φ = θ = 20 deg line that we have assumed for this example.

The ratio of the slip (R lines) on the upper flat relative to the lower flat can now be read off the diagram, which in this case is about 0.87; therefore, the slip along the upper flat must be 0.87 of the slip along the lower flat. Field geologists have often observed

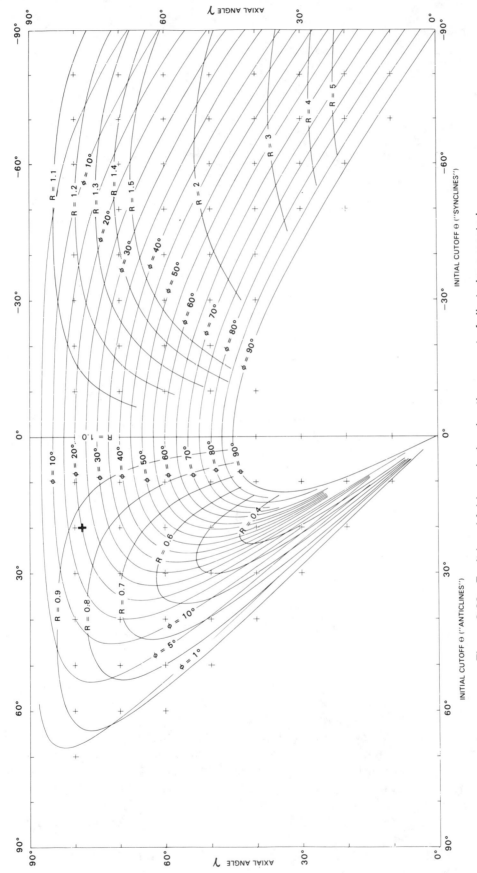

Figure 9-22 Fault bend fold graph showing the amount of slip to be expected along different portions of a fault surface. The R lines indicate the slip ratio along the upper flat relative to slip along the lower flat. (From Suppe 1983. Published by permission of the American Journal of Science.)

that the slip on faults dies or decreases within the cores of folds. This exercise is most useful when experimenting with structures that exhibit unusual geometries or complicated shapes. Another check on the solution would be to measure bed lengths on more than one structural level.

Fault Propagation Folds

Fault propagation folds are the most common fold type to be observed in outcrop and in seismic (Figs. 9-23 and 9-24), and like fault bend folds are known to be good producers. Fault propagation folds, as the name implies, possess the particular characteristic that as the fold grows the deformation advances at the tip of a propagating thrust fault (Fig. 9-24), hence the name fault propagation fold (Suppe 1985). As long as the structure has not been faulted through (i.e., been subject to breakthrough), the slip is consumed by bedding plane displacements located along the frontal limb of the fold (Fig. 9-25c).

Fault propagation folds typically have higher cutoff angles than fault bend folds, in the range of about 20 deg to 40 deg, which causes these fold types to possess steeply dipping to overturned frontal limbs that are not imaged on seismic sections, along with a characteristic *asymmetry* (Fig. 9-26). This striking asymmetry, when imaged in seismic sections, gives the appearance of a striking snake, giving rise to the expression *"snake head."*

The kinematics of fault propagation folds are as follows: a fault, propagating upward from a decollement, causes the beds in front of the propagating fault tip to bend forward and to move up the ramp (Fig. 9-25a). As in fault bend folding, the beds will also roll up the ramp created by the propagating thrust fault and through axial surface B, creating the back dip panel outlined by axial surfaces B and B' (Fig. 9-25a).

The exercise supplement to this book describes how this style of folding tends to accommodate the slip along the fault surface by an increase in the amount of deformation within the core of the fold. Therefore, the beds near the end of the thrust fault bend forward, often at steep angles (Fig. 9-25a). This intense deformation within the more steeply dipping beds, as well as bedding plane slip between the beds, *consumes* the slip along the thrust fault. Thus, the slip dies out within the core of the fold. In addition, the more steeply dipping beds between the front and the top of the structure form two axial surfaces A and A' (Fig. 9-25a).

As the fold grows in amplitude and the propagating fault moves forward, it incorporates more material into the frontal limb of the structure. Consequently, as the fault moves forward and as axial surface A' moves away from axial surface A, axial surface A' incorporates point 2 of Fig. 9-25b into the steeply dipping frontal limb. With increasing deformation, the dip panels defined by axial surfaces AA' and BB' broaden (Fig. 9-25c). Also notice that axial surface B' is an *active* surface in that it modifies the dip of axial surface A as axial surface B' moves forward. This has the effect of increasing the amplitude of the fold, subjecting the structurally lower layers to tight folding and incorporating more material into the back limb of the structure (Fig. 9-25c). The active axial surface also has the effect of fracturing the rock.

Fault propagation folding can exhibit a variety of structural styles depending upon the cutoff angle (Fig. 9-26) and the amount of slip. As the cutoff angle increases, and for the same amount of slip, the folding will appear to be more symmetric on seismic sections even though the amount of slip remains unchanged. If the fold forms according to the processes described in Fig. 9-25, the cutoff angle can be determined directly from the dip of the beds within the back dip panel as these beds parallel the ramp.

Given additional amounts of slip, the fault propagation may find a weak or incom-

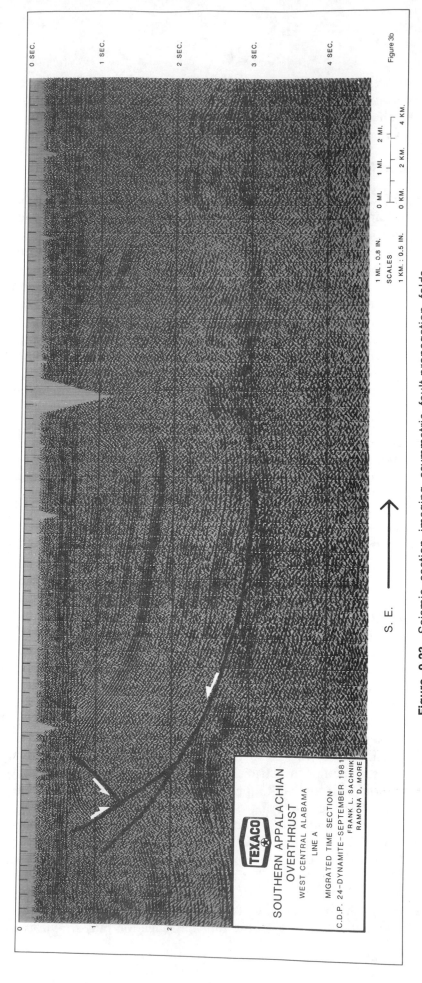

Figure 9-23 Seismic section imaging asymmetric fault-propagation folds, Southern Appalachians, Alabama. (Interpretation by Dick Bischke. After Sachnik and More. From *Seismic Expression of Structural Styles* by Bally 1988. Published by permission of the American Association of Petroleum Geologists.)

S. E.

SOUTHERN APPALACHIAN
OVERTHRUST
WEST CENTRAL ALABAMA
LINE A
MIGRATED TIME SECTION
C.D.P. 24-DYNAMITE-SEPTEMBER 1981
FRANK L. SACHNIK
RAMONA D. MORE

TEXACO

SCALES
1 MI. : 0.8 IN.
1 KM. : 0.5 IN.

0 MI.   1 MI.   2 MI.
0 KM.   2 KM.   4 KM.

0 SEC.
1 SEC.
2 SEC.
3 SEC.
4 SEC.

Figure 3b

430

petent horizon that parallels bedding and become a hybrid fault bend fold (Fig. 9-27c). Alternatively, the structure can break through the anticlinal, the synclinal, or the overturned limb portions of the fold, creating more complex geometries (Figs. 9-27a, b, and d).

As with fault bend folds, fault propagation folds can be balanced utilizing formula or graphs (Suppe and Medwedeff 1984; Suppe 1988) (Fig. 9-28). In outcrop or on seismic sections, fault propagation folds can be balanced by observing either the ramp angle (θ) or the back limb dip of the fold. Remember to depth correct the seismic or to choose sections that are roughly on a scale of one to one. This can be readily accomplished by digitizing the more prominent seismic horizons and by depth correcting the digitized profiles utilizing interval or stacking velocities. We have found this procedure to be adequate for most cases.

We shall study a simple case for balancing fault propagation folds. For example, you may observe beds on a depth converted seismic section, dipping at 30 deg and overlying flat beds. As the back limb beds are determined to dip at 30 deg, the corresponding axial angles γ_p and γ_p^* can be read off of Fig. 9-28 (about 53 deg and 38 deg, respectively). The kink method can now be employed as follows. First construct a 30-deg dipping ramp. At the tip of the thrust fault, construct the structurally lower γ_p axial surface which dips at 53 deg (Fig. 9-29a). The tip of the thrust fault also determines the position of the branch point along the γ_p and γ_p^* axial surfaces (Fig. 9-28). As the dip of the front limb (β) is defined by

$$\beta = \theta + 2\,(\gamma_p^*)$$

Figure 9-24 Fault propagation fold, Appalachians, Tennessee. The frontal limb dips more steeply than the back limb. (From Suppe 1985.)

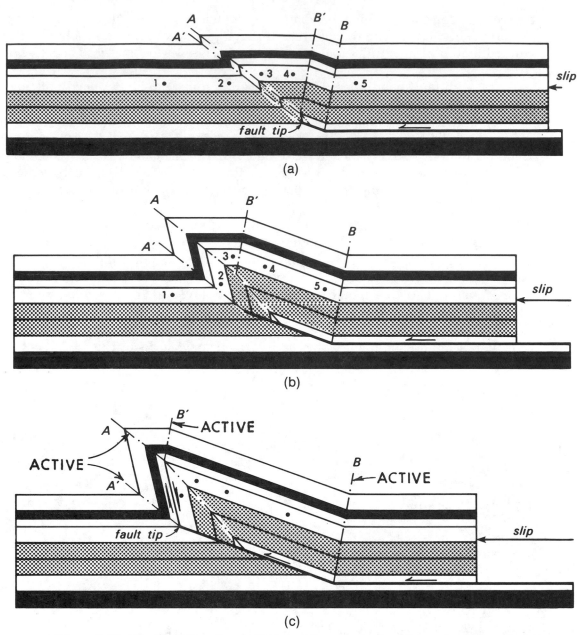

Figure 9-25 Fault propagation fold kinematics, illustrating the progressive development of beds deforming at the tip of a propagating thrust fault. (Modified after Suppe 1985.)

(see Fig. 9-28, 9-29b), the frontal limb dip in this case is 106 deg. The position of the branch point is located by projecting the horizon, which is on the level of the fault tip, in from the right. This horizon is bent upward by the active axial surface (surface B, Fig. 9-25), is bisected (105 deg, Fig. 9-29b), and then is projected upward to where it terminates at the intersection with the front dip panel. The length of the frontal dip panel will now determine where the slip terminates along the decollement. Next, the structurally higher γ_p axial surface can be drawn in, along with the lower γ_p^* axial surface (Fig. 9-29b). The elements of the structure are now complete, additional layers can be projected throughout the structure, and the line lengths can be measured for area confirmation (Fig. 9-29c).

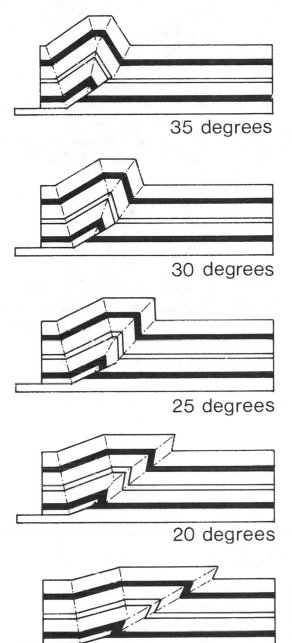

35 degrees

30 degrees

25 degrees

20 degrees

15 degrees

Figure 9-26 Fault propagation folds at different cutoff angles. Frontal limb dips increase as the cut-off angle decreases. At low cutoff angles, the limb dips are too high to be imaged on conventional seismic sections. (Published by permission of John Suppe.)

Imbricate Structures

As the thrust belt moves progressively over the foreland, there is a tendency for new thrust faults to form near the toe (front) of the thrust wedge and for these thrusts to seek a lower level. Thrust faults which form at a higher structural level have been, perhaps inappropriately, called *out-of-sequence* thrusts. When a thrust fault forms below a pre-existing fault(s), motion along the deeper fault will cause the shallow fault(s) to bend or fold, and for the deformation to produce some rather interesting and complex geometries (Fig. 9-30).

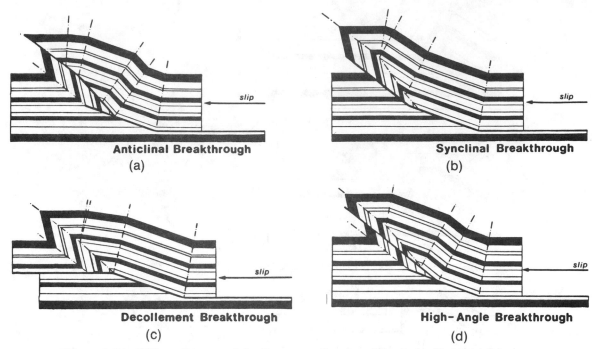

Figure 9-27 Different types of fault propagation breakthrough. (Published by permission of John Suppe.)

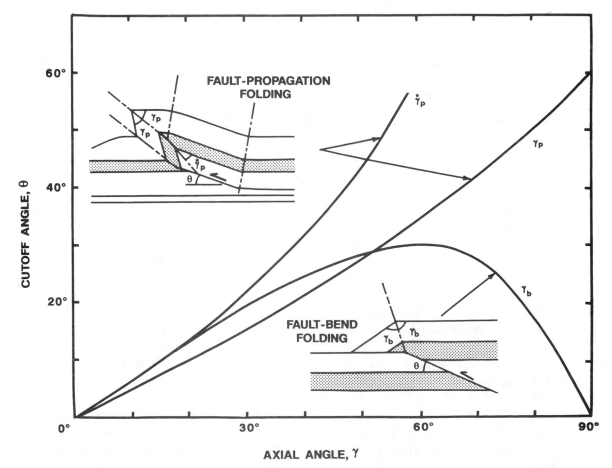

Figure 9-28 Fault propagation fold graph for a simple step-up from a decollement surface. This simple theory applies to cutoff angles greater than 20 to 30 deg. (Modified from Suppe 1985.)

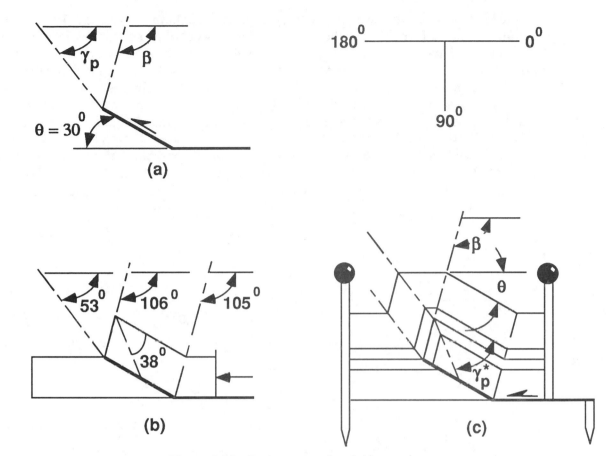

Figure 9-29 Fault propagation fold exercise.

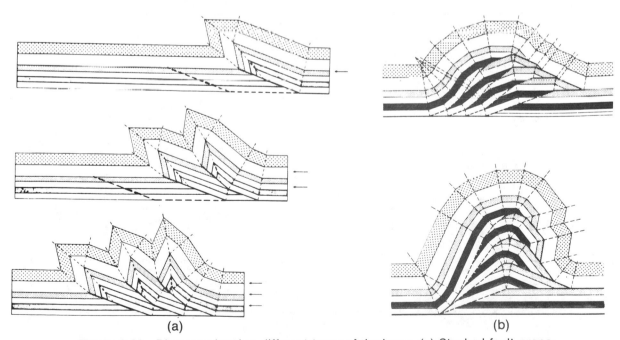

Figure 9-30 Diagram showing different types of duplexes. (a) Stacked fault propagation folds (b) Foreland and anticlinal stacked duplexes (From Mitra 1986. Published by permission of the American Association of Petroleum Geologists.)

This complex process is best described through example. We shall first assume that a fault bend fold formed near the front of a thrust wedge, as shown in Fig. 9-31a. In this example we have assumed that the cutoff angle is (as before) 20 deg, and that the fault formed ramp AB. We can now determine the frontal dip panel angles utilizing the methods developed in the section on Fault Bend Folds. We shall also assume for purposes of demonstration that, at a particular point in time, another ramp forms in front of the ramp that formed the fault bend fold. The new ramp which forms along the lower decollement is ramp CD (Fig. 9-31b).

This wedge-shaped structure, which is completely surrounded by ramps AB and CD and flats BD and AC, is called a **horse** (Boyer and Elliott 1982). If several horses move up their ramps, then they form a *duplex* of folded imbricate thrusts (Fig. 9-30).

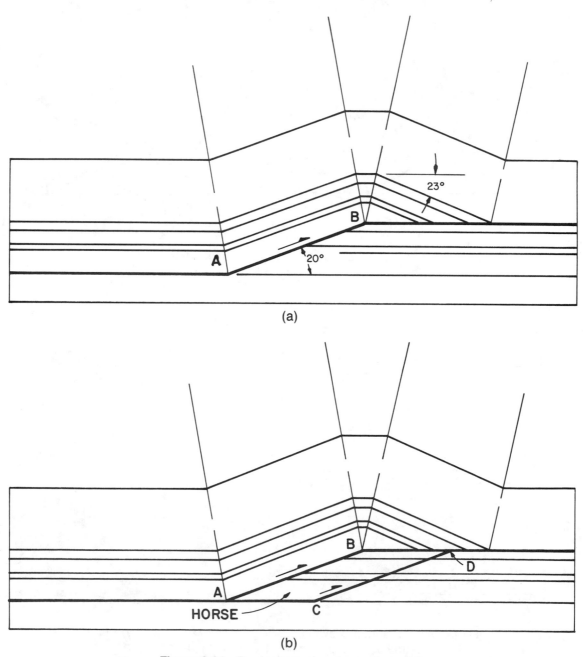

(a)

(b)

Figure 9-31 Duplex exercise, forward model.

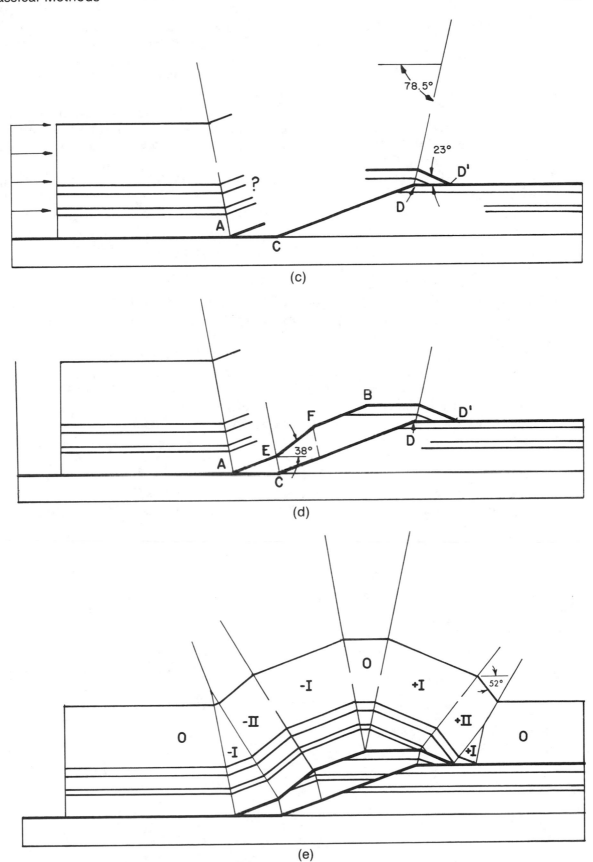

(c)

(d)

(e)

Figure 9-31 *(continued)*

Next we assume, again for purposes of demonstration, that the thrusting has progressed to a stage that the original distance between points A and C has been exactly halved, which means that only a portion of the rhomb-shaped horse has moved onto the upper flat (Fig. 9-31c). As the frontal portions of the rhomb-shaped horse move up the ramp, its frontal potions will bend in the same manner as layers deform during fault bend folding. Thus, for this portion of the deformation, we are able to determine the deformed shape of the horse utilizing the kink method and the techniques developed in the section on Fault Bend Folds. It now follows that as the horse moves up a 20-deg ramp and onto an upper flat, it will have the same frontal dip angle (B) as the fault bend fold had when it moved up ramp AB in Fig. 9-31a. Therefore, frontal dip angle can be determined to be 23 deg from Fig. 9-20, or from Table 9-1, and the amount of slip consumed by the bending of the layers within the frontal dip panel of the horse can be determined from Fig. 9-22, which in this case is about 0.87 of the total slip, or 0.87 of ½ AC (Fig. 9-31c). Therefore, after deformation, the distance DD′ will be equal to 0.87 of the original distance AC in Fig. 9-31b.

Next we bisect the angle between the portions of the horse that were bent (where it rode up the ramp) and its undeformed portions (in this case 78.5 deg), and then project the top (upper most) layers of the horse to the left. As bed length consistency requires that the layers be the same length before and after deformation, length BD before deformation (Fig. 9-31b) should be equal to length BD′ after deformation (Fig. 9-31d).

The problem of the deformed horse can now proceed as follows. The portion of ramp AB located near point B (Fig. 9-31b) rode up ramp CD without being deformed. As the cutoff angle is 20 deg, the upper portions of ramp AB can be projected downward at a 20-deg angle in Fig. 9-31d (i.e., line segment FB). Similarly, the lower portions of ramp AB (located near point A of Fig. 9-31b) slid along the lower flat without being deformed, and thus the lower portions of the ramp AB can be projected upward at a 20-deg angle (see line AE in Fig. 9-31d).

The central portions of ramp (i.e., line segment EF in Fig. 9-31d), however, have been subject to deformation as the horse moved up ramp CD and through axial surface

TABLE 9-1

Dip Spectral Analysis

Foreward dips (+)							Fundamental cutoff angle° θ	Back dips (−)						
VII	VI	V	IV	III	II	I		I	II	III	IV	V	VI	VII
61.6°	52.5°	43.0°	34.0°	25.2°	16.6°	8.2°	8°	8°	15.9°	23.4°	30.6°	37.3°	43.5°	49.3°
70.2°	59.2°	48.6°	38.3°	28.3°	18.6°	9.2°	9°	9°	17.8°	26.2°	34.0°	41.3°	47.9°	53.9°
80.6°	67.6°	55.2°	43.3°	31.9°	20.9°	10.3°	10°	10°	19.7°	28.9°	37.4°	45.1°	52.0°	58.2°
93.1°	77.3°	62.6°	48.8°	35.7°	23.3°	11.4°	11°	11°	21.6°	31.5°	40.6°	48.7°	55.9°	62.2°
109°	88.8°	71.0°	54.8°	39.8°	25.8°	12.6°	12°	12°	23.5°	34.1°	43.7°	52.1°	59.5°	65.9°
128°	102°	80.5°	61.5°	44.3°	28.5°	13.8°	13°	13°	25.4°	36.7°	46.7°	55.4°	62.9°	69.4°
160°	119°	91.3°	68.6°	48.9°	31.2°	15.0°	14°	14°	27.2°	39.1°	49.5°	58.4°	66.1°	72.5°
—	146°	104°	76.3°	53.6°	33.9°	16.2°	15°	15°	29.1°	41.5°	52.3°	61.4°	69.0°	75.5°
	—	124°	85.9°	59.0°	36.8°	17.4°	16°	16°	30.9°	43.9°	54.9°	64.1°	—	—
		—	99.2°	65.6°	40.2°	18.8°	17°	17°	32.7°	46.2°	57.5°	—		
		—	123°	73.1°	43.7°	20.2°	18°	18°	34.4°	48.4°	59.9°	—		
			—	82.2°	47.4°	21.6°	19°	19°	36.2°	50.6°	—			
			—	97.6°	52.0°	23.2°	20°	20°	37.9°	52.7°	—			
				—	57.0°	24.8°	21°	21°	39.6°	—				
				—	63.6°	26.6°	22°	22°	41.3°	—				
				—	72.0°	28.4°	23°	23°	42.9°	—				
					—	30.4°	24°	24°	—					

Published by permission of the American Journal of Science (Suppe 1983).

EC (Fig. 9-31d). Axial surface EC is pinned to the lower footwall cutoff at point C in Fig. 9-31d. As ramp CD has a 20-deg initial cutoff angle, axial surface EC dips at 80 deg.

We have now determined that, as fault AB moves through axial surface EC, it must be deformed (bent upward); and as ramp AB dips at 20 deg before deformation, it must dip at an *even higher angle* after deformation! As a 20-deg dipping line (line AB) ran up a 20-deg dipping ramp (line CD of Fig. 9-31d), one might *incorrectly* conclude that the central deformed portion of fault AB (i.e., line segment EF in Fig. 9-31d) presently dips at 40 deg. In the section on fault bend folding we learned that in order to maintain line lengths, the angle β or the dip of the frontal flap must be greater than the (initial) cutoff angle (θ). This relationship must be maintained on every structural level within imbricate structures. Therefore, as the frontal portions of the horse cause the overlying beds above it to dip forward at a higher angle (+ II region in Fig. 9-31e), the dips at a higher structural level will experience a *quantum increase* in dip (Suppe 1980, 1983). In other words, as a 20-deg ramp caused a 23-deg frontal dip (Fig. 9-31a), the insertion of the horse onto the upper flat will cause the uppermost beds (+ II region in Fig. 9-31e) to dip at an angle which is greater than twice 23 deg, or in this case 52 deg (Table 9-1). In order to maintain line length and formation thickness, the strata above the horse will shear in such a manner than an *increase* in dip in the frontal panel is accommodated by a *decrease* in dip in the back panel. Thus, the *compensating dip* of line segment EF in Fig. 9-31d is 38 deg and not 40 deg (Table 9-1).

Although this exercise may at first seem to be an annoying complication, we shall use these small changes of dip to our advantage in what is called **dip spectral analysis** (Suppe 1980, 1983). This procedure can be utilized to develop the often poorly imaged subthrust plays.

Our problem can now be completed by bisecting the angles along the deformed upper ramp (e.g., angles AEF, EFB, etc. in Fig. 9-31d) and projecting the dip domains toward the surface (Fig. 9-31e). At the top of the upper ramp located at point B (Fig. 9-31a) exists the rollover associated with the original fault bend fold, and a frontal dip panel that dips at 23 deg. The frontal dip panel of the original fault bend fold was subsequently deformed by the frontal portions of the horse (between points D and D' in Fig. 9-31c), and thus the beds above the deformed portion of the horse will dip at an even higher angle, which can be determined from Table 9-1 to be 52 deg (Fig. 9-31e). Projecting in the upper strata yields the final solution shown in Fig. 9-31e.

Consider the geometry present in Fig. 9-31e which has several implications concerning petroleum exploration. Assume that the thin horizon above the lower flat is a productive reservoir horizon. Notice that this reservoir can be intersected on two structural levels resulting in two potential plays. The first play is the closure associated with the original fault bend fold present in Fig. 9-31a. The second play is a partial closure located within the horse, and thus its prospectivity would depend on the trapping mechanism, or the permeability of the beds above the thin reservoir horizon.

Dip Spectral Analysis. Next we review how to locate potential subthrust plays in practice. We learned from the imbricate structure exercise (previous section) that, as one moves up the structural pile, the dips at the front of the imbricate structures increase at an increasing rate. Furthermore, the back dips, while exhibiting a corresponding increase, do so at a decreasing rate. Therefore, the frontal dips exhibit a *unique quantum increase* in dip (in our case 52 deg is greater than twice 23 deg), while the back dips exhibit a *unique quantum rate decrease* in dip (Table 9-1). These unique changes in dip allow us to determine the number of subthrusts and their approximate position.

Notice that in Fig. 9-31e the final structure exhibits three frontal dip panels or domains (labeled $+I+II$, and $+I$) and three back dip panels (labeled $-I$, $-II$, $-I$) that are separated by a region of initial dip (labeled 0). In nature, detailed mapping across the structure shown in Fig. 9-31e could result in the topographic section shown in Fig. 9-32a. In this figure, the following dips occur from left to right: 0, -38, -20, 0, 23, 52, 0, -20, -20, -20, and 0 deg. If a regional back dip of 5 deg were to exist, then the corresponding dips would be -5, -43, -25, -5, 18, 47, -5, -25, -25, -25, and -5 deg; therefore, a brief analysis of these numbers suggests that a regional dip of 5 deg

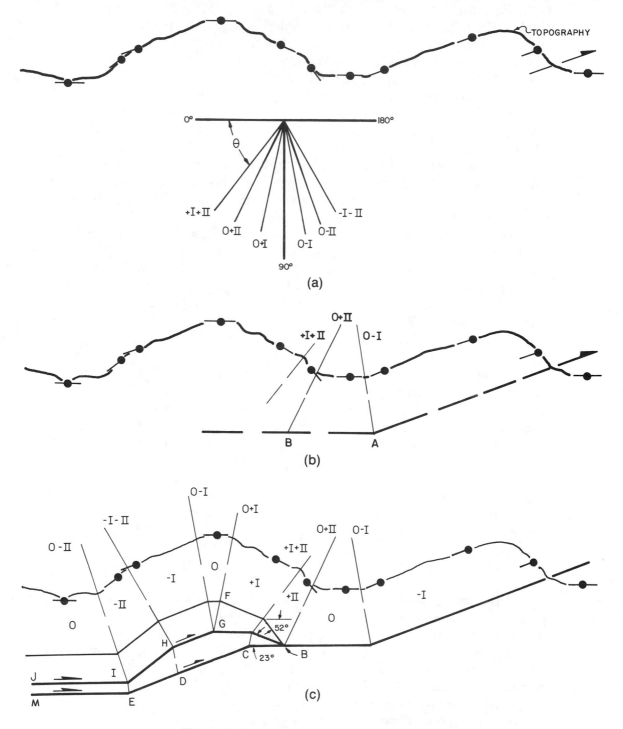

Figure 9-32 Duplex exercise, inverse model.

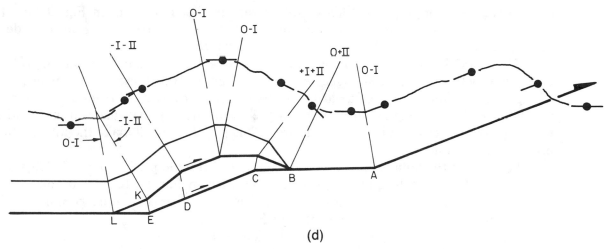

(d)

Figure 9-32 (*continued*)

should be removed prior to analysis. The resultant dips corrected for regional dip are 0, −38, −20, 0, 23, 52, 0, −20, −20, −20, and 0 deg. We have shown that the higher dips exist at the front of the structure, while lower dips occur at the back (Fig. 9-31c). Thus, the 52-deg dip and its associated 23-deg dip are *forward dips*, while the negative −38-deg dip and its associated −20-deg dip are *backward dips* (see Table 9-1). As there are two forward dips, the 52-deg dip represents a +II domain, while the 23-deg dip represents a +I dip domain. These numbers, each of which represents an individual dip domain, can be compared to Table 9-1 (line 13) to indicate that there are two thrusts ramping at 20 deg causing second-order frontal and back dips of 52 deg and −38 deg, respectively, and first-order frontal and back dips of 23 deg and −20 deg, respectively. If three faults are present, then we would expect an additional forward dip of about 98 deg and a back dip of about 53 deg. As only two forward dips exist in our example, Table 9-1 suggests that only two faults are present. Therefore, surface and subsurface data can be compared to Table 9-1 in order to determine the number of imbrications that exist.

Let us proceed to solve the structure presented in Fig. 9-32a, which is based on Fig. 9-31e. The first task to be accomplished is to determine the dips of the axial surfaces which separate every one of the observed dip domains. Once the *related* dips within each of the dip domains have been averaged as was described in the section on Kinking, the axial surface can be determined from the following formula:

$$\theta = (\text{Dip}_1 + \text{Dip}_2 + 180°)/2$$

where θ = dip of axial surface taken counterclockwise from the horizontal
Dip_1 = average dip in dip domain 1
Dip_2 = average dip in adjacent domain 2

Dip is taken to be *negative* if it is in the 90 deg to 180 deg quadrant (see Fig. 9-32a for sign convention).

The dips of the axial surfaces for the appropriate dip domains presented in Figs. 9-32a and 9-31e are given in Table 9-2. For example, at the top of the hill a 0-deg dip exists adjacent to a 23-deg dip (Fig. 9-32a). These two dips represents a 0 and a +I dip domain (Fig. 9-31e and Table 9-1), and at the intersection of the two domains they form a dip domain boundary, which by definition is the 0+I axial surface. Thus, the 0+I dip domain boundary dips at $(0 − 23 + 180)/2 = 78.5$ deg (Table 9-2). Notice that Fig. 9-32a does not include the −20-deg dip panel above the lower flat that is present in Fig. 9-31e. This should make our solution more interesting and realistic. The 0−II data are included in Table 9-2 for reasons which will soon become apparent.

These axial surface dip calculations are best applied to the solution of problems utilizing a method which was suggested by John Suppe, and shown in the lower portion of Fig. 9-32a. The dip of each axial surface is projected downward from a central point, and labeled. Two triangles can now be aligned with the axial surface dip data and slid into any position that the interpreter desires. This procedure will make the interpretation process more rapid.

Possessing the data set presented along the topographic profile in Fig. 9-32a and the knowledge of the solution to the problem presented in Fig. 9-31e will not allow us to arrive at a unique solution to our problem. Trial and error and some guessing will be required to solve the problem presented in Fig. 9-32a.

The first step in the solution of our problem is to examine the data from a geometric point of view. Two observations are critical to an accurate interpretation. First, the observed dips tend to follow the topographic slope, which is often the case in nature. Second, the beds above the thrust fault dip at about the same angle as the thrust fault, which is also commonly observed, suggesting that a ramp is responsible for the 20-deg tilt to these beds. Therefore, the thrust fault can be projected downward to where it intersects the adjacent (0-deg) dip domain. The change in topography can then be utilized to position and project the O−I axial surface downward to the point where we predict the thrust fault to ramp (Fig. 9-32b, point A). Accordingly, we have also predicted the structural level of the upper flat or decollement (Fig. 9-32b).

We now apply the following reasoning. Table 9-1 was utilized to suggest that there are two 20-deg ramping thrusts that created the observed 23-deg and 52-deg forward dips, and thus these two thrusts must be imbricated (i.e., stacked) in order to produce the observed quantum increase in dip. The data suggests that a 52-deg dip domain adjoins a 0-deg dip domain and therefore a 64-deg axial surface (the 0+II domain of Table 9-2) is positioned at the appropriate change in topographic slope (Fig. 9-32b). We know that two things happen to the left of the intersection of the 0+II axial surface and its intersection with the upper flat (Fig. 9-32b, point B). First, the structurally higher beds dip at 52 deg to the right, and second, a 23-deg deformed thrust must exist beneath the 52-deg dipping beds. This follows from a direct application of Table 9-1 and the theory that we presented earlier.

Consequently, at the intersection of the 0+II axial surface and the upper flat, we draw in the 52-deg dipping beds and the deformed 23-deg dipping fault (Fig. 9-32c). The 52-deg dipping beds are then projected up to the +I+II axial surface (Fig. 9-32c, which

TABLE 9-2

Axial Surface Calculations

Dip domain boundary*	$(Dip_1{}^* + Dip_1{}^* + 180°)/2$	Dip of axial surface (in degrees)
−I − II	(38 + 20 + 180)/2	119.0
0 − II	(0 + 38 + 180)/2	109.0
0 − I	(0 + 20 + 180)/2	100.0
0 + I	(0 − 23 + 180)/2	78.5
0 + II	(0 − 52 + 180)/2	64.0
+I + II	(−23 − 52 + 180)/2	52.5

* Dips and dip domains are taken from Figure 9-32a.

is positioned at the break in topographic slope). At this point, the kink method is applied to the deformed horse block and a 78.5-deg (0+I) axial surface is drawn downward to a point where it intersects the upper flat (Fig. 9-32c, point C). The frontal portions of the *deformed* horse block have now been properly bisected (i.e., we have bisected the undeformed and the deformed regions of the horse block). This point, labeled C in Fig. 9-32c, not only marks the position where the horse block has ridden onto the upper flat, but also determines where the structurally lower ramp can be projected downward at a 20-deg angle (Line EC, Fig. 9-32c).

Proceeding with the construction: the 52-deg dipping beds being bent by the +I+II axial surface can be projected at a 23-deg angle up to the 0+I axial surface, or to where the beds go horizontal (point F, Fig. 9-32c). Notice that the 0+I and 0−I axial surfaces converge at point G, which defines the edge of the horse block or the upper footwall cutoff to the original fault bend fold (Fig. 9-31a, point B). At this point, the upper fault can be projected downward at a 20-deg angle up to the −I−II axial surface (point H, Fig. 9-32c), which is defined by the data and by a minor break in the topographic slope. At this intersection, the kink method is applied once again and the deformed fault is projected downward at a 38-deg angle, as defined by the dip data, until the upper fault intersects the 0−II axial surface (point I, Fig. 9-32c). Bisecting the angle at the 0−II axial surface projects the upper fault to level out on the IJ structural level. At point I, the beds between the upper and the lower fault are bisected (a 100-deg axial surface) and a 0−I axial surface is drawn in to predict the position of the lower fault's footwall cutoff (point E, Fig. 9-32c).

What is the result? We have apparently obeyed all of the rules with a result that is somewhat complex and confusing! The upper and lower faults have not merged on the same structural level (compare I to E, Fig. 9-32c). How can we improve the solution and what features should we look for when attempting to arrive at a more satisfactory solution? First, the distance ED that the horse block has ridden up the lower ramp is not compatible with the distance CB that the horse block has ridden onto the upper flat. We learned previously that CB must be 0.87 of ED (Fig. 9-21 and 9-22). Thus, an error was made in this portion of our analysis that represents not only a clue to the proper solution, but also the approximate position of our difficulties. Second, returning to Table 9-1 we re-examine the possible dip domains that are associated with two 20-deg ramping thrusts. As there are obvious problems at the back of the structure, perhaps the problem lies in this region. Table 9-1 suggests that a −I dip domain can exist between a 0 and −II back dip domain. Labeling the dip panels on our solution reveals that a −I-deg dip domain was not included into the back area portions of our solution.

We now proceed to backtrack, modifying the first solution by inserting a −I (back) dip domain between the −II and the 0 back dip panels. We know that the distance ED in Fig. 9-32d should be equal to CB/0.87, and this distance is entered on the figure. Therefore, from the *calculated* position of point E we project a 0−I axial surface upward to the base of the deformed upper thrust (point K in Fig. 9-32d). From this point we project a −I−II axial surface upward toward the surface. It also follows from this reasoning that to the left of point K exists a 20-deg back dip panel and a −I−II axial surface that intersects the decollement at point L. This solution, although slightly different from Fig. 9-32c, creates a deformed horse block.

From this complex exercise we conclude: (1) duplexes produce more rounded structures; (2) structural balancing even under ideal situations can be nonunique; (3) the interpretation process can be rigorous, but in this age of domestic energy shortfalls the rewards could be substantial. The better we understand the detailed geometry of structures, the more likely we are to extract oil from these structures.

Box and Lift-Off Structures

Box and lift-off structures represent a particular class of folds, which when viewed relative to the regional dip are roughly *symmetric* but *angular* structures (Figs. 9-33 and 9-34). Both structural types form in association with a zone of weak detachment located at depth, and possess the characteristic that the decollement is isoclinally folded into the hanging wall (Laubscher 1961; Namson 1981). In the Jura Mountains, this zone of weakness consists of evaporites (i.e., gypsum), although over-pressured shales are likely to produce a similar deformational style. Box and lift-off structures differ from diapiric structures in that there is less mass transport or flow into the cores of these folds. This causes the box and lift-off structures to have almost vertically dipping limbs at the lower structural level. In addition, diapiric structures result from a gravity instability, whereas box and lift-off folds result from compression.

Box and lift-off structures can be recognized in outcrop or seismic from their bilateral symmetry, but also from their angular geometry. If broad zones of vertically dipping beds are encountered in outcrop (e.g., 70–80 deg), these structural styles are to be suspected (Fig. 9-34).

Box folds have nearly flat tops and vertically dipping limbs and axial surfaces which dip at about 45 deg (Fig. 9-33). If two axial surfaces intersect at nearly right angles, then you should consider the possibility that box folds are present. On seismic, the vertically dipping beds would not be imaged, and thus a pattern of symmetrically dipping reflectors separated by two zones of noncoherent reflectors, representing the almost vertically dipping beds, may be an indication of this style of deformation. However, zones of noncoherent reflectors on seismic sections can result from other causes such as wrench faulting, rock type or poorly collected data. Seismic reflection analysis (Payton 1977; Sheriff 1980) could resolve this problem as the sedimentary sequences on the flanks of box and lift-off folds are elevated within the cores of these structures.

Lift-off folds differ from box folds in that the shallow limbs of the lift-off fold types

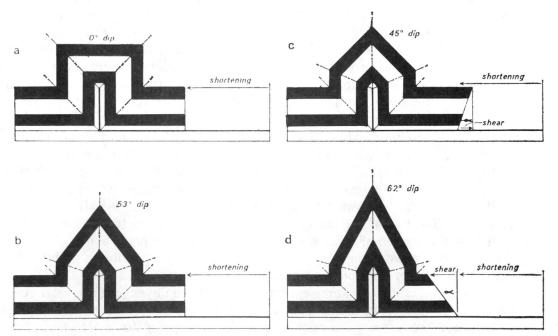

Figure 9-33 Box and lift-off structures. (From Namson 1981. Published by permission of the Chinese Petroleum Institute.)

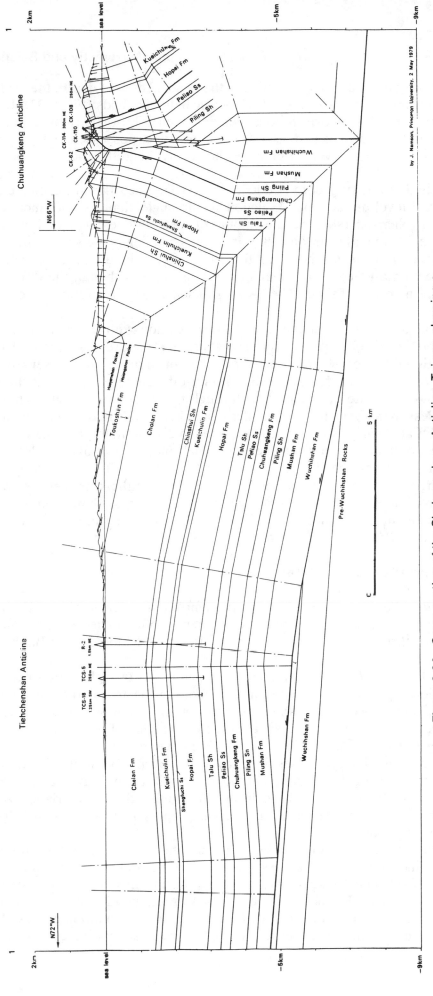

Figure 9-34 Cross section of the Chuhuangkeng Anticline, Taiwan, showing a broad region of near vertical dips. (From Namson 1981. Published by permission of the Chinese Petroleum Institute.)

445

tend to dip in the 45-deg to 60-deg range relative to the regional dip. At depth, the shallow steeply dipping limbs merge into a zone of nearly vertical dipping beds (Fig. 9-33). If you observe lift-off structures just above the decollement, then a lift-off structure may be impossible to distinguish from a box fold, as both structural types possesses 45 deg dipping axial surface on this structural level. In practice, however, this difference may be academic.

If the lift-off fold is not subject to bedding plane shear, then the limbs of the fold at a higher structural level dip at about 53 deg. This dip angle changes with increasing bedding plane shear (Namson 1981) and the amount of shear can be calculated from the dip of the fold limbs (Mitra and Namson 1989). If there is bedding plane shear within the structure, then Mitra and Namson show that this shear affects the depth to detachment presented in the section on Depth to Detachment Calculations, although the difference is not major for small amounts of shear. Their work should be consulted for more accurate depth to detachment calculations.

Box and lift-off folds are commonly found in association with each other. In the Pre-Alps, Mosar and Suppe (1988) have observed that lift-off structures form in the leading or the trailing position relative to fault propagation folds. In addition, they observed that fault propagation folds may transform laterally into lift-off structures, and that the two structural styles may be related to each other through the local cutoff angle. Thus, at low cutoff angles (less than about 18 deg to 20 deg), fault bend folds may form in an area, whereas if the cutoff angle is greater than about 20 deg to 25 deg, fault propagation folds usually form instead of fault bend folds. If the cutoff angle increases to over 60 deg along the strike of a structure, then the structure may transform into a lift-off or box fold.

When mapping box or lift-off structures, one should utilize the kink method. When applying this approximate method to these symmetric structures, remember that the hanging wall decollement is assumed to rise vertically above the basal detachment and to roll back upon itself (Figs. 9-33 and 9-34).

Triangle Zones and Wedge Structures

Triangle zones and wedge structures are types of complex structures that exhibit both a lower and an upper detachment. The basal detachment is often called the *sole thrust*, while the uppermost thrust is called the *roof thrust* (Boyer and Elliott 1982; Banks and Wharburton 1986).

Gordy et al. (1975, 1977) initially utilized the concept of a triangle zone to explain the complex relationships associated with an anticlinorium located at the front of the Canadian Rockies. Jones (1982) later refined the concept, and showed that the structure contained a duplex and that it was responsible for the termination of the eastern directed thrusting along the Rocky Mountain Thrust Front. We have learned that during the orogenic process the deformation progresses (i.e., advances) toward the foreland and therefore, at a previous time during the formation of the thrust belt, the old "frontal portions" would presently exist hinterland of the thrust front. This implies that fossil triangle zones can exist within the cores of mountain ranges, perhaps representing the frontal edge of the deformation at a previous time.

A simple triangle zone is illustrated in Fig. 9-35 that utilizes the concept of a ramping monocline. Notice that the deformation *terminates* where the roof thrust meets the sole thrust, creating a half syncline. This syncline with only one limb lies foreland of the thrust belt. Jones (1982) mapped a half syncline along the Rocky Mountain Front and concluded that a wedge-shaped body of material must be thrust underneath the dipping beds of the half syncline. A seismic section of a complex triangle zone is shown imaged in Fig. 9-

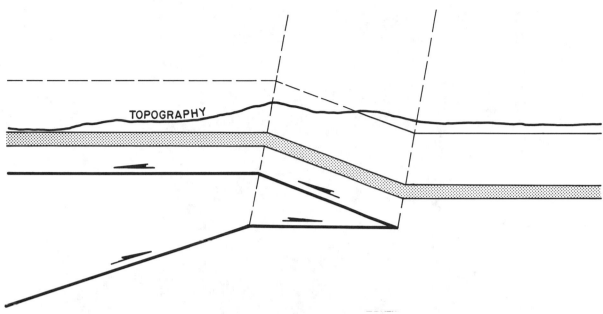

Figure 9-35 Simple triangle zone.

36. Notice that the wedge-shaped body represented by the duplication of the reflection located at between sp 190 to sp 240 and at 1.2 sec to 1.5 sec appears to correlate to the flat reflection at 1.5 sec in front of the structure. This section also exhibits characteristics of a *wedge structure* to the left of sp 260 at 0.4 sec to 0.8 sec.

Medwedeff (1988, 1989) has extended the concept of interactive sole and roof thrusts to single structures, and he has called these interactive thrusts *wedge structures*. Figure 9-37a is an example of a wedge structure that has two cutoff angles on its sole thrust and a single bend in its roof thrust. As the deformation progresses [Fig. 9-37(b)], motion along the sole thrust deforms the roof thrust, as back and frontal dip panels form over what is essentially a fault bend fold that also has an upper roof detachment. Notice, however, that the beds riding up the roof thrust, which is also a back thrust, will also form frontal and back dip panels associated with a fault bend fold that forms as a result of the ramp in the upper detachment. As this structurally higher fold is caused by the bends in the roof thrust, its dip panels terminate at the upper detachment (Fig. 9-37b). The result is two folds for the price of one that are slightly offset from each other. As the deformation progresses (Fig. 9-37c), the axial surfaces interfere and annihilate each other as they form branch points. This example illustrates that the deformation process can be very transient, and that the introduction of additional fault bends results in folds that have more rounded tops (Fig. 9-37c). Medwedeff has utilized wedge structures to model the complex stratigraphic relationships present at Wheeler Ridge, California. Restored and present well logs and the positions of the wells along the structure are shown in Figs. 9-38 and 9-39, respectively. These figures demonstrate how well logs can be used to define the complex relationships that exist within some structures. Precise correlations and balancing can be effectively integrated to locate prospects that may not be recognized by normal mapping techniques.

Interference Structures

Would you believe that anticlines can form over synclines with no evidence of an intervening fault or evidence of more than one deformation? Nevertheless, we have seen clear evidence for this seemingly contradictory relationship on proprietary seismic lines from

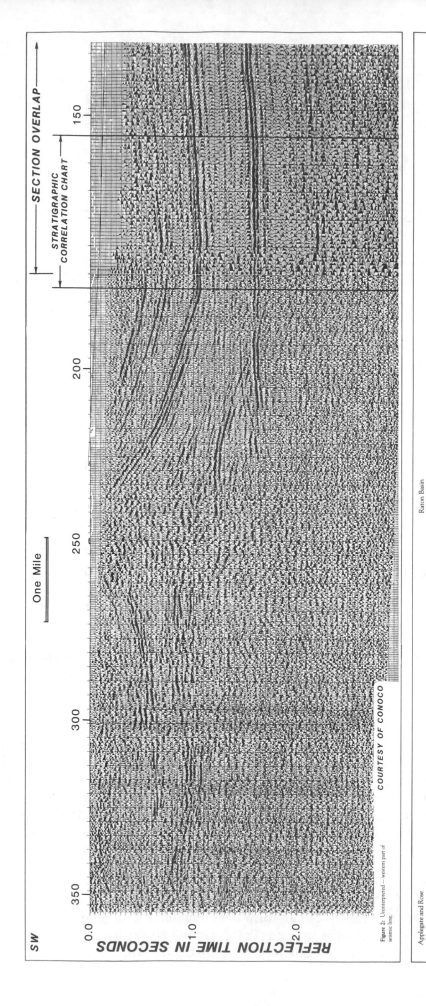

Figure 9-36 Seismic section imaging a triangle zone to the right of sp 250, Raton Basin, Colorado. (After Applegate and Ross. From *Seismic Exploration of the Rocky Mountain Region* by Gries and Dyer 1985. Published by permission of the Rocky Mountain Association of Geologists and the Denver Geophysical Society.)

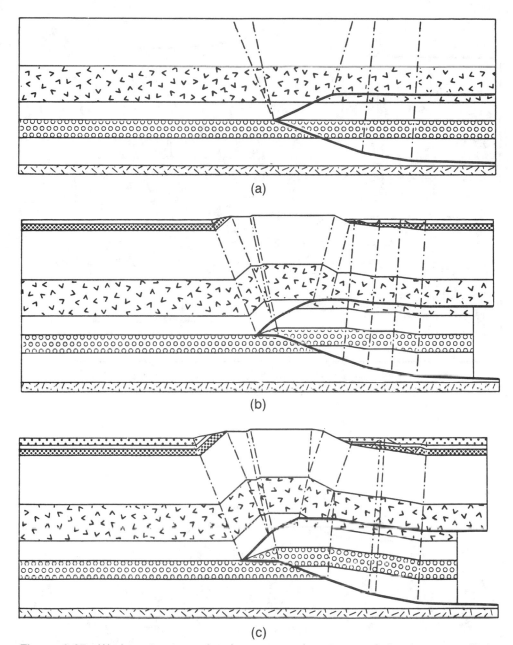

Figure 9-37 Wedge structure showing progressive stages of development. (Published by permission of Don Medwedeff 1988.)

several major oil companies. In the last section on wedge structures, we saw that deformation on a lower level could modify the shape and the form of dip panels of structures located on a higher structural level. When structural modification results from a single deformation along one thrust surface, as illustrated in Fig. 9-40a, the resulting structures are called *interference structures* (Suppe 1988).

Interference structures are commonly present when the spacing between ramps is relatively narrow, causing the back dip panel of the leading structure to interfere with the frontal dip panel of the trailing structure (Figs. 9-40a and b). The interference tends to produce chevron folds and conjugate kink structures (Weiss 1972; Suppe 1988).

The resulting interference patterns that are created by the deformation are dependent on ramp spacing, initial cutoff angle, and the total amount of slip. Two patterns are useful

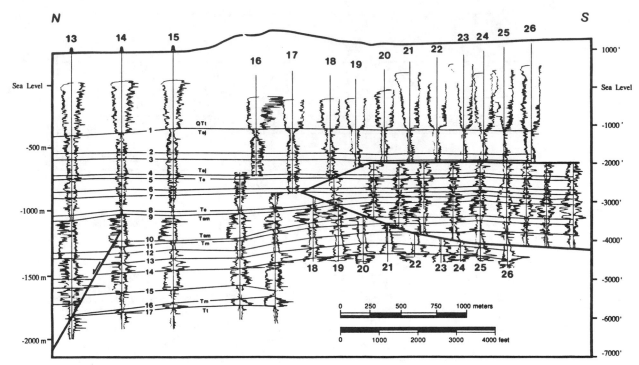

Figure 9-38 Wedge structure in its initial or restored state as defined by well logs, Wheeler Ridge, California. (Published by permission of Don Medwedeff 1988.)

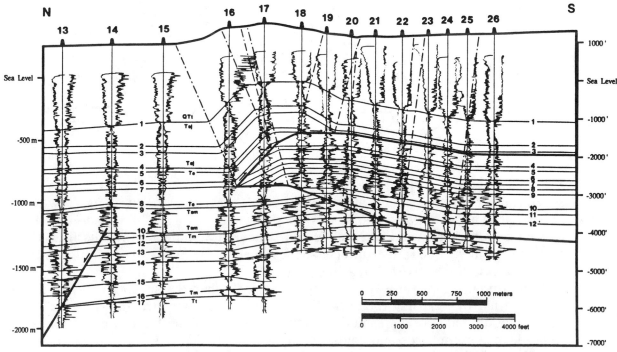

Figure 9-39 Wedge structure in its present state as defined by well logs, Wheeler Ridge, California. (Published by permission of Don Medwedeff 1988.)

in mapping these types of structures, although this does not deny the usefulness of other types of patterns. In the first example, the leading fault bend fold has run up a ramp while the frontal dip panel of the trailing fold (a monocline) occupies a portion of the lower flat (Fig. 9-40a). The resulting deformation creates a structure in which a frontal anticline lies beneath a structurally higher anticline formed by the trailing fold. Flat dips and a syncline exist above the lower flat directly in front of the trailing monocline.

As the deformation progresses, the frontal portions of the trailing monocline will start to run up the leading ramp (Fig. 9-40b). If both ramps have about the same cutoff angle, then as the trailing monocline runs up the second ramp, the beds in the frontal dip panel of the trailing monocline will flatten (Fig. 9-40b). One result of the deformation is to create a region of nearly flat dips and a narrow syncline over the leading ramp as the monocline *unfolds*. This exercise once again stresses the progressive nature of the deformation. **Structures are not cast in a mold**; they move, bend, and rebend strata. Knowledge of which regions of the fold have been subject to refolding should aid in the prediction of fracture porosities and better well site location. In our example, some of the strata in the back dip panel of the leading fault bend fold was first bent backwards and then forward by the advancing monocline. As the active axial surfaces sweep through

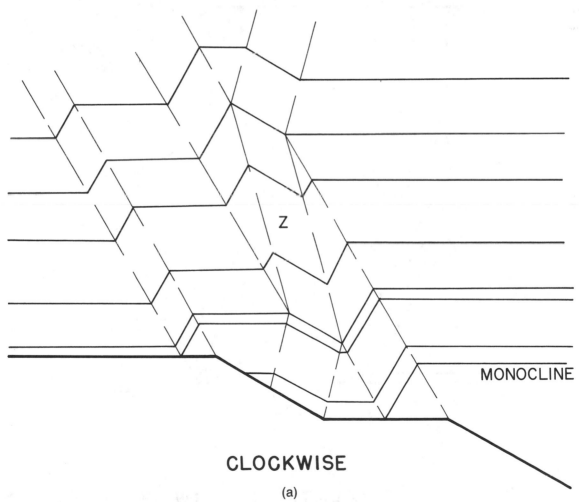

CLOCKWISE

(a)

Figure 9-40 Interference structures (a) and (b) for clockwise deformation and increasing slip (c) for counter clockwise shear. (Modified after Suppe 1988. Published by permission of John Suppe.)

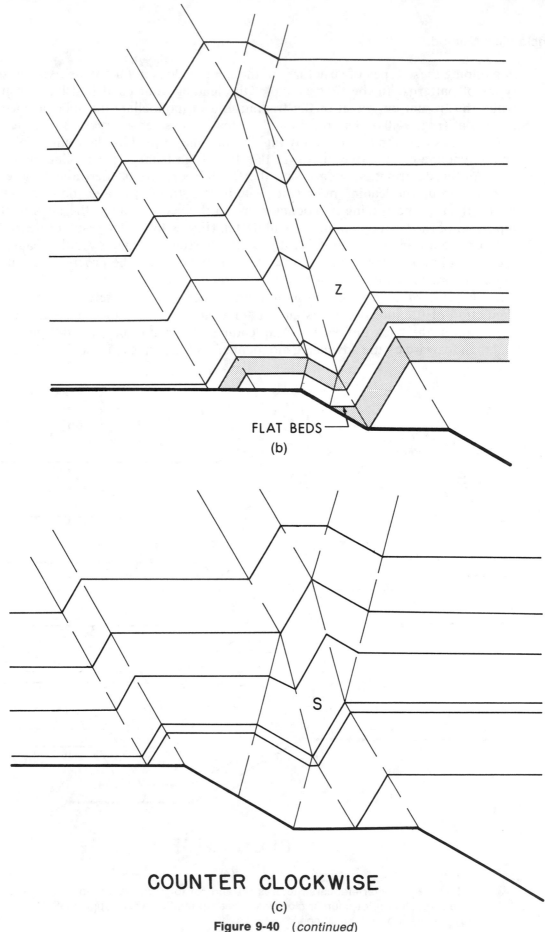

FLAT BEDS

(b)

COUNTER CLOCKWISE

(c)

Figure 9-40 (*continued*)

452

the structure, particular regions within these folds will be subject to repeated deformation. One can study the refolding by *utilizing the kink method* and by forward modeling increasing amounts of slip (see section on Strategy for Balancing Cross Sections).

We shall make two more points before leaving this subject. First, as the initial cutoff angle decreases, the structurally higher anticline will move vertically away from the structurally lower syncline, but at the same time shift to a position where it is located almost directly above the syncline. Second, the two examples presented thus far were for a clockwise shear within the interfering frontal and back dip panels. In other words, the beds within the interfering dip panels exhibit a Z vergence (Suppe 1988). An example of an S vergence (counterclockwise shear), in which the frontal dip panel of the trailing monocline passes through the upper anticline, is shown in Fig. 9-40c.

A Strategy for Balancing Cross Sections

We previously emphasized that balancing is not unique. It is subject to errors of interpretation. Mount et al. (1990) have proposed a straightforward cross section balancing procedure that utilizes the concept of trial and error forward modeling, which should help *minimize interpretation errors*. The fundamentals of the forward modeling strategy are as follows (Fig. 9-41):

1. *Compile and examine the data:* All data should be compiled on a scale of one to one, perpendicular to the strike direction of the structure under study (Chapter 6). The data are then subject to an initial interpretation.
2. *Develop a working hypothesis of the data set:* This stage of the interpretation process involves the generation of a ''bright idea'' or, following Chamberlain (1897), the

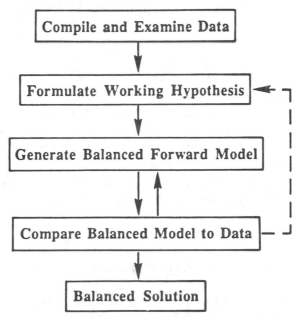

Figure 9-41 Flowchart for forward modeling strategy. (Published by permission of Mount, Suppe, and Hook 1990.)

generation and testing of all "reasonable" bright ideas. Each idea or interpretation can then be tested in accordance with the data. The largest scale, or the longest wavelength, features should be studied first (see section on Long Wavelength Mapping), and secondary or smaller features analyzed later. If the hypothesis is *incorrect*, the forward modeling process is likely to result in a negative result, but the incorrect hypothesis should be filed for future reference and use.

3. *Generation of balanced forward models:* Fault bend and fault propagation and growth fault bend fold theory (Medwedeff and Suppe 1986) are utilized to generate a balanced forward model that approximates the major features present in the data.

4. *Compare the model to observed features:* If the generated model does not match the known features, then a new hypothesis is tested, or following Chamberlain, you may wish to test all of your reasonable ideas. If a particular model tends to duplicate most of the observed features, then structural refinements are incorporated into the preferred model.

5. *Balanced solution:* "After a number of iterations, a balanced solution is converged upon that adequately honors the observed structural and stratigraphic relationships" (Mount et al. 1990). These procedures exhibit the additional advantage in that forward models are more readily balanced than the actual field, well, or seismic data.

EXTENSIONAL STRUCTURES: BALANCING AND INTERPRETATION

Introduction

The balancing of extensional structures is presently in the initial stages of development; therefore, the concepts and techniques derived from this work are new and have been subject to limited testing. We have applied the following concepts to several regions in the Gulf Coast, particularly to the Brazos Ridge region, offshore Texas (Vogler and Robison 1987), with good results. Until these techniques and concepts are successfully applied to other regions of the world in different extensional settings, we cannot guarantee the unversality or the generality of our methods. Therefore, in this section we document our procedures and conclusions so that you can more critically assess our work. Nevertheless, many of our readers will be active in the Gulf Coast region where our study and others were initiated.

This section is divided into two parts: (1) the study of compaction effects along growth normal faults and how sandstone/shale ratios are utilized to predict fault shape and position, or conversely how fault shape is used to predict sandstone/shale ratios; (2) the origin of rollover, antithetic, and synthetic faults and keystone structures. We address such subjects as (a) why some growth faults die in both the upward and downward directions and what this means physically, and (b) where rollover structures are likely to exist or be positioned along major listric normal faults (i.e., faults which exhibit large vertical expansion across the fault surface).

Compaction Effects Along Normal Faults

In this section we present a theory by Xiao and Suppe (1989) which attempts to explain why some growth normal faults are listric (concave upward) or anti-listric (concave downward). Generally, about 20% to 30% of the faults in the Gulf Coast region are mapped as anti-listric, whereas in the Brazos Ridge area of our studies the majority of the compensating (antithetic) faults are anti-listric. We should stress that the following techniques for calculating fault shape and position are only intended to apply to syndepositional faults that formed at the same time as the growing sedimentary package (i.e., are active when the sediments are being deposited), and not to faults within the pre-growth sediments which were deposited prior to the normal faulting, nor to faults within the growth sediments which formed after the sediments are deposited. In other words, these techniques apply only to the growth faults or to those *portions* of a fault that experience growth.

Consider two small columns of material, column A located at the sea floor, and column B located at depth (Fig. 9-42). Column A has just been deposited and has not been subject to compaction, whereas column B has been buried and is subject to compaction. Initially, column A has width ΔX and an initial porosity of ϕ_o. If the height or depth of column A is defined as ΔY, then the initial dip of the growth normal fault relative to the *footwall* is defined by

$$\tan\theta_o = \Delta Y/\Delta X \qquad (9\text{-}1)$$

where

θ_o is the initial fault dip

After burial, column A will dewater and take on the shape of column B (Fig. 9-42),

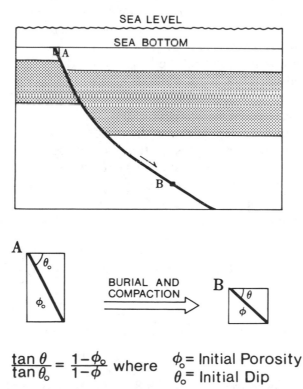

$$\frac{\tan\theta}{\tan\theta_o} = \frac{1-\phi_o}{1-\phi} \quad \text{where} \quad \begin{array}{l} \phi_o = \text{Initial Porosity} \\ \theta_o = \text{Initial Dip} \end{array}$$

Figure 9-42 Compaction and burial along a growth normal fault. Element A at the sea floor compacts to the geometry present at B. These relationships are taken relative to the compacted footwall. (From Xiao and Suppe 1989. Published by permission of the American Association of Petroleum Geologists.)

where $\Delta X'$, $\Delta Y'$, θ, and ϕ are the buried width, height, fault dip, and porosity, respectively. Therefore, the fault dip at B is defined as

$$\tan\theta = \Delta Y'/\Delta X' \tag{9-2}$$

However, the compaction primarily affects the height and not the width; therefore

$$\Delta X = \Delta X' \tag{9-3}$$

and thus, utilizing Eqs. (9-1) and (9-2):

$$\tan\theta/\tan\theta_o = \Delta Y'/\Delta Y \tag{9-4}$$

As porosity controls the extent of compaction (Baldwin and Butler 1985), the mass conservation formula written in terms of solid volume is

$$\Delta X' \cdot \Delta Y'(1 - \phi) = \Delta X \cdot \Delta Y(1 - \phi_o) \tag{9-5}$$

and from Eqs. (9-3) and (9-4), we derive

$$\tan\theta/\tan\theta_o = (1 - \phi_o)/(1 - \phi) \tag{9-6}$$

Therefore, if we are able to predict the initial fault dip (θ_o) and porosity (ϕ_o), along with the porosity (ϕ) of the rock at any given depth level, then we can determine the fault dip (θ) at the working level (Fig. 9-42).

As dewatering and compaction is a rapid process that occurs within about the upper 700 ft (Baldwin and Butler 1985), any compaction equation will be controlled primarily by the porosity, which is in turn related to the sandstone/shale ratio. Thus, if the position of a normal fault is known in a single well and if the sandstone/shale ratios are known from local or adjoining footwall well logs, then the fault shape and its location can be extrapolated between adjoining wells using Eq. (9-6).

Before applying Eq. (9-6) to an area, we must be able to calculate the amount of porosity to be expected in sandstone and shale horizons at any given depth. Porosity/depth equations are dependent on the region being studied (Baldwin and Butler 1985), and as these relationships represent the average porosity of sandstone or shale at any given depth, you should determine these relationships from local data. As the porosity/depth equations represent averages, your results will also represent averages.

For the Gulf Coast, which contains many over-pressured shales, the following empirically derived sandstone and shale porosity/depth equations can be applied (Xiao and Suppe 1989). For shale horizons,

$$\phi_{sh} = 0.2684 - 0.1972 \times \log_{10}(z/3300 \text{ ft}) \tag{9-7}$$

where

$$\phi_{sh} = \text{shale porosity}$$

$$z = \text{depth in ft}$$

For sandstone horizons, Eqs. (9-8) to (9-10) may apply:

$$\phi_{ss} = 0.3198 - 0.0327 \times (z/3300 \text{ ft})$$

$$\text{below 8000 feet,} \tag{9-8}$$

and

$$\phi_{ss} = 0.3933 - 0.0635 \times (z/3300 \text{ ft})$$

$$\text{above 8000 ft,} \qquad\qquad\qquad (9\text{-}9)$$

or, alternatively, you may test the following equation (Atwater and Miller 1965):

$$\phi_{ss} = \phi_o - 0.01265 \times (\text{depth}/1000 \text{ ft}) \qquad\qquad (9\text{-}10)$$

where $\qquad\qquad \phi_{ss}$ = sand stone porosity

$\qquad\qquad\qquad \phi_o$ = the initial porosity

Porosity/depth equations for areas other than the Gulf are presented by Baldwin and Butler (1985) and Sclater and Christie (1980).

The technique can now be applied as follows. First, the sandstone portions of a well are determined by locating the sandstone and the shale base lines from SP logs. Common industry practice suggests that sandstones exist wherever the SP log exceeds 25% of the distance between the sandstone and shale baselines (Fig. 9-43). Values of less than 25%

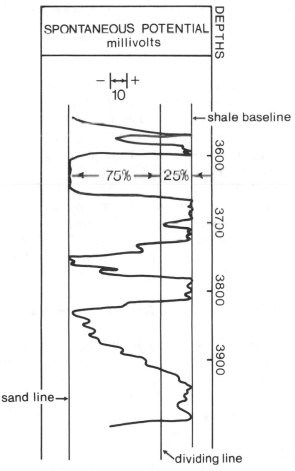

Figure 9-43 Relationships utilized to determine sand-shale ratios. Sand is present wherever the SP log deviates more than 25% off the shale base line. (From Xiao and Suppe 1989. Published by permission of the American Association of Petroleum Geologists.)

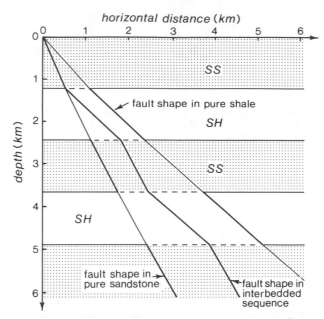

Figure 9-44   The "splicing" method for estimating fault dips from sand-shale horizons determined from SP logs. Fault dip in a sand horizon is taken to be pure sand, and in a shale horizon, to be pure shale. The dips in the alternating sand and shale horizons are then spliced or added together. (From Xiao and Suppe 1989. Published by permission of the American Association of Petroleum Geologists.)

on the divided SP log are taken to be shale (Schlumberger 1987). Second, the "splicing method" is employed (Fig. 9-44). For example, Eqs. (9-8) to (9-10) are utilized to describe those regions of the SP log that have been determined to be sandstone, and Eq. (9-7) is applied to those portions of the log that represent shale.

Equation (9-6) is also dependent on the initial fault dip and porosities which have been determined to be 67.5 deg, and 39% and 68% for sandstone and shale, respectively (Xiao and Suppe 1989). You may wish to confirm these initial values from local data prior to applying Eq. (9-6) to your area of interest. We have determined the sandstone and the shale portions of the sedimentary section from the SP logs and can calculate their average porosity at any given depth from Eqs. (9-7) to (9-10). Therefore, we can now calculate the average fault dip (θ) at any depth from Eq. (9-6) and the "splicing method." These formulae were tested on several wells in the Gulf with good results (Fig. 9-45).

Inverting Fault Dips to Determine Sand/Shale Ratios

In the Brazos Ridge area, many of the growth normal faults at first dip at shallow angles (40 deg to 45 deg) and with greater depths, dip at higher angles (50 deg to 60 deg). These anti-listric normal faults steepen into a known section of higher sandstone/shale ratios (approximately 1.0 to 1.5), and thus these normal faults present a method of utilizing the fault dips to determine sandstone/shale ratios (Bischke and Suppe 1990a). The basic theory for accomplishing this task has been presented in the previous section, and is implicit in Eq. (9-6).

For example, if the initial fault dip (θ_o) and porosities (ϕ_o) are known (as has been defined previously), and if the average porosities (ϕ) are known at all depth levels [e.g., Eqs. (9-7) to (9-10)], then the fault dip (θ) at the working level can be calculated from Eq. (9-6). This is called the forward modeling procedure.

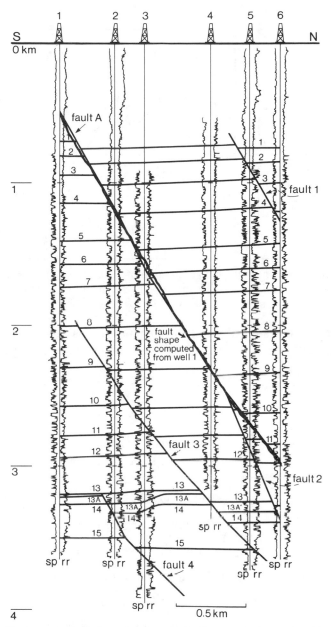

Figure 9-45 Test example for projecting fault dips between wells. The SP log from Well No. 1, the fault dip (Fig. 9-42), and the porosity/depth equations are utilized to predict the shape and position of fault A with depth. (From Xiao and Suppe 1989. Published by permission of the American Association of Petroleum Geologists.)

If, however, the fault dip (θ) is known from depth corrected seismic sections, then the porosities (ϕ) can be calculated from Eq. (9-6) at any working level. This is called the inverse problem. As porosities are dependent on lithology [see Eqs. (9-7) to (9-10)], we can determine the sand–shale/depth curve in any area, provided that we have information on local porosity depth curves. This information can be derived from the nearest wells, which in frontier areas could be many miles from the area of interest.

Next, we must calculate the sandstone/shale ratio, which can be accomplished utilizing Fig. 9-46. In practice, the observed height of the figure is the top and the bottom of the digitized trace of an interpreted normal fault. This distance, which must be depth

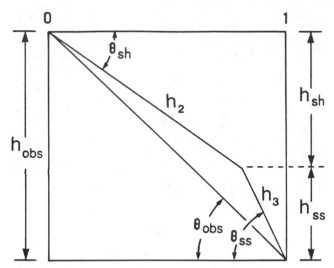

Figure 9-46 An alternating sequence of sand and shale horizons can be generalized into a shale horizon of thickness h_{sh}, and a layer of sandstone of thickness h_{ss}. (Published by permission of Dick Bischke.)

corrected, represents a column of interbedded sandstone and shale horizons. Utilizing a "deck of cards" analogy, the sand layers are removed from the deck and placed at its bottom.

We now calculate

$$h_2 = h_{sh}/\sin(\theta_{sh})$$

$$h_3 = h_{ss}/\sin(\theta_{ss})$$

where:

$$h_{sh} = \text{thickness of shale}$$
$$h_{ss} = \text{thickness of sandstone}$$
$$\theta_{sh} = \text{fault dip in shale unit}$$
$$\theta_{ss} = \text{fault dip in sandstone unit}$$
$$\theta_{obs} = \text{observed fault dip at working level}$$
$$h_2 = \text{length of fault in shale horizon}$$
$$h_3 = \text{length of fault in sandstone horizon}$$

and utilizing the law of sines:

$$\sin(\theta_{ss} - \theta_{obs})/h_2 = \sin(\theta_{obs} - \theta_{sh})/h_3$$

and substituting and rearranging yields

$$\bar{r}_{calc} = h_{ss}/h_{sh} = [\sin(\theta_{obs} - \theta_{sh})\sin(\theta_{ss})]/[\sin(\theta_{ss} - \theta_{obs})\sin(\theta_{sh})]$$

where

$$\bar{r}_{calc} = \text{average calculated sandstone/shale ratio}$$
$$\theta_{obs} = \text{observed fault dip calculated from an interpreted and depth corrected section}$$

As θ_{sh} and θ_{ss} can be obtained from Eq. (9-6) and the porosity from Eqs. (9-7) to (9-10), we can calculate the average sand/shale curve with increasing depth.

Before proceeding to a test case, we will outline a number of factors that affect the accuracy of our methods.

1. *The method only applies to growth normal faults,* and thus to areas which are experiencing both active sedimentation and extension.
2. Although the method can be applied wherever velocity information is available, the velocity function is critical to obtaining reliable results. Velocity determinations are discussed in Chapter 5.
3. The theory in its present stage of development only involves compaction, and thus deformation of normal faults by other processes is not taken into account.
4. The method is restricted to the depth to which the seismic reflections can be resolved.
5. The method only applies to sandstone/shale lithologies, and not to limestones.

In order to compare observed well log results (which resolve horizons to within feet) to seismic data, in which the resolution is a function of depth, the well log data must be transformed to longer wavelengths. This is accomplished by first determining the sand and shale horizons from SP logs, as was discussed earlier (Fig. 9-43), and then measuring the wavelength of coherent reflectors directly off the seismic section. The frequency content of the seismic section is thus determined. As the frequency content decreases with depth, the wavelength of the reflections increases with depth. This length is then depth corrected and passed over the SP data as a moving average, the length of the operator increasing with the depth. The result of this averaging process from a sample well is the sandstone/shale depth curve shown in Fig. 9-47. The data are plotted at a frequency of two seismic wavelengths. This well log data can be directly compared to the normal fault calculations shown next. Typical results from a nearby growth fault, that was processed utilizing *stacking velocities*, is shown in Fig. 9-48 for comparison. The normal fault, which terminates at 7000 ft, has predicted the higher sandstone/shale ratios present between the 5000-ft to 6000-ft level.

This method has broad application to petroleum exploration, and we have found that the method is robust in that useul results are obtained from imprecise velocity data, such as stacking velocities (Bischke and Suppe 1990a). Normally, when a well is drilled, the least known parameters are the depth and the sand/shale ratio of the reservoir. As the section contains shale from 0 to 4600 ft, Fig. 9-48 also produces a positive prediction as to seal. Such information can be utilized by the geologist-engineer to better determine the target depths for wells and to improve the calculations of potential reserves prior to drilling. As the method can be applied to the third dimension, three-dimensional sand/shale ratio maps can be constructed to better select well locations. All that is required to apply the method is an array of small growth normal faults. The usefulness of the method increases as well control becomes limited and the distance between wells increases.

Origin of Rollover

Rollover structures, like anticlines, have been successfully drilled for many years, yet many questions remain concerning their origin. A major insight into the origin of rollover was initiated by Hamblin (1965) when he recognized that these strange "reverse drag

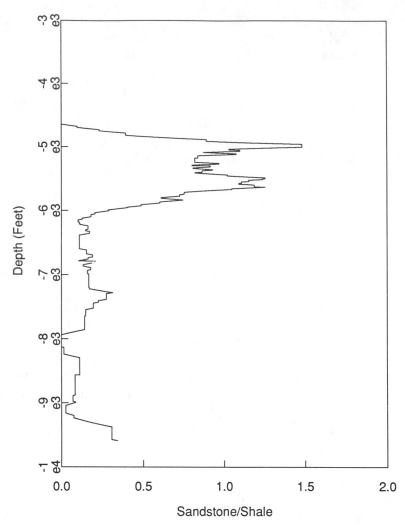

Figure 9-47 Processed sandstone/shale depth curve obtained by averaging sand/shale horizons taken from an SP log. The data are on the frequency of two seismic wavelengths and range in depth from 3000 ft to 10,000 ft. (Published by permission of Dick Bischke.)

folds'' are the natural consequence of motion along listric normal faults. He reasoned that if the hanging wall block separates from the footwall block, a hole would open up between the two blocks, and as the hanging wall block gravitationally collapses onto the footwall block, a reverse drag structure would form. However, Hamblin did not exactly specify how this gravitational collapse occurs. Gibbs (1984) later recognized from North Sea data that extensional structures (as in analogy with compressional structures) seemed to form duplexes, complete with horses and transfer zones, etc. Yet a number of questions remain to be fully answered, such as:

1. Can the shallow rollover structures be utilized to predict the deeper structure (e.g., subfault structure) and to extrapolate into regions of poor or nonexisting data (e.g., into poorly imaged seismic zones)?
2. What is the precise origin of the rollover and is it possible to predict the amplitude, style, and position of the rollover from known geologic features?

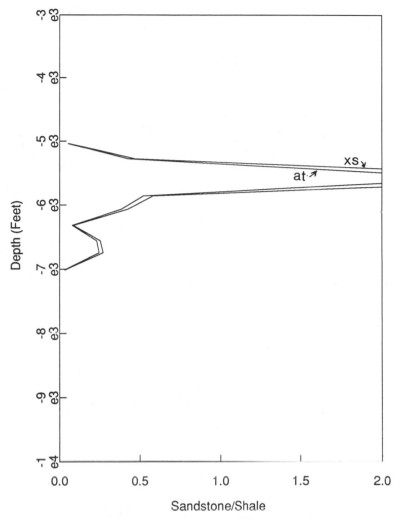

Figure 9-48 Processed growth fault showing predicted sandstone/shale depth curve derived from fault dips. The fault is adjacent to the well data shown in Fig. 9-47. XS curve uses the Xiao-Suppe compaction equation, while at denotes the Atwater equation. (Published by permission of Dick Bischke.)

3. What are the origins of the antithetic or compensating faults and keystone structures and how do these structures influence rollover geometry?

4. What causes some (perhaps most) compensating growth faults to exhibit displacements along their surfaces which die in both the upward and downward directions in a seemingly contradictory relationship?

Answers to at least some of these questions should aid explorationists in reducing the time and expense of locating rollovers, generating better maps, and isolating trapping mechanisms. Our research and that of others suggests that compaction (see section on Compaction Effects), fault shape, antithetic and synthetic faulting, and cross structures affect the geometry of the rollover structures.

When developing a theory for a structure that is as complex as a rollover, the initial theories are likely to be overly simplistic. Our position is that a theory is only as good as its ability to quantitatively explain the observations. Therefore, we expect that in the future, modifications and refinements are likely to make these theories more exacting.

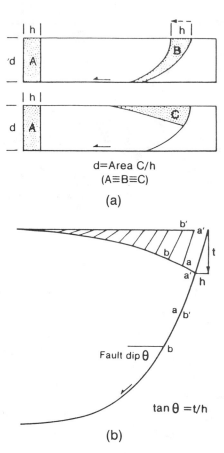

(a)

(b)

$\tan\theta = t/h$

Figure 9-49 Method for estimating listric fault dips. (From Gibbs 1983. Published by permission of the Journal of Structural Geology.)

Vertical Breakup Theory

The first attempt at predicting the shape of rollover structures was presented by Gibbs (1983). Following Hamblin (1965), Gibbs and others reasoned that as the hanging wall block moved over the footwall block, the hanging wall block would separate from the footwall block and then *collapse vertically* onto the fault surface (Fig. 9-49). It follows that the (collapsed) vertical displacements at the top of the hanging wall block (t) are related to the horizontal component of slip (h) by the relationships shown in Fig.9-49. Thus, the fault dip is defined as the tangent of the angle between the vertical drop (t) and the horizontal distance (h), which are taken relative to the dip of the normal fault.

If the shape of the rollover is known, then the rollover can be segmented at a uniform interval h with respect to a horizontal line that is drawn tangent to the top of the rollover. Vertical lines are projected downward from the top of the horizontal line for every h increment, and vectors (e.g. aa', bb', etc.) are then constructed for every t-h component. The vectors are added to determine the position of the normal fault at depth (Fig. 9-49). This exercise results in a listric normal fault. Problems with the vertical breakup theory are that it does not account for a graben-like structure that is often present near the front of the rollover (Fig. 9-60, between C and D). Furthermore, the vertical breakup theory is not very sensitive to rollover dips, so the theory may predict listric faults that are too deep relative to their rollover dips. Normally, structures observed to have high rollover dips also tend to have shallow faults (i.e., the fault dip decreases rapidly with depth). Rowan and Kligfield (1989) have shown, however, that the vertical collapse theory may generalize the complex displacements that are associated with some rollover structures.

0 _____ 5 km   scale: horizontal=vertical

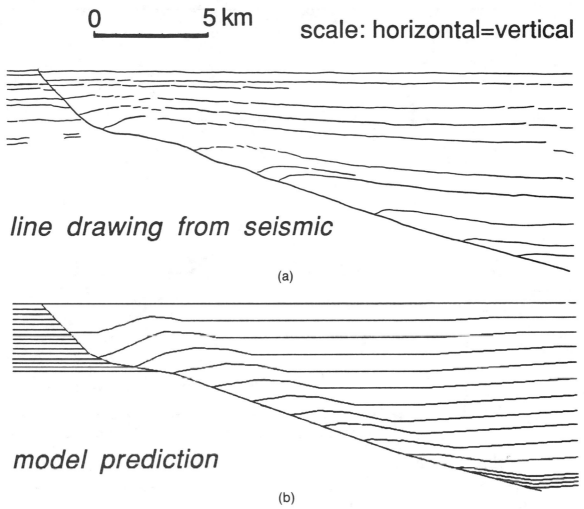

line drawing from seismic

(a)

model prediction

(b)

Figure 9-50   Comparison of Brazos Ridge seismic data (a) to computer model (b) utilizing the Coulomb break up theory. (Published by permission of Xiao and Suppe.)

Coulomb Collapse Theory

A theory utilizing a *coulomb* or frictional breakup of the hanging wall onto the footwall block has been advanced by Xiao and Suppe (1988), and elements of this theory are also present in work by Groshong (1989). According to this theory, the rock is assumed to fail along the internal or coulomb friction surfaces oriented at about 30 deg to the maximum principal stress (σ_1) (Billings 1972), which in the case of extensional deformation is subvertical. This angular breakup is in contrast to the vertical breakup surfaces assumed by the **Chevron Method**. This theory qualitatively describes many of the features observed on seismic sections, and has in simple cases mimicked the observed rollover geometry (Fig. 9-50).

The elements of the coulomb theory are shown in Figs. 9-51a to 9-51e, where Fig. 9-51a represents the initial or pre-faulted state. This simple example shows a major fault that has, for purposes of demonstration, one single concave upward bend. In nature, major listric normal faults are normally curved, and this more realistic case can be duplicated by introducing several concave upward bends into a model.

As the hanging wall block moves over the footwall block, a small hole opens up

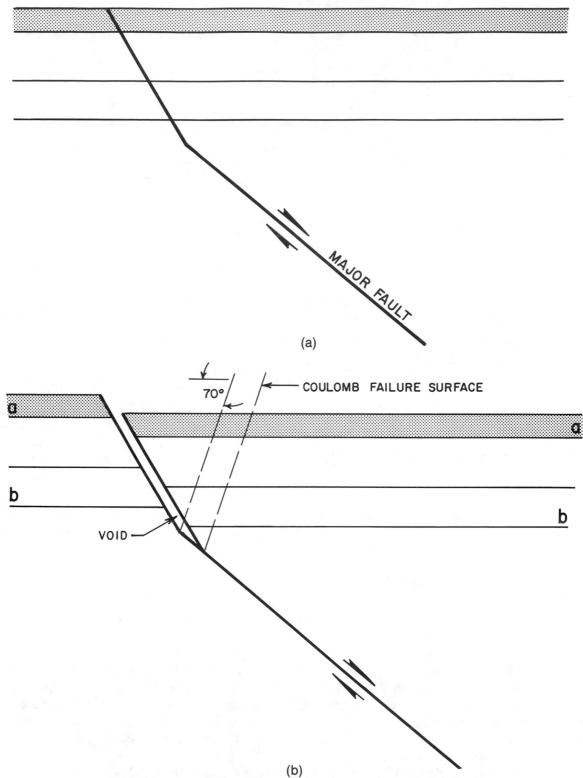

(a)

(b)

Figure 9-51 Coulomb failure or shear collapse model, showing different stages in the development of a simple rollover. A bend in the major fault subjects the hanging wall to deformation at the active axial surface, which is fixed to the bend in the footwall. Increased slip causes the inactive axial surface to migrate away from the surface of active deformation. (Modified after Suppe 1988. Published by permission of John Suppe.)

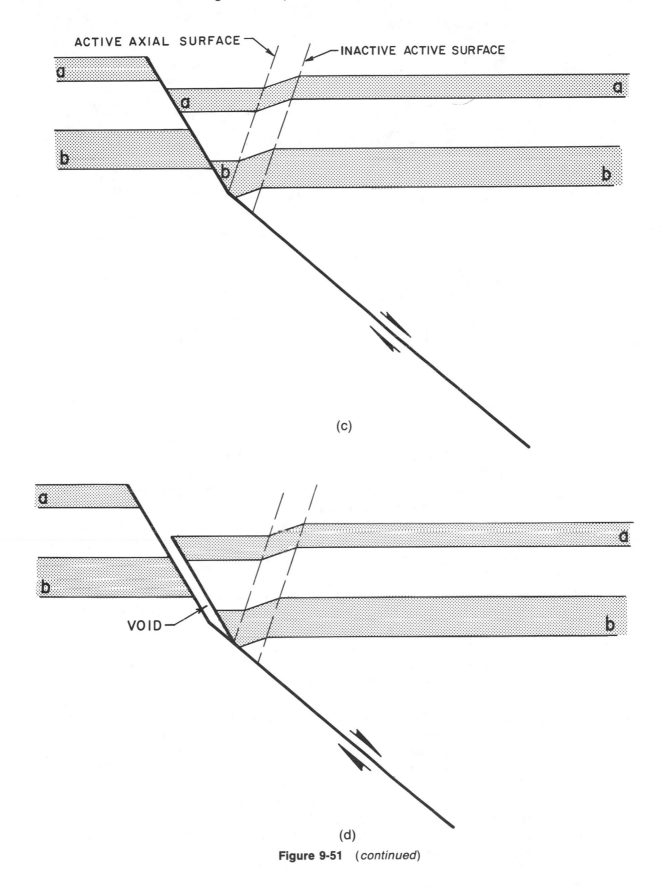

(c)

(d)

Figure 9-51 (*continued*)

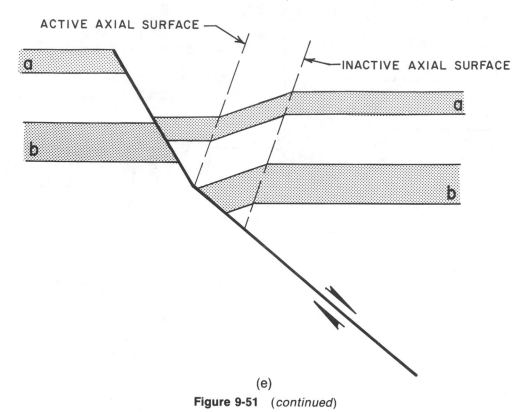

(e)

Figure 9-51 *(continued)*

between the hanging wall and the footwall blocks, as has been described by Hamblin (1965) (Fig. 9-51b). In the coulomb shear case, the hanging wall block will gravitationally collapse into the void (created by the sliding) along the coulomb failure surfaces, which in this case are assumed to be 70 deg, rather than along the vertical breakup surface assumed by the Chevron method (Fig. 9-49). As material fills the hole, the beds which experience deformation are assumed to shear parallel to the coulomb failure surfaces, which results in their extension (Fig. 9-51c). Gibbs (1984) was first to realize that the frontal limbs of rollover structures would experience large extension.

Observe in Figs. 9-51b and 9-51c that the *slice* of material which is passing through the coulomb shear surface, and which is fixed to the concave upward bend in the major fault, is the material that is experiencing distortion. For this reason, this surface is called the *active axial surface* (Fig. 9-51c). Therefore, deformation only occurs along active axial surfaces which are affixed to the bends of listric faults.

The initial material which passed through the bend was deformed by the active axial surface and was translated basinward. This material presently lies adjacent to the *inactive axial surface* (Fig. 9-51c).

This process can be more readily understood if the fault is subject to another increment of sliding (Fig. 9-51d). Of course in nature, these increments are infinitesimal. The sliding opens up another hole between the hanging wall and the footwall blocks and the active axial surface that formed in Fig. 9-51c is immediately translated basinward (Fig. 9-51d). However, the void is instantaneously closed by the gravitational shear failure pinning the active axial surface to the footwall portion of the bend. The resultant structure shown in Fig. 9-51e contains a graben-like structure adjacent to the steepest portion of the major fault and a monoclinally shaped rollover structure. Sedimetary compaction within the basin would close the structure.

You can better understand these processes by redrawing Fig. 9-51e and cutting the hanging wall block from the footwall block with a scissors. Then subject the major fault to another increment of motion and collapse the suspended material onto the footwall parallel to the coulomb failure surfaces.

Most major normal faults in the Gulf of Mexico, for example, are *active growth faults*. This means that sedimentation is occurring as the major fault is slipping, and that the sediments are subject to vertical expansion across the fault surface (i.e., the sediments thicken on the downthrown side of the growth fault). How does this syndepositional sedimentation affect the problem?

Referring back to Fig. 9-51c, let us assume that a layer of sediments is deposited across the hangwall and footwall portions of our structure (Fig. 9-52a). As the graben is deeper than the top of the rollover, the sediments will be thickest over the graben and will thin over the footwall and the crestal portions of the monoclinal rollover. We saw in Figs. 9-51b through 9-51e that the active axial surface is fixed to the bend in the major fault and that in the pre-growth sediments that the inactive axial surface (which is no longer subject to deformation) migrates basinward. Notice in Fig. 9-52a that as the inactive surface is not subject to deformation, it cannot affect the recently deposited sediments. The active axial surface, however, which is a zone of recent deformation, will deform the growth sediments. Consequently, a *growth axial surface* must connect the active axial surface to the inactive axial surface (Fig. 9-52a).

You can convince yourself of this statement by visualizing that a sliver of growth sediments is deposited above layer 1. These sediments will be deposited horizontally. Another small increment of sliding along the major fault will deform this sliver of recently deposited sediments in the region where the active and the growth axial surfaces converge (Fig. 9-52a). An additional increment of sliding combined with growth sedimentation will produce the geometry observed in Fig. 9-52b. The growth sedimentary wedge expands as the growth sediments move through the active axial surface (Fig. 9-52b).

Another interesting feature of this model is the following. For larger rollover structures, sandy sediments are more likely to be deposited in the graben, while suspended pelagic sediments are more likely to be dominant across the top of the rollover. Therefore, facies changes and stratigraphic traps are predicted to occur in the vicinity of the growth axial surface. Some listric faults are observed to initially flatten (to dip more gently with depth) and then to possess steeper dips with increasing depth (Fig. 9-56). This creates a concave downward bend in the major fault. If this occurs, the coulomb collapse theory predicts that the deformation takes place along the basinward dipping *conjugate shear surface*. Shear can occur along two conjugate surfaces (Billings 1972), and in our examples this shear is in the clockwise direction (Fig. 9-53a).

The Brazos Ridge can be utilized to demonstrate the coulomb breakup theory for a more realistic case. In the most general sense, the Corsair Fault, the major listric normal fault in the Brazos area, at first dips at about 45 deg, flattens to about 10 deg, and then steepens to approximately 20 deg. The Brazos region is presently subject to both active deformation and sedimentation and thus the sediments deposited upon the hanging wall block are growth sediments. If the simple Brazos fault model described above is subject to gravitational (basinward) sliding, two holes would be created, one above the concave up portion and the other adjacent to the concave down portion of the master fault. Clockwise collapse above the deeper concave downward bend (Fig. 9-53a) and counter-clockwise collapse above the concave upward bend (Fig. 9-53b) would produce the geometry observed in Fig. 9-53c. If the deposition rates are high (as in the Brazos), then sediments will fill the graben-like feature adjacent to the shallowest portions of the major fault, thin

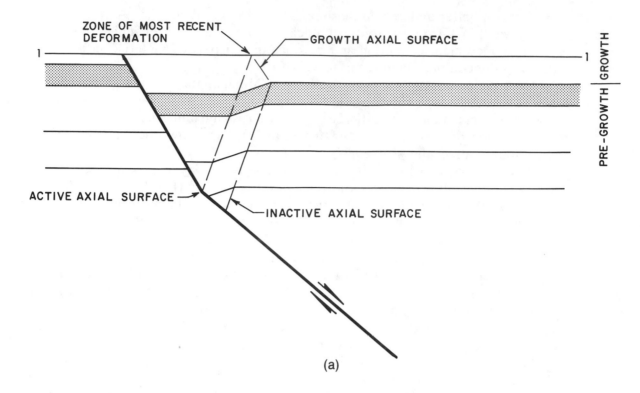

(a)

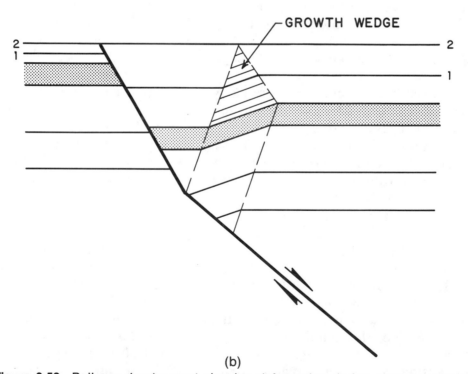

(b)

Figure 9-52 Rollover development showing deformation during growth sedimentation. The most recently deposited sediments are deformed at the point where the active and growth axial surfaces converge. (Modified after Suppe 1988. Published by permission of John Suppe.)

over the top of the rollover structure, and then thicken basinward. The vertical expansion observed in this example clearly demonstrates why the sediments which were deposited during the period of time that the deformation is active are referred to as growth sediments. Also notice that the nondeformed and the deformed growth sediments in Fig. 9-53c are separated by a growth axial surface (compare to Fig. 9-52). Figure 9-53c contains rollover amplitudes that are higher than are observed along the Brazos Ridge (Fig. 9-50). However, compaction combined with synthetic and antithetic faulting could reduce the rollover amplitudes.

Provided that information exists on the fault dips that created a rollover, the coulomb breakup theory can be utilized to predict the rollover angle (θ) or bed dips. In the coulomb theory, the rollover angle (θ) can be related to the change in fault dip (ϕ) through a complicated set of trigonometric formulae which can be represented by graphs (Figs. 9-54 and 9-55). The assumptions used in the graphs are that the material in the hanging wall is subject to coulomb collapse (as discussed previously), and that the structure has *not* experienced sedimentary compaction, as described in the section on Compaction Effects Along Normal Faults. Thus, these diagrams strictly apply to the nongrowth phase of the sedimentation and will *overestimate* the rollover angle within growth sediments. Let us assume that a seismic section images the shallow and deeper portions of a rollover structure but poorly images the bed dips. If the abovementioned conditions are met and if the initial fault dip (β) and the fault dip at the working level (α_1) can be observed on a depth converted seismic section, then the rollover dip at the working level can be estimated from Fig. 9-54 or Fig. 9-55. Figure 9-55 assumes a coulomb breakup angle of

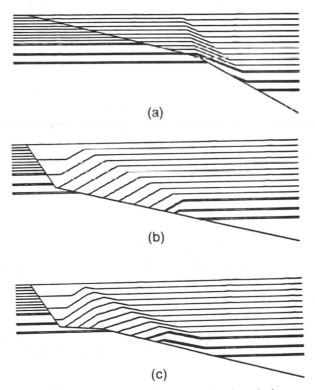

(a)

(b)

(c)

Figure 9-53 Deformation of growth sediments at the bends in normal faults. (a) A downward bend in the fault activates basinward dipping or synthetic shear. (b) An upward bend activates landward or antithetic shear. (c) Simple generic model of Brazos Ridge rollover development, not including compaction. (Published by permission of Xiao and Suppe.)

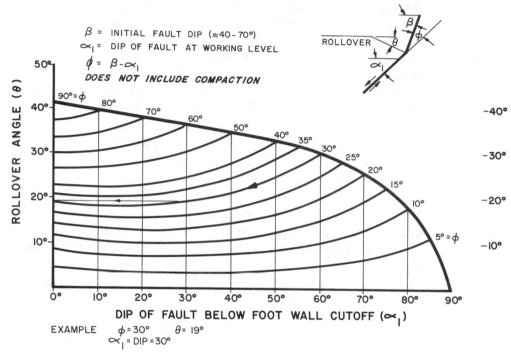

Figure 9-54 Theoretical prediction of rollover angle (θ) from fault dips (φ and α₁) for a coulomb shear angle of 60 deg. Diagram does not include compaction. (Published by permission of Dick Bischke.)

70 deg, whereas Fig. 9-54 assumes a 60-deg breakup angle. For example, the Brazos Ridge generally has a major fault which initially dips at about 45 deg and flattens to about 15 deg. Therefore, φ = 30 deg. From Figs. 9-54 and 9-55 the predicted bed dips at the front of the rollover structures are estimated to be 18 deg to 22 deg. Even though the Brazos Ridge growth sediments have been subject to compaction, these results compare favorably to observed bed dips located along the *flat or base* of the rollover, which average 20 deg. Figures 9-54 and 9-55 can also be utilized to generate simple noncompacted generic or idealized models of rollover structures.

Origin of Synthetic, Antithetic, and Keystone Structures; and Downward Dying Growth Faults

In the Brazos Ridge region, as in other areas, most of the antithetic or compensating faults are not associated with the active axial surface that forms at the concave upward bend in the major normal fault (e.g., Fig. 9-56). Surprisingly, few antithetic faults are present in the region directly above the concave upward bend in the major fault. As this region is subject to extension, as was discussed previously, this is contrary to what might be expected. Instead, most of the antithetic faults terminate at and lie above and basinward of a synthetic fault that appears to be associated with the concave downward bend in the major fault (Fig. 9-56) (Bischke and Suppe 1990b). For lack of a better term, this synthetic fault will be called the **"master synthetic."** In Fig. 9-56, the master synthetic fault is imaged between sp B and sp C at 0.2 sec to 2.0 sec. Figure 9-56 also depicts that

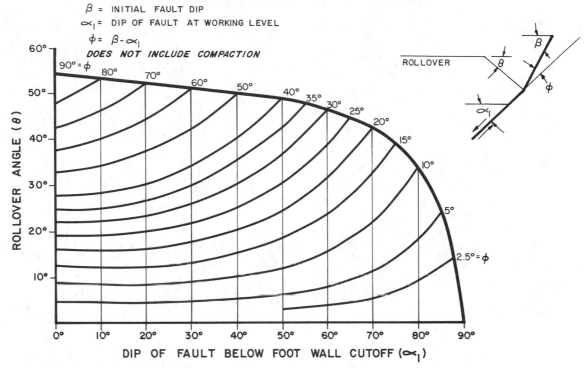

Figure 9-55 Same as Fig. 9-54, but for a coulomb shear angle of 70 deg. (Published by permission of Dick Bischke.)

the antithetics terminate along the master synthetic fault and do not generally exist beneath this synthetic.

From Figs. 9-53c and 9-56 and our discussion of simple rollover structures, it appears likely that the master synthetic fault is related to the active axial surface that forms at the concave downward bend in the master fault (i.e., is caused by the clockwise or synthetic shear produced by the convex bend). Other factors could cause this fault to take on a more listric shape. More importantly, the master synthetic can be observed to offset reflectors between the 1.0-sec to 2.0-sec level, but not reflectors at 2.1 sec to 2.4 sec (near sp C). Notice that the bold reflector at sp C and at 2.1 sec is coherent (i.e., not displaced). A closer inspection of the master synthetic fault shows that not only do the **displacements die downward** to zero with depth, but also that the displacements die upward toward the sea floor. The fact that the displacements are negligible near the sea floor is not surprising, as little time has lapsed since the most recent sediments were deposited. Therefore, our examination of the Brazos Ridge seismic lines has resulted in two seemingly confusing relationships concerning the slip along the antithetics and the master synthetic fault: (1) the slip along the antithetics generally terminates at the master synthetic fault; (2) the slip along the master synthetic fault in Fig. 9-56 at first increases from shallow depths but then dies at greater depths. Where does all this slip go, and does it really go to zero? The conservation of fault size rule discussed in Chapter 8 suggests otherwise.

In previous sections we made the point that slip along a thrust fault need not remain constant and that the slip could totally die in the cores of fault propagation folds. We did, however, account at all times for the changes in slip, and how the slip was consumed

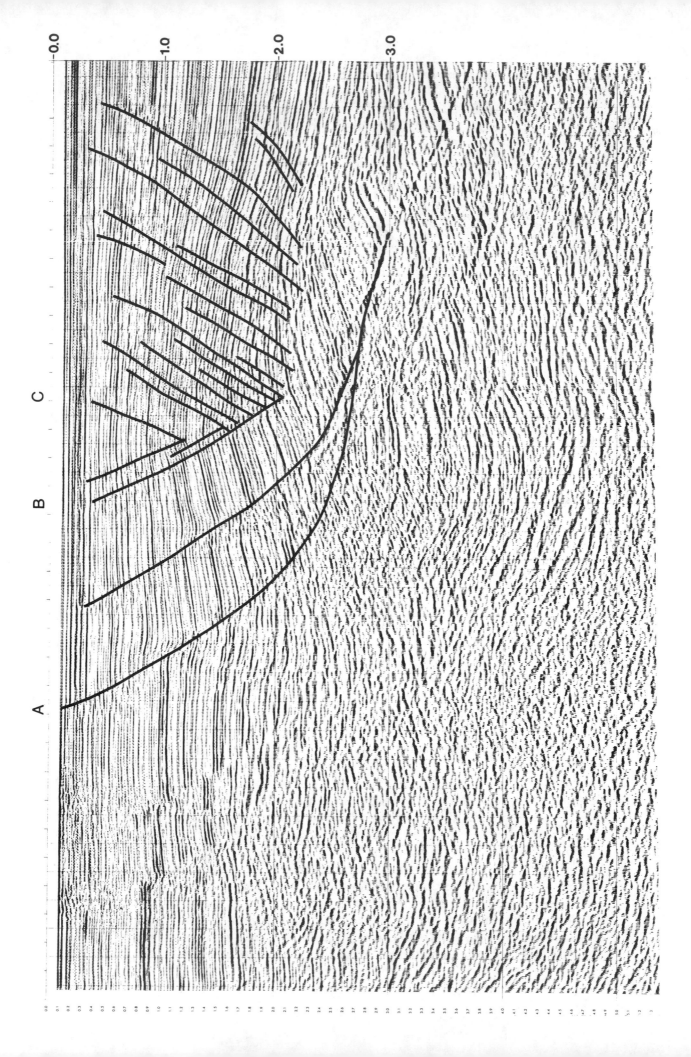

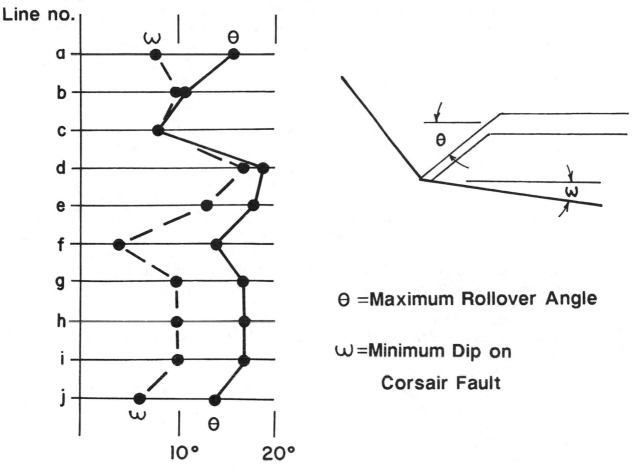

Figure 9-57 Minimum dip (ω) on Corsair Fault vs maximum bed dips (θ) at the front of the rollover and above its base. The frontal limb dip on the rollover structures generally exceeds the gentle dips on the Corsair Fault. (Published by permission of Dick Bischke.)

or dissipated. In the case of the Brazos Ridge, this slip, which appears to vanish so abruptly, apparently follows the *bedding planes*. How does this process work?

Backsliding Process. One can envision the Corsair fault to be a large, slowly moving landslide (Xiao and Suppe 1990). As the fault slips, a void opens up between the hanging wall and the footwall, and the overlying material instantaneously collapses along coulomb shear surfaces producing the rollover (Figs. 9-51 and 9-53). We learned from Figs. 9-54 and 9-55 that the greater the concave bend (φ) in the master fault, the greater the rollover angle (θ). If the dip along the major fault shallows with depth, then eventually the rollover angle may increase to an extent where it (θ) exceeds the dip on the major fault (Fig. 9-57). Frictional failure could then occur along the bedding surfaces that form the frontal portions (limb) of the rollover structure and not just along the coulomb shear surfaces. The higher the dip of the rollover structure, the closer the sedimentary beds

Figure 9-56 Jebco Seismic Inc., Houston, TX, line from Brazos Ridge imaging Corsair Fault sp A, rollover structure, and downward dying antithetic and synthetic faults. Antithetics terminate along a master synthetic fault. (Interpretation by Dick Bischke. From Xiao and Suppe 1990. Published by permission of John Suppe.)

come into parallelism with the antithetic coulomb failure direction. At some angle of inclination, the bedding planes are likely to present less frictional resistance than a sequence of sedimentary layers, and thus frictional failure along the bedding surfaces would be favored over the coulomb shear surface (of 60 deg to 70 deg) which cuts across the layering at a high angle (Fig. 9-51b). These potential bedding plane slip surfaces are likely to occur along *overpressured shale zones* which would present the least possible frictional resistance.

We therefore conclude that backsliding along bedding surfaces is a mechanism which can account for the apparent termination of the slip along the downward dying synthetic and the antithetic faults. Backsliding is analogous to reverse thrusting in fold and thrust belts, where the reverse faulting is opposed to the general motion. On the Brazos Ridge, the backsliding appears to initiate when the rollover angle exceeds about 10 deg and is a second-order effect relative to the total amount of slip along the major normal fault—one part of bedding plane slip to about every seven parts of the slip along the Corsair fault. Although the bedding plane slip within the frontal limb of the rollover structure is small compared to the total slip, we shall see that it can have a marked effect upon the amplitude of the rollover structure.

Backsliding Model. Exactly how does this backsliding mechanism operate? We present two examples, the first of which outlines the backsliding process; and a second, more realistic model, which assumes that the backsliding consumes about one part in ten of the amount of slip that occurs along the major fault. Depending upon geologic conditions, however, backsliding may at times be the dominant process.

Figure 9-58 illustrates the backsliding mechanism. Initially, slip along the master fault must occur to an extent that a *critical rollover failure angle* is exceeded (Fig. 9-58a). This critical failure angle should depend upon local geologic conditions, such as the pore pressures, and is likely to vary from area to area. As the pore pressure is unlikely to be large in pre-growth sediments, backsliding is most likely to occur within the growth sedimentary package. At some point in time, an overpressured shale moves through the concave upward bend. When this occurs, an increment of backsliding along the bedding surfaces that comprises the frontal portions of the rollover structure fills the void created by the basinward sliding (Fig. 9-58b). Initially, the plane of detachment propagates along the bedding plane surface and (in the case of the Brazos) up to the master synthetic. At this location, the clockwise shear associated with the concave downward bend in the major fault turns the strata downward and forms the upper portions of the rollover (Figs. 9-53c and 9-56). At the position of the master synthetic fault, the shear within the weak horizon can follow one of four possible paths. First, the propagating bedding plane slip could also turn downward and follow the bedding surfaces across the top of the rollover structure. This possibility seems unlikely, as large amounts of energy would be required to move the detached block. Second, the propagating slip could cut across the more gently dipping bedding surfaces located near the top of the rollover to a pre-existing antithetic fault (plane of weakness), and then follow the older antithetic fault to the surface. This would create small triangular-shaped structures. Third, and at the position of the master synthetic, the growing bedding plane fault could follow a coulomb failure surface up to the sea floor, creating a new antithetic fault (Fig. 9-58c). This deformation path appears to be the mechanism of least resistance, and the path most favored by the Brazos Ridge compensating faults. This mechanism would also result in antithetic faults that appear to terminate along the master synthetic, which is what normally is observed on our Brazos Ridge seismic sections. Fourth, the hanging wall block above the bedding plane detachment and the major and master synthetic fault could detach and slump toward

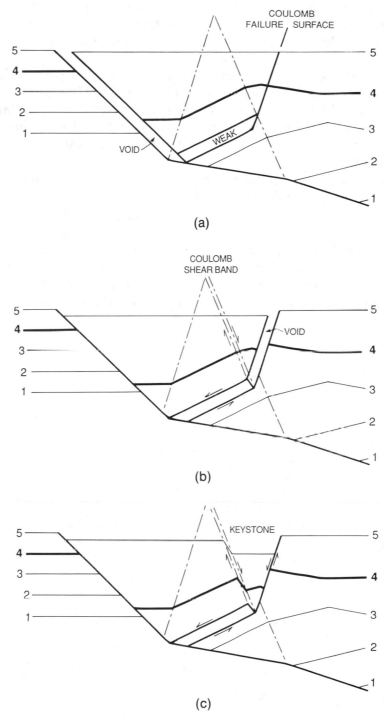

Figure 9-58 Backsliding model in growth sedimentary section. (a) Uniform shear along bedding surfaces causes the hanging wall to detach from the undeformed (foot-wall wedge. (b) The backsliding forms an antithetic fault and opens a hole along the newly formed fault surface. (c) The upper void is filled by synthetic coulomb collapse along a basinward dipping shear band. The balanced model forms a keystone and a downward dying antithetic and synthetic fault. (Published by permission of Dick Bischke.)

the major fault (not shown in Fig. 9-58). This mechanism would produce a surface of detachment along the master synthetic. This mechanism is not a dominant process in our area of study, but it could be present in other areas. Therefore, we presently favor the third and to a lesser extent the second mechanism to explain the observed antithetic faulting.

The third mechanism would cause another hole to open up along the newly formed antithetic fault (Fig. 9-58b). This hole can only be filled by the basinward dipping coulomb failure surface (Fig. 9-58c) forming a keystone structure. According to the balanced deformation model just described, another increment of backsliding will result in the formation of another antithetic which forms above the master synthetic and the previously formed antithetics.

Another consequence of this model is that the bedding plane slip is likely to occur along shale horizons that could be related to fluctuations in sea level (Payton 1977). As a result, the intervening horizons are more likely to contain *sands or reservoir rock*.

Example from Corsair Trend.   A more realistic or generic example for the Brazos Ridge is illustrated in Fig. 9-59. In this example, the major fault initially dips at 45 deg, decreases to 10 deg on the flat, and then steepens basinward to 20 deg. This fault configuration can be considered to be the first approximation for the Corsair fault. Also, in

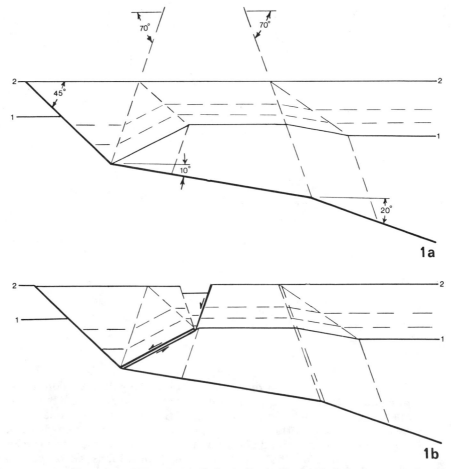

Figure 9-59 Generic model of the Corsair Fault illustrating the progressive development of downward dying antithetic and synthetic faults through growth stages 1 to 7. (Published by permission of Dick Bischke.)

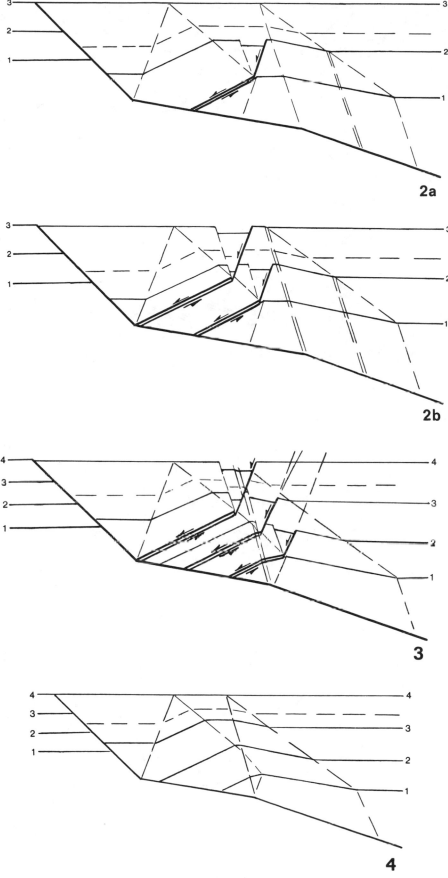

Figure 9-59 (*continued*)

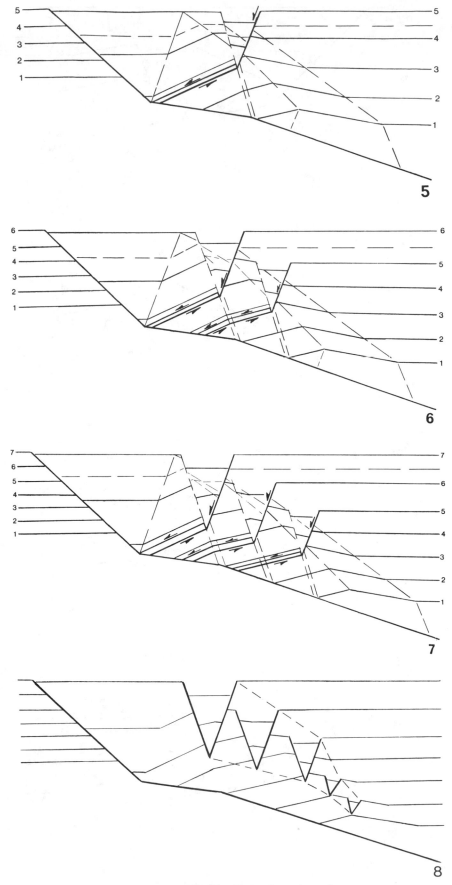

Figure 9-59 (*continued*)

Fig. 9-59-1a, the coulomb failure angle is assumed to be 70 deg and the growth phase of the sedimentation to begin at horizon 1. In Chapter 6 we described how to distinguish the growth interval from the nongrowth interval using expansion indices. Since the sediments deposited beneath horizon 1 are pre-growth sediments, and as the backsliding mechanism is likely to initiate in an over-pressured horizon, the backsliding process is not likely to begin until a critical angle is reached within the growth sediments. At this stage, the first increment of backsliding occurs (Fig. 9-59-1b).

Additional sliding along the major fault produces the geometry present in Fig. 9-59-2a, and another increment of backsliding results in Fig. 9-59-2b. The backsliding has the effect of creating keystones that **young** upward. The basinward sliding along the major fault has deactivated the antithetic fault that formed in Fig. 9-59-1b, and this antithetic fault stopped growing after horizon 2 was deposited (Fig. 9-59-2a). The backsliding also has the effect of reducing the rollover amplitude. After the deposition of layer 4, the geometry appears as shown in Fig. 9-59-3. At this stage, the antithetics which formed during the interval of time that horizons 1 to 3 were deposited have moved through the active axial surface that is associated with the concave downward bend in the major fault. The clockwise shear associated with this deformation has the effect of slightly rotating the antithetic faults clockwise (Fig. 9-59-3). During the interval of time that horizon 5 is deposited, slip along the major fault has advanced to the stage that the newly forming antithetics initiate to the left of the clockwise shear-active axial surface, and thus the antithetics which form after the deposition of layer 4 will not be rotated clockwise by the concave downward bend in the major fault (Fig. 9-59-5). Additional increments of backsliding are shown in Figs. 9-59-6 to 9-59-7.

Figure 9-59-8 is a generalization of all of the deformation that occurred during the deposition of layers 1 through 7. Dashed lines are drawn at the base of the antithetic faults where the slip entered the bedding planes, and at the top of the antithetic faults where they ceased to grow and became inactive (Fig. 9-59-8). These lines can be considered *axial surfaces* that are associated with the growth phase of the antithetic faulting and the formation of the keystones.

Again notice that the more recent antithetic faulting and keystones form to the left of the older rollovers. If this model is correct, then it can be tested by the data. Figures 9-60 and 9-61 are two seismic lines from the Brazos Ridge and Brunei (S.E. Asia), respectively. On both of these lines, the antithetic faults have a tendency to young upward and to age with depth into the deeper sediments.

Let us briefly review the consequences of the above described deformation.

1. Growth antithetic faults form basinward and above synthetic faults located near the crests of rollovers.

2. The antithetics young upward with the older antithetics being positioned basinward and terminating at a deeper level.

3. Slip along the master synthetic may die with depth, and slip along the antithetic faults appears to terminate at the position of the master synthetic.

4. The deformation mechanism forms a keystone structure or a central graben, which along with the antithetics has the effect of reducing the amplitude of the rollover structure.

5. The deformation mechanism forms faults which die in both the upward and downward directions.

6. The process may be controlled by the sedimentary Fary cycle.

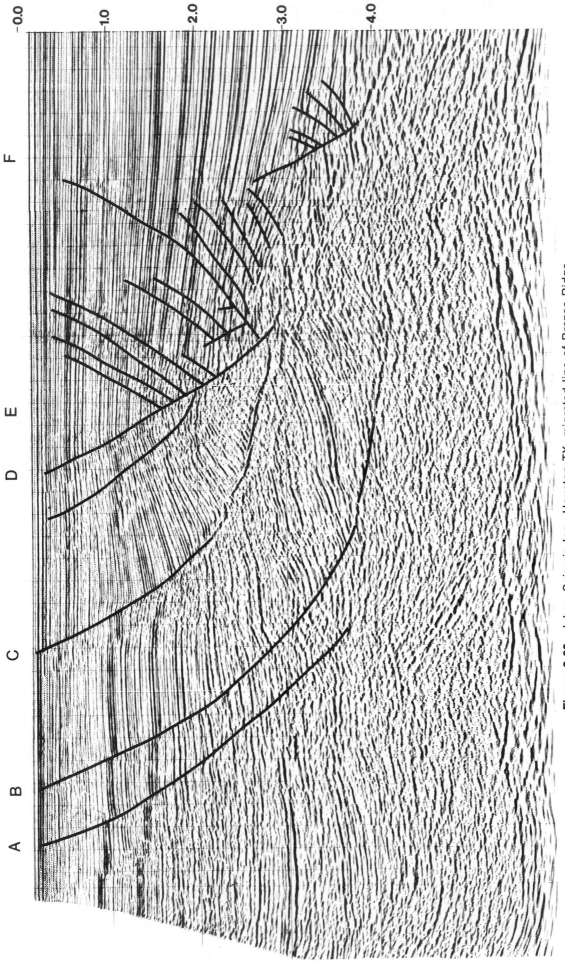

Figure 9-60 Jebco Seismic Inc., Houston TX, migrated line of Brazos Ridge showing antithetic faults that die with depth into the sedimentary basin. Corsair Fault is imaged at sp C and two other less active faults at sp A and B. (Interpretation by Dick Bischke. From Xiao and Suppe 1990. Published by permission of John Suppe.)

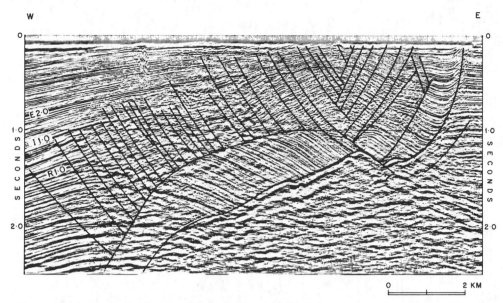

Figure 9-61 Seismic line from Brunei (Borneo) showing antithetic faults that appear to terminate along bedding surfaces. The antithetic faulting also dies with depth into the basin. (Published by permission of Muzium Brunei.)

Three-Dimensional Effects and Cross Structures

Our observation of the Brazos Ridge tie lines revealed the existence of previously unreported **cross or transverse structures**. As these structures determine the position of the rollovers, cross structures may play a fundamental role in locating future rollovers. In fact, all of the Brazos Ridge rollovers that we studied could be located directly from the *tie or strike lines*, or from the *fault surface maps*.

Many geophysicists, however, mistrust strike lines, and this is probably unfortunate. They correctly reason that collecting data along (parallel to) a steeply dipping surface such as the Corsair fault is like collecting data from a roof. For example, seismic energy from the upslope portion of the roof returns to the receiver, while the energy returning from directly beneath the receiver is deflected downslope. Thus, in regions of structural dip, the energy recorded on strike lines comes from out of the plane of the seismic section and is called **"side-swipe."** Consequently, reflections on strike lines are difficult to depth convert (Chapter 5).

Nevertheless, our experience has led us to conclude that *strike lines can be used to interpret data*. Strike lines may not image the structures in their correct location, but we take the position that often more can be learned from strike lines than from dip lines, even though strike lines are generally more poorly imaged relative to dip lines. Indeed, strike lines look like dip lines with one additional advantage. Strike lines also image the *cross structures*.

We make this point for the following reason. The location of all of the rollover closures along a 25-mi section of the Brazos Ridge that we studied, and this includes five rollover structures and four major gas fields, *can be predicted from a single strike line*.

The Brazos Ridge strike lines clearly image the Corsair fault and a number of faults on a deeper structural level. In Fig. 9-62, the Corsair fault is imaged as a distinct **boomer** (strong reflection) between the 2.3-sec to 2.7-sec level to the right of sp C. **The fault**

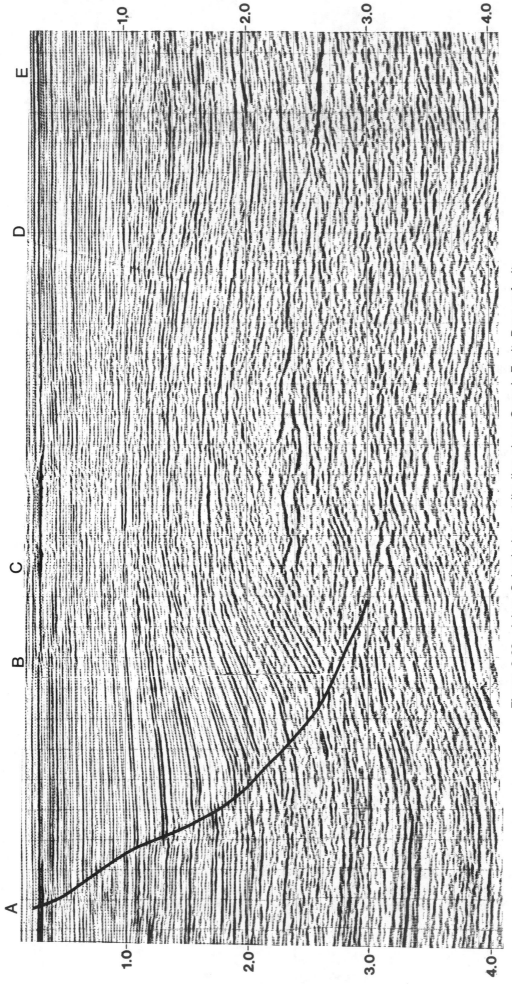

Figure 9-62 Jebco Seismic Inc., strike line along Corsair Fault. Deeper faults have deformed the Corsair into a series of highs and lows that create cross structures on the fault surface map (Fig. 9-63). The low on the Corsair Fault at sp B creates Chute 1 on Fig. 9-64. (Published by permission of Jebco Seismic Inc., Houston, TX.)

MAJOR FAULT SURFACE MAP
(DEPTHS NOT SHOWN)

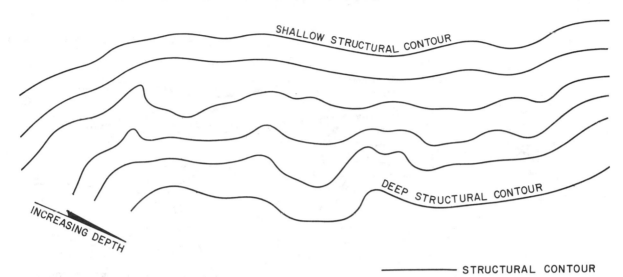

Figure 9-63 Fault surface map of a portion of the Brazos Ridge. The structural contours deepen toward the lower portion of the diagram. The cross structures segment the fault surface map into a series of low gradient areas (bows) and high gradient areas (chutes). (Prepared by W. L. Keyser. Published by permission of Texaco USA, Eastern E & P Region.)

LOCATION OF BOWS & CHUTES

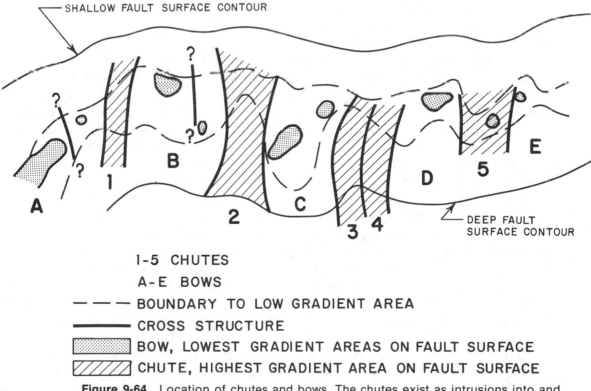

Figure 9-64 Location of chutes and bows. The chutes exist as intrusions into and the bows as protrusions upon the fault surface map (Fig. 9-63). Five bows and chutes exist along trend, with Chutes 3 and 4 being a double chute. (Published by permission of Texaco USA, Eastern E & P Region.)

485

GENERALIZED TIME HORIZON, PARTIALLY RESTORED

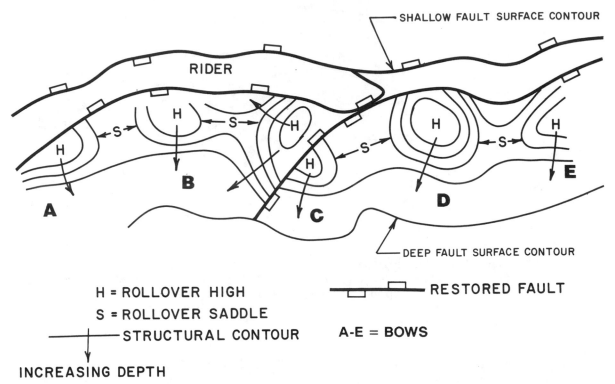

Figure 9-65 Partially restored time horizon map. The rollover highs correspond to the position of bows, while the saddles correlate to the chutes. (Published by permission of Texaco USA, Eastern E & P Region.)

surface is not planar. The Corsair is level between sp C to D, but then it deepens to 2.7 sec at sp E. The Corsair fault is seen to surface in Fig. 9-62 at about sp A (0.2 sec). You can follow this fault downward to about the sp B area, where another *deeper* fault is observed to *deform* (rollover) the Corsair. This deeper fault continues beneath the Corsair to perhaps the 3.55-sec level, where it levels off (at sp D). On the dip lines, these deeper faults can be observed to surface to the north (landward) of the Corsair fault, to dip beneath the Corsair fault, and to be presently active, although at much lower slip rates (Fig. 9-60, sp A and B). In Fig. 9-60, the Corsair fault is seen to surface at sp C. On the strike lines, these deeper faults can be observed to intersect the Corsair fault (Fig. 9-62, sp B), to offset the fault, and to form cross structures.

These cross structures trend subperpendicular to the Corsair fault and **deform and partition** the footwall into a series of high and low gradient areas (Fig. 9-63). In certain places, these deeper intersecting faults propagate upward into the hangwall block and form tear faults. On a regional scale, the series of highs and lows on the Corsair fault, which are caused by the *major subfault deformation*, are bounded by the cross structures (Fig. 9-64). The low gradient, shelf-like, or "flat" areas on the Corsair fault produce "bows" in the fault surface. Bows are known in the Gulf to be a key petroleum trap indicator. We propose the term **"chute"** for the higher gradient regions of indentation into the fault surface. The strike line (Fig. 9-62) images chute 1 (Fig. 9-64) at sp B.

Finally, we demonstrate that the general position of the rollover structures can be predicted from the fault surface map of the Corsair fault (Fig. 9-63), with the bows or lowest gradient areas on the fault surface corresponding to the position of the rollover highs, and the chutes or lows corresponding to the position of the saddles between the structural highs. These relationships are shown in Fig. 9-65, which is a partially restored time map of a horizon that was rolled over by the Corsair fault. The map was restored to a common level by simply closing the faults. A comparison of Fig. 9-65 to Figs. 9-64 and 9-63 demonstrates that the bows correlate to the position of the closures, while the chutes correlate to the saddles. Thus, rollovers can be located from fault surface maps, which further demonstrates the value of constructing these maps (Chapter 7).

CHAPTER 10

ISOCHORE/ISOPACH
MAPS

INTRODUCTION

An *isopach map* is a map which shows by means of contour lines the distribution and thickness of a specific mapping unit (Bishop 1960) as shown in Fig. 10-1. The terms isochore and isopach are often used synonymously in the petroleum industry, but they are different. **An isochore map delineates the true vertical thickness of a rock unit, while an isopach map illustrates the true stratigraphic thickness of a unit**.

An isochored or isopached unit may be as small as an individual sand only a few feet thick or as large as several thousand feet and encompassing a number of sand units. An isopach map is extremely useful in determining the "tectonic framework," or the structural relationship responsible for a given type of sedimentation. The shape of a basin, the position of the shoreline, areas of uplift, and under some circumstances the amount of vertical uplift and erosion, can be recognized by mapping the variations in thickness of a given stratigraphic interval (Bishop 1960).

Isochore and Isopach maps are used for a number of purposes by the petroleum geologist, including: (1) depositional environment studies, (2) genetic sand studies, (3) growth history analyses, (4) depositional fairway studies, (5) derivative mapping, (6) the history of fault movement, and (7) calculation of hydrocarbon volumes.

In this chapter, we discuss several different types of isochore/isopach maps important to the evaluation of petroleum potential. These include **interval isopachs, net sand, and net pay isochore maps**. An *interval isopach* map delineates the true **stratigraphic thickness** of a specific unit. A *net sand isopach* map is an isochore map which represents the total aggregate **vertical thickness** of porous reservoir quality rock present in a particular stratigraphic interval, as illustrated in Fig. 10-2. The techniques and calculations to derive

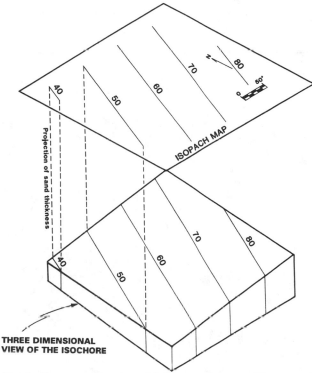

Figure 10-1   Block diagram showing thickness of a specific mapping unit. Upper portion of the figure is an isopach map of the unit. Modified from Appelbaum. Geological & Engineering mapping of Subsurface: A workshop course by Robert & Appelbaum.

maps. An Interval Isopach map delineates the true **stratigraphic thickness** of a specific unit. A *net sand isochore map* represents the total aggregate **vertical thickness** of porous reservoir quality rock present in a particular stratigraphic interval, as illustrated in Fig. 10-2. The techniques and calculations to derive vertical thickness are explained in detail in this chapter. The fluid contained in an isochore interval may be hydrocarbons or water or any combination of two. Figure 10-3 shows a net sand isopach map of the 10,500-ft

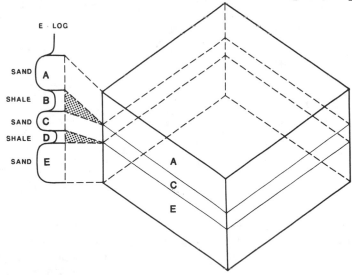

Figure 10-2   Net sand consists of porous reservoir quality rock. All shale and non-reservoir quality rock are removed. (From Tearpock and Harris 1987. Published by permission of Tenneco Oil Company.)

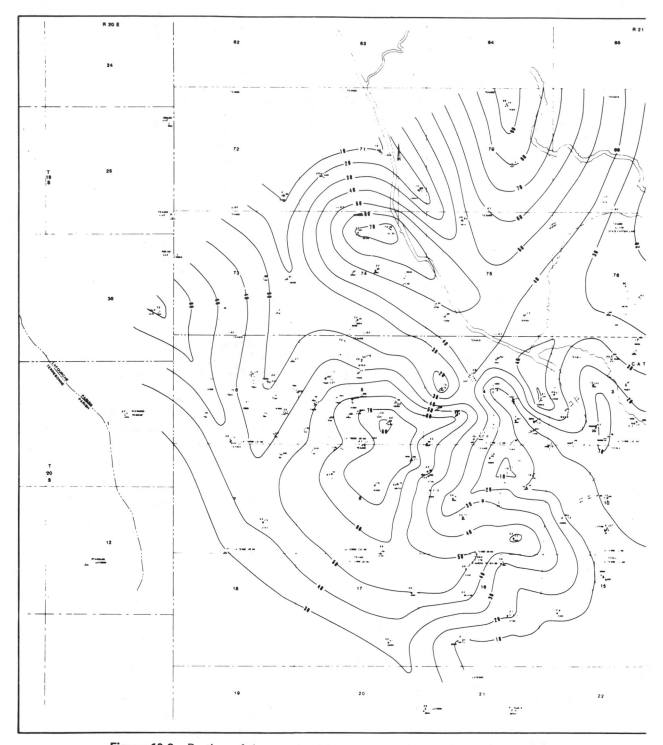

Figure 10-3 Portion of the net sand isochore map of the 10,500-ft Sand in Golden Meadow Field, Lafourche Parish, Louisiana. (Published by permission of Texaco, USA.)

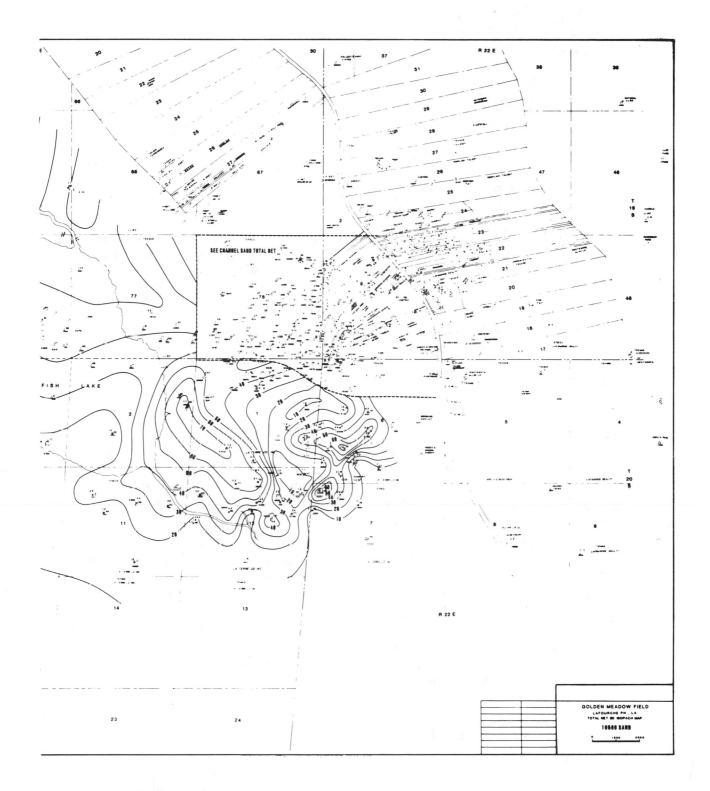

GOLDEN MEADOW FIELD
LAFOURCHE PH., LA
TOTAL NET SD ISOPACH MAP
10500 SAND

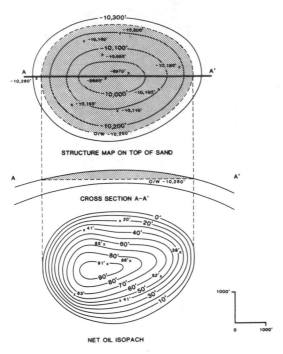

STRUCTURE MAP ON TOP OF SAND

CROSS SECTION A-A'

NET OIL ISOPACH

Figure 10-4 Structure map, cross section, and net oil isochore map for a bottom water reservoir. Net oil isochore map is a special map showing the thickness of reservoir quality rock containing hydrocarbons. (From Tearpock and Harris 1987. Published by permission of Tenneco Oil Company.)

Sand in Golden Meadow Field, Lafourche Parish, Louisiana. *A net pay isochore map* that delineates the thickness of reservoir quality sand which contains hydrocarbons (gas, oil, or both), is shown in Fig. 10-4.

Net sand and net pay isochore maps of subsurface units are usually prepared from well log data. Interval isopach maps may be constructed from well log, as well as seismic data, where coverage is adequate. As with structure maps, the completeness and accuracy of an isochore/isopach maps depends upon the amount and accuracy of data available. Even in isochore/isopach mapping, we cannot get away from the importance of log correlation work. Well log data, particularly correlations, should be studied very carefully in order to prepare an accurate and precise isochore map.

For volumetric reserve calculations, we are interested in obtaining the volume of a reservoir in terms of *acre-feet*. To many people, an acre-foot is an abstract measurement, but the concept is relatively simple. One acre-foot is defined as that volume of fluid contained in an area one acre in size with a thickness of one foot. How big is an acre? There is a very easy way to visualize the size of an acre. One acre contains just about the same area as a football field from goal line to goal line. A football field 300 ft long and 160 ft wide is equal to 48,000 sq ft, while one acre is equal to 43,560 sq ft. If we fill a football field one foot deep with oil, the volume of oil is just about equal to one acre-foot. In terms of barrels of oil, there are 7758 barrels of oil in one acre-foot.

SAND/SHALE DISTRIBUTION

Most individual sand bodies do not consist exclusively of sand; shale and other nonreservoir quality rock material are commonly distributed throughout the sand as interbedded shale or impervious zones. The percentage and distribution of shale members or impervious zones varies from sand to sand. Net sand and net pay isochore maps are drawn on net effective sand (porosity) only; therefore, shale and other nonreservoir quality rock must be subtracted from the total sand interval to determine the net effective sand for isochore mapping.

The net effective sand in a well to be used for isochore mapping is normally determined by detail analysis of a 5-in. electric log. In Chapter 4, in the section Annotation and Documentation, we outlined a method for distinguishing and annotating the percentage and distribution of sand and shale that is present within a particular sand member (Fig. 4-37). Once the net sand is determined for each well, a net sand map can be prepared for that sand. The aggregate of net sand for any particular well may contain water or hydrocarbons; net pay is that portion of the net sand that contains hydrocarbons.

BASIC CONSTRUCTION OF ISOCHORE MAPS

The procedure used in constructing an isochore map depends on whether the reservoir being mapped is a *bottom water* or an *edge water* reservoir. A bottom water reservoir is a reservoir which is completely underlain by water, and an edge water reservoir is one not completely underlain by water, with some portion of the reservoir completely filled with hydrocarbons from the top to base of the sand or unit.

It is very important to be able to visualize a hydrocarbon reservoir in three dimensions. Your ability to understand the configuration of a reservoir can impact the location of development wells, completion practices, and planned production.

Bottom Water Reservoir

Figure 10-4 illustrates in both map and cross-sectional views a bottom water reservoir. Looking at the cross section in the center of the figure, we see that this is an oil reservoir. Notice that the hydrocarbons are completely underlain by water. This bottom water reservoir contains a wedge of oil, therefore nowhere in the reservoir is there a constant or full thickness of oil.

Figure 10-5 is a three-dimensional model of a bottom water reservoir. This hydrocarbon accumulation consisting of oil and gas is trapped on an anticlinal structure. Notice how the oil and gas are segmented within the reservoir and completely underlain by water.

3-DIMENSIONAL MODEL

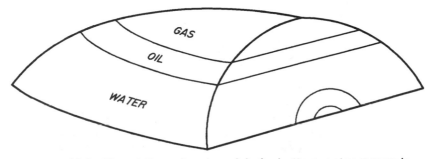

Figure 10-5   Three-dimensional model of a bottom water reservoir.

Net Pay Isochore Map Construction.   The construction of a *net pay isochore map* for a bottom water reservoir requires a structure map on the top of porosity (reservoir top) and net pay values for each well in the reservoir. The following procedure is used to construct the net pay isochore map for a bottom water reservoir (Fig. 10-4).

1. Post the net pay values for each well on a blank base map. The net pay values must be corrected, if required, to true vertical thickness.

2. Overlay the isochore base map onto the structure map for the reservoir being mapped and draw the outer limit of the hydrocarbon bearing reservoir of which is any boundary such as an oil/water contact, fault, pinchout, permeability barrier, etc. This outer limit of the productive reservoir area becomes the *zero line* on the net pay isochore map, as shown in Fig. 10-4. In this case, the outer limit of the reservoir is an oil/water contact at a subsea depth of 10,250 ft.

3. Contour the net pay isochore map, which is contained within the area outlined by the zero line on the base map. Be sure to honor all posted net pay values. If the well control is limited, additional points of contour control may be obtained by using a method which is called **"walking wells."** This method is explained in detail later in this chapter. The net pay isochore contours generally parallel the structure contours and are often drawn as being equal spaced. Because of variations in net sand thickness around the reservoir, the isochore contours need not be equally spaced. Basically, the net pay isochore map for a bottom water reservoir illustrates a reservoir that is totally underlain by water (an entire wedge of hydrocarbons), as shown in Fig. 10-4.

Edge Water Reservoir

Figure 10-6 illustrates in map and cross-sectional views an edge water reservoir containing oil. The cross section shows that there are at least **two** different wedge zones involved in this type reservoir: (1) a water wedge, and (2) an oil wedge. From the oil/water contact

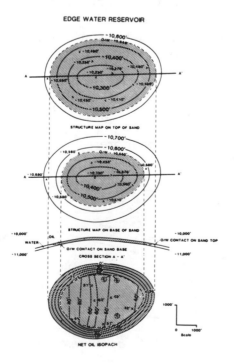

Figure 10-6 Structure maps, cross section, and net oil isochore map for an edge water reservoir. The procedure for constructing a net oil isochore map is summarized in this figure.

3–DIMENSIONAL MODEL

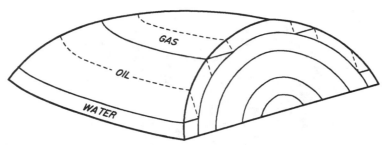

Figure 10-7   Three-dimensional model of an edge water reservoir.

on the top of the sand to the oil/water contact on the base of the sand, there is an oil wedge sitting on top of a water wedge. Updip of the oil/water contact on the base of sand, the reservoir is full of oil from the top to base of sand.

Edge water reservoirs are obviously more complex than bottom water reservoirs. An edge water reservoir can become extremely complex if it contains oil and gas and is cut by one or more faults. When mapping a reservoir cut by faults, in addition to the hydrocarbon wedges, consideration may be given to mapping one or more fault wedges, resulting in very complex isochore maps. Fault wedges are discussed in detail later in ·this chapter.

Figure 10-7 is a three-dimensional model of an edge water reservoir containing oil and gas. The hydrocarbons are trapped on an anticlinal structure similar to that shown in Fig. 10-6. Compare the configuration of this reservoir with the bottom water reservoir model in Fig. 10-5. It is obvious that this type reservoir is more complex and harder to visualize in three dimensions.

An understanding of the reservoir type and configuration is very important in such decisions as the location of development wells, completion practices, and production plans. Take a few minutes and review the figure, especially the areas of multiple wedge zones.

Net Pay Isochore Map Construction-One Hydrocarbon. The generally accepted method for construction of a net hydrocarbon isochore map for an edge water reservoir is called the **Wharton Method** after Jay B. Wharton (1948). The data needed to construct a net hydrocarbon isochore map for this type reservoir are:

1. a structure map on the formation top (top of porosity);
2. a structure map on the formation base (base of porosity);
3. a net sand isochore map;
4. net pay values for all available wells; and
5. depth or elevation of all fluid contacts (oil/water, gas/water, oil/gas).

We first outline in detail the procedure for construction of a net hydrocarbon isochore map for an edge water reservoir containing one hydrocarbon. For this example, we consider the formation to be sandstone and the hydrocarbon to be oil. The procedure is illustrated in Fig. 10-8.

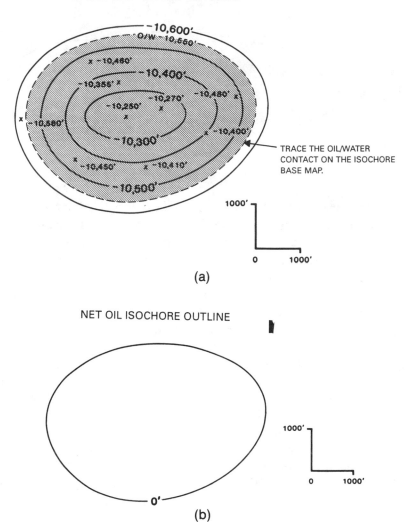

STRUCTURE MAP ON TOP OF SAND

TRACE THE OIL/WATER
CONTACT ON THE ISOCHORE
BASE MAP.

(a)

NET OIL ISOCHORE OUTLINE

(b)

Figure 10-8 (a) An overlay of the structure map on top of the sand onto a blank base map. The oil/water contact is traced. (b) The oil/water contact becomes the zero line on the net oil isochore map. (c) Overlay the isochore base map onto the structure map on the base of sand and trace the oil/water contact. (d) Isochore base map delineating the two major areas comprising the net oil isochore map: (1) wedge zone from the zero line to the inner limit of water (oil/water contact on the base of the sand), and (2) the area of full hydrocarbon thickness (area inside the inner limit of water). (e) To contour the full thickness area, superimpose the net oil isochore base onto the net sand isochore map and trace the net sand isochore contours inside the inner limit of water (dashed line). (f) Net oil isochore outline with contours drawn for the 100% oilfilled area. (g) Ail full thickness area contours that intersect the inner limit of water must connect with contours of equal value in the wedge zone. (See text for procedure.) (h) Completed net oil isochore map with important points of isochore construction highlighted. (From Tearpock and Harris 1987. Published by permission of Tenneco Oil Company.)

STRUCTURE MAP ON BASE OF SAND

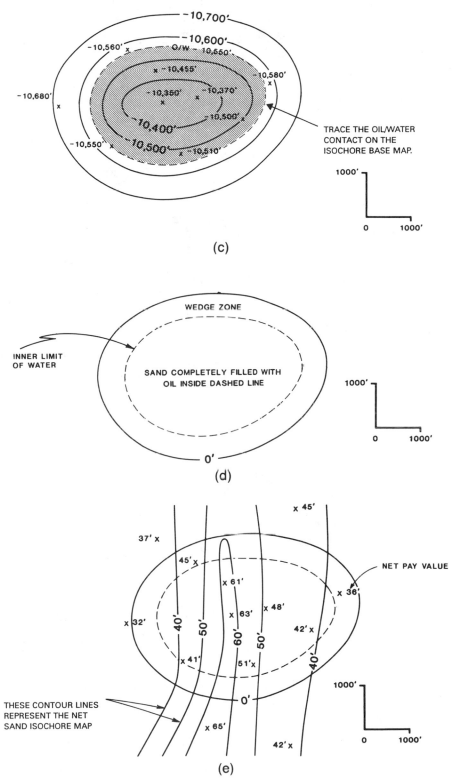

(c)

(d)

(e)

Figure 10-8 (*continued*)

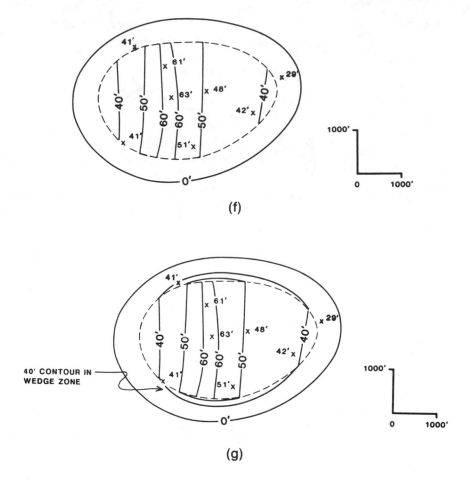

(f)

(g)

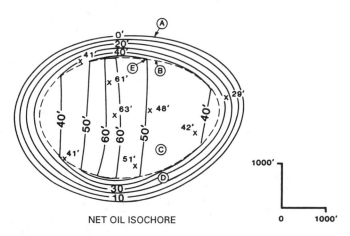

NET OIL ISOCHORE

A. "0" LINE OR OUTER LIMIT OF HYDROCARBON/WATER CONTACT
   FROM STRUCTURE MAP ON TOP OF SAND.

B. INNER LIMIT OF WATER IS HYDROCARBON/WATER CONTACT ON
   BASE OF SAND.

C. AREA UPDIP OF "B" IS TOTALLY FILLED WITH HYDROCARBONS.

D. WEDGE ZONE.

E. ABRUPT ANGLE TOWARD NEXT NUMERICALLY LARGER CONTOUR.

(h)

Figure 10-8   (*continued*)

1. Start with a blank base map with all the well control spotted.

2. Place the base map over the structure map on the top of sand (Fig. 10-8a) and trace the outer limit of the productive reservoir. This is the oil/water contact in this example and corresponds to the *zero line* on the net oil isochore map, just as in the previous example. The zero line which delineates the reservoir limit is shown in Fig. 10-8b. From this point forward, we refer to the blank base map as the net oil isochore map.

3. Place the net oil isochore map over the structure map on the base of sand (Fig. 10-8c) and trace the oil/water contact onto the isopach map using a *dashed line*. This dashed line represents the *inner limit of water* for the reservoir. The area inside this dashed line is filled with oil from top to base of the sand (Fig. 10-8d) and this area is referred to as the *full thickness area*. The area between the two oil/water contacts (top of sand and base of sand) is the reservoir wedge zone.

4. Post net pay values for all wells within the reservoir, corrected to true vertical thickness, on the net oil isochore base map.

5. In the full thickness area (inside the dashed line drawn in step 3) the net oil contained in the reservoir equals the total sand as interpreted on the net sand isochore map. This is so because the entire sand is full of oil in this area. Therefore, to contour the full thickness area, place the net oil isochore map over the net sand isochore map as shown in Fig. 10-8e (see introduction for definition of net sand isochore map), and trace the contours which fall within the dashed line onto the net oil isochore map. The full thickness area of the net oil isochore map is now finished, as illustrated in Fig. 10-8f.

6. The next step is to contour the wedge zone. The wedge zone is the area of the isochore between the oil/water contact on the top of sand and the oil/water contact on the base of sand as shown in Figs. 10-8d and 10-8f. The wedge zone contains oil and water and is wedge shaped (see cross section in Fig. 10-10). All well data in the wedge zone must be honored. Unlike the full thickness area contours which were controlled by the net sand, the major influences on contours in the wedge zone are the structural attitude of the sand and the shale distribution within the sand body. The wedge edge isochore contours generally parallel the structure contours, but may not necessarily be equally spaced (variations in contour spacing are discussed later). The 40-ft and 50-ft isochore contour lines in the wedge zone are shown in Fig. 10-8g.

7. The final step is to connect the contours in the full thickness area of the isochore to those in the wedge zone. Notice in Figs. 10-8g and 10-8h that when the full thickness contour lines are connected with the wedge edge contour lines of the same value, the full thickness contour lines make an abrupt turn at the inner limit of water in the direction of increasing sand thickness (or in the direction of the next higher contour).

The completed net oil isochore map is shown in Fig. 10-8h. This figure highlights five important points of the net pay isochore construction.

The method of connecting the full thickness contours to those in the wedge zone is extremely important and deserves special attention. The inner limit of water shown as the dashed line on the net pay isochore map separates the area where the entire sand is full of hydrocarbons from that in the wedge zone which contains both hydrocarbons and water.

Why cannot the contours in the full thickness area continue uninterrupted past the

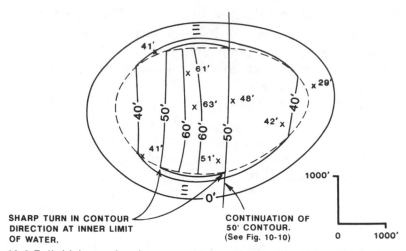

Figure 10-9 Full thickness isochore contours make an abrupt turn at the inner limit of water, toward the next higher contour. (Modified from Tearpock and Harris 1987. Published by permission of Tenneco Oil Company.)

inner limit of water, downdip, into the wedge zone? Look at Fig. 10-9, which is similar to Fig. 10-8g. Let's discuss the construction of the easternmost 50-ft contour highlighted in the figure. Why must this contour line sharply change direction at the inner limit of water on the net oil isochore map rather than continue straight into the wedge zone? Figure 10-10 is a diagrammatic cross section running parallel to the 50-ft net sand isochore contour line. Everywhere along the cross section there is exactly 50 ft of net sand. In the portion of the reservoir that is updip to the oil/water contact on the base of the sand (inner limit of water), the entire 50 ft of net sand is full of oil from top to base. However, if we go one foot downdip of the inner limit of water, we are in the wedge zone where the sand contains both oil and water. Therefore, anywhere outside the inner limit of water, there is both oil and water; and, since the total net sand is still 50 ft thick; there must be less than 50 ft of oil. The 50-ft net pay isochore contour, therefore, cannot continue along the 50-ft total net sand contour downdip of the inner limit of water.

Where must the contour from the full thickness area be drawn in the wedge zone? Since contour lines must close, there must be an area of 50 ft of net oil in the wedge zone. This area exists where the net sand is greater than 50 ft. In Fig. 10-8e, notice that

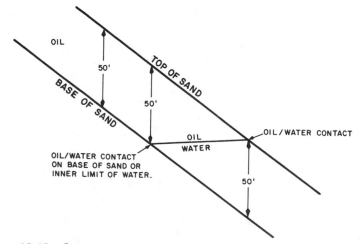

Figure 10-10   Cross section parallel to the 50-ft net sand contour line.

the net sand increases in thickness west of the 50-ft contour to a maximum of 63 ft. Therefore, in order to close the 50-ft contour and correctly isochore the wedge edge, the contour must turn sharply at its intersection with the inner limit of water, toward the area of thicker sand, and connect with the other 50-ft full thickness contour line on the map (Fig. 10-8g).

This procedure must be undertaken for all contour lines contained within the full thickness area of the net pay isochore map. The correct application of this technique is most important. If the 50-ft contour is carried incorrectly into the wedge zone, the volume of hydrocarbons determined for the reservoir, by planimetering, will be over-estimated.

Figures 10-11a and 10-11b present a summary of the method for preparing a net pay isochore map for an edge water reservoir containing one hydrocarbon.

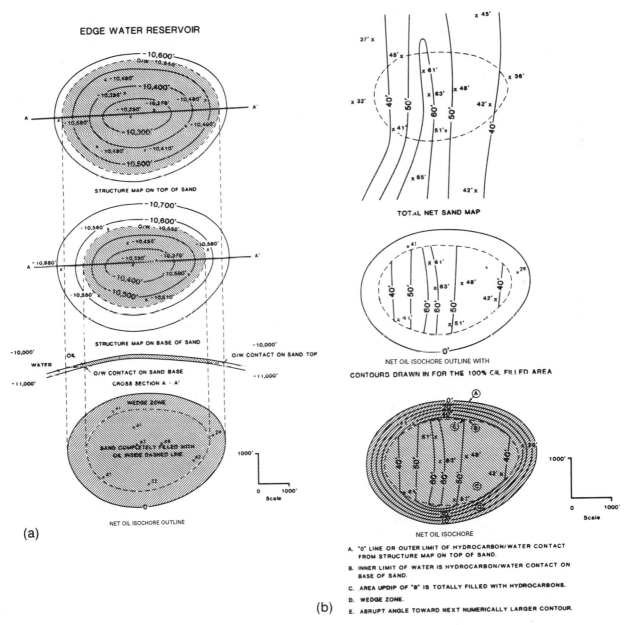

Figure 10-11 Summary of method for constructing a net hydrocarbon isochore map for a reservoir containing one hydrocarbon (oil or gas). (Modified from Tearpock and Harris 1987. Published by permission of Tenneco Oil Company.)

Net Pay Isochore Map Construction—Oil and Gas. There are two ways to determine the volumes of oil and gas in a reservoir containing both hydrocarbons. The simplest and quickest method is to construct a total net hydrocarbon isochore map and a net gas isochore map, calculate the volumes of each, and subtract the gas volume from the total hydrocarbon volume to determine the oil volume. This method is appropriate when only one lease or owner is involved or when the total volume of oil and gas is required without an interest in the actual distribution of oil and gas within the reservoir. When a reservoir is underlain by two or more separate leases, it is very important to know the volume of gas and oil under each lease. In this case, net gas and net oil isochore maps must be constructed, preferably using the procedure outlined in this section.

First, draw the basic maps used in the Wharton method, which are the structure map on top of porosity, structure map on base of porosity, and net sand isochore map. Using these maps, construct the net gas isochore map as shown in Figs. 10-12a and 10-12b. The net gas isochore map is constructed using the same procedure explained in the previous section on edge water reservoirs containing one hydrocarbon. The only difference in this case is that the gas/oil contact forms the downdip outer (zero) limit of the gas reservoir, whereas in the previous cases the limit was a hydrocarbon/water contact.

The last map to be constructed is the net oil isochore map. This map differs from the previous maps in that it has two wedges, an inner wedge zone (gas/oil) and an outer

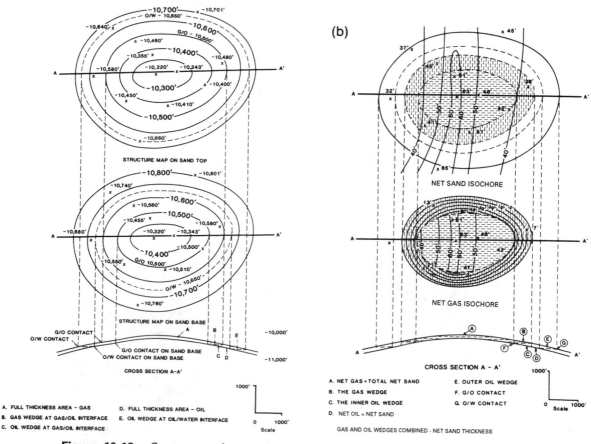

Figure 10-12   Summary of procedure to construct a net gas isochore map for a reservoir containing both oil and gas. (Modified from Tearpock and Harris 1987. Published by permission of Tenneco Oil Company.)

wedge zone (oil/water), as shown on the cross section in Fig. 10-12a. The outer wedge zone and any full thickness areas are constructed using the Wharton method, as previously discussed; the inner wedge requires additional steps.

Figure 10-13a shows the base map for the oil isochore with an inner and outer wedge of oil and an area of full thickness in between. The base also shows the area containing full thickness of gas and no oil. First, contour the area containing a full thickness of oil.

OIL ISOCHORE OUTLINE

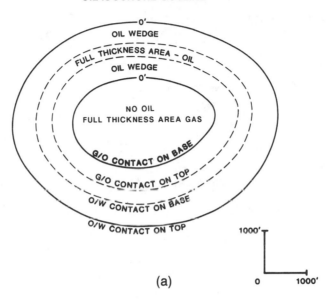

(a)

OVERLAY OF NET OIL ISOCHORE OUTLINE WITH NET SAND ISOCHORE

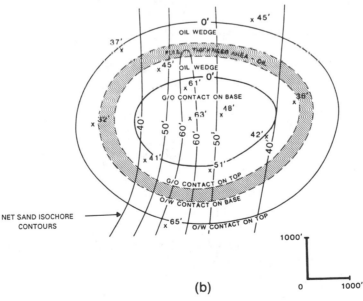

(b)

Figure 10-13 (a) Outline of oil isochore base showing the inner and outer wedge zones and area of full thickness. (b) Overlay of net oil isochore base map onto the net sand isochore map. The contours in the area of full oil thickness are equal to the net sand contours. (c) Full thickness area is contoured. (d) Overlay of net gas and net sand isochore maps used to aid in the construction of the inner oil wedge contours. (e) Completed net oil isochore map. (From Tearpock and Harris 1987. Published by permission of Tenneco Oil Company.)

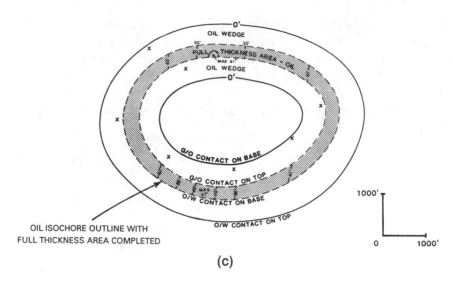

OIL ISOCHORE OUTLINE WITH
FULL THICKNESS AREA COMPLETED

(c)

OVERLAY OF NET GAS ISOCHORE AND TOTAL NET SAND ISOCHORE

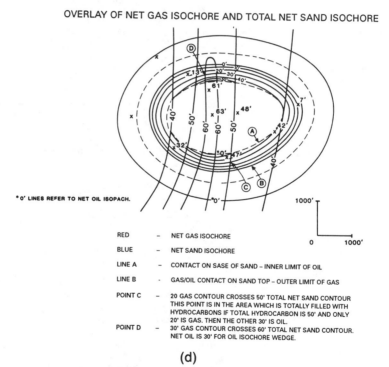

*0' LINES REFER TO NET OIL ISOPACH.

RED	–	NET GAS ISOCHORE
BLUE	–	NET SAND ISOCHORE
LINE A	–	CONTACT ON SASE OF SAND – INNER LIMIT OF OIL
LINE B	–	GAS/OIL CONTACT ON SAND TOP – OUTER LIMIT OF GAS
POINT C	–	20 GAS CONTOUR CROSSES 50' TOTAL NET SAND CONTOUR THIS POINT IS IN THE AREA WHICH IS TOTALLY FILLED WITH HYDROCARBONS IF TOTAL HYDROCARBON IS 50' AND ONLY 20' IS GAS. THEN THE OTHER 30' IS OIL.
POINT D	–	30' GAS CONTOUR CROSSES 60' TOTAL NET SAND CONTOUR. NET OIL IS 30' FOR OIL ISOCHORE WEDGE.

(d)

Figure 10-13   (*continued*)

Post the net oil values next to each well and overlay the isochore base map onto the net sand isochore map, as shown in Fig. 10-13b. Trace the net sand contours onto the base map in the full thickness area of oil as shown in Fig. 10-13c and this portion of the isochore is complete.

The outer and inner wedge zones on the net oil isochore map remain to be constructed. By referring to the cross section in Fig. 10-12b, we see that within the inner oil wedge zone, **the sum of net oil and net gas sand equals the total net sand**. Therefore, in order to determine the amount of net oil in the inner wedge zone, the following procedure is used. Overlay the net gas isochore map over the net sand isochore map and mark each location where the contours for the two separate maps cross. The net oil sand value at each contour

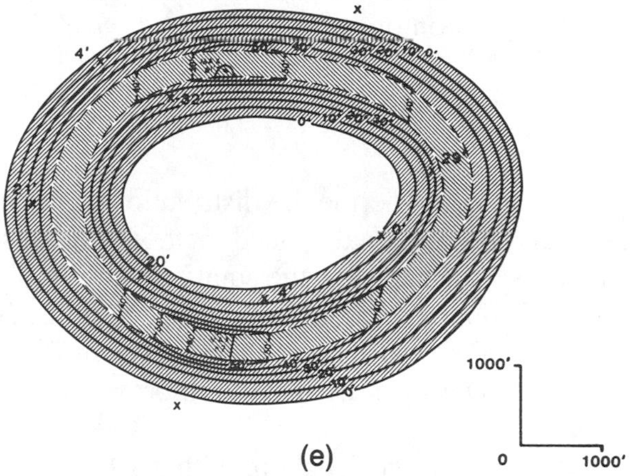

COMPLETED NET OIL ISOCHORE

(e)

Figure 10-13 (*continued*)

intersection is equal to the difference in values of the two contours. For example, at point C in Fig. 10-13d, the 20-ft contour line on the net gas isochore map crosses the 50ft contour line on the net sand isochore map. By subtracting the 20 ft of net gas from the 50 ft of net sand, a value of 30 ft of net oil is obtained for this point. Take a minute and review the data at point D. As indicated, a known value is established wherever a contour line on the net gas isochore crosses a contour line on the net sand isochore. The net oil sand value at each intersection is the difference in values of the two contours. Figure 10-13d shows that there are 48 calculated points of control plus data from four wells to aid in contouring the inner wedge zone of the oil isochore map.

The net oil isochore map should be overlain on both the net gas and net sand isochore maps while contouring the inner wedge. This allows the inner oil wedge to be mapped with the assurance that **the sum of the net gas and net oil does not exceed the total net sand**. This is one of the most complex, and therefore most difficult, isochore maps to construct.

Finally, contour the outer wedge zone just as shown in the previous section on contouring the wedge zone in a single phase reservoir (Fig. 10-8h). We now have constructed the inner and outer wedges, and the full thickness area for the oil isochore. The completed net oil isochore map containing two contoured wedge zones and an area of full thickness is shown in Fig. 10-13e.

METHODS OF CONTOURING THE WEDGE ZONE

Limited Well Control or Evenly Distributed Shale

With limited well control in the wedge zone of a reservoir, or a fairly even distribution of shale within the sand body, the most common method for contouring the wedge is to proportionally space the isochore contours, while honoring the available well control. This is how the outer oil wedge and gas wedge in the last example were constructed.

The configuration of the contours within a wedge zone is primarily controlled by the structural attitude of the formation and the distribution of sand and shale within the overall sand member. If the distribution of shale or other nonreservoir quality rock is fairly uniform, the primary influence on the contours in the wedge zone is the structural attitude of the sand. In such a case, the contours should be equal spaced, and more or less parallel the structure contours.

Walking Wells

At times, we may wish to better define the distribution of net pay sand within the wedge zone instead of just using the proportional spaced method. Perhaps the reservoir is being mapped for a unitization, there is a dispute as to the equity of various leases overlying the reservoir, the sand/shale distribution is not uniform, or both thick and thin shale layers occur within the sand. In such cases, a more detailed estimate of the reserves in the wedge zone may be required. A technique called *walking wells* can be used to improve the accuracy of the contouring in the isochore wedge zone.

We wish to construct a net gas isochore with a 10-ft contour interval for a reservoir with limited well control, and would like additional control in the oil wedge; we decided to walk a well through the wedge zone. Any well to be walked can be located in the reservoir itself or downdip of the hydrocarbon/water contact in the water leg. **The key point in walking a well is to choose a well or wells that can be walked parallel to the nearest contour line on the net sand isochore map.** This is a key point because when walking a well through the wedge zone, the assumption is made that the amount and distribution of sand and shale, as seen in the well, are the same in a direction parallel to the net sand isochore contours. For example, if a well has 50 ft of net sand, it will fall on the 50-ft total net sand contour line. This 50-ft contour line is constructed with the assumption that there is exactly 50 ft of net sand all along this contour line and not just at the well location (Fig. 10-14). If a series of wells were drilled along this contour line, each well should encounter exactly 50 ft of net sand. We further make the assumption that along the contour line (at least for any limited distance), the distribution of sand and shale is the same in the formation or unit being mapped. This assumption regarding the sand/shale distribution may not be true along the 50-ft contour over long distances or on opposite limbs of a fold because of changes in depositional environments, variable structural growth histories, and other factors; however, it is a reasonable assumption to make for a limited distance from the well, parallel to the nearest net sand contour line.

Procedure for Walking a Well.   To walk a well through the wedge zone, lay the net hydrocarbon isochore base map over the structure map on the top of the sand and

NET SAND ISOCHORE

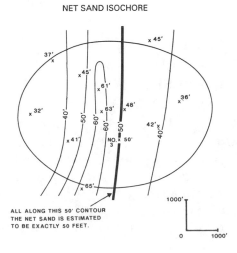

ALL ALONG THIS 50' CONTOUR
THE NET SAND IS ESTIMATED
TO BE EXACTLY 50 FEET.

Figure 10-14 Each net sand contour line is constructed with the assumption that the amount of net sand is constant along the contour line.

the net sand isochore map as shown in Fig. 10-15a. We wish to walk Well No. 2, which is located near the crest of the structure, through the wedge zone. The well has 48 ft of net sand.

1. First use a detailed 5-in. electric log to determine the net sand in the well and the distribution of sand and shale. The 5-in detailed electric log for Well No. 2 is shown in Fig. 10-15b.

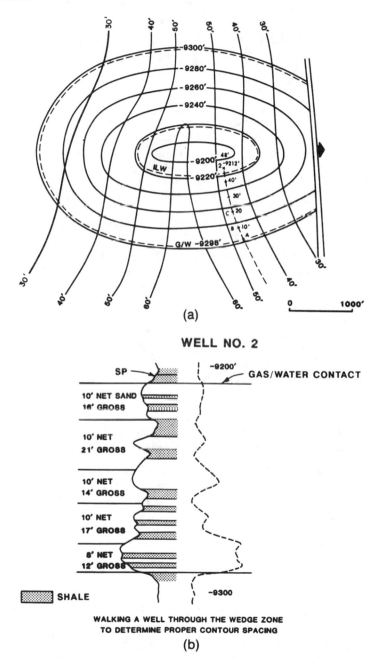

(a)

WELL NO. 2

WALKING A WELL THROUGH THE WEDGE ZONE
TO DETERMINE PROPER CONTOUR SPACING

(b)

Figure 10-15 (a) Structure map and net sand isochore map overlaid onto the net gas isochore base map. Well No. 2 is walked through the wedge zone parallel to the nearest (5Gft) net sand contour line. (b) Five-inch detailed log for the 9200-ft Sand showing the net feet of gas sand per gross feet of interval. (c) Completed gas isochore map for the 9200-ft Sand. The contour spacing in the southeastern portion of the wedge zone was improved by walking Well No. 2. (From Tearpock and Harris 1987. Published by permission of Tenneco Oil Company.)

ISOCHORE MAP CONSTRUCTED BY "WALKING WELLS THROUGH THE WEDGE ZONE.

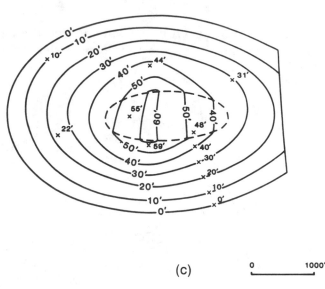

(c)   0 _____ 1000'

Figure 10-15 (*continued*)

2. Move the location of the well from its actual structural position, parallel to the nearest net sand isochore contour line, and place it such that the top of the sand is at the zero (0') line or outer limit of the net gas isochore map (Point A in Fig. 10-15a). A well drilled at this location would: (1) encounter the sand top at the gas/water contact, (2) have no gas pay, and (3) contain 48 ft of net sand.

3. Starting at the top of the sand on the electric log, determine the number of vertical gross feet of section required to obtain 10 ft of net reservoir quality sand. In Fig. 10-15b, a total of 16 ft of gross section are required to obtain 10 ft of net pay. From Point A, move up-structure 16 ft parallel to the nearest total net sand contour line (50-ft line) to locate Point B. This Point B becomes a 10-ft net gas data point for contouring the gas wedge. A well drilled at Point B would encounter the sand 16 ft above the gas/water contact and contain 10 ft of net gas sand.

4. To determine the location of the 20-ft net gas point next, start at the base of the previous 10-ft section and repeat the procedure. In this example (Fig. 10-15b), it requires 21 vertical gross ft to obtain the next 10 ft of reservoir quality sand. Move up-structure 21 ft from Point B, parallel to the 50-ft net sand contour, to locate Point C or the 20-ft net gas data point. Continue the same procedure until the well is back at its original structural position.

Using this method, the well may be walked completely across the wedge zone, resulting in a more accurate contour spacing than using the arbitrary equal-spaced method. Figure 10-15c shows the net gas isochore map constructed for this reservoir incorporating the data obtained from walking Well No. 2.

We emphasize strongly that a well must be walked parallel to the nearest contour line on the net sand isochore map. Significant contouring errors can occur if this procedure is not followed. Consider Well No. 5 in the western portion of the reservoir (Fig. 10-16). If we wish to develop a more accurate contour spacing in this area of the reservoir, can Well No. 5 be walked from the water level updip to the inner limit of water, along the dashed line? The answer is no. Well No. 5 has 28 ft of net sand. If the

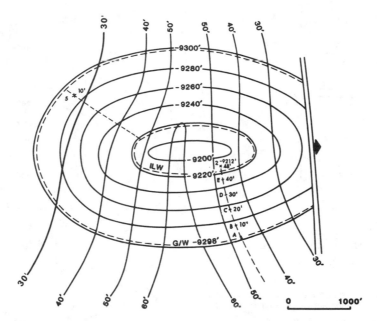

Figure 10-16 Structure map superimposed onto the net sand isochore. Well No. 5 in the western portion of the reservoir **cannot** be walked through the wedge zone to improve the isochore contours. (Modified from Tearpock and Harris 1987. Published by permission of Tenneco Oil Company.)

well is walked from the water contact updip, it will be walked into an area of greater net sand than is actually present in the well, based on the net sand map. Therefore, Well No. 5 cannot be walked to improve the contour spacing in the wedge zone of the reservoir in this area. Caution must be taken when walking wells to be sure that the assumptions made in choosing a well to walk are geologically reasonable and can be supported by a net sand isochore map, and if necessary, additional sand study.

Reservoir Sands with Significant Shale Intervals   When a reservoir is encountered with one or more significant shale intervals between net sand, such as that shown in Well No. 2 in Fig. 10-17a, the accurate construction of the net hydrocarbon isopach map wedge zone may depend upon the walking of wells through the wedge zone. It is obvious from a review of the 5-in. detail log for Well No. 2 that the net sand and shale are not evenly distributed throughout the gross interval. Therefore, the use of the equal-spaced contour method for contouring the wedge zone would result in significant error.

The map in Fig. 10-17b shows the location of Well No. 2 on structure and the placement of the 10-ft and 20-ft net gas values used for contouring the net gas isochore map based on walking Well No. 2 through the wedge. Notice that the first 10 ft of net sand are obtained in 16 ft of gross interval; however, it takes another 52 ft of gross interval to obtain another 10 ft of net pay sand.

Figure 10-18a is a net gas isochore map prepared for this reservoir using an equalspaced method of contouring for the wedge zone, while honoring the net pay values assigned to each well. Figure 10-18b is a net gas isochore map prepared by walking Wells No. 2 and 3 through the wedge zone. Observe the significant difference between the two net gas isochore maps. If there were several leases involved in this gas reservoir, the net gas isochore map prepared by walking the wells would provide a more accurate map for assigning equities to each lease.

There are at least three other methods for constructing accurate net pay isochore maps. First, there is a more accurate method of walking wells. Second, there is

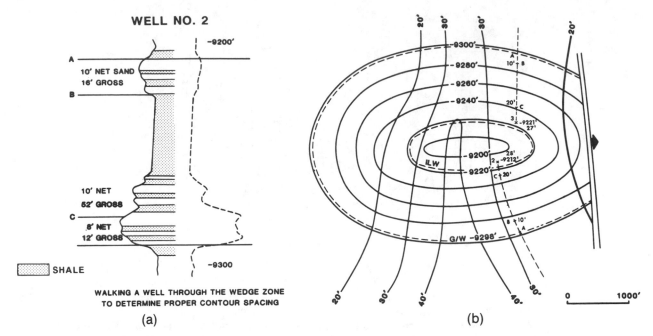

Figure 10-17   (a) Five-inch detailed electric log for Well No. 2. The sand and shale are not evenly distributed throughout this sand member. (b) Structure map superimposed onto the net sand isochore map. The dashed lines indicate paths along which Wells No. 2 and 3 were walked through the wedge zone to improve the net gas isochore contour spacing. (From Tearpock and Harris 1987. Published by permission of Tenneco Oil Company.)

a method that employs the construction of a net-to-gross ratio map for the entire reservoir which is used to aid in the construction of the net pay isochore map. Finally, there is the construction of an $S_o\phi h$ map. In this section of the chapter, we review the more accurate method of walking wells.

Using the same reservoir as in Fig. 10-17, we illustrate a more detailed method of walking wells. Figure 10-19 shows a north-south diagrammatic cross section along the path used to walk Well No. 2. On the right side of the figure, Well No. 2 is positioned so that the top of the sand is at the gas/water contact (-9298 ft). On the left side of the figure, the well is positioned at the inner limit of water (-9218 ft). Well No. 2 must be walked 16 ft up-structure from the gas/water contact to -9282 ft to obtain 10 ft of net pay sand. From this point updip to -9248 ft, no additional pay is added since the section being raised above the water level contains all shale. At -9248 ft, the net pay is still 10 ft. At this point, the lower sand member is now at the gas/water contact; therefore, up-structure from this position, additional pay is added to the reservoir. Continuing to walk the well updip, we must go to -9230 ft before an additional 10 ft of net pay is added to the reservoir from the lower member. Finally, at the gas/water contact on the base of sand, all the sand (28 ft net) is above the water contact. From this point to the actual well location, shown on the far left in the figure at a depth of -9212 ft, the net pay sand is a constant 28 ft.

On the cross section, there are two locations that have 10 ft of net gas sand (-9282 ft and -9248 ft), and the area between these points has a constant 10 ft of net gas sand. The accuracy of the net gas isochore map can be improved by constructing the isochore honoring the two 10-ft net pay values. Well No. 3 was also walked through the wedge zone, as shown in Fig. 10-17b, to aid in the construction of the net pay isochore map. The resulting net gas isochore map is shown in Fig. 10-20.

At first glance, it may appear as if an important contouring rule has been broken in the construction of the isochore map; contours cannot merge or split. However, no rules are broken. The two 10-ft contour lines that appear to merge represent the limits of a **very wide** 10-ft contour line. Everywhere within the area of the wide contour, the net gas has a constant value of 10 ft.

One may, ask since the sands are so far apart and separated by such a thick shale break, why not map each sand separately and construct two isopach maps. In the western part of the reservoir, the two sand members coalesce into one member with no consistent shale break within the gross interval. Therefore, the thick shale wedge is localized in the eastern section of the reservoir. The fact that the shale wedge interval decreases to the west is evidenced in the rapid decrease in width of the 10-ft contour line to the west. If such a shale interval were known to be continuous over the entire reservoir, it would be necessary to prepare a structure map for each sand member and construct a separate net pay isochore map for each sand.

The procedure outlined in this section is more involved than the two previous methods shown, but provides further accuracy in the construction of a net hydrocarbon isochore map. The method chosen to prepare a net hydrocarbon isochore map depends upon a number of factors, including the available time, detail, and accuracy required.

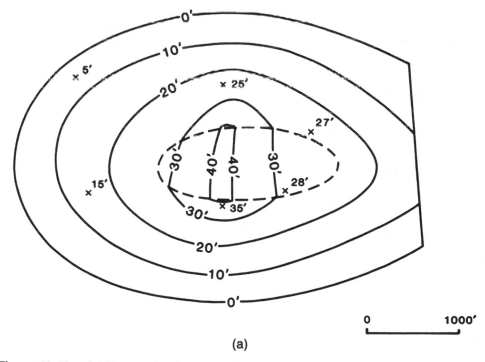

ISOCHORE MAP CONSTRUCTED USING PROPORTIONALLY SPACED CONTOURS

(a)

Figure 10-18   (a) Net gas isochore map based on equally spaced contours. (b) Net gas isochore map with the contour spacing based on walking Wells No. 2 and 3 through the wedge zone. Compare this isochore to that shown in Fig. 10-18a. (From Tearpock and Harris 1987. Published by permission of Tenneco Oil Company.)

ISOCHORE MAP CONSTRUCTED BY
"WALKING" WELLS THROUGH THE WEDGE ZONE

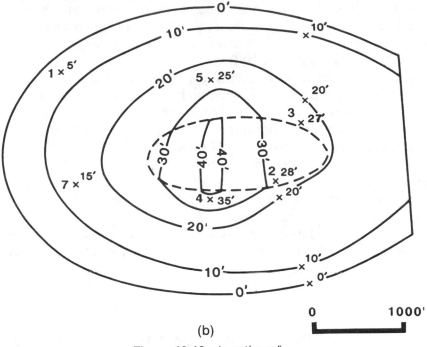

(b)

Figure 10-18 (*continued*)

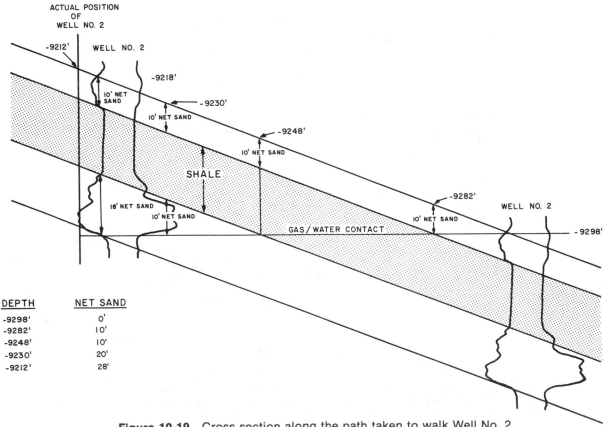

DEPTH	NET SAND
-9298'	0'
-9282'	10'
-9248'	10'
-9230'	20'
-9212'	28'

Figure 10-19 Cross section along the path taken to walk Well No. 2.

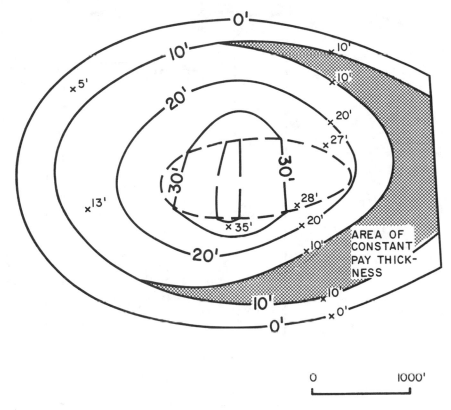

Figure 10-20 Net gas isochore map using a more accurate method of contouring the wedge zone based on the results of walking Wells No. 2 and 3.

VERTICAL THICKNESS DETERMINATIONS

True vertical thickness is the thickness of a bed when measured in a vertical direction. As mentioned in several sections of the text (see Chapter 4), vertical log thickness is a very important measurement. It is this tertical thickness that is required to measure the vertical separation of a fault; it is also the thickness required to count net sand and net pay from 5-in. detailed logs, and it is the thickness used to construct net sand and net pay isochore maps.

In a vertical well, the actual thickness seen on the electric log is the vertical thickness. In the case of a directionally drilled well, however, a correction factor must be applied to correct the exaggerated or diminished log thickness to true vertical thickness (**TVT**).

If a reservoir is horizontal (zero bed dip), the thickness of the reservoir for net sand or net pay isochore mapping is defined by the true stratigraphic thickness (Fig. 10-21). However, if the same reservoir is rotated to some angle, such as 45 deg, the thickness of the reservoir required for net sand and net pay isochore mapping is no longer the stratigraphic thickness. Figure 10-21 illustrates the cross-sectional area of a reservoir with a fixed width. The horizontal reservoir (zero dip) in the lower portion of the figure has a cross-sectional area of 50,000 sq ft. The length of the reservoir as seen in map view is 500 ft, and the thickness is 100 ft. Since the dip on the reservoir is zero, the vertical and stratigraphic thicknesses are the same (100 ft). If the same reservoir is now rotated to an angle of 45 deg, as shown in the upper portion of the figure, notice that the length of the reservoir, in map view, has now been shortened to 354 ft. Since the area of the reservoir has not changed and is still 50,000 sq ft, the thickness must be something greater than 100 ft. If we measure the vertical thickness of the dipping reservoir, it is 141.25 ft (141.25 ft x

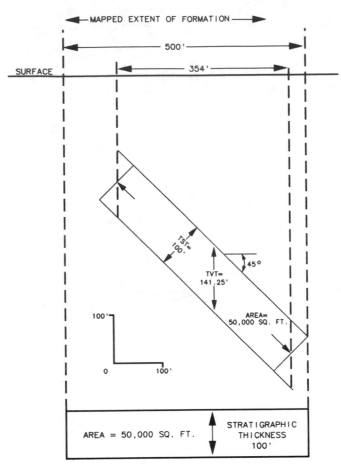

Figure 10-21   Cross-sectional area of two reservoirs of equal volume and a constant stratigraphic thickness of 100 ft. One reservoir is horizontal, the other is dipping at 45 deg.

354 ft = 50,002.5 sq ft). A measurement of the stratigraphic thickness, which is measured perpendicular to the bed dip, reveals that it is still 100 ft. We conclude from this example that as a reservoir of fixed length rotates from the horizontal, the areal extent in map view shortens; therefore, in order to maintain the same area or volume of the reservoir, the shortened length must be multiplied by the vertical thickness.

Figure 10-22 illustrates the effect of dipping beds and directional wells on the apparent thickness of a sand encountered by wells drilled in an updip and downdip direction. True vertical thickness (TVT) of the sand unit shown is 150 ft, which is the thickness represented in a vertical well. If a straight hole were drilled into this sand, the thickness on the electric log would be 150 ft.

For directionally drilled wells, the log thickness can be thicker, the same as, or thinner than that seen in a vertical well drilled through the same stratigraphic section. A correction factor must be applied to the log thickness seen in most deviated wells to convert the borehole thickness to true vertical thickness. There are two parts to the correction factor: the first is the correction for borehole deviation angle, and the second is for the formation or bed dip. Equations (4-3), (4-4), (4-5), or (4-6) shown in Chapter 4 can be used to calculate this correction factor. In Chapter 4, the equations were used to estimate the true vertical thickness of missing or repeated section in a well as the result of a fault. Remember, the vertical separation of a fault is defined in terms of the **true vertical thickness** of the stratigraphic section missing or repeated in a wellbore. In this

TRUE VERTICAL THICKNESS CORRECTIONS

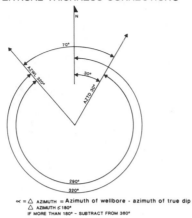

∝ = △ AZIMUTH = Azimuth of wellbore - azimuth of true dip
△ AZIMUTH ≤ 180°
IF MORE THAN 180° - SUBTRACT FROM 360°

Figure 10-22 Azimuth is measured from 0 deg to 360 deg in a clockwise direction from true north. A Δ azimuth is the azimuth of the wellbore minus the azimuth of true bed dip.

chapter, we look at the same correction factor equations to convert deviated borehole thickness to true vertical thickness for use in net sand and net pay isochore mapping.

For convenience, we repeat correction factor equation (4-6). Equation (4-6), which is a three-dimensional equation, is the preferred equation to use for correction factors because this one equation can be used to calculate the thickness correction factor regardless of the direction of wellbore deviation, and the true dip of the beds is used instead of the apparent dip required in the two-dimensional equations.

$$\text{TVT} = \text{MLT} \left[\cos \psi - (\sin \psi \cos \alpha \tan \phi)\right] \qquad (4\text{-}6)$$

where

TVT = True Vertical Thickness

MLT = Measured Log Thickness

   ψ = wellbore deviation angle

   ϕ = true bed dip

   α = Δ Azimuth—acute angle between the wellbore azimuth and the azimuth of true bed dip

Figure 10-22 illustrates the measurement of azimuth and Δ azimuth for use in Eq. (4-6). The Δ azimuth is always the acute angle between the azimuth of the wellbore and the azimuth of the true bed dip. The maximum Δ azimuth is 180 deg.

In order to more closely examine the two directionally drilled wells shown in Fig. 10-23, look first at the well drilled to the east in a downdip direction (Fig. 10-23a). The measured thickness of the sand as seen in the well log is 466 ft. By applying the correction factor for wellbore deviation only, the thickness is reduced to 357 ft, shown in the figure as the true vertical depth thickness. This thickness is still highly exaggerated, because the correction for wellbore deviation does not take into account the dip of the beds. The thickness of the sand corrected for wellbore deviation only is called the *True Vertical Depth Thickness* (*TVD*). The true vertical depth thickness is that thickness obtained from a **TVD log,** and for dipping beds, the TVD thickness **is not** equivalent to true vertical thickness (see Chapter 4). With the final correction for bed dip, the thickness is converted

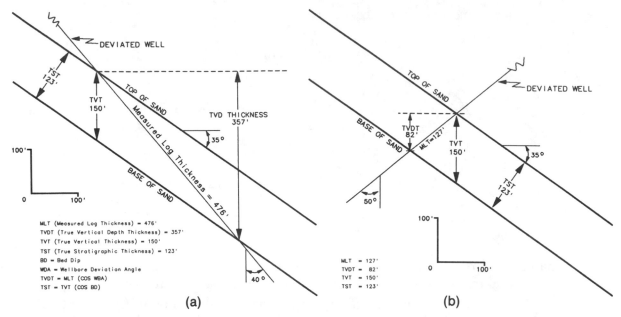

Figure 10-23 (a) True vertical thickness correction for a well drilled in a downdip direction. (b) True vertical thickness correction for a well drilled in an updip direction.

to its true vertical thickness equal to 150 ft, shown in the figure at the penetration point of the wellbore in the top of the sand. Note that the true stratigraphic thickness (thickness perpendicular to the bed dip) is 123 ft. The true stratigraphic thickness (**TST**) is calculated by multiplying the TVT by the cosine of the angle of bed dip (35 deg).

The well in Fig. 10-23b is deviated updip to the west. The log thickness is actually less than true vertical thickness; it is 127 ft. A correction factor for the well deviation angle alone, which is a correction to true vertical depth thickness, actually reduces the thickness even more, to 82 ft. When the correction factor for bed dip is applied, however, the thickness converts to true vertical thickness, again in this case equal to 150 ft. As an exercise, use Eq. (4-6) to verify the TVT for the two wells shown in Fig. 10-23 to confirm the results shown.

There are various computer systems that can be used to create *TVD*, *TVT*, and *TST* (*True Stratigraphic Thickness*) logs from measured depth (*MD*) logs for use in mapping. The actual deviated log can be placed into a computer system along with the directional survey for the well and bed dip information. The bed dip information can be obtained either from completed structure maps or a dipmeter. The log curve data may be obtained from the logging company tapes or digitized from the actual log. The directional survey data are furnished by the directional company that worked the well. The output logs can be in standard presentation or any scale desired. The logs in Fig. 10-24 were created using what is called IEPS (Integrated Exploration and Production System). The log sections for Well MP-D5 shown from left to right in the figure represent the (1) measured depth log, (2) true vertical depth log, (3) true vertical thickness log, and (4) true stratigraphic thickness log. This log is from the Main Pass 296 Salt Dome Field in an area of significant bed dip. Notice that there is very little difference between the measured depth (MD) and the true vertical depth (TVD) logs. This is so because the TVD log is only corrected for wellbore deviation and not bed dip. However, the true vertical thickness (TVT) log shows a considerable reduction in thickness from the measured depth log because this log thickness has been corrected for both wellbore deviation and bed dip.

MAIN PASS 296 SALT DOME

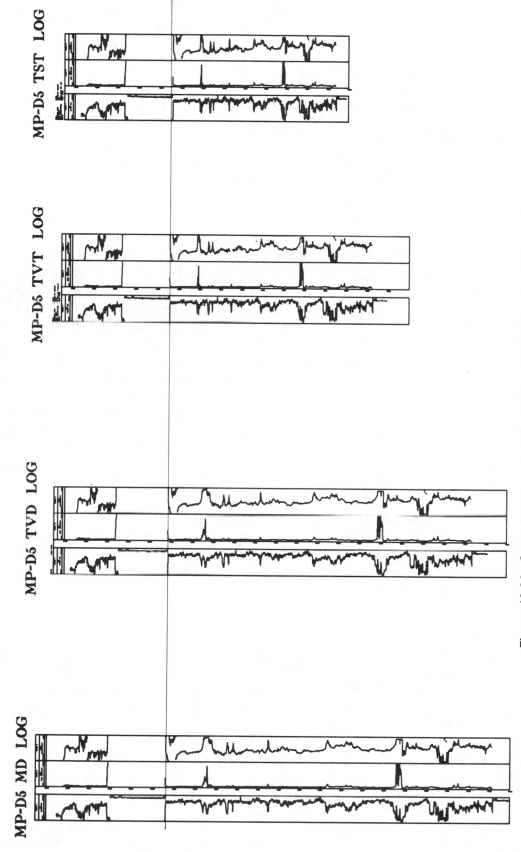

Figure 10-24 Computer generated electric logs illustrating the difference in thickness between measured depth, true vertical depth, true vertical thickness, and true stratigraphic thickness logs generated for the same well. (From Tearpock and Harris 1987. Published by permission of Tenneco Oil Company.)

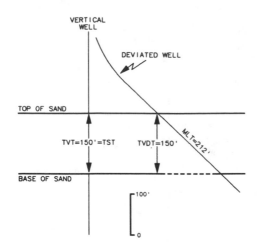

Figure 10-25 True vertical depth thickness is equal to true vertical thickness when the formation is horizontal.

We caution here that too often, TVD logs are made as standard practice when logging a deviated well and then used for purposes that are not applicable with this log. There is a **widespread misunderstanding** that a TVD log, prepared from a measured depth log, can be used to: (1) correlate with other well logs, (2) determine the vertical separation fault, and (3) count net sand or net pay for isochore mapping. Remember, a TVD log is only corrected for wellbore deviation and not bed dip. In areas of flat lying beds (no bed dip), a TVD log is equivalent to a TVT log because the only correction factor required is for wellbore deviation (Fig. 10-25). When dealing with dipping beds, however, particularly over 10 deg, a TVD log, in most cases, does not represent the log thickness required to aid in correlation work, determine the vertical separation for a fault, or use for net sand or pay counting. For these purposes a deviated well **must** be corrected so that the log thickness represents the true vertical thickness as seen in a vertical well. Look again at Fig. 10-24 and observe the significant difference in thickness between the TVD and the TVT logs. To determine the vertical separation of a fault by correlation with a deviated well log, and to count all net sand and net pay from a deviated well log, a *True Vertical Thickness Log (TVT)* or its equivalent **must** be used. By equivalent, we mean that one or more correction factors must be determined for each deviated well and these correction factors applied to the thickness of a measured depth log to convert the log thickness to true vertical for use in such determinations as the vertical separation for faults and net sand and net pay counts for isochore mapping.

The Impact of Correction Factors

Figure 10-26 presents an example of two separate net pay isochore maps prepared for a reservoir on the flank of a salt dome in the Offshore, Gulf of Mexico. Notice that there are two platforms from which wells were drilled. Platform D is located over the up-structure position with the D Platform wells directionally drilled downdip. The A platform is located over the flank of the structure with most of the A Platform wells directionally drilled updip.

Figure 10-26a is a net pay isochore map prepared for the T-1 Sand, Reservoir A. The net pay values posted on the net pay isochore map were corrected for borehole deviation but not for bed dip, which is about 35 deg at this location on the dome. In addition to the error of failing to correct the net pay values for bed dip, there are several other serious isochoring problems that are not discussed here.

Figure 10-26b is a net pay isochore map for the same reservoir with new net pay values which have been corrected for borehole deviation and bed dip. This new isochore map was prepared solely to compare the effect of the correction factor for bed dip on

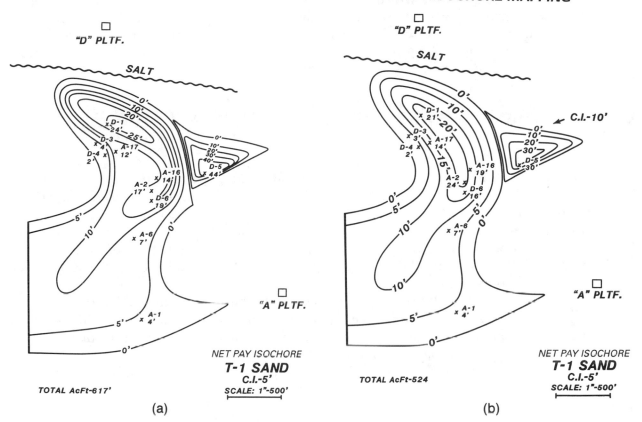

CORRECT ISOCHORE MAPPING

(a)

NET PAY ISOCHORE
T-1 SAND
C.I.-5'
SCALE: 1"-500'

TOTAL AcFt-617'

(b)

NET PAY ISOCHORE
T-1 SAND
C.I.-5'
SCALE: 1"-500'

TOTAL AcFt-524

Figure 10-26 (a) Net pay isochore map for the T-1 Sand, Reservoir A. The net pay values for the deviated wells from Platforms A and D were corrected only for wellbore deviation. (b) Net pay isochore map for the T-1 Sand, Reservoir A. The net pay values for the deviated wells from Platform A and D were corrected for wellbore deviation and bed dip. Compare the net pay value for each well with those shown in Fig. 10-6a. (From Tearpock and Harris 1987. Published by permission of Tenneco Oil Company.)

the total volume of the reservoir. Therefore, the isochore errors made in Fig. 10-26a are incorporated in the net pay isochore map in Fig. 10-26b. The planimetered volume for the isochore corrected for bed dip is 524 acre-feet; the incorrect map has a volume of 617 acre-feet. Therefore, the estimated hydrocarbon volume for the map prepared without correcting the net pay thickness values for bed dip is **18%** greater than the isochore map prepared taking into account the correction factor for bed dip. This means that the reserves, based on the incorrect map, are overstated by 18%.

In a situation like this, we would expect the error factor to be larger than 18%, and it would be in most cases; however, look at Wells No. A-2 and D-5. Well No. A-2 was corrected upward from 17 ft net pay to 24 ft net pay, while Well No. D-5 was corrected down from 44 ft net pay to 30 ft net pay. It so happens that the D Platform wells, drilled downdip, result in a reduction in net pay values when the correction factor for bed dip is considered; while the A Platform wells, drilled updip, result in an increase in net pay. Therefore, a significant part of the error is negated because of the manner in which the wells were drilled.

This reservoir is only one of a number of oil and gas reservoirs that are productive within this field. A complete remapping of the field was undertaken when several major mapping errors, such as the one shown here, were identified. The remapping of the field resulted in a significant writedown of hydrocarbon reserves that were overestimated because of mapping errors such as the failure to incorporate the proper correction factors in determining net pay values for isochore mapping. This example illustrates the impact that correction factors can have on estimated hydrocarbon volumes determined from volumetric calculations using net pay isochore maps.

VERTICAL THICKNESS AND FLUID CONTACTS IN DEVIATED WELLS

We have already discussed that the net pay values required for isochore mapping must be expressed as the true vertical thickness for each well penetrating the sand. The mathematical equations that are used for converting log thicknesses to true vertical thicknesses were reviewed in Chapter 4, as well as in the previous section in this chapter. However, some additional discussion is required regarding these correction factors when dealing with deviated wells with fluid contacts. The mathematical treatment in these situations is not as straightforward.

Normally, net sand and net pay values for isochore mapping are determined at the position where a well penetrates the top of the formation or sand. For a vertical well, the penetration points with all encountered sands are at the same location in map view, directly under the surface location of the well (Fig. 10-27). For a directionally drilled well, the intersection of the well with the top and base of sands at various depths is in

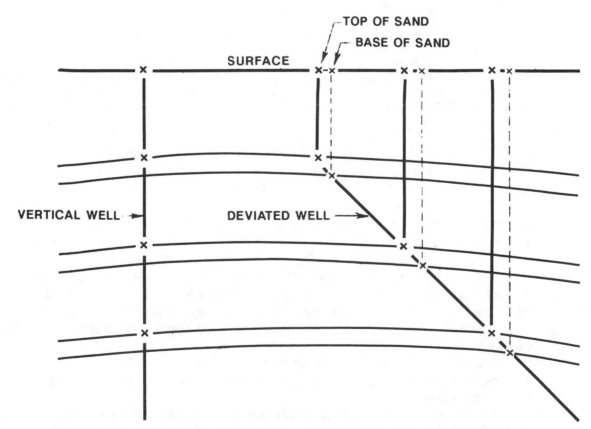

Figure 10-27 A deviated well penetrates a formation top and base at different positions in the subsurface.

different locations with respect to the horizontal, as seen in map view, along the path of the deviated well (Fig. 10-27).

In many cases, due to the low angle of wellbore deviation or the minimal thickness of a sand, the calculation and positioning of the net sand or pay values at the point where the well penetrates the top of the sand is sufficient. There are situations involving highly deviated wells, fluid contacts, dipping beds, or thick sands, however, where a single data point for net sand or net pay at the penetration point of the well at the top of the sand may be insufficient, as well as incorrectly calculated or posted for isochore mapping. These special conditions are discussed here.

Equation (4-6) is designed to calculate the correction factor for sand thickness when there are no fluid contacts in the well, as shown in Fig. 10-28. In this figure, the deviated well has penetrated a sand with a true vertical thickness of 150 ft. The measured log thickness from the depth where the well penetrates the top of the sand to its penetration at the sand base is 219 ft. Using the data given in Fig. 10-28 and Eq. (4-6), the 219 ft of log thickness are converted to 150 ft true vertical thickness. Observe that there is an oil/water contact just downdip from the well's penetration with the base of the sand. The entire wellbore penetration within the sand is therefore confined to the full thickness area of the reservoir sand, with no fluid contacts present in the well.

Consider the following situation. A well penetrates the top of a sand directly over the oil/water contact with the following data:

1. wellbore deviation is 30 deg due west;
2. bed dip is 20 deg due west;
3. log length through the sand from top to base is 219 ft; and
4. log length from the top of sand to the oil/water contact is 115 ft.

What is the thickness of the oil column vertically under the penetration point of the well at the top of the sand? Draw a cross section of the well and sand relationship and, using Eq. 4-6, calculate the true vertical thickness of the oil sand vertically under the penetration point of the well at the top of the sand. Is the correct answer 150 ft, 100 ft, or 79 ft? If you calculated 150 ft, this thickness is equal to the total true vertical thickness of net sand directly under the penetration point of the well at the top of the sand (Fig. 10-29). If you calculated 79 ft, this is the vertical thickness of net oil pay directly above the point where the well penetrates the oil/water contact. A review of Fig. 10-29 shows that the oil/water contact is a horizontal surface. If we consider the penetration point of

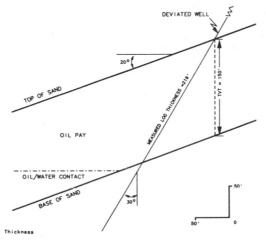

Figure 10-28 Cross section illustrates a deviated well penetrating a formation above the oil/water contact. The conversion of measured log thickness to true vertical thickness is accomplished by use of Eq. (4-6).

the well at the top of the sand as a point, we do not have to consider any bed dip correction factor to calculate the net pay directly under the penetration point of the well at the top of the sand. This is so because there is no bed dip effect on the thickness of the net oil sand since the oil/water contact is a horizontal surface and the penetration point at the top of the sand is a point **directly over** the oil/water contact. Therefore, the calculation of net oil sand directly beneath the penetration point of the well at the top of the sand reduces to the correction factor for wellbore deviation multipled by the log thickness from the top of the sand to the oil/water contact, which is 115 ft. The net oil pay is equal to

$$\text{TVT} = \cos 30 \text{ deg} \times 115 \text{ ft}$$

TVT = 0.866 × 115 ft = 100 ft *net oil sand at the position where the well penetrates the top of the sand.*

If we use the entire equation to determine the correction factor for this case, we are calculating the correction factor used to determine the true vertical thickness of **net sand** at the penetration point of the well at the top of the sand, as well as the correction factor for the **net oil sand directly above the penetration point of the well at the oil/water contact** (Fig. 10-29).

$$\text{CF} = [\cos \psi - (\sin \psi \cos \alpha \tan \phi)]$$

$$(4\text{-}6)$$

$$\text{TVT} = \text{CF} \times \text{MLT}$$

Using the data provided in Fig. 10-29, the correction factor is:

$$\text{Wellbore deviation } (\psi) = 30 \text{ deg}$$
$$\text{True bed dip } (\phi) = 20 \text{ deg}$$
$$\text{Delta azimuth } (\alpha) = 0 \text{ deg}$$

$$\text{CF} = [\cos 30° - (\sin 30° \cos 0° \tan 20°)]$$

$$\text{CF} = 0.866 - (0.5)(1)(0.364)$$

$$\text{CF} = 0.684$$

1. The true vertical thickness of the net sand at the penetration point of the well at the top of the sand is:

$$\text{TVT} = (\text{CF})(\text{measured log thickness in sand})$$
$$\text{TVT} = (0.684)(219 \text{ ft})$$
$$\textbf{TVT = 150 ft}$$

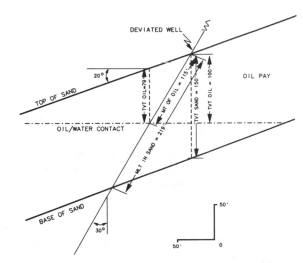

MLT (Measured Log Thickness) in Sand = 219'
MT (Measured Thickness) of Oil in Well = 115'
TVT (True Vertical Thickness) of Sand = 150'
TVT of Oil below Penetration of Well in Sand Top = 100'
TVT of Oil above Penetration of Well at O/W Contact = 79'

Figure 10-29 Deviated well penetrates the sand directly above the oil/water contact. TVT of oil sand directly below the penetration point of the well at the top of the sand is calculated by simply multiplying the measured log thickness of oil by the cosine of the wellbore deviation angle.

2. The true vertical thickness of the net oil pay sand above the penetration point of the well at the oil/water contact is:

TVT = (CF)(log thickness from top of sand to the oil/water contact)

TVT = (0.684)(115 ft)

TVT = 79 ft

We have now introduced a special condition for which only the correction factor for wellbore deviation is required to convert log thickness to true vertical thickness, even though the beds are dipping at a significant angle. We make an important conclusion here: *Whenever the penetration point of a well in the top of a sand is over a fluid contact (in the wedge zone of a reservoir) as in Fig. 10-29, the essential dip of the beds can be considered zero for determining the* **vertical net pay sand** *thickness above the fluid contact at a point vertically under the penetration of the well at the top of the sand.*

Figure 10-30 illustrates a situation in which a reservoir containing oil and gas is penetrated by a deviated well. In this case, the vertical thickness of the oil and gas columns must be determined for isochore mapping. Using the data provided in the figure and Eq. (4-6), verify the true vertical thickness for the oil and gas at two separate locations: (1) directly beneath the penetration point of the well at the top of the sand, and (2) where the well penetrates the gas/oil contact. Finally, calculate the TVT of water, oil, and gas at the point where the well penetrates the oil/water contact (dashed line).

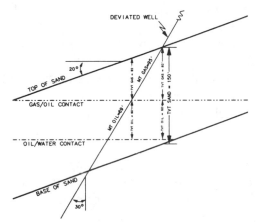

Figure 10-30 The deviated wellbore penetrates the sand directly above two separate fluid contacts (gas/oil and oil/water). See text for explanation for calculating the TVT values for gas and oil.

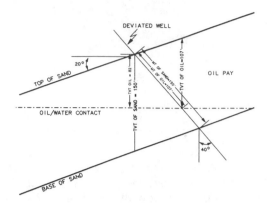

Figure 10-31 Deviated well drilled in an updip direction. True vertical thickness calculations are the same as those for a well drilled in a downdip direction.

Figure 10-31 illustrates a well deviated in an updip direction penetrating an oil reservoir in the oil/water wedge zone. The same procedures as previously discussed are used to calculate the TVT of oil at the locations where the well penetrates the top of the sand and at the oil/water contact. If desired, the TVT of oil can also be calculated at the position where the well penetrates the base of the sand. This is done by first calculating the TVT of the water column at this location and subtracting this value from the TVT of the sand to arrive at the vertical thickness for oil.

By having a good understanding of the geometric relationship of the sand, wellbore, and fluid contacts, we can use this knowledge to our advantage. The calculation of the net gas or net oil pay at various points in a well, such as the well's penetration point at the top of the sand, base of sand, or at fluid contacts, can provide additional net pay values. These values can be used in the preparation of the net gas or net oil isochore maps, providing additional control in the wedge zones.

The detailed calculations shown in this section are not always required or justified. However, where detailed accurate mapping is needed for some specific reserve estimate, development plan, enhanced recovery program, unitization, or litigation, the use of these techniques may prove to be very important.

STRUCTURE TOP VERSUS POROSITY TOP MAPPING

We discussed the effect of mapping on a structure top versus porosity top with regard to structure mapping in Chapter 8. We now review this special condition as it relates to net pay isochore maps. We mentioned in Chapter 8 that the upper portion of a reservoir sand unit may be composed of nonreservoir quality rock. This nonreservoir quality rock is usually referred to as a tight zone or tight streak. Although the top of the sand may represent the actual stratigraphically equivalent top, it does not constitute reservoir quality rock. Therefore, the structure map prepared to interpret the structure may not be useful to evaluate the reservoir itself.

Once the structure mapping is complete, the question arises as to whether a separate map on the *top of porosity* is required for accurate delineation of the reservoir and use in the construction of net hydrocarbon isochore maps. Two parameters are considered in evaluating the importance of the nonreservoir quality rock: (1) the thickness of the tight zone, and (2) the relief of the structure. A thick tight zone has a greater effect than one that is thin. Low relief structures introduce greater error in delineating the limits of a reservoir than steeply dipping structures, particularly if the low relief structure contains a reservoir with bottom water. This is true because a steeply dipping reservoir is associated with a relatively small wedge zone when compared to the total area of the reservoir.

On a low relief structure, the wedge zone of a reservoir can represent a significant portion of the total reservoir area (Fig. 10-32).

Figure 10-32a shows in map and cross-sectional views a low relief bottom water reservoir mapped on the top of a sandstone which consists of nonreservoir quality rock in the upper 75 ft of the sand. The same reservoir is mapped on the top of porosity in Fig. 10-32b. The net oil isochore map prepared from each structure map is shown in Fig. 10-32c. The same net pay values are assigned to each well in both isochore maps. In this case, because the reservoir is on a low relief structure, the difference in reservoir volume between the incorrectly constructed isochore map (Fig. 10-32a) and the correctly constructed isochore map (Fig. 10-32b) is a significant **32%** (Fig. 10-32d). Consequently, the volume of recoverable hydrocarbons based on the isochore map in Fig. 10-32a is overestimated by 32%, which is equal to 637 acre-feet.

The decision to prepare a separate map on the top of porosity, when the upper portion of a sand unit is tight, needs to be made on a reservoir by reservoir basis. Depending upon the geometry of the reservoir and thickness of the tight zone, the difference in volume between a map on the top of the unit and one on the top of porosity may be too insignificant to warrant additional mapping.

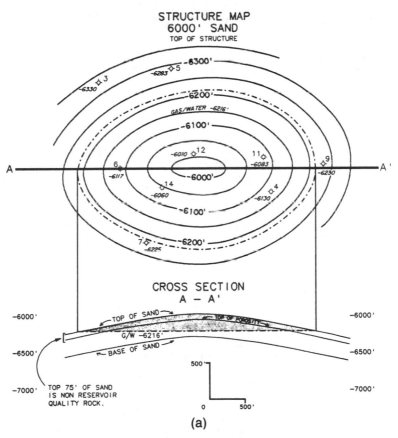

Figure 10-32 (a) Structure map on the top of the 6000-ft Sand and cross section AA'. Upper 45 ft of the sand unit contains nonreservoir quality rock. (b) Map on top of porosity (6000-ft Sand) and cross section A-A'. (c) Two separate net pay isochore maps: (1) the upper isochore is based on the structure map on the top of sand, (2) the lower isochore is based on the porosity top map. Net pay values for all the wells are the same for each map. (d) There is a 32% reduction in reservoir volume for the net pay isochore map constructed from the top of porosity map vs the net pay isochore constructed from the top of structure. This is a significant reduction in volume.

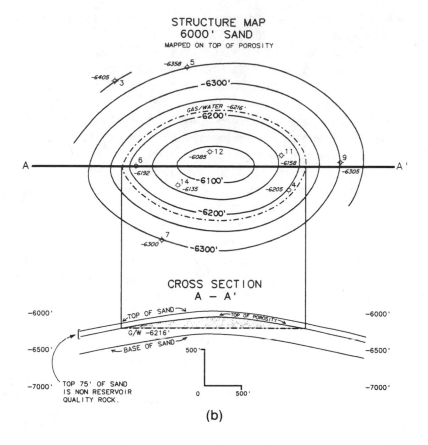

(b)

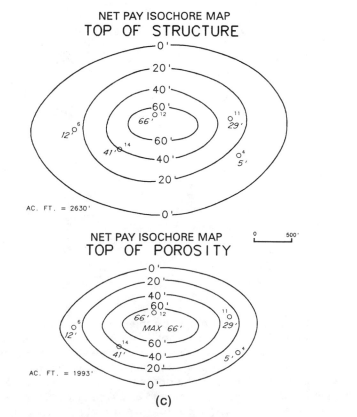

(c)

Figure 10-32 (*continued*)

STRUCTURE TOP VERSUS POROSITY TOP

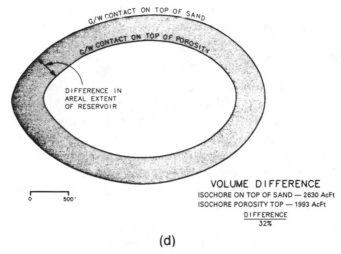

VOLUME DIFFERENCE
ISOCHORE ON TOP OF SAND — 2630 AcFt
ISOCHORE POROSITY TOP — 1993 AcFt
DIFFERENCE
32%

(d)

Figure 10-32 (*continued*)

FAULT WEDGES

A *fault wedge* is defined as a wedge-shaped section of strata bounded by a fault. It is often just as important to map the fault wedge of a productive reservoir as it is to map the water wedge. If a productive sand is thin or the bounding fault is at a high angle, the reservoir volume affected by the fault wedge may be insignificant and can be ignored for all practical purposes. In cases where the reservoir sand is thick or the fault is at a low

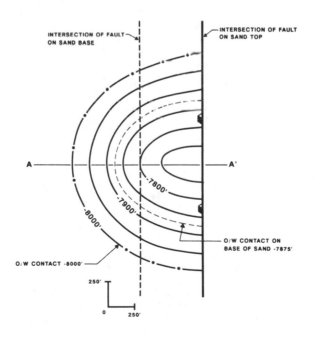

STRUCTURE MAP ON TOP OF SAND

Figure 10-33 Structure map on top of sand shows the intersection of the fault with the top and base of sand. These two intersections are required to isochore the fault wedge.

angle, the reservoir volume affected by the fault may be significant and must therefore be considered when constructing the net pay isochore map. There are several ways to handle the mapping of a fault wedge. It may be contoured in the conventional way, or a mid-point method can be used.

Conventional Method

The conventional method of contouring the fault wedge is the most accurate and should be employed whenever possible. With this method, the wedge is actually contoured using all control points in the same manner as contouring a water wedge. As with a water wedge, when the impermeable zones (shale, etc.) are fairly evenly distributed throughout the sand, the isochore fault wedge contours can be evenly spaced in the wedge zone. All the control points must be honored, however, even though it may cause an uneven spacing of the contours. Figure 10-33 is an example of a structure map on the top of a productive sand. The reservoir is bounded on the east by a west dipping fault, with the remainder of the reservoir bounded by an oil/water contact at - 8000 ft. The structure map shows the intersection of the top of the sand with the fault, as well as the intersection of the fault with the base of the sand (shown as a dashed line). The area between these two intersections is the area of fault wedge. In this case, the fault is dipping to the west at 45 deg and the sand is dipping at 30 deg. It is readily apparent from the fault intersections

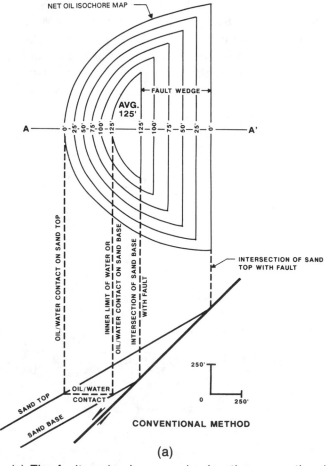

Figure 10-34 (a) The fault wedge is mapped using the conventional method. See cross-section view below map. (b) Mid-trace method for mapping the fault wedge (see cross section).

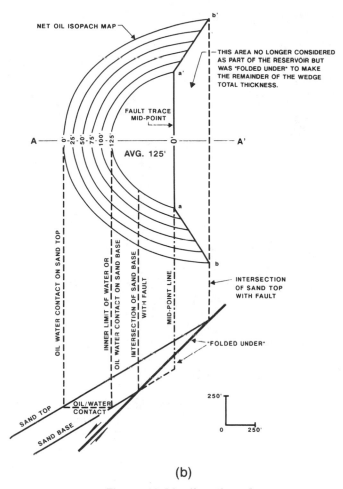

(b)

Figure 10-34   (*continued*)

with the top and base of sand that the fault wedge affects a very large portion of this reservoir. For simplicity, the reservoir is assumed to have 50% sand and 50% shale evenly distributed throughout the gross sand interval. Because of this even shale distribution, the contours in both the water and fault wedge zones can be evenly spaced. *The control required to map the fault wedge in this example is the intersection of the* **top and base of the sand** *on the downthrown side of the fault.* The upthrown trace of the fault, and therefore the fault gap as seen on a structure map, plays no part in isochoring the fault wedge.

Figure 10-34a illustrates the net oil isochore map for this reservoir with both the water and fault wedges conventionally contoured. The cross section A-A' drawn below the net oil isochore map depicts certain key control points in the reservoir, including the oil/water contact on the sand top, the inner limit of water, which is the oil/water contact on the sand base, the intersection of the sand base with the fault, and the intersection of the sand top with the fault. These key control points play an important part in the construction of the wedge zones for this net oil isochore map.

Mid-Trace Method

The use of the conventional method of contouring the fault wedge can be tedious and at times not justified. In such cases, there is a shortcut method for contouring the fault

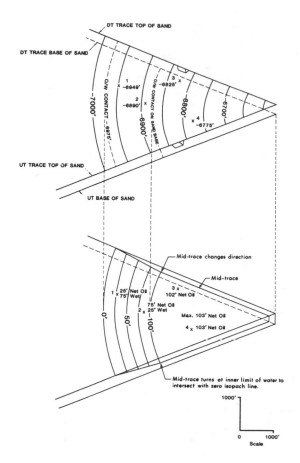

Figure 10-35 Construction of two fault wedges involving intersecting faults. Note the change in direction of the mid-traces at the inner limit of water.

wedge referred to as the **Mid-Point Method** (Fig. 10-34b). To construct an isochore map using the mid-point method, a line is drawn through the full thickness area of the isochore at the mid-point between the intersections of the sand top and sand base with the fault. This line intersects and stops at the inner limit of water shown as points a and a' in Fig. 10-34b. Next, extend this line from the inner limit of water straight to the intersection of the oil/water contact on the sand top as indicated by a-b and a'-b' on the figure. This line becomes the boundary of the reservoir. Any reservoir area outside this boundary line in the fault wedge zone is considered as being *"folded under"* to convert the wedge zone inside this line to an area of full thickness. This is illustrated in the cross section below the net oil isochore map in Fig. 10-34b. All contours are then extended through the water wedge to intersect with line segments a-b and a'-b' as illustrated in the figure.

Figure 10-35 illustrates the use of the mid-point method of contouring a fault wedge with a reservoir bounded by two intersecting faults. Note that the mid-point extension from the inner limit of water to the oil/water contact on the top of the sand is in opposite directions on the individual fault wedges. Remember, the mid-fault trace must always intersect with the zero isochore contour line on the top of the sand.

NONSEALING RESERVOIRS

Although faults play a very important role in the trapping of hydrocarbons, studies have shown that at times faults are nonsealing, thereby permitting the migration of hydrocarbons from one fault block to the next. One of the most common situations resulting in a nonsealing fault occurs when part of a single sandstone is juxtaposed across a fault within the hydrocarbon column (Smith 1980). This occurs when a fault does not have sufficient displacement to separate an entire sand body from one fault block to the next.

With nonsealing faults it is important to map fault wedges because they can contain significant amounts of hydrocarbons. The method for isochoring the fault wedge is basically the same as presented in the previous section with one exception. Since there is hydrocarbon pay upthrown and downthrown to the fault, there are two reservoirs and two fault wedges that must be mapped to account for all the hydrocarbon volume. The isochore map for the two reservoirs can be constructed individually or contoured as one.

Figures 10-36 and 10-37 illustrate an example of a hydrocarbon bearing sand juxtaposed across a nonsealing fault. The example is the 9500-ft Sand on the southern flank of a piercement salt structure. Figure 10-36a is the structure map on the top of the 9500ft Sand, Reservoirs A and B. The reservoirs are limited to the north by salt, to the east and west by faults, and in the downdip direction by a gas/water contact at - 10,550 ft. The interval from top to base is in excess of 200 ft thick and it is all sand. Fault B, with a vertical separation of 200 ft, is of insufficient size to completely separate the sand; therefore, part of the sand is juxtaposed across the fault. The same gas/water contact in both Reservoirs A and B indicates that the fault is nonsealing and that the two reservoirs are in communication. Figures 10-36b, 10-36c, and 10-36d are the structure map on the base, fault map, and net sand map, respectively. These maps are required for isochore construction.

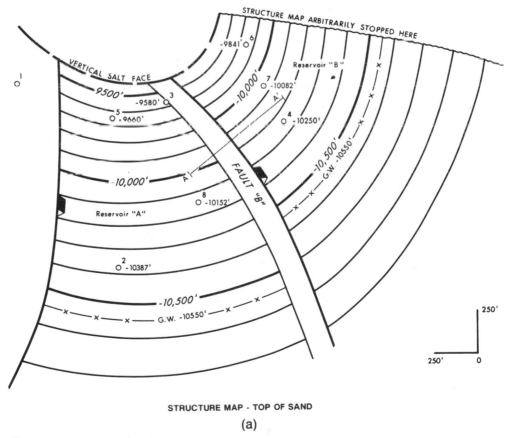

STRUCTURE MAP - TOP OF SAND

(a)

Figure 10-36 (a) Structure map. Top of sand Reservoirs A and B. Salt intrusion is vertical at this depth. (b) Structure map—base of sand. (c) Fault surface map—Faults A and B. (d) Net sand isochore map for 9500-ft Sand.

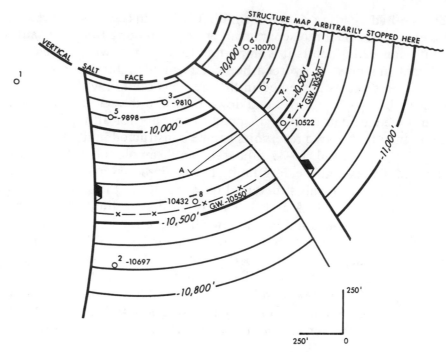

STRUCTURE MAP - BASE OF SAND

(b)

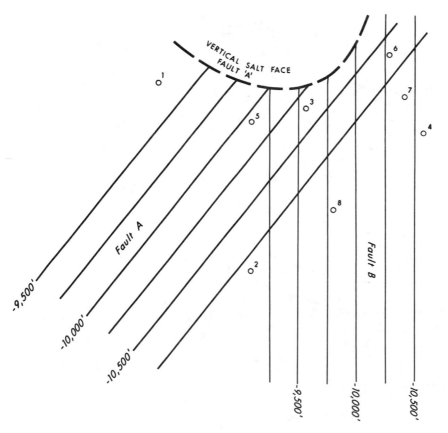

FAULT MAPS - FAULTS A & B

(c)

Figure 10-36 (*continued*)

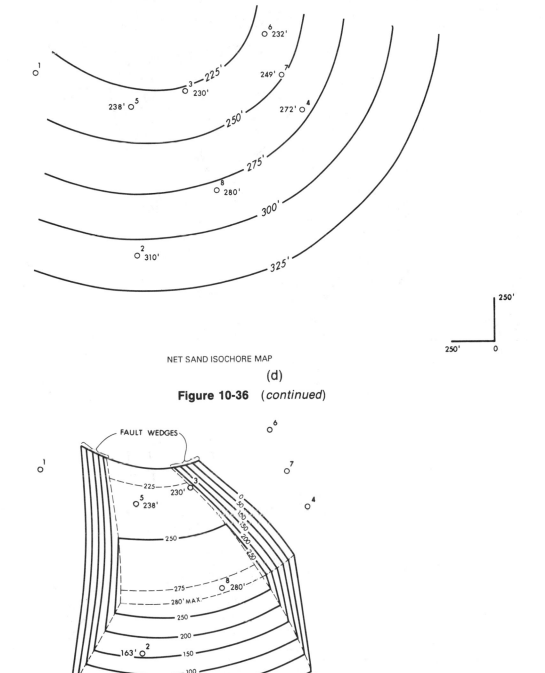

NET SAND ISOCHORE MAP

(d)

Figure 10-36   (*continued*)

NET GAS ISOCHORE- RESERVOIR A

(a)

Figure 10-37   (a) Net gas isochore map for the 9500-ft Sand, Reservoir A. Isochore includes the mapping of two fault wedges and a water wedge. (b) Net gas isochore map for the 9500-ft Sand, Reservoir B. Fault and water wedges are a significant portion of the reservoir volume. (c) Composite net gas isochore map for Reservoirs A and B. Notice the complexity in the mapping of the two fault wedges created by Fault B when both reservoirs are mapped together.

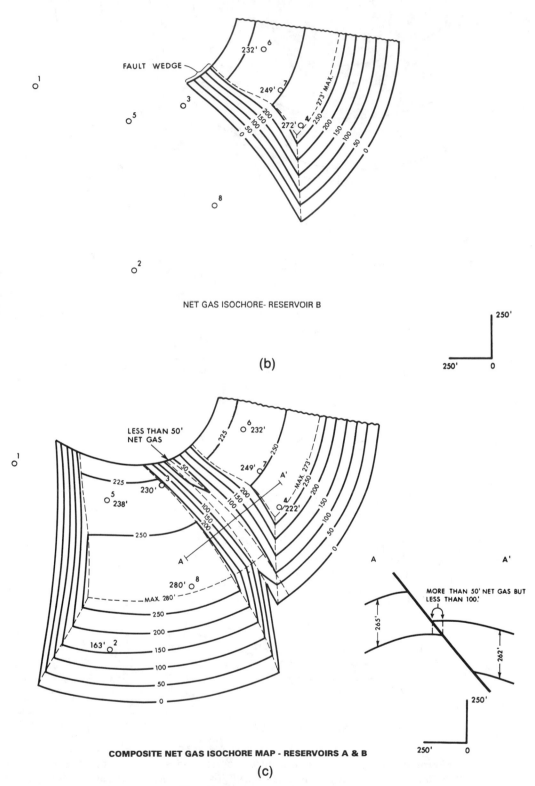

NET GAS ISOCHORE- RESERVOIR B

(b)

COMPOSITE NET GAS ISOCHORE MAP - RESERVOIRS A & B

(c)

Figure 10-37   (*continued*)

The net gas isochore map for the 9500-ft Sand, Reservoirs A and B, can be constructed separately or as one single isochore map. The isochores in Figs. 10-37a and 10-37b show the individually constructed net gas isochore maps for Reservoirs A and B. Notice that the fault wedges for Faults A and B cover a significant portion of the reservoirs. Figure 10-37c illustrates how the two reservoirs can be isochored together combining the wedges for Fault B. Cross section A-A' shows about 75 ft of sand juxtaposed across the fault at this position in the reservoir.

Isochore maps in which the fault wedges are combined, such as the one shown in Fig. 10-37c, are difficult to construct and can easily result in error. We recommend for simplicity of construction and planimetering that even with nonsealing faults, each reservoir and fault wedge be isopached separately as shown in Figs. 10-37a and 10-37b.

CONFIGURATION OF A NET PAY ISOCHORE MAP

So far, we have discussed a number of methods used to contour net pay isochore maps including the full thickness area and all wedge zones. We have not as yet discussed the geometry of a net pay isochore map as it compares to that of the original structure. In preparing a net pay isochore map, the net pay sand is completely rearranged to the extent that the isochore does not at all resemble the structural configuration of the reservoir.

Figure 10-38 relates a net gas and net oil isochore map to the actual structural configuration of a reservoir. The structure section in the center of the figure shows the actual structural configuration of the reservoir. The cross sections in the lower portion of the figure are sections through the net gas and net oil isochore maps. It is important to notice that the isochore is not a true representation of the net pay sand in its actual structural configuration; it has been artificially flattened or piled up (*referred to as isochore piling*) to represent the same volume in a configuration that can be used for planimetering and volume determinations.

RESERVOIR VOLUME DETERMINATIONS FROM ISOCHORE MAPS

There are two methods commonly used to determine reservoir volume from net pay isochore maps. These are the *Horizontal Slice* and *Vertical Slice Methods*.

Horizontal Slice Method

Two equations are generally used to determine the volume of a net pay isochore map which has been planimetered (Craft and Hawkins 1959). The first is the volume of the frustum of a pyramid:

$$\textbf{Volume} = \tfrac{1}{3}h(A_n + A_{n+1} + \sqrt{A_nA_{n+1}}) \tag{10-1}$$

where

$$h = \text{Interval between isochore lines in feet}$$
$$A_n = \text{Area enclosed by lower isochore line in acres}$$
$$A_{n+1} = \text{Area enclosed by upper isochore line in acres}$$

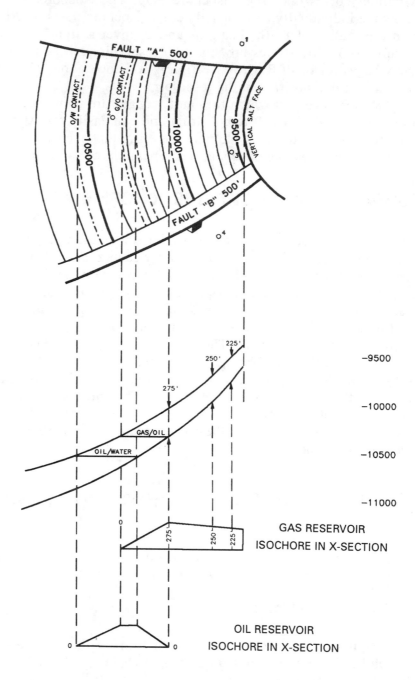

CROSS-SECTIONS OF GAS AND OIL RESERVOIR ISOCHORES
TO SHOW ISOCHORE "PILING"

Figure 10-38   In the preparation of a net hydrocarbon isochore map, the configuration of the sand is completely rearranged and the base is artificially flattened. Compare the configuration of the net gas and oil cross sections in the lower portion of the figure to that of the structure cross section in the center.

This equation is used to determine the volume between successive isochore lines, and the total volume is the sum of these separate volumes.

The second equation is the volume of a trapezoid:

$$\text{Volume} = \tfrac{1}{2}h(A_n + A_{n+1})$$

or, for a series of successive trapezoids:

$$\textbf{Volume} = \tfrac{1}{2}h(A_0 + 2A_1 + 2A_2 \ldots 2A_{n-1} + A_n) + t_{\text{avg}}A_n \qquad (10\text{-}2)$$

where

$$A_0 = \text{Area enclosed by the zero isochore line in acres}$$

$$A_1, A_2 \ldots A_n = \text{Areas enclosed by successive isochore lines}$$

$$t_{\text{avg}} = \text{Average thickness above the top or maximum thickness isochore line in feet}$$

The pyramidal equation provides the most accurate results; however, because of its simplicity, the trapezoidal equation is commonly used. Since the trapezoidal equation introduces an error of about 2% when the ratio of successive areas is 0.5, there is a common convention used to employ both equations. Wherever the ratio of the areas of

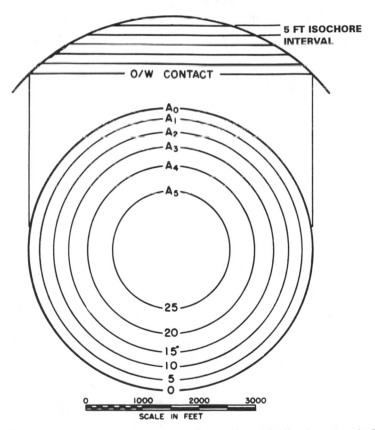

Figure 10-39 Cross section and net pay isochore of an idealized reservoir. The cross section shows the isochore divided into 5-ft horizontal slices. (B. C. Craft/M. F. Hawkins, *Applied Petroleum Reservoir Engineering*, © 1959, p. 28, Reprinted by permission of Prentice-Hall, Inc., Upper Saddle River, N.J.)

TABLE 10-1

Area	Planimeter Area Sq. In.	Area Acres	Ratio of Areas	Interval h Feet	Equation	ΔV Ac. Ft.
A_0	19.64	450	—	—	—	—
A_1	16.34	375	0.83	5	Trapezoid	2063
A_2	13.19	303	0.80	5	Trapezoid	1695
A_3	10.05	231	0.76	5	Trapezoid	1335
A_4	6.69	154	0.67	5	Trapezoid	963
A_5	3.22	74	0.48	5	Pyramid	558
A_6	0.00	0	0.00	4	Pyramid	99
					Total ac. ft.	**6713***

* The percentage difference in acre feet between the horizontal and vertical slice methods is less than 1%.

any two successive isochore lines is smaller than five-tenths, the pyramidal equation is applied. Whenever the ratio of the areas of any two successive isochore lines is larger than five-tenths, the trapezoidal equation is used.

Figure 10-39 and Table 10-1 outline the volume determination using the horizontal slice method. Take a few minutes and review this example to obtain a good understanding of the procedure.

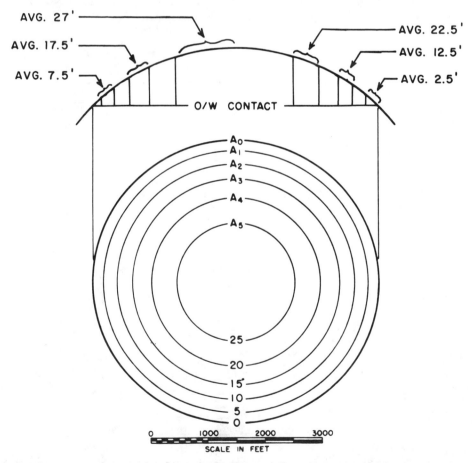

Figure 10-40 Cross section and net pay isochore of an idealized reservoir. The cross section shows the isochore divided into vertical slices. (B. C. Craft/M. F. Hawkins, *Applied Petroleum Reservoir Engineering*, © 1959, p. 28. Reprinted by permission of Prentice-Hall, Inc., Upper Saddle River, N.J.)

TABLE 10-2

Area	Planimeter Area Sq. In.	Area Acres	Difference in Areas $A_{n-1} - A_n$	Average Thickness Feet	V Ac. Ft
A_0	19.64	450	—	—	—
A_1	16.34	375	75	2.5	187
A_2	13.19	303	72	7.5	540
A_3	10.05	231	72	12.5	900
A_4	6.69	154	77	17.5	1347
A_5	3.22	74	80	22.5	1800
A_6	0.00	0	74	27.0	1998
				Total ac. ft.	6772*

* The percentage difference in acre feet between the horizontal and vertical slice methods is less than 1%.

Vertical Slice Method

The vertical slice method is sometimes referred to as the donut method because the individual areas used to determine the reservoir volume fall between successive isochore lines that often appear as being donut shaped. This method is considered by many to be less confusing than the horizontal slice method, particularly if isochore maps have a number of thick and thin areas. The equation for the vertical slice method is:

$$\textbf{Volume} = h(A_0 - A_1) + h(A_1 - A_2) + \ldots h(A_{n-1} - A_n) + h_{avg}(A_n) \quad (10\text{-}3)$$

where

h = Average thickness of the isochore between successive contour lines

A_0 = The zero isochore line

A_1 = Next thicker or successive isochore line

A_n = Thickest or last isochore line

h_{avg} = Average thickness within the last or thickest isochore contour line

Figure 10-40 and Table 10-2 illustrate the procedure for volume determinations using the vertical slice method. The isochore used for this example is the same one used for the horizontal slice method so that the results can be compared. The difference in calculated volume between the horizontal and vertical slice methods, for the example in Figs. 10-39 and 10-40, is less than 1%.

Choice of Method

The choice of using the horizontal or vertical slice methods is usually based on individual preference, since both methods are reasonably accurate. The choice is less important than the assurance that the method is used correctly, avoiding planimetering pitfalls. It is therefore of utmost importance that anyone doing actual planimetering be thoroughly familiar with the mathematics of each method and the pitfalls that can be encountered when planimetering.

A geologist may spend months working on a prospect with the end result being a net pay isochore map prepared to estimate the hydrocarbon volume for the prospect. If the isochore map is planimetered incorrectly, as a result of carelessness or a lack of understanding of the planimetering procedures, a viable project can be mistakenly rejected. We highly recommend that all planimetered work be spot checked by the geologist or reservoir engineer who prepared the isochore map.

INTERVAL ISOPACH MAPS

As discussed in the beginning of this chapter, an isopach map is one in which the thickness of a single unit is contoured, such as a sand body or formation, or the thickness of a section between two markers, which is referred to as an *interval isopach map*. Interval isopach maps are particularly useful in determining the history of movement along faults, of saltdome uplift, or fold development, as well as in the interpretation of depositional environments.

The full application of interval isopach maps is beyond the scope of this text. More specifically, the determination and use of the proper interval thicknesses obtained from well logs or seismic data used for construction of interval isopach maps are addressed here. Interval isopach maps reveal the **true stratigraphic thickness of units** and not the vertical thickness. The stratigraphic thickness or true thickness is the thickness measured perpendicular to the surface of the bed or formation. Where strata are flat lying, or have only gentle dips, the two types of thicknesses are very similar; but with increasing dip, the differences become substantial. The relationship is given by:

$$\text{Stratigraphic Thickness} = (\text{TVT})(\cos \phi) \tag{10-4}$$

where

$$\text{TVT} = \text{True vertical thickness}$$
$$\phi = \text{True structural or bed dip}$$

For example, with a 10-deg bed dip, a bed with a true stratigraphic thickness of 100 ft has a vertical thickness of 101.5 ft; at a 20-deg dip, the vertical thickness is 106.5 ft. At a 45-deg dip, however, the vertical thickness becomes 141 ft, which is significantly different than the stratigraphic thickness. If the true vertical thickness is the value contoured, the map is more correctly referred to as an *isochore map*. Therefore, net sand and net pay maps are technically isochore maps.

Well Logs

The determination of stratigraphic thickness from well logs presents a few complications. In areas of nearly flat lying beds, the vertical thickness is very close to true stratigraphic thickness. The determination of stratigraphic thickness does, however, become more complicated around steeply dipping structures. Figure 10-41 shows the effect of changing bed dip on log thickness although the stratigraphic thickness is constant. With zero bed dip, the vertical and stratigraphic thickness are the same. At a 40-deg bed dip, the vertical thickness is equal to 1.30 times stratigraphic thickness, and at 60 deg, the vertical thickness is twice as thick as the stratigraphic thickness. Assuming vertical wells, if the upper and lower markers chosen for interval isopaching are parallel or nearly so (that is, they are at or near the same dip), Eq. (10-4) can be used to convert vertical thickness to

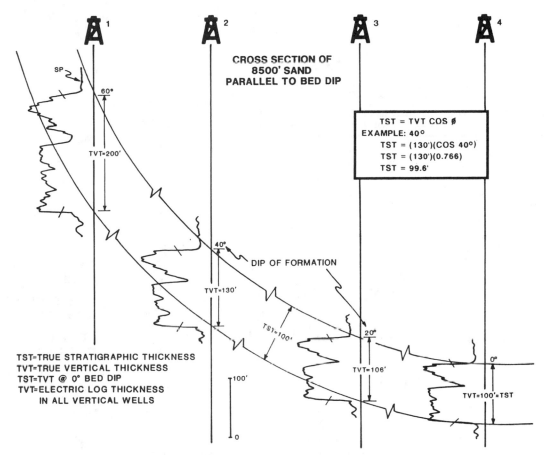

Figure 10-41 Effect of changing bed dip on true vertical thickness, although stratigraphic thickness is constant.

stratigraphic thickness. If the dips of the upper and lower markers are different, the correction factor for the cosine of a single bed dip, either the upper or lower, will not be accurate. For such cases, another equation is required (Tearpock and Harris 1987).

$$\text{Stratigraphic Thickness} = \frac{\text{TVT}}{(\sin \alpha \tan \phi) + \cos \alpha} \tag{10-5}$$

where

$$\text{TVT} = \text{True Vertical Thickness}$$
$$\alpha = \text{Angle of upper horizon}$$
$$\phi = \text{Angle of lower horizon}$$

The equation and an example problem are shown in Fig. 10-42. This equation takes into account the dip of the upper and lower markers and the interval's vertical thickness.

Finally, in the case of deviated wells, the measured log thickness must be first corrected to vertical thickness and then corrected to true stratigraphic thickness. The procedure for vertical thickness conversions was previously discussed in this chapter and Chapter 4.

To avoid making laborious stratigraphic thickness calculations, the nomogram in

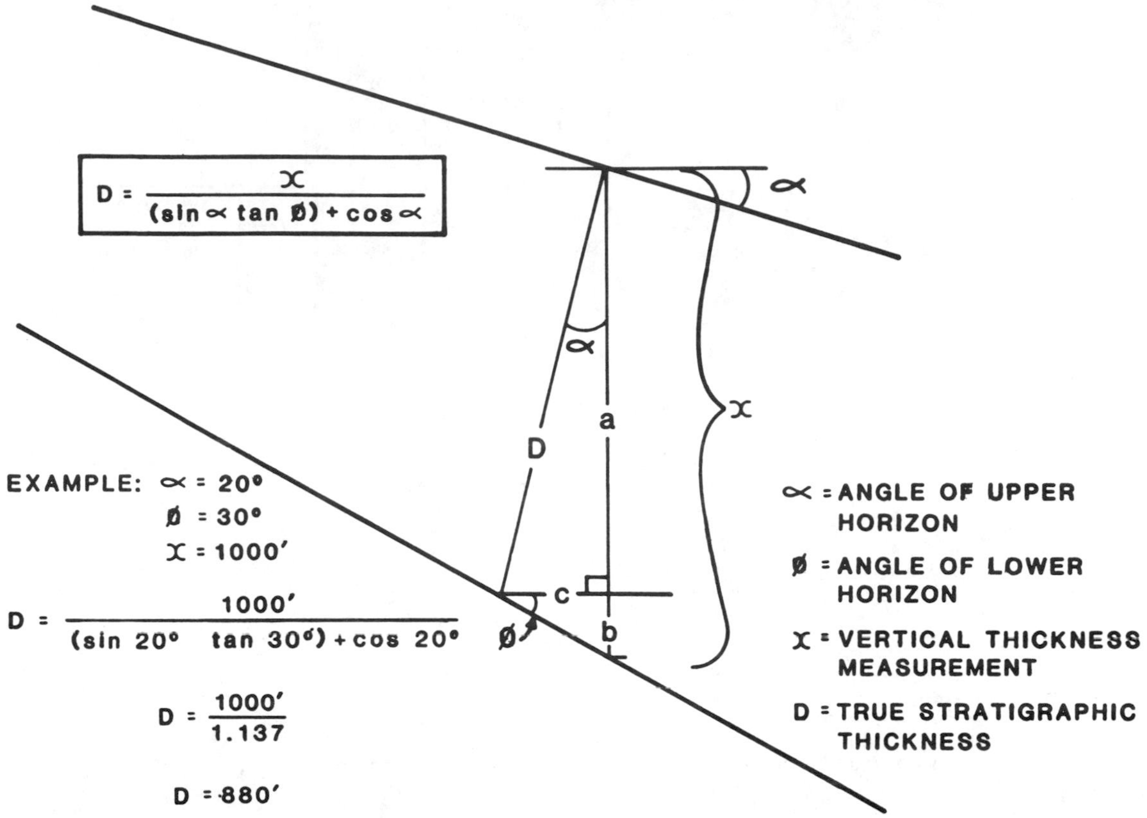
D = x / ((sin α tan Ø) + cos α)
α
α
a
D
x
c
b
Ø
EXAMPLE: α = 20°
Ø = 30°
x = 1000′
D = 1000′ / ((sin 20° tan 30°) + cos 20°)
D = 1000′ / 1.137
D = 880′
α = ANGLE OF UPPER HORIZON
Ø = ANGLE OF LOWER HORIZON
x = VERTICAL THICKNESS MEASUREMENT
D = TRUE STRATIGRAPHIC THICKNESS

Figure 10-42 Cross section shows the geometric relationship between two markers that have different dip rates. Equation (10-5) is used in this type of situation to convert true vertical thickness to true stratigraphic thickness. (Prepared by C. Harmon. Modified from Tearpock and Harris 1987. Published by permission of Tenneco Oil Company.)

Fig. 10-43 can be used to calculate stratigraphic thickness if the vertical thickness and the dips of the upper and lower markers are known. The horizontal axis represents the dip of the top bed, the vertical axis represents the correction factor, and the curves within the nomogram represent the difference in degrees between the lower and upper beds. Consider the following example:

Data:

Dip of upper bed = 20 deg

Dip of lower bed = 30 deg

Vertical thickness = 1000 ft

1. To use the nomogram, first subtract the dip of the upper bed from the lower bed. This value determines which of the curved lines to use for the correction factor.
2. Enter the chart on the horizontal axis at 20 deg and move vertically upward until you intersect the curve equal to the difference in degrees between the lower and upper beds. In this case it is the +10-deg curve.
3. From the intersection with the curve, move laterally to the left to intersect the vertical axis which is the correction factor. In this case it is 0.88.

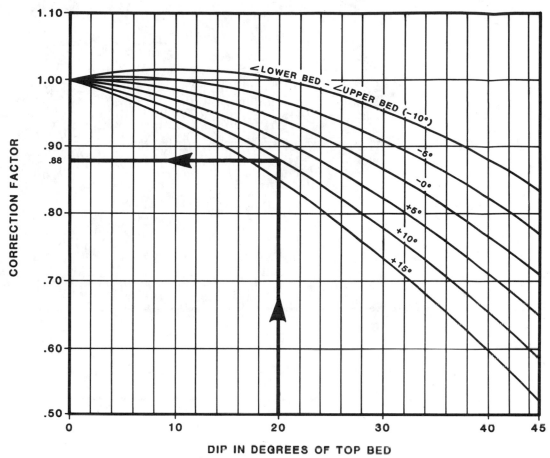

Figure 10-43 Nomogram derived from Eq. (10-5) used to determine the correction factor for converting true vertical thickness to true stratigraphic thickness when the upper and lower beds or markers dip at different angles. (Prepared by C. Harmon. From Tearpock and Harris 1987. Published by permission of Tenneco Oil Company.)

Therefore:

$$\text{True Stratigraphic Thickness (TST)} = (1000 \text{ ft})(0.88)$$

$$\text{(TST)} = 880 \text{ ft}$$

When Is a Perpendicular Not a Perpendicular?

Using seismic data for interval isopach construction can give you many additional data points between well control. In areas of relatively low dip (10 deg or less) and parallel horizons, the vertical thickness calculated from seismic is a close approximation of the true stratigraphic thickness. The procedure in this type of area is straightforward: the time-converted depth of the upper horizon is subtracted from the time-converted depth of the lower horizon to arrive at an interval thickness. The basic requirement is an accurate time-depth function.

In areas of steeper dip and nonparallel horizons, you should be aware of some visual pitfalls inherent in seismic sections. The basic point to remember is that *a time section is not a cross section*. It is distorted because of the two very different dimensions displayed on a section: time along the vertical axis and distance along the horizontal axis.

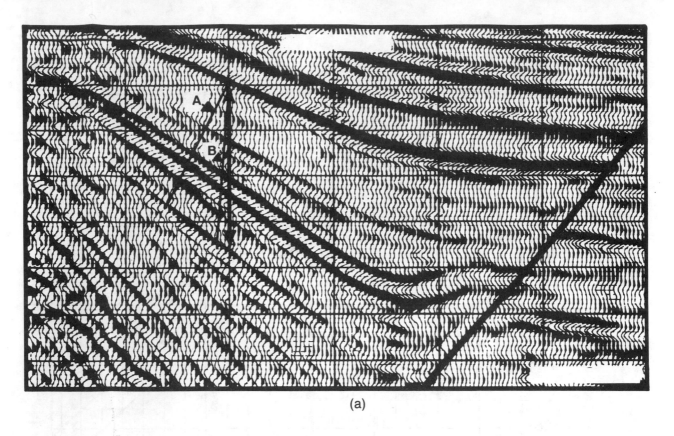

(a)

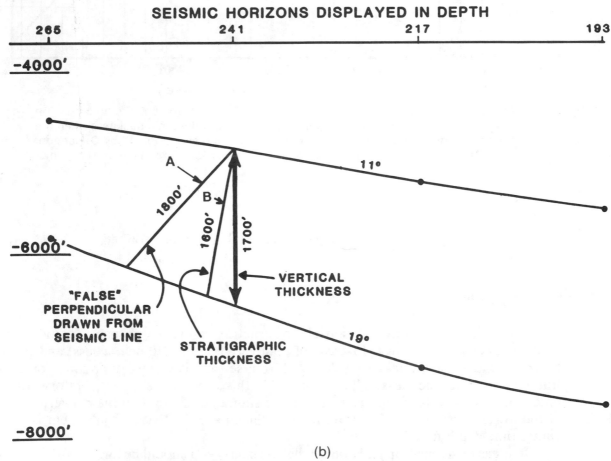

SEISMIC HORIZONS DISPLAYED IN DEPTH

265   241   217   193

−4000'

A

1800'

B

1800'

1700'

11°

−6000'

VERTICAL THICKNESS

"FALSE"
PERPENDICULAR
DRAWN FROM
SEISMIC LINE

STRATIGRAPHIC
THICKNESS

19°

−8000'

(b)

Figure 10-44 (a) Which line (A or B) represents the true stratigraphic thickness of the designated interval? (b) True 1:1 scale cross section of the seismic interval shown in (a). (Prepared by C. Harmon. From Tearpock and Harris 1987. Published by permission of Tenneco Oil Company.)

These dimensional differences often introduce some very pronounced vertical exaggeration.

To illustrate this, observe the two horizons marked in Fig. 10-44a. They obviously diverge from one another as the interval thickens into a fault. How is the stratigraphic thickness measured? The first inclination is to draw a perpendicular line, shown as A on the figure, from the top horizon to intersect the bottom horizon, and calculate trigonometrically the stratigraphic thickness using the time converted depths at both points along with the lateral distance between the two points. **THIS IS WRONG**.

To see graphically what is really present, look at Fig. 10-44b, which shows the sections' horizons converted to depth and displayed at a true 1:1 scale. Line A drawn earlier as perpendicular to the top horizon on the seismic line is in fact a longer segment than the true perpendicular, which is line B on the seismic line and cross section. The reason for this pitfall is that the seismic line, at this depth, has about a 2:1 vertical exaggeration. In this case, you would post a larger thickness for the interval than is actually present. To get corrected data points, you need to apply the correction factor in Eq. (10-5), which uses the dip of the top and bottom beds, and the thickness of the interval measured vertically. The nomogram in Fig. 10-43 can also be used to calculate the stratigraphic thickness.

In summary, seismic information can be a valuable source of interval thickness data, as long as you are aware of the visual distortion inherent in seismic data, and properly account for it in the calculation of stratigraphic thicknesses.

CHAPTER 11

FIELD STUDY
METHODOLOGY

INTRODUCTION

Up to this point in the text, we have presented many subsurface mapping and cross section techniques with both idealized and actual examples. In this chapter, we show the methodology and application of these techniques in conducting detailed field studies. In Chapter 1, we mentioned that the techniques and methods of mapping can be designed and modified to apply to any type geologic study. There are three basic phases of prospecting in which various mapping techniques and methods are required.

1. The first is the initial exploration phase of a property. Generally, few, if any wells have been drilled on or near the prospect and therefore, prospect evaluation relies on seismic data, limited well control, a good geologic concept, and comparisons or analogies to nearby properties. For exploration purposes the emphasis may be on defining major sand units, locating the larger potential fault traps, and preparing regional subsurface maps. The aim of this work is to identify undiscovered economic hydrocarbon accumulations.

2. The second phase of prospecting occurs after a field has been discovered and several wells have been drilled. Continuously accumulated well log, core, seismic, and performance data are used to fully develop the field and obtain a good estimate of the volume of recoverable hydrocarbons. In these newly discovered fields, the geologic study may include the mapping of all recognized faults, the preparation of structure contour maps, interval isopach maps, and net sand and net pay isopach maps for all known hydrocarbon bearing reservoirs, in addition to a variety of cross sections. These maps and cross sections may be used to estimate reserves, justify additional development drilling, or establish a field depletion plan.

546

3. The third and final phase of prospecting occurs in maturely developed properties. With increasing amounts of log and performance data, volumetrics are compared to performance to refine the geological maps and ultimate reserve estimates. All the geologic, geophysical, and engineering information is used to help identify any hydrocarbons that remain. In any integrated study being conducted in a mature area with the purpose of identifying economic reserves not capable of being recovered by existing producing wells, to identify additional potential overlooked in previous studies or to plan secondary or enhanced recovery projects, *the most detailed maps are required, as well as the analysis of reservoir performance data.* This detailed work requires a team effort consisting of team members experienced in various disciplines, including geology, geophysics, and petroleum engineering. *It is during this final phase of prospecting that many companies have failed to recognize and approve the kind of integrated, detailed interdisciplinary work required to find all the remaining oil and gas.*

In this chapter we present the methodology and guidelines that illustrate the synergistic team approach to prospecting. This requires the organization of an interdisciplinary team with members from the following disciplines: geology, geophysics, and engineering. In addition, there should be a technical support staff. We believe that if you are capable of undertaking such detailed work, you can certainly modify the amount of detail needed to conduct other geologic and engineering studies such as those required in newly discovered fields or for wildcat exploration.

For the first time, we introduce the role of the petroleum engineer and technical assistant as integral members of a detailed study team. The buzz word used today to describe an integrated team approach is **"synergy,"** which comes from the Greek word *synergos* meaning *"working together."* When undertaking major integrated field studies, each member of a team must work toward a common goal with specific objectives. The team members must have diversified backgrounds and areas of specialized training.

We present a summary of the methodology required for conducting detailed geologic, geophysical, and engineering studies in maturely developed fields. The primary objective of such studies is to identify previously unrecognized development and exploratory hydrocarbon potential. These detailed studies are unique. They require technically integrated synergistic work teams specialized in conducting very detailed geologic and engineering work; and, more importantly, they require individual members capable of working together in an atmosphere that encourages a free flow of ideas, data, assistance, and work across all disciplinary boundaries. If detailed field studies were simple to undertake and required no special experience, there would be no need for anyone to conduct studies in these mature fields, because all the hydrocarbons would have already been found; however, we know this is not the case.

There are *real opportunities* in maturely developed fields. We discuss the methodology needed to optimize the efficiency of such work, to maximize the prospect potential.

BACKGROUND

To conduct a detailed field study from scratch, all data generated from the field are required for study. These data include all available data from the company owning the field, log libraries, production reporting services, regulatory agencies, in addition to internally generated data. An example of such data are all company correspondence, lease and well files, unitization orders, all company correlated logs, all previously generated maps, a clean set of logs, new base map, directional surveys, core analyses and descriptions, drill stem tests, all production data (including oil, gas, and water), well tests, BHP's, completion schematics, well histories, and seismic lines and a velocity survey. The previously correlated logs, generated maps, and reserve calculations provide an opportunity to quickly gain experience and knowledge about the field.

In prospecting in old mature fields, you must conduct a complete evaluation of all known reservoirs for a number of reasons, including: (1) estimation of good recovery factors, (2) identification of attic potential, (3) recognition of poor completion practices, (4) evaluation of well drainage efficiencies, and (5) the identification of infill development potential. Future potential may be found in a number of areas, including producing reservoirs, reservoirs with proved reserves behind pipe, accumulations (defined as attic, infill, untested fault blocks, and miscellaneous development potential), wildcatting in and around the field, and in deeper stratigraphic sections than the current field limit. All future potential must be supported by sound geologic, geophysical, and engineering procedures. Reserve estimates, where possible, should be determined by at least two methods in producing or previously produced reservoirs (volumetrics and analyses of performance histories) and by volumetrics and field analogies for all other potential. The specific method or combination of methods for estimating reserves depends upon the stage of reservoir development, availability and quality of data, and judgement of team members. The work must demonstrate with reasonable certainty that the reserves are economically recoverable under existing economic and operating conditions.

FIELD STUDY METHODOLOGY

Objectives

The methodology used in conducting any field study is primarily dictated by (1) the specific results expected, (2) size of field, (3) state of depletion, (4) timetable, and (5) manpower. The primary objectives for conducting any field study are:

1. to ascertain remaining proved developed producing reserves;
2. to identify all future recompletion, workover, development, and exploratory drilling potential;
3. to evaluate any enhanced recovery opportunities;
4. determine the fair market value of the field; and
5. to conduct a reserve audit.

The following discussion presents a summary of the phase by phase procedures required to conduct the most detailed studies in maturely developed areas. The methodology and procedures are shown in Fig. 11-1.

Phase I—Data Collection

The data used in each study are obtained from a number of sources, and much are internally generated. A general list of these data are shown here. All these data must be collected, cataloged, and organized into files. During the study, the organized data are updated and new files generated as needed.

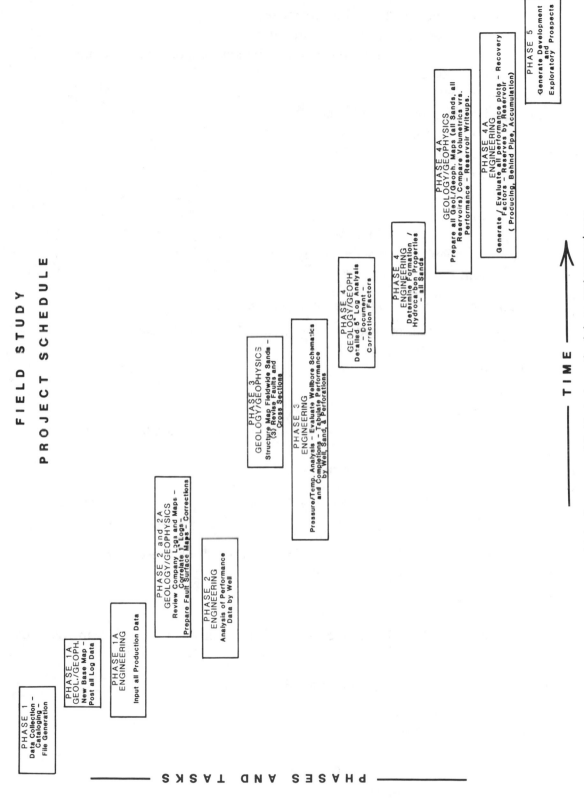

Figure 11-1 Summary of field study methodology and procedures.

549

Geologic Data—Available or Generated

1. One-in. logs, one clean and one correlated
2. Five-in. logs, one clean and one correlated
3. Place all core data, perforations and production data on all 5-in. detailed logs
4. Directional surveys
5. Well histories
6. Production summary for all wells
7. Well data inventory
8. Current completion data
9. All fault cut data
10. All sand data
11. New base map
12. All previously prepared maps
13. Seismic profiles and velocity survey
14. Paleo data, if available

Engineering Data—Available or Generated

1. Monthly tabulation of production by well (oil, gas, and water)
2. Gauge data for all wells
3. PVT data
4. Formation water analysis
5. Bottom hole pressure data
6. Pore pressure plots for wells—determine the pressure gradient
7. Volumetric parameters

Temperature	Viscosity of oil
Porosity	FVF
Water saturation	Recovery factors
Gas/oil ratio (GOR)	Condensate yields
Specific gravity	Pressure

8. Cumulative production by well, perforations, sand, and reservoir
9. Rate vs time plots (by well, by reservoir)
10. Rate vs cumulative production plots
11. Oil cut vs cumulative production by well
12. Summary of well problems such as mechanical problems, sanded up wells, extraneous water, communication problems, etc.

Geophysical Data—Available and Generated

1. Seismic base map
2. All seismic sections, full- and half-scale, if possible
3. Good velocity function
4. Previously interpreted seismic sections
5. Gravity and magnetic data

Internally Generated Data—Phases I through IV.   During the life of any project, additional information may be necessary to conduct the study. Also, data internally generated during the life of the project must be organized in a usable format. Examples of such data are shown here.

1. A tabulation of all the productive sands and the wells in which they are productive.
2. A tabulation of available well data such as location, type of logs run, elevation of Kelly Bushing, mud weights, core reports, total depth, etc.
3. A tabulation of test data and cumulative production for all perforated intervals in each well.
4. Computer generated production plots.
   a. Tabulation of production by month.
   b. Rate vs time plots.
   c. Rate vs cumulative production plots.
   d. Oil cut vs cumulative production plots.
   e. P/Z vs cumulative production.
   f. Flowing tubing pressure vs cumulative production.

Flow of Geologic and Geophysical Work

Phase IA

1. *Base Map Preparation.* A new base map is often required for each individual field study. Commercially available base maps and even company prepared bases often contain errors sufficient to effect the geologic and geophysical effort in detailed studies. A new base map assures the accuracy of all well locations and seismic lines, and permits the detailed subsea plotting of all directionally drilled well paths.
2. *Posting Data on Well Logs.* Post all test, completion, and production data on all 1-in. logs. Detailed completion and core data should be posted on 5-in. logs (refer to Chapter 4).

Phase II

3. *Initial Seismic Analysis.* Data from wells falling directly on any seismic lines are incorporated into the lines. Seismic sections are initially interpreted to identify major markers, sands, amplitudes, and all recognized faults.
4. *Correlation of Electric Logs.* Prior to log correlation, one or more type logs are prepared as standard correlation logs. The 1-in. logs are correlated to identify major markers, sands, and all recognized fault cuts. All data are documented and correction factors made to the data where applicable.
5. *Fault Surface Map Construction. Fault surface maps* are prepared for all recognized faults. A reasonable structural interpretation in faulted areas must begin with the construction of accurate fault maps. The data for the fault surface map interpretations come from well log correlations and seismic interpretations. The fault map interpretations must be geologically reasonable, have three-dimensional geometric validity, and tie to seismic data, if available. Normally, fault maps are not finalized until the later stages of the study because of a continuous evolution during structure and reservoir mapping. Fault data are documented and all fault cut correction factors made for any deviated wells.
6. *Cross Section Construction.* During the early stage of a field study, *problem-solving cross sections* are laid out to begin evaluating the gross structural framework. Also, they help resolve questionable correlations and aid in solving subsurface fault, struc-

ture, and stratigraphic problems. Normally, straight line cross sections are prepared during this phase of the study. Later in the study (Phase III), *final illustration cross sections and special correlation sections* can be laid out to evaluate individual reservoirs and membered sands. If possible, all cross sections should be checked for balance.

Phase III

7. *Fieldwide Structure Maps—At Least Three Horizons*. At least *three fieldwide structure maps* should be prepared from well log and seismic data. These maps, prepared on shallow, intermediate, and deep horizons, are integrated with the fault maps. Integrated multiple horizon mapping, on a fieldwide basis, is essential to the development of a geologically sound and accurate interpretation of a structure. Depending upon the size of the field and number of pay horizons, additional fieldwide maps may be required.

Phase IV

8. *Five-in. Log Analyses*. Five-in. logs must be correlated and analyzed to obtain all the required information for conducting the reminder of the geologic and engineering effort. The following is a partial list of basic data obtained and recorded on the logs and documented on data sheets.

   a. correlative sand tops   f. fluid contacts
   b. porosity sand tops   g. porosity
   c. correlative bases   h. permeability
   d. net pay   i. perforations
   e. resistivity—R_t and R_o j. production data

   The sand data must be documented on some type of data sheet for each productive sand. All correction factors must be made, including subsea and net sand thickness calculations.

9. *Preparation of All Required Geological Maps*. Various geological maps are prepared on all productive or potentially productive horizons to evaluate the entire field. These include structure, porosity top, interval isopach, net sand, net hydrocarbon, original limit, current limit, log, attic, and a variety of special maps. During this phase, the cross sections are refined if necessary and correlation sections or three-dimensional RAMs (Reservoir Analysis Models) prepared for complex multimembered pay sands.

10. *Reserve Comparisons—Volumetrics vs Performance*. Net pay isopach maps are planimetered to determine the productive acre-feet for volumetric calculations. The geologist and engineer conduct comparison studies of reserve estimates calculated at least two different ways, if possible. Discrepancies are analyzed and work changed or modified to resolve reserve problems. (Reservoir maps may require revision, additional geologic work may be needed, or recovery factors, formation, or hydrocarbon properties might be revised.) At times, regardless of the integrated work effort, discrepancies cannot be adequately explained and still honor the data. Such facts must be taken into consideration when reviewing the results.

Phase V

11. *Generate Development and Exploratory Prospects*. A complete geologic and engineering packet is prepared for each prospect generated. The packet should contain

all supporting material, including logs, fault, structure and isopach maps, seismic data, cross sections, reservoir engineering data (production plots, reserve calculations—volumetrics and performance), and economic analyses.

Flow of Engineering Work

There is a significant amount of test, production, and pressure data to document, sort in various ways, and use to prepare standard tabulations and production plots (tabulation or production by month, rate vs time plots, oil cut vs cumulative production, P/Z vs cumulative production, and others). Computer and word processing equipment and software are required to undertake these tasks.

Phase IA

1. *Preparation of Database.* Initially, a computer or word processor database for all production related data should be prepared. Establish guidelines for data input, so these data can be used and manipulated during the course of the study. In addition to performance plots, production data may require sorting by perforations, leases, wells, sands, or reservoirs.

Phase II

2. *Evaluate Wellbore Schematics and Initial Performance Data.* It is best to conduct an initial engineering evaluation of performance prior to the completion of the geological mapping. All wellbore schematics and completion practices are evaluated during this phase. Initial production plots by *well*, or by *completion*, are assembled and analyzed. In single well reservoirs, these data are used to estimate reserves by performance.

Phase III

3. *Analysis of Subsurface Formation Pressures.* A detailed study of formation pore pressures can be made for the entire field area, from the surface through the deepest penetrated stratigraphic section. Initial bottom-hole pressure measurements, if available, are plotted versus depth. If there is a geopressured section, the depth to the top of geopressure should be identified and the section evaluated.

4. *Formation Temperature Analysis.* Formation temperatures obtained from bottom-hole measurements are used to define the geothermal gradient for the field. The temperatures are plotted vs depth. If possible, generate an equation from the plotted data which can later be used to estimate temperatures at any depth.

5. *Determination of Formation Properties.* Estimate the average porosity for each sand and reservoir using porosity values reported from analyses of whole core or side wall core samples. It may be necessary to determine a millidarcy cutoff for the definition of permeable sand. (Example: for Gulf of Mexico Miocene Sands, permeabilities of less than 10 millidarcies are usually considered tight and excluded from analysis.)

   Connate water saturations (S_w) are determined from electric log analyses of resistivity curves. Water saturation calculations are estimated for each sand and reservoir to be used in volumetric analyses.

6. *Determination of Hydrocarbon Properties.* Three properties of hydrocarbons are used directly in estimating reserves by the volumetric method.

a. Specific gravities of oil and condensate are obtained from reported test data and laboratory analyses of produced hydrocarbons.
b. Specific gravity of separator gas is determined from laboratory analyses.
c. Gas-oil ratios are determined from test data and production records.

Phase IV

7. *Estimation of Recovery Factors*
   a. *Oil:* An understanding of the type of reservoir drive mechanism is important in volumetric reserve estimates. Recovery factors used in volumetric calculations are directly related to the type of drive. For oil reservoirs, there are five basic drive mechanisms: (1) water, (2) dissolved gas, (3) gas-cap, (4) gravity, and (5) combination. Standard equations are available for estimating recovery factors for each related drive mechanism. These standard equations often assume a 100% sweep efficiency. The actual performance data, where possible, from depleted or nearly depleted oil reservoirs should be used to aid in the estimation of recovery factors.
   b. *Gas:* There are three basic drive mechanisms for gas reservoirs: (1) gas expansion (also called a depletion or volumetric drive), (2) water, and (3) combination. There are standard equations available for calculating recovery factors for each of the drive mechanisms.

8. *Performance Evaluation—Reserve Estimates.* Using all the available test and production data, various performance methods are available to estimate reserves by sand and by reservoir, including: (1) production decline curves, (2) cumulative production curves, and (3) material balance. These methods are also used to project production trends into the future. Computer plots are usually generated for performance work.

9. *Volumetric Evaluation—Reserve Estimates.* Reserves are estimated by the volumetric method as geologic maps are prepared defining reservoir limits and volumes. Volumetric analyses should be used to estimate reserves for all reservoirs including producing, behind pipe, and various accumulations.

10. *Volumetric vs Performance.* Where possible, reserves calculated by volumetrics and performance should be compared. If necessary, additional geologic and/or engineering work may be undertaken to attempt to resolve discrepancies in reserves determined from the various methods. The comparison of estimated reserves (volumetrics vs performance) is very important when evaluating previously produced reservoirs for additional potential.

Phase V

11. *Generate Development and Exploratory Prospects.* Prepare a complete geologic and engineering packet for each prospect generated. The packet should contain all supporting material, including logs, fault, structure, and isopach maps, seismic data, cross sections, reservoir engineering data (production plots, reserve calculations), and economic analyses.

ORGANIZATION OF A SYNERGISTIC STUDY TEAM

There are a number of ways in which study teams can be organized and manpower allocated for conducting a major field study. It often depends on the individual philosophy of the company, the degree of compartmentalization within the company, immediate goals and objectives, and the number of geologists, geophysicists, engineers, and technical support staff experienced in this kind of detailed and integrated study. It is a **specialized area of expertise, often requiring additional training and organization to be successful**.

Positive economic results are the benefit that all companies expect to achieve through their efforts, not only in maturely developed areas, but also in exploration. Some companies are more successful than others. Many factors influence success, including better technology, aggressive management, experience, and even serendipity. But a significant underlying cause of success, often overlooked or taken for granted, is the quality of the subsurface geological mapping.

The main emphasis of this textbook is to improve your skills in the preparation of subsurface geological maps by presenting a wide variety of mapping techniques. With improved techniques you can accomplish the following:

1. provide the most accurate geologic picture with the available data;
2. generate more quality prospects;
3. improve the accuracy of reservoir volume determinations;
4. increase the efficiency and reduce the risk of drilling and recompletion programs; and
5. employ the best and most accurate methods of finding and developing reserves at the lowest cost per net equivalent barrel.

Therefore, one of the main requirements for this kind of work is a team that is knowledgeable and experienced in detailed mapping techniques such as those covered in this text.

Manpower Requirements

For purposes of establishing the manpower needed to undertake a detailed field study, we make the following assumptions as to the field parameters and composition of the study teams:

Typical Field Parameters

100 wells within the field
25 producing horizons
±50 individual reservoirs (active and inactive)

Composition of an Integrated Study Team

1 geologist
1 petroleum engineer
1 part-time geophysicist
1 technical assistant/computer operator
1 secretary/computer operator/word processor

Each member of an integrated study team must have specific expertise and responsibilities in his or her particular area of study, as well as an understanding of the basic fundamentals of the other disciplines. The following are general guidelines for each technical team member.

GEOLOGIST

I. Technical Responsibilities

A. Geologic

Experience in exploration and exploitation geology

Base map preparation

Geologic data collection and cataloging

Log correlation
  Preparation of correlation type log
  General (1-in.) and detailed (5-in.) log correlation
  Vertical and deviated wells
  Fault cut determinations and correction factors
  Stratigraphic analysis
  Net sand/net pay determinations
  Water saturation calculations
  Evaluation of core analyses
  Directional survey data and calculations

Good understanding of structural geology of tectonic setting

Fault surface, salt, and unconformity mapping

Integrated detailed mapping experience
  Structure and porosity top mapping
  Salt/sediment, salt/fault
  Structure derivative mapping

Isopach mapping
  Gross interval isopach/depositional modeling
  Net sand
  Net hydrocarbon (single/multiphase reservoirs)
    Original, current, behind pipe, and attic

Cross section construction and application
  Electric log, stick, and seismic
  Problem solving and illustration type sections
  Correlation sections
  Balanced cross sections

Deviated wells
  Vertical and stratigraphic correction factors
  Fault cut correction factors

Depositional environment modeling from log analyses

Planimetering—vertical/horizontal slice methods

Prospect generation and evaluation

Recompletion/workover evaluation

Property acquisition evaluation

B. Engineering

Volumetric reserve estimation

Volumetric versus performance analysis

Performance evaluation
  Production decline curve analysis

B(1). *Familiarity with the Following Engineering Functions*

Application of pressure and temperature data
Recovery factor calculations
Evaluation of partially depleted reservoirs
Gas deliverability studies

C. *Logging*

Application and use of open and cased hole logs
Performance of basic quality evaluation of all logs
Familiarity with quantitative and qualitative interpretation of well logs
Supervision of logging operations on location

D. *Planning*

Assist in preparation of field depletion plans
Prepare development plan for drilling field wells
Consider any competitive situation
Prepare directional drilling plans and targets, and monitor during drilling
Coordinate plans with partners on joint interest leases
Coordinate logging operations

E. *Completions, Recompletions, and Workovers.* Provide all geologic data and recommendations regarding all completions, recompletions, and workovers.

II. Administrative and General Responsibilities

A. *Regulations*

Have knowledge of statewide rules and regulations affecting oil and gas operations
Have knowledge of proration or unitization rules affecting the studied areas

B. *Procedures.* Initiate and follow through to completion the following:

Location letters
Geologic contribution to AFE
Certified completion plats
Lost hole reports
Geologic contribution to plans of development
Preparation and distribution of master logs

C. *Filing.* Secure and properly file all maps, logs, and reports in a logical and orderly manner.

D. Budget Planning. Prepare, nominate, present, and recommend drilling projects to reflect the use of all geologic, geophysical, engineering, production, and economic data.

III. Personal Qualifications

A. Basic Communication Skills—can express points of view to:

management and supervisors
outside partners
investors
state, federal, or other jurisdictional agencies
other team players

B. Team Player (synergistic team approach)—understand team member responsibilities and activities

Self-motivator
Positive attitude
Open minded—can take suggestions (and criticism) as positive
Task oriented

PETROLEUM ENGINEER

I. Technical Responsibilities

A. Engineering

Application and use of all open and cased hole logs
Perform quantitative and qualitative log interpretations
Net sand/net pay count determination
Shaley sand analysis—thin bed corrections

Volumetric reserve evaluation (in place and recoverable)
   Use of physical rock properties
   Use of hydrocarbon properties
   Recovery factor calculations—all drive mechanisms
Volumetric vs performance evaluation
Analyses of depleted or partially depleted reservoirs
Residual oil and gas saturation calculations
Planimetering—vertical and horizontal slice methods
Material balance calculations

Performance methods of reserve evaluation
   Z-factor calculations
   Production decline curve analysis
   Cumulative production curve analysis
   P/Z vs cumulative production
   BHP and flowing tubing pressure curves

Analysis of individual completion performance
Formation pore pressure analyses including geopressured
Evaluation of wellbore schematics
Evaluation of perforation and completion practices
Formation damage analysis
Evaluation of cross fault drainage
Reservoir modeling studies
Detailed economic analyses—new drills, recompletions, and workovers

B. Geological

Prepare net pay isopach maps—Wharton Method
Prepare detailed cross sections—correlation sections

B(1). Familiarity with the Following

Basic understanding of geologic principles
Basic understanding of applied subsurface mapping techniques

C. Logging

Application and use of open and cased hole logs
Performance of qualitative and quantitative evaluation of all logs
Supervision of logging operations on location (especially cased hole logs)

D. Computer Operations

Hands-on experience and application of computer hardware and software
Organization and coordination of production and test data base
Application of computer generated economic analyses
Computer graphics programs
Familiarity with word processing programs to prepare reports and manipulate production and test database

E. Planning

Preparation of field depletion plan
Support preparation of developing drilling plan
Preparation of AFEs

F. Monitoring of Offset Operations

Monitor the operation of offset operators
Advise and make recommendations regarding the status of offset operations

G. Completions, Recompletions, and Workovers. Provide engineering data and economics regarding all completions, recompletions, and workovers.

II. Administrative and General Responsibilities

A. Regulations

Have knowledge of federal and statewide rules and regulations effecting oil and gas operations

Have knowledge of proration or unitization rules affecting the field being studied

B. Procedures.   Initiate and follow through to completion the following:

Location letters

Engineering contribution to AFE and project economics

Certified completion plats

Engineering contribution to plans of development

All payout economic evaluations

C. Production Data Base

Organization of internal engineering data base

Familiarity with computer application for production and test data

Tabulation of cumulative production and test data

Sorting and manipulation of production data base

D. Budget Planning

Assist in the preparation, nomination, and presentation of drilling projects to reflect the use of all geologic, geophysical, production, and economic data.

III. Personal Qualifications

Same as those presented for the geologist.

DEVELOPMENT GEOPHYSICIST

I. Technical Responsibilities

A. Geophysics

Proficient in **detailed** interpretations

Tie geophysical data with geologic well log data

    Fault cuts—understand vertical separation vs throw

    Major mapping horizons

    Fault surface mapping

Bright spot analysis

Familiarity with seismic processing and reprocessing

Physics of logging (sonic/density logs for synthetics)

Model pay sands—reservoir characteristics

Deal with velocity problems

3-D seismic interpretation experience helpful

B. Geology

Educational background in geology. A degree in geology is preferred in order to have a good understanding of structural geology.

Proficiency in correct subsurface mapping techniques

Integrate geophysical/geologic data to prepare maps: fault surface, salt, unconformity, interval isopach, and fault/structure integrated maps.

II. Personal Qualifications

Same as those presented for geologists.

TECHNICAL SUPPORT

I. Technical Responsibilities

A. Engineering Support

Prepare individual well, lease, and field files

Prepare inventory to accompany well files

Prepare individual well histories

Coordinate assemblage of engineering performance data:
    Production well histories
    Assemble and prepare individual well test data
    Assemble and prepare bottom hole pressure data
    Assemble and prepare core data

Prepare computer generated performance curves:
    Rate vs time
    Rate vs cumulative production
    P/Z vs cumulative production
    Percent oil cut vs cumulative production
    Flowing tubing pressure vs cumulative production
    Shut-in tubing pressure vs cumulative production
    FTP/Z vs cumulative production

B. Geologic Support

Assist in recording basic sand data

Prepare TVD, TVT, TST computations

Post data on 1-in and 5-in logs:
    All production and completion data on 1-in. logs
    All production, test, core, and completion data on 5-in. logs

Place subsea fault cuts and sand tops on all base maps

Prepare cross section bases

Collect, catalogue, and file all maps

Computer Requirements

Computer hardware applicable for use in detailed field evaluations is available today in the form of desktop microcomputers which are relatively inexpensive. Considering the size of a production database per field study and the requirements for manipulating data, plotting graphics, and preparing reports, one set of hardware is recommended for each evaluation team. The software can be used on all computer systems. The following computer hardware and software are recommended:

Computer Hardware

286 (minimum) microprocessor
80-megabyte hard drive
Extended RAM 1.0 megabyte
3.5-in. internal floppy (1.4 mgbyte)
5.25-in. internal floppy (1.2 mgbyte)
Mouse 2400 Baud (minimum) internal modem
VGA color monitor
VGA adapter card
Wide carriage graphics printer
6–8 expansion slots on processor
6- or 8-pen color plotter

Computer Software

Database/word processor
Spreadsheet program
Graphics analysis programs
   Economic analysis program
   Systems analysis program
   Utility programs
   Basic and basic compiler

INTRODUCTORY RESERVOIR ENGINEERING

Reservoir engineering must be involved in all phases of oil and gas prospect evaluation from the initial phases of exploration through field development. Prior to any wells being drilled on a prospect, the most likely reservoir engineering parameters must be assumed in a preliminary evaluation to estimate the potential hydrocarbons. Following initial drilling, offset information from analogous fields should be used to predict the performance of the new discovery. Finally, during the field's productive life, performance data needs to be constantly evaluated to allow revisions to the original assumptions, and develop and modify the field's depletion plan. The last phase is especially critical to optimize drainage of recoverable reserves yielding maximum productivity and revenue.

An excellent review of the history of reservoir engineering is found in *Applied Petroleum Reservoir Engineering*, by Craft and Hawkins (1959). One of the major benefits to the geologist of becoming familiar with basic reservoir engineering is to understand what a reservoir engineer requires for the evaluation of hydrocarbon reserves. Refer to Table 1.1 in Craft and Hawkins (1959), which is a list of the symbols and nomenclature adopted by the Society of Petroleum Engineers for use in reservoir engineering. It is essential for a geologist to become familiar with the basics of petroleum engineering and to work closely with the petroleum engineer to obtain the necessary data to evaluate hydrocarbon reserves.

Reservoir Characterization

After the identification of hydrocarbons in a porous, permeable media, a plan must be established to estimate the in-place volumes of these hydrocarbons and what might be expected to be recovered. Recoverable hydrocarbons are referred to as *hydrocarbon reserves*. Because the productive characteristics of a reservoir or field may not be fully known until it is maturely developed, offset analogous reservoirs need to be considered as models to the development of any new discovery. With a model to follow, various reservoir engineering parameters such as reservoir porosity, permeability, and water saturation can be estimated to allow volumetric analyses of the reservoir or field to estimate potentially recoverable reserves.

Estimation of Reserves

Reservoir bulk volume is calculated from net hydrocarbon isopach maps as discussed in Chapter 10. From detailed log analysis, rock properties can be identified. As an example, interstitial water saturation must be estimated for reserve determinations. Detailed reservoir petrophysics for use in calculating reserves is beyond the scope of this book; however, a presentation of general reservoir engineering may be helpful and is outlined here.

Using the letter symbols G and N, initial in-place volumes of oil and gas can be determined by the following equations.

$$G = (43,560)(\phi)(1 - S_{wi})(B_{gi})(\text{acre-feet}) \tag{11-1}$$

and

$$N = \frac{(7758)(\phi)(1 - S_{wi})(\text{acre-feet})}{B_{oi}} \tag{11-2}$$

where

$$G = \text{original gas in place in cu ft}$$
$$N = \text{original oil in place in barrels}$$
$$\phi = \text{effective porosity, fraction}$$
$$S_{wi} = \text{interstitial water saturation, fraction}$$
$$B = \text{formation volume factor, dimensionless}$$
$$43,560 = \text{cu ft per acre-foot}$$
$$7758 = \text{barrels per acre-foot}$$
$$\text{subscript } o = \text{oil bearing zone}$$
$$\text{subscript } g = \text{gas bearing zone}$$
$$B_{gi} = \text{scf/reservoir cu ft}$$
$$B_{oi} = \text{reservoir barrels/stock tank barrels}$$

Gas Reservoirs. Equation 11-1 is used to estimate original gas in place. There are several unknown factors in the equation that must be determined. The formation volume factor (FVF) is defined as the relationship of gas volumes from surface conditions to reservoir conditions. For gas reservoirs, B_{gi} is expressed in standard cubic feet per cubic foot, SCF/cu ft. The porosity (ϕ) is expressed as a fraction of the bulk volume, and the interstitial water (S_w) is a fraction of the pore volume.

To determine the unit recovery for a gas reservoir, the final reserve volume per acre-foot is determined based on the reservoir drive mechanism. This fractional recovery or *recovery factor* represents the difference between the initial unit-in-place gas and the final or abandonment unit-in-place gas.

$$\text{Recovery Factor } (RF) = \frac{100(G - Ga)}{G}$$

or

$$RF = \frac{100(B_{gi} - B_{ga})}{B_{gi}} \text{ percent} \qquad (11\text{-}3)$$

where

B_{gi} = formation volume factor at initial conditions

B_{ga} = formation volume factor at abandonment conditions

This recovery factor is indicative of depletion drive reservoirs where interstitial water saturation remains unchanged and conversely, gas saturation remains constant. The other end of the spectrum with regard to drive mechanisms is a strong water drive where produced gas is being replaced by encroaching water (there is no appreciable pressure loss and $B_{gi} = B_{ga}$). The recovery factor for a water drive gas reservoir, which is representative of the change in gas and water saturations in the reservoir due to production, is shown in the following equation.

$$RF = \frac{100(1 - S_{wi} - S_{gr})}{(1 - S_{wi})} \text{ percent} \qquad (11\text{-}4)$$

where

S_{wi} = interstitial water saturation, decimal

S_{gr} = residual gas saturation, decimal

Oil Reservoirs. Oil reservoirs are often more difficult to analyze for a number of reasons, including the presence, at times, of both phases of hydrocarbons (oil and gas). If an oil reservoir is found without free gas, the oil is said to be *undersaturated*. An oil reservoir with a free gas cap is indicative of a saturated oil reservoir.

Equation 11-2 is used to volumetrically determine original oil in place. The variables in Eq. 11-2 are very similar to those discussed for Eq. 11-1. If we consider an undersaturated oil reservoir under a strong water drive, the recovery factor is based on the following equation.

$$RF = \frac{100(1 - S_{wi} - S_{or})}{1 - S_{wi}} \text{ percent} \qquad (11\text{-}5)$$

where

$$S_{or} = \text{residual oil saturation, decimal}$$

When there is an initial gas cap, the oil is saturated. In such cases, the reservoir can be produced under drive mechanisms other than water drive. These include dissolved gas, gas cap, or a combination drive. The opposite end of the recovery factor spectrum from a water drive reservoir is that of a dissolved gas drive reservoir.

$$RF = \frac{(1 - S_{wi} - S_{ga})(B_{oi})}{(1 - S_{wi})(B_{oa})} \text{ percent} \tag{11-6}$$

where

$$S_{wi} = \text{interstitial water saturation, decimal}$$

$$S_{ga} = \text{gas saturation at abandonment, decimal}$$

$$B_{oi} = \text{initial formation volume factor}$$

$$B_{oa} = \text{formation volume factor at abandonment}$$

Field Production History

Once reservoir or field production has been established and sufficient quantities of hydrocarbons produced performance data can be used to estimate original and remaining reserves, and to forecast future performance. It is good practice to monitor both production and reservoir pressure data. These points of reference are invaluable in evaluating reservoir heterogeneities and are applicable to material balance equations.

There are various performance curves that can be used to evaluate reserves and forecast future production trends. The most common performance curves are those which plot the production of oil, gas, and water vs time. In many instances, performance evaluation is by far the most accurate method for estimating the original in-place volume of hydrocarbons, estimating recoverable reserves, and forecasting future performance. The following is a partial list of the types of performance curves that can be plotted.

1. Monthly production (oil or gas) vs time
2. Flowing tubing pressure vs time
3. Bottom-hole pressure data vs cumulative production
4. Percent oil cut vs cumulative production
5. Water yield vs cumulative gas production

A detailed discussion on these performance curves is beyond the scope of this book. We again refer you to Craft and Hawkins (1959).

CASE HISTORY

**Geological and Engineering Evaluation
of Texaco Inc. Interests
in Golden Meadow Field
Located in Lafourche Parish, Louisiana
As of January 1, 1982**

SUMMARY OF FIELD STUDY

In 1982 a detailed geologic and engineering field study of the Golden Meadow Field located in Lafourche Parish, Louisiana was undertaken by Texaco, Inc. The purpose of the study was:

1. to determine the remaining reserves effective January 1, 1982, of Texaco's holdings in the field;
2. to evaluate the field for recompletion opportunities, development, and exploratory potential; and
3. to evaluate enhanced recovery opportunities.

The field study was undertaken by an integrated (synergistic) team composed of three geologists, one geophysicist, one reservoir engineer, and two technicians. The study took 15 months to complete and resulted in three separate reports. The first report summarized the entire field study and detailed the evaluation of reserves as of January 1, 1982; the second report presented the evaluation of future development and exploratory potential identified in the study; and the third report covered the methods, procedures, and tasks undertaken during the study. The data presented here are only a summary of the actual field study. In addition to the three reports, the study resulted in the construction of over 300 geological maps and over 30 (3-in.) three-ring binders of geologic and engineering data on 268 studied reservoirs. **The purpose of presenting this case history is to outline the methodology required to undertake a complete, integrated, study of a large mature oil and gas field.**

The reserves shown in this study are gross to Texaco's working interests. As of January 1, 1982, Texaco Inc., had 107 producing wells in the field: 96 oil wells and 11 gas wells. The monthly production was 73,000 barrels of oil and 168 MMcf of gas from 68 separate reservoirs. Table 11-1 summarizes the proved producing reserves, proved reserves behind pipe and hydrocarbon accumulations estimated for the Texaco Inc., interests in Golden Meadow Field.

As part of this study, the future development and exploratory potential of the field was reviewed. *Fifteen development prospects, three development prospects with deeper exploratory potential, and ten exploratory prospects were identified and evaluated. Table 11-2 summarizes the results from this part of the study.*

A preliminary review of the future enhanced oil recovery potential was undertaken during the study. Screening guidelines were prepared and used to select reservoirs as

TABLE 11-1

Summary of the Proved Producing Reserves, Proved Reserves Behind Pipe and Hydrocarbon Accumulations

Classification of reserves	Reserves	
	Oil/cond.-MBbls.	Gas-MMcf
Proved producing	3,146.3	10,872.5
Proved behind pipe	1,774.3	7,146.4
Hydrocarbon accumulations	3,178.0	14,275.0
Total reserves	**8,098.6**	**32,293.9**

Source: Published by permission of Texaco, USA.

TABLE 11-2

Summary of the Identified Development and Exploratory Field Potential

	Reserves	
Classification of reserves	Oil/cond.-MBbls.	Gas-MMcf
Development prospects	2,045.8	11,374.0
Exploratory prospects	3,823.8	30,115.5
Development prospects with deeper exploratory potential	1,269.8	6,868.4
Total reserves	**7,139.4**	**48,357.9**

Source: Published by permission of Texaco, USA.

potential candidates. Through the initial screening process, ten reservoirs were identified which appeared to be prospective for enhanced oil recovery. The estimate of recoverable oil from these ten reservoirs is 15,326,000 barrels.

The total identified potential recoverable hydrocarbons is the summation of reserves from Table 11-1, the exploratory reserves from Table 11-2, and the enhanced recovery potential. The development potential shown in Table 11-2 is accounted for in the reserve numbers summarized in Table 11-1. Therefore, the estimate of Texaco's proved, probable, and possible (risked) reserves and enhanced recovery potential for the Golden Meadow Field was 28,095,600 barrels of oil and condensate and 68,193 MMcf of gas as of January 1, 1982.

INTRODUCTION

In addition to data in Texaco's files, data from log libraries, Dwight's Energydata, Inc., and Adam's and Roundtree Technology Inc. were used. To estimate reserves, all data available were studied, including correspondence, lease and well files, Office of Conservation unitization orders, electric logs, directional surveys, core analyses and descriptions, drill stem tests, performance histories, well test data, and selected seismic profiles. Geologic and engineering work previously conducted by Texaco was not used for the study so that the field study would be completely independent of previous work.

Proved reserves for the study were defined as the estimated quantities of crude oil, condensate, natural gas liquids, and natural gas that geologic and engineering data demonstrated with reasonable certainty to be economically recoverable from known reservoirs under existing economic and operating conditions. Reserves were assigned to some reservoirs where productivity with reasonable certainty was indicated by favorable core analyses or electric log interpretations which were similar to other reservoirs which had produced in the field.

For this study, *proved producing reserves* were defined as the estimated quantities of hydrocarbons to be produced from completion intervals open to production in existing wells, with existing equipment and operating methods at the time of the study. *Proved reserves behind pipe* were defined as the estimated quantities of hydrocarbons behind the casing in existing wells that are expected to be produced by plugback recompletions. Accumulations were assigned to estimated quantities of hydrocarbons that are reasonably expected to be recovered by plugdowns or the drilling of new wells. The accumulations were supported by favorable geologic and engineering data and could be classified as either proved behind pipe or undeveloped reserves.

All reserves were estimated by the volumetric method or by analyses of performance histories using standard industry geologic and engineering procedures. The method or combination of methods used for each reservoir was determined by the stage of development of the reservoir, the availability and quality of the data, and the judgment of the study team.

FIELD DEVELOPMENT AND HISTORY

The Golden Meadow Field is located approximately 45 mi south-southwest of New Orleans, Louisiana, as shown on the field location map in Fig. 11-2. The field was discovered in December, 1938, with the completion of the Texas Company (Texaco, Inc.) LaTerre No. 1. The initial completion in the 8600-ft Sand through perforations from 8500 ft to 8520 ft flowed 1094 barrels of 38.2 API gravity oil per day with a flowing tubing pressure of 2450 psig on a $\frac{3}{8}$-in. choke. **Approximately 130 separate sands and more than 300 reservoirs have proved productive.** The shallowest production has come from a gas well in the 1500-ft Sand, through perforations from 1472 ft to 1482 ft. The deepest production has occurred in the Texaco Dusenbury No. 1, in the 16,800-ft Sand through perforations from 16,802 ft to 16,824 ft. The only penetration of salt is at 15,355 ft in the Gulf Oil Corporation Lafourche Realty No. 1 in Section 21, T19S-R22E.

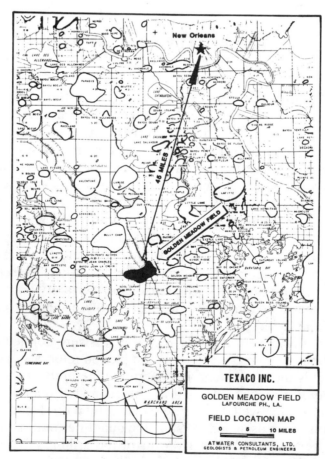

Figure 11-2 Location map for the Golden Meadow Field. (Published by permission of Texaco USA.)

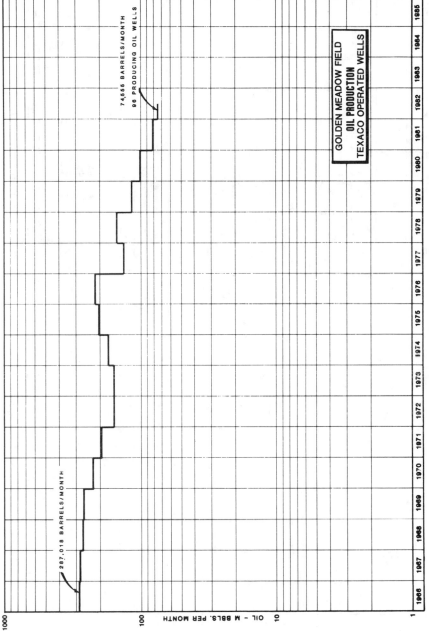

Figure 11-3 Golden Meadow Field—Oil decline curve. (Published by permission of Texaco USA)

569

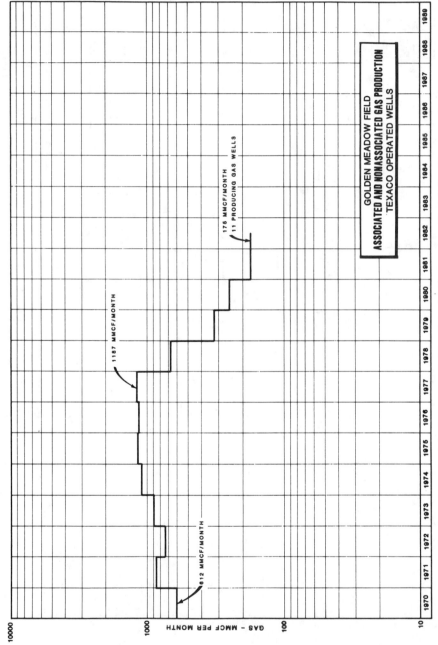

Figure 11-4 Golden Meadow Field—Gas decline curve. (Published by permission of Texaco USA.)

As of January 1, 1982, Texaco had drilled a total of 608 wells in the field. Texaco had 107 wells producing as of January 1, 1982, including 96 oil wells and 11 gas wells. The monthly production rate was 73,000 barrels of oil and 168,000 Mcf of gas from 68 separate reservoirs. The decline curves for Texaco's gross oil and gas production to June 1982 is shown in Figs. 11-3 and 11-4.

FIELD STUDY METHODOLOGY

Objectives

The methodology used in conducting a field study is primarily dictated by the specific results expected from the study. The charge for the Golden Meadow Field Study was centered around three major objectives:

1. determine the remaining Texaco working interest reserves as of January 1, 1982;
2. evaluate future recompletions and development and exploratory drilling potential; and
3. evaluate enhanced recovery opportunities.

Data Collection, Cataloging, and Generation of Data Files

The following is a general list of the data used in the study.

Geologic Data

1. One-in. electric logs, 608 Texaco operated wells
2. One-in. electric logs, 103 outside operated wells
3. Five-in. detailed logs, available for most Texaco operated wells
4. Directional surveys for all deviated wells
5. Core analyses and descriptions
6. Well histories
7. Paleontological reports
8. Seismic velocity survey and selected lines
   Note: Of the total 711 well logs reviewed, 389 logs were of wells deeper than 5000 ft.

Engineering Data

1. Production data—Texaco operated wells
   a. Six years' monthly production on magnetic tapes
   b. Texaco production reports
   c. Texaco gauge books: 1/1/75 through 7/1/81
   d. Production by perforations through 3/1/81
2. Copies of Louisiana Office of Conservation unitization orders for the field
3. Gross and net interests for Texaco leases and units
4. Laboratory analyses of formation water samples
5. Bottom-hole pressure and temperature data
6. PVT analyses of recombined fluid samples
7. Shale resistivity versus depth plots for pore pressure determination

Internally Generated Data.   During cataloging of all furnished data, additional information deemed pertinent or necessary to conduct the study was identified. Additional data files were compiled from the available information. The following is a list of the internally generated files and data.

1. Field Sand Summary File—tabulation of the wells and the productive sands in each well.
2. Well Data Inventory File—tabulation of well data such as name, location, serial numbers, type of logs provided, elevation, log depth, mud weight, etc.
3. Tabulation of cumulative production and test data for each well and perforated interval sorted by lease and well, and by sand and reservoir.
4. Computer generated production data.
   a. Tabulation of production data by months—1976 through 1981
   b. Rate vs time plots
   c. Oil cut vs cumulative production plots
   d. Flowing tubing pressure vs cumulative production plots

Flow of Work—Geologic

Following the collection, cataloging, and sorting of data, the geologic work was undertaken as summarized here.

Base Map Preparation.   A new base map was required for the field study because of location discrepancies and the detailed work to be undertaken. All surface locations of wells were checked and spotted on the base. Wells with a course deviation greater than 50 ft from the surface location were plotted as directional wells. The course for each directionally drilled well was plotted by coordinate position for each 1000 ft of true vertical depth.

The base map was prepared on a scale of 1 in. equals 1000 ft. Exceptions were made for those reservoirs occurring in the high well density areas in the northeast portion of the field. For mapping here, a separate base was prepared at a scale of 1 in. equals 600 ft, in order to provide greater detail and clarity of mapping.

Porosity Top vs Structure Top Mapping.   All reservoirs were isopached using top of porosity maps rather than top of structure maps. In those cases where top of porosity and top of structure were identical, only one map was drawn. In cases where minor variances occurred between top of porosity and structure, only a porosity top map was prepared. If the top of porosity and top of structure were significantly different, both a porosity and structure map were drawn; the structure map to verify continuity of structure and fault patterns, and the porosity map to add accuracy to isopach mapping and volumetrics.

Reservoir Names and Codes.   The field reservoirs were named after the faults that limited the reservoir or the nearest fault. For example, a reservoir named E indicates either that Fault E acts as a reservoir trap or that it is the nearest fault. A reservoir named D/E indicates that both Faults D and E are bordering the reservoir.

Because of the large number of reservoirs, in addition to the reservoir names, each reservoir was given an individual nine-digit code number. The first six digits of the code were determined by the sand in which the reservoir was located. The last three digits were determined by the location of the reservoir in the field.

Map Datum. During the early stages of the study, a decision was made to perform all mapping based on true vertical depth rather than true vertical subsea depth. Reasons for this decision include lack of elevation data on some wells, and discrepancies from different sources of data. Since derrick elevations were rather consistent on most wells throughout the field, the overall effect on mapping was considered minimal. We recommend, however, whenever possible prepare all structure maps on a subsea basis.

Study of Competitor Wells. One hundred and three competitor wells were incorporated into the study. Only 1-in. logs were obtained for most of these wells. Many were old logs which were of rather poor quality. Since low resistivity oil pay is common in this area, accurate interpretation of these logs was difficult, and in some cases, impossible.

Production data for the competitor wells were generally unavailable. If no record of production was found in a sand exhibiting a hydrocarbon show, it was given a show symbol on the map. It is likely, however, that some of these wells have produced. The lack of production data made it difficult to calculate remaining reserves based on volumetrics or performance for these areas.

Deviated Wells. A limited number of intentionally deviated wells were drilled in the field. These were dealt with as outlined in previous chapters in the text. True vertical depths were computed using the directional surveys. True vertical thickness corrections were made where necessary. Any directional well with a surface location and bottom-hole location less than 50 ft apart was treated as a vertical well.

Posting Data On Logs. All completion, production, and test data were posted on 1-in. electric logs. Completion and core data were posted on all detailed 5-in logs. **A total of 1546 completion and test intervals were recorded on the well logs.**

Correlation of Electric Logs. Electric logs with a scale of 1 in. equals 100 ft were correlated to identify fieldwide shale resistivity markers, all productive sands, and all fault cuts. In addition to the sand correlation nomenclature, a numbering system was established for the major producing sands. **A total of 711 1-in. logs were correlated during the course of the study.**

Fault Surface Map Preparation. Fault surface maps were constructed to depict the geometry of the faults from recognized fault cuts in the well logs, as well as seismic data. Customary fault mapping techniques were used with strategically located straight line cross sections providing valuable help in the more complex areas. **A total of 39 primary and secondary normal faults were identified and mapped.**

The size of the fault cuts posted on the fault surface maps represents, in most cases, an average amount of missing or repeated section (vertical separation) as determined by the initial correlation with two or more wells. A fault cut tabulation was prepared, which lists all fault cuts by well and also denotes the fault to which each cut has been assigned. Some of the original fault cut information was revised during the fault and structure mapping.

No attempt was made to evaluate fault contour offset patterns in the immediate vicinity of fault intersections where both faults continue through the zone of intersection. Both intersecting faults were contoured across the intersection without offset of either fault. This technique results in minor errors in fault trace construction on the sand structure maps in the immediate vicinity of the intersecting faults.

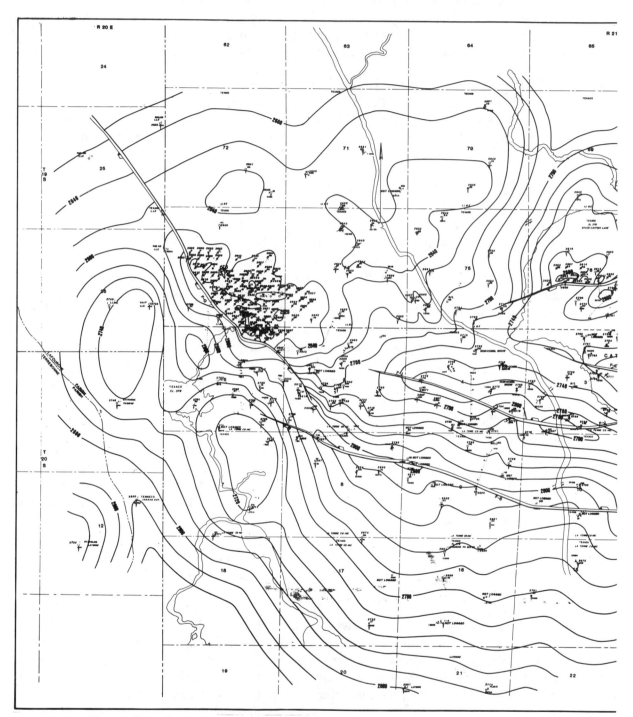

Figure 11-5   Structure map—Top of 2700-ft Sand. (Published by permission of Texaco USA.)

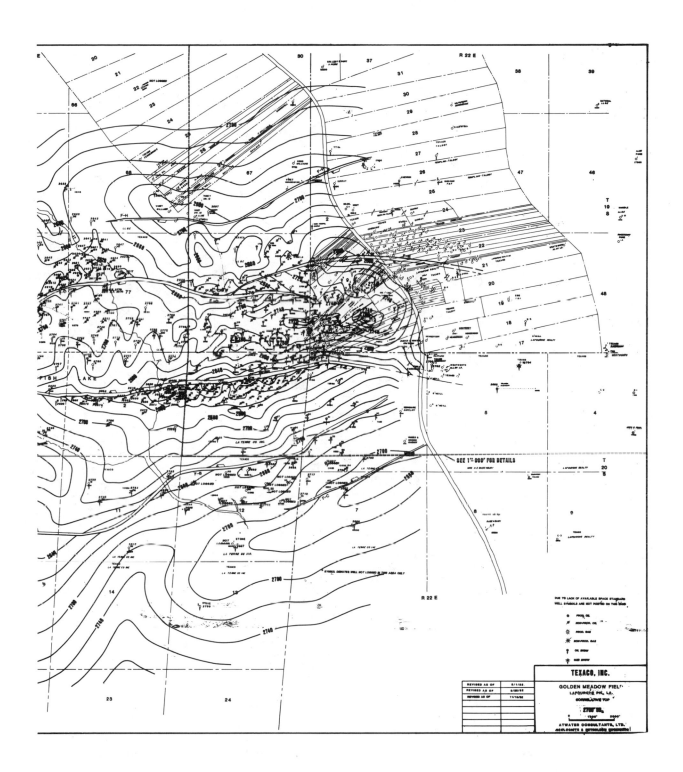

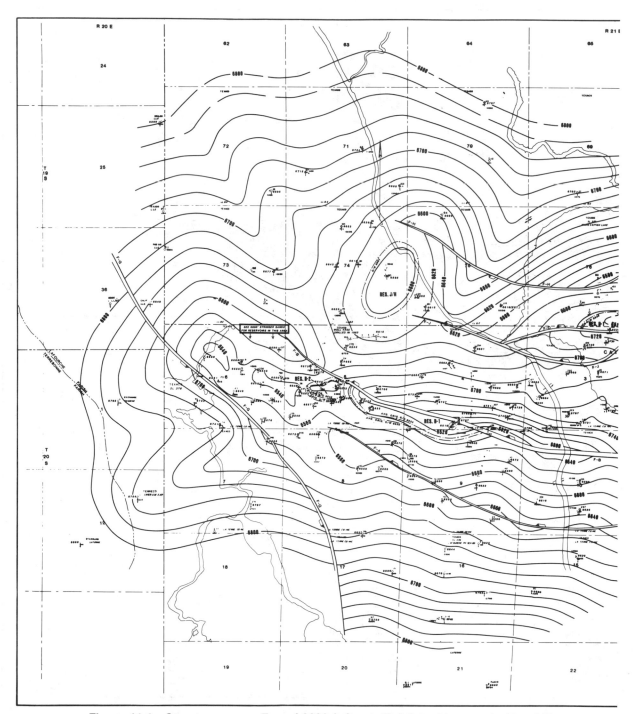

Figure 11-6   Structure map—Top of 6600-ft Sand. (Published by permission of Texaco USA.)

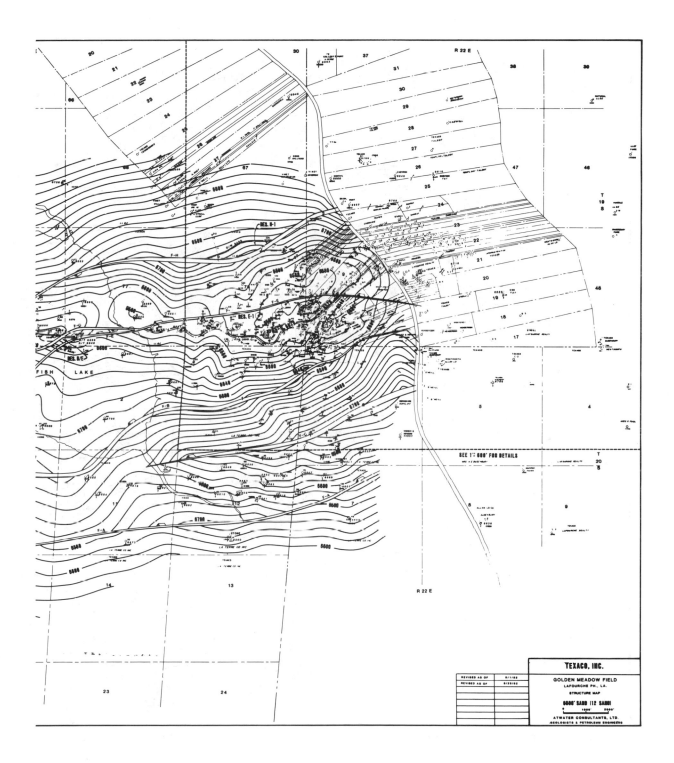

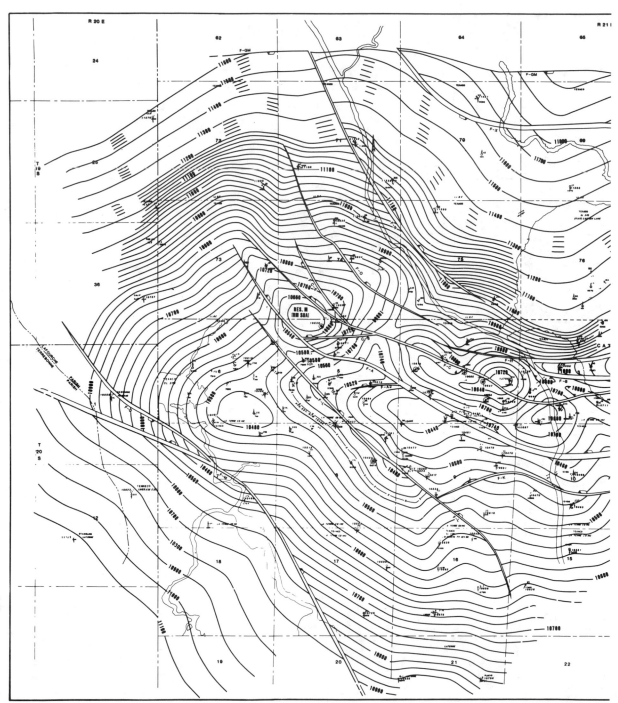

Figure 11-7   Structure map—Top of 10,500-ft Sand. (Published by permission of Texaco USA.)

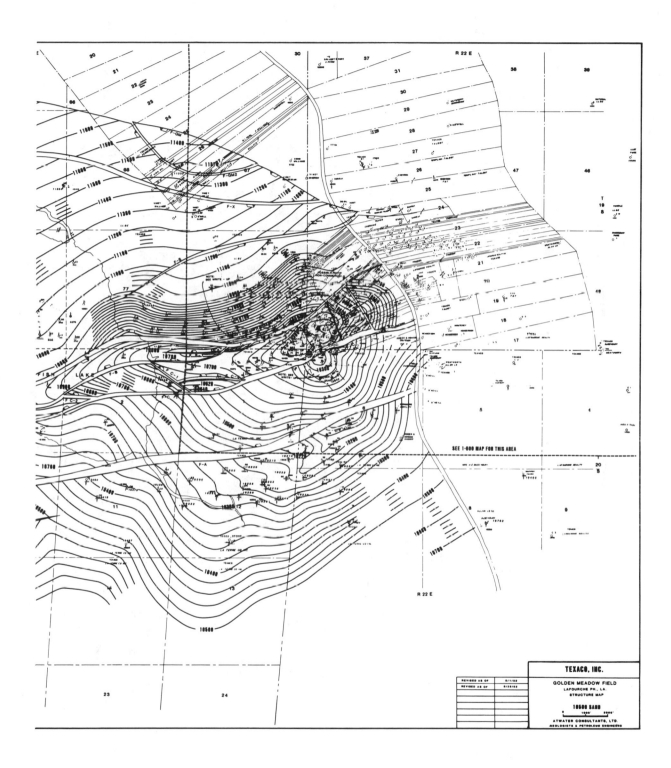

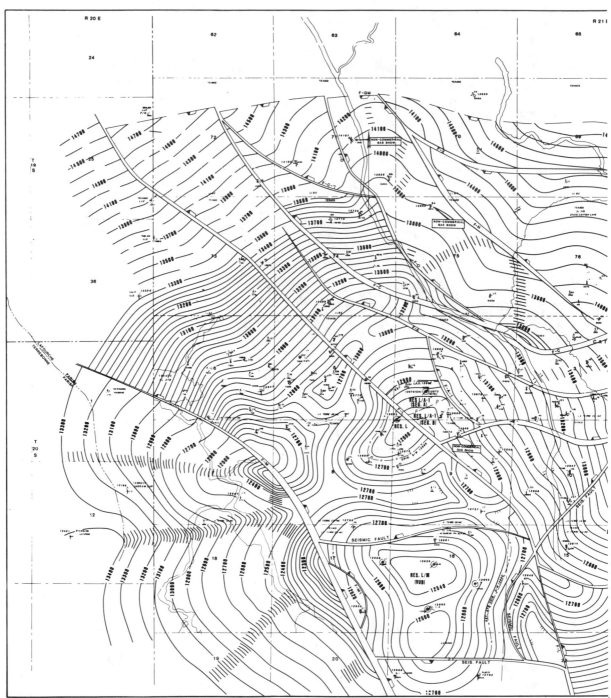

Figure 11-8 Structure map—Top of McGowan (13,000 ft). (Published by permission of Texaco USA.)

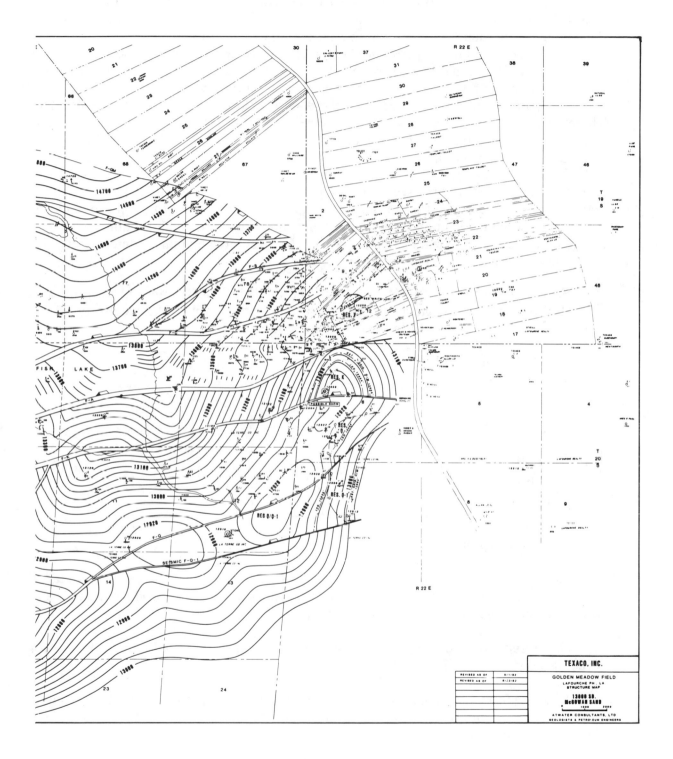

Cross Section Construction.   Straight line problem-solving cross sections were prepared to provide data for resolving and supporting the fault interpretations and sand correlations. All cross sections were constructed using the same horizontal and vertical scales. Various scales were used ranging from 1 in. equals 20 ft to 1 in. equals 400 ft, depending on the depth of wells and the horizontal distance involved.

A total of 20 electric log cross sections were constructed. Nine of these sections are in a general north-south direction approximately perpendicular to the east-west structural axis and the strike direction of the major faults in the primary graben fault system.

The primary purpose of most sections was to help determine the fault geometry. Therefore, the correlation lines included were limited to a sequence of sand tops and shale markers for which the correlation was considered reliable and which were most needed in resolving the fault pattern.

Correlations and faults on most sections were shown as straight lines. Therefore, it must be emphasized that the sections do not depict the true structure or fault geometry between wells. The straight line problem-solving cross section construction proved very helpful in resolving and supporting the fault interpretations, but such sections could not be used for accurate delineation of reservoir limits, structure and fault geometry between wells, or analysis of cross fault drainage.

Sand Data Sheet Preparation.   The sand data sheet is a statistical form used to record all pertinent geologic and engineering data obtained from the electric logs and core analyses. A summary of the data is shown here.

a. correlative sand tops
b. porosity sand tops
c. correlative bases
d. net pay sand
e. fluid contacts
f. resistivity (R_t and R_o)
g. porosity
h. permeability
i. remarks
j. miscellaneous data

Sand data sheets were prepared for each sand or reservoir studied using detailed electric logs and core, test and production data. **Approximately 130 productive sands and sand stringers were identified during the study.**

Dipmeter Evaluation.   All available dipmeters were reviewed and evaluated for structural dip.

Seismic Data.   All available seismic sections were interpreted and tied to well data where possible. Due to the effect of a massive shallow channel fill covering most of the field, the seismic data had limited use.

Preparation of All Appropriate Geological Maps.   Seven fieldwide structure maps at various depths were prepared to assist in the following:

1. evaluation and analysis of the gross structural framework of the field;

2. analysis and resolution of the gross fault interpretation; and

3. construction of structure maps for all productive sands between the sands mapped fieldwide.

The fieldwide structure maps were integrated with the preliminary fault maps and cross sections. The integration process allowed for revisions of fault maps, structure maps, and cross sections, as necessary, to aid in evaluating possible alternative interpretations. Revisions were made to produce the best interpretation with the available data before continuing with additional mapping on a reservoir basis.

For the Golden Meadow Field Study, fieldwide maps were prepared on seven sands, including the 2700-ft, 6000-ft, 8500-ft, 10,000-ft, 10,500-ft, 11,800-ft, and 13,000-ft Sands, four of which are shown in Figs. 11-5 to 11-8. Various types of geological maps were prepared for the evaluation of hydrocarbon accumulations, including structure, porosity top, total net sand, net gas and net oil isopach maps, and special maps as required. Where net oil or gas isopach maps were prepared, the Wharton Method outlined in Chapter 10 was used. The maps were constructed on a scale of 1 in. equals 1000 ft, except for the northeast portion of the field, where a scale of 1 in. equals 600 ft was used because of the well density.

Flow of Work—Engineering

This section summarizes the engineering effort performed concurrently with the geologic activity.

Processing of Production Data

1. Computer software was used to generate standardized production plots and tabular data from magnetic tapes. These data were used to analyze the performance of individual completion intervals over the last five years and provided the basis for the estimation of reserves from performance data.

2. Production data by perforations were stored on magnetic discs for use with word processing equipment to generate various tabulations. The word processor was used to sort this material by leases and wells in order of descending depth in the stratigraphic section, and also by sand and reservoir.

VOLUMETRIC ANALYSIS—GENERAL VOLUMETRIC PARAMETERS

Analysis of Subsurface Pressures. A comprehensive study of formation pore pressures was made of the entire Golden Meadow Area, from the surface through the deepest penetration. The details of the subsurface pressure analysis go well beyond the scope of this text; only a brief summary of the analysis is presented here.

Hydropressured (Normal) Section: The hydropressured section extends from the surface down to approximately 11,300 ft. All available initial bottom-hole pressure measurements were plotted vs depth. Analysis of the plotted data led to the conclusion that a pressure gradient of 0.455 psi per foot of depth best fit the data and was used for predicting initial pressures where actual measurements were unavailable. This gradient is slightly different from the gradient of 0.465 considered average for the Texas-Louisiana Gulf Coast. The plot of pressure vs depth used to determine initial reservoir pressures in the normal or hydropressured stratigraphic section is shown in Fig. 11-9.

Geopressured (Abnormal) Section: There is an obvious break in the normal pressure trend below 11,300 ft which indicates entry into the geopressured (abnormal pressured) section. A comprehensive study relating to the occurrence and magnitude of geopressures in the field was conducted. Abnormal pressures are encountered in and below a 200–300-ft shale section which occurs between the 11,200-ft Sand and the 11,600-ft Sand. This shale section provides the "geopressure seal" necessary for the creation of abnormally high formation pressures.

To define pore pressures in the geopressured section, a method involving the analysis of shale resistivity readings from electric logs was used, since this was the tool providing the most complete coverage in the area. A total of 88 logs from wells drilled deeper than 11,000 ft were examined. Shale resistivity values were recorded from 6000 ft to total depth for each well, and the data were computer processed.

The resulting pore pressure versus depth profile was adjusted to incorporate measured bottom-hole pressure data, pressures calculated from surface readings, and mud kick data from wells. The resistivity values plotted on semi-log paper with R_{sh} on the logarithmic horizontal axis and depth on the linear vertical axis are shown in Fig. 11-10.

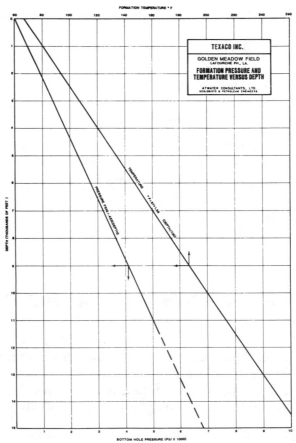

Figure 11-9   Plot of temperature and pressure vs depth for the Golden Meadow Field. (Published by permission of Texaco USA.)

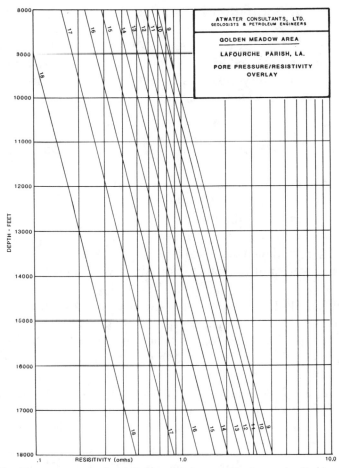

Figure 11-10 Pore pressure/resistivity overlay plotted vs depth for the Golden Meadow Field. (Published by permission of Texaco USA.)

Formation Temperature Analysis

The formation temperature values obtained from bottom-hole pressure measurements define the geothermal gradient for the Golden Meadow Field. The temperatures were plotted vs depth, and an equation for estimating temperatures was developed based on a least squares curve fit analysis. From these data, a curve of the form $Y = A + B(X)$, which is the equation for a straight line, provided a curve fit with the highest correlation coefficient (0.9831) for the four types of curves examined. The value of the constant A was 66.9, which represents the temperature at $X = 0$ (surface) or the ambient temperature. The value of the constant B was 0.0133, which is the slope of the line in degrees per foot. The plot of temperature vs depth is shown in Fig. 11-9.

Reservoir temperatures were calculated from the following equation:

$$T = 67° + [(1.33)\text{Datum}/100]$$

where

T = temperature in degrees Fahrenheit

Datum = reservoir datum in feet

Determination of Formation Properties

Porosity. Porosities were obtained by averaging the values reported from analyses of sidewall core samples where available. No values were included in the average where the reported permeability was less than 10 millidarcies. In the absence of sidewall core analyses, the porosities were determined as a function of depth from the following equation:

$$\text{Porosity}(\phi) = .41 - [(0.01285)\text{Datum}/1000]$$

This correlation is based on a plot of porosity vs depth (Atwater and Miller, 1965) for South Louisiana miocene oil and gas fields. This calculation was made for comparison even when sidewall core data were available.

The porosity equation was not used to estimate porosity in the geopressured section, since sidewall core data were available for most of the reservoirs. In a few cases where sidewall cores were not available, porosities were determined by analogy.

Interstitial Water Saturation. Connate or interstitial water saturations were obtained from analyses of electric log resistivity curves. Values of true resistivity (R_t) were determined from the electric logs for each interval of the hydrocarbon-bearing zones and related to the net feet of pay for each comparable segment within the interval. The resistivity of the water-bearing sands (R_o) was obtained either from a representative wet section below the hydrocarbon contact or the closest clean water sand. The water saturation for each section of net pay was calculated from the following equation:

$$\text{Water Saturation } (S_{wi}) = \left[\frac{R_o}{R_t}\right]^{\frac{1}{1.73}}$$

where

$$R_o = \text{formation resistivity with 100\% water}$$

$$R_t = \text{true formation resistivity}$$

The average interstitial water saturation for each reservoir was obtained by multiplying the net feet of pay by the interstitial water for each section and then dividing the summation of these results by the total footage of the sections included.

Determination of Oil Zone Parameters

Gas-Oil Ratios. The initial gas oil ratio for each reservoir was obtained by averaging the results from well test data reported early in the producing life of the reservoir. If this information was unavailable or not considered representative, the initial solution gas-oil ratio was calculated from equations developed by Standing (1977). The determination of solution gas-oil ratio in this manner assumes that the oil is at or near saturation pressure and requires input on initial reservoir pressure, formation temperature, API gravity, and separator gas gravity.

The average gas-oil ratio used to predict future solution gas production or reserves for a reservoir may not necessarily be the same as the initial gas-oil ratio used to determine initial formation volume factors. The average gas oil ratio was selected after consideration of such factors as averaged produced ratio from production data, anticipated pressure behavior of the reservoir, the possibility of gas-cap production, and any other pertinent factors which could influence future gas production from reservoir completions.

Specific Gravity of Oil. The specific gravity of produced oil for each reservoir expressed in A.P.I. degrees (API) was obtained from an average of reported test data for the completions in the reservoir. Data from Reservoir Fluid Studies were also used.

Specific Gravity of Separator Gas. The specific gravity of the separator gas was based on average data reported in Reservoir Fluid Studies, and the value 0.62 (AIR = 1.0) was used throughout this study.

Viscosity of Oil. The viscosity of oil for each reservoir was determined from the correlation of Beal (1946) and from those of Chew and Connally (1958). These correlations were programmed for a handheld calculator and required the input of separator pressure and temperature, oil gravity, separator gas gravity, gas-oil ratio, and reservoir temperature and saturation pressure.

Viscosity of Salt Water. The viscosity of formation salt water was obtained from a correlation of viscosity vs depth developed from actual water sample analyses obtained from completions in this field. Using the correlations for viscosity vs percent **NACL** at various temperatures, the following equation was developed to provide this information.

$$u_w = 3.582 - 0.342 ln \text{ (datum)}$$

where

$$u_w = \text{viscosity of salt water at reservoir conditions}$$

A graphical representation of the correlation is shown in Fig. 11-11.

Formation Volume Factor. The oil formation volume factors were obtained from equations published in Standing Correlations, which were programmed for a handheld calculator. The initial gas-oil ratio (or initial reservoir pressure), formation temperature, API gravity of oil, and specific gravity of separator gas were input to determine the formation volume factor.

Oil Recovery Factor. The actual performance analyses of 36 depleted or nearly depleted oil reservoirs in the field indicated average recoveries by 1000-ft depth intervals ranging from 31% to 51% of original oil-in-place. The determination of the oil recovery factor for use in the volumetric calculation for each reservoir was obtained using the correlations presented in API Bulletin D-14 with the application of a discount factor in an effort to approach the level of recovery factors determined from experience. The discount factor was determined considering such points as departure from optimum reservoir geometry, well location, sand quality, and operating and economic conditions. The discount factors, referred to in the calculations as *"sweep efficiency,"* ranged from 0.5 to 1.0 and were assigned after consideration of all available data.

As previously noted, 36 depleted or near depleted oil reservoirs in the field were reviewed to develop recovery data based on actual performance. The results of these

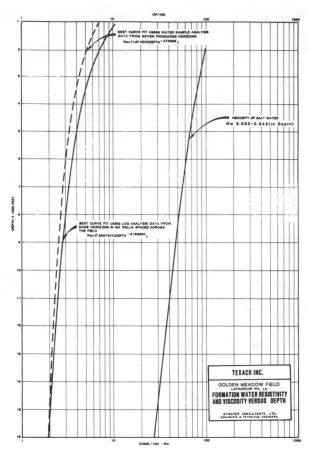

Figure 11-11 Formation water resistivity and viscosity vs depth for Golden Meadow Field. (Published by permission of Texaco USA.)

analyses are summarized in Table 11-3. Only those reservoirs for which reservoir limits and production data were reasonably well defined were included in the review. The estimated ultimate recovery from these 36 reservoirs represents approximately one-third of the estimated ultimate recovery for the field. The average for the hydropressured reservoirs was 44%.

Determination of Gas Zone Parameters

Condensate Yields. Condensate yields are expressed as barrels of separator liquid per million cubic feet of separator gas. The initial condensate yield for each gas reservoir was obtained by averaging results of well test data reported early in the producing life of the reservoir. The average condensate yield used may be different than the initial yield used to determine the initial gas deviation factor (Z factor). Condensate yields were determined after consideration of such factors as average produced yield from production data, anticipated pressure behavior of the reservoir, possibility of oil rim production, and any other pertinent factors which could influence future liquid hydrocarbon yields.

Specific Gravity of Condensate. The specific gravity of condensate expressed in API degrees was obtained from the average of reported test data for the gas completions in each reservoir.

TABLE 11-3

Average Recovery Factors Determined for 36 Depleted or Nearly Depleted Reservoirs from 2000 ft to 14,000 ft.

Depth range	No. reservoirs	Average recovery % of in place	Est. ultimate recovery MB
2,000–3,000	2	40	7,648
3,000–4,000	1	30	48
4,000–5,000	2	34	322
	5	40	8,014
5,000–6,000	4	44	1,754
6,000–7,000	3	47	712
7,000–8,000	4	51	4,577
8,000–9,000	5	50	1,045
	16	49	8,088
9,000–10,000	—	—	—
10,000–11,000	8	43	5,404
11,000–11,300	1	41	13
	9	43	5,417
11,300–14,000	6	49	13,954
Total	**36**	**46**	**35,474**

Source: Published by permission of Texaco, USA.

Gas Deviation Factor (Z) and Shrinkage Factor. The following equations and procedures were used to determine the gas deviation factors for associated and nonassociated gas reservoirs.

$$G_c = \frac{141.5}{°API + 131.5}$$

$$V_c = (2938.81)(1.03 - G_c)$$

$$G_m = \frac{G_s + 4493G_c}{1 + V_c/R_c}$$

$$SF = \frac{R_c}{R_c + V_c}$$

where

G_m = specific gravity of full well stream

G_s = specific gravity of separator gas

G_c = specific gravity of condensate

R_c = gas condensate ratio (ft^3/Bbl)

V_c = condensate vaporizing volume (ft^3/Bbl)

SF = shrinkage factor (fraction of dry gas in full well stream)

The calculations used in the above equations were used in gas calculations at the Louisiana statutory pressure base of 15.025 psia.

After calculation of the full well stream gas gravity, the critical pressure (P_c) and the critical temperature (T_c) were obtained from a well-known and widely used plot of these critical constants vs full well stream gas gravity (Brown et al. 1948). The pseudoreduced pressures (P_r) were calculated by dividing the reservoir pressure by the critical pressure. The pseudoreduced temperatures (t_r) were calculated by dividing the reservoir temperature by the critical temperature. The pseudoreduced pressure and temperature were used to determine the appropriate gas deviation factor (Z factor or compressibility factor) for each gas reservoir from a well-known and widely used correlation of compressibility vs pseudoreduced pressure for various pseudoreduced temperatures (Standing and Katz 1942). Refer to Craft and Hawkins (1959) for detailed explanation of the gas deviation factor and its application in reserve determinations.

Gas Recovery Factor.   Equation (11-4) was used to determine gas recovery factors for the hydropressured section.

$$RF = \frac{100(1 - S_{wi} - S_{gr})}{1 - S_{wi}} \qquad (11\text{-}4)$$

S_{gr} was obtained from a correlation of Residual Gas Saturation vs Porosity (Katz et al. 1966; McKay 1974).

For the geopressured section, Eq. (11-4) yielded recovery factors that were usually too optimistic. These recovery factors were reduced to approach recoveries indicated by analysis of performance data.

For strong water drive reservoirs, sand heterogeneity, reservoir geometry, well locations that were less than optimum, and the consideration of the perforation interval (i.e., perforations close to a gas/water contact), were taken into account in applying a discount factor to the recovery factors. This discount factor was designed to account for the coning effect of water or the watering out of a producing well before normal recovery of recoverable gas.

PERFORMANCE ANALYSIS

Rate vs Time Analysis

Monthly production vs time plots were generated by computer from production data stored on magnetic tapes. The data provided production by month for the years 1976 through 1981 and cumulative totals for each completion interval. Plots were generated on semi-log paper for all completion intervals which reported production during this period. Intervals producing as of January 1, 1982, were analyzed and the data extrapolated to determine remaining reserves using the basic equation and procedures for exponential decline extrapolation.

$$Q = \frac{(P_1 - P_{el})(12)}{ln(1 - d)} \qquad (11\text{-}7)$$

$$T = \frac{ln(P_l)}{ln(1 - d)} \qquad (11\text{-}8)$$

where

Q = remaining reserves (Bbls or MMcf)

P_1 = initial producing rate per month with the effective data January 1, 1982

P_{el} = final producing rate per month

d = slope of decline curve or percent decline per year

$$\text{percent decline} = \frac{P_1 - P_2}{P_1}$$

T = time in years to produce Q

The same equations and procedures were used to estimate both oil and gas reserves. In this study, P_{el} was not an economic limit, but a mechanical or terminal limit. In the case of oil, a 100-Bbls per month limit was used based on observations of actual performance histories which indicated that production was seldom sustained below this level. Oil reserves for reported production at rates below 100 Bbls per month were usually assigned for one additional year at the current rate. Figure 11-12 is an example of a plot of rate vs time from a reservoir in the 11,800-ft Sand. In the case of gas, a value of 3000 Mcf per month was used for the P_{el} based on similar observations. In fields such as

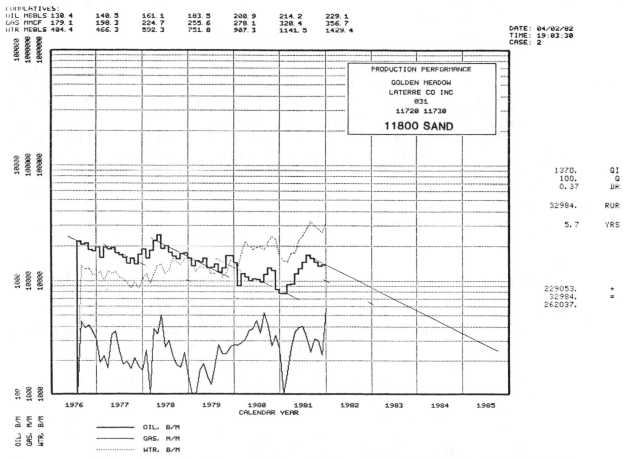

Figure 11-12 Monthly production rate (oil, gas, water) vs time. (Published by permission of Texaco USA.)

Golden Meadow with large multi-well leases, the calculation and use of true economic limits for individual wells is considered unrealistic.

Percent Oil vs Cumulative Production.   The percent oil vs cumulative production plots were also generated by computer from the production data provided. In some cases, the data permitted extrapolation to ultimate recoveries which were used to support the reserves based on the plot of rate vs time. In most instances, the percent oil production declined rapidly and erratically to a level where artificial lifting was required and then leveled off at low rates which could not be extrapolated with any degree of reliability.

Bottom-hole Pressure vs Cumulative Production.   Initial bottom-hole pressure records were available for a number of completions. The pressure data, where two or more pressures were available for the same reservoir taken at different times, indicated excellent pressure maintenance. From these observations and other related data, it was generally concluded that most of the reservoirs in the hydropressured section were producing under the influence of strong water drives. Where the data indicated otherwise, it was used to estimate abandonment pressures.

GENERAL OVERVIEW OF GOLDEN MEADOW GEOLOGY

Structural Geology

The Golden Meadow Field is on an east-west trending major anticlinal complex with an associated crestal graben system and a deep-seated salt ridge. Salt was encountered at 15,355 ft in the Gulf Oil Corporation, Lafourche Realty No. 1 well on the eastern flank of the field in Section 21, T19S-R22E. The east-west trending anticline plunges to the southwest and gradually migrates southward with depth. It becomes more sharply defined at depths of 5000 ft to 8000 ft than at shallower depths. The major regional south dipping "Golden Meadow Fault" north of the field also trends east-west. The anticline originates with a deep-seated salt ridge and is amplified by rollover caused by the Golden Meadow Fault.

Approximately 130 sands have been produced by Texaco operated wells from the 1500-ft Sand of Plio-Pleistocene age to the 16,800-ft Sand of Middle Miocene age. Four major types of trapping mechanisms are responsible for the significant hydrocarbon accumulations encountered in the field: (1) low relief anticlinal highs in the western part of the field and those associated with the east-west trending ridge; (2) fault closures in downthrown blocks or the graben area; (3) fault closures in upthrown fault blocks; and (4) stratigraphic traps resulting from lateral changes in depositional environments. Most of the faults appear to be sealing with very few occurrences of drainage across faults.

The seven fieldwide structure maps show the structural geology of the field with depth and the significance of the crestal graben system. Most of the field production is associated with the crestal graben system.

The primary fault pattern of the field is that of a compensating (antithetic) fault system with central graben blocks in the crestal area, and normal faults on both the north and south flank positions. Figure 11-13 (Faults A, S, and X) and Fig. 11-14 (Faults H, K, L, and Q) show examples of the faults mapped in the field. As would be expected, all faults are normal resulting from tensional forces. The primary faults are those that have a general east-west strike and are associated with the antithetic fault system. Dips of the faults vary from approximately 40 deg to 70 deg. The maximum observed fault displacements (vertical separation) range from 350 ft to 650 ft. The secondary faults are those with other than east-west strike, such as Fault L in Fig. 11-14 and Faults G and O in Fig. 11-15. The dips of these faults are similar to those of the primary faults, but displacements rarely exceed 300 ft.

Thirty-nine primary and secondary faults were identified. There are seven primary down-to-the-south faults, one of which is the Golden Meadow Fault, which served as the northern limit of the field study area. Eleven primary down-to-the-north faults were also identified. Several of the primary faults bifurcate within the field area.

At least two ages of fault activity are indicated. The first is related to the main graben system and appears to have been initiated relatively early and continues active through very shallow depths. The second relates to the cross faults and the deep-seated faults which also appear to have originated early, but die out upward at depths of 11,000 ft to 9000 ft (e.g., Faults K, L, and Q shown in Fig. 11-14 and Fault O shown in Fig. 11-15). The fault interpretations for the northernmost faults such as Faults S and X are speculative at best. Correlation problems and sparse well control place several limitations on the reliability of the fault contour maps for these faults.

Most of the east-west striking faults related to the graben system exhibit growth with depth. The upward extent of these faults is documented to depths as shallow as 1000 ft to 2000 ft. The compensating faults terminate downward against opposite-dipping faults at depths between 5000 ft and 12,000 ft. Those faults which carry through the vertical section without terminating against another fault continue downward through the deepest well control to depths of 12,000 ft to 15,000 ft. Lateral changes in vertical separation are present in numerous faults; some faults become smaller or die to the east, and others show similar decrease toward the west or in both directions.

Stratigraphy of the Field

The stratigraphy of the Golden Meadow Field is that of a predominant prograding deltaic system of alternating sands and shales resulting from cyclic trangressive and regressive depositional episodes. As a result of this progradational system, the vertical depositional sequence reflects several diachronic facies (i.e., alluvial, upper delta plain, lower delta plain, and active delta front) which can be identified from well logs and are supported in general by paleontological data (see composite stratigraphic type log—Fig. 11-16).

The recognition of the overall depositional system and the interpretation of the related depositional environments are important to the understanding and economical exploitation of stratigraphically trapped hydrocarbon accumulations. Numerous stratigraphically trapped hydrocarbon accumulations have produced in the field and additional stratigraphically trapped hydrocarbon accumulations were identified in this study.

The stratigraphic section was divided into depth zones, generally corresponding to the different deltaic subenvironments, as shown on the stratigraphic type log. These subdivisions provide a convenient way of studying and discussing the field geology in greater detail. The subdivisions, although not arbitrary, are interpretive and based on electric log characteristics and limited paleontological data.

Stratigraphic Zone from the Surface to 5000 Feet. Depositionally, this zone appears to be in an upper delta plain. This zone is characterized primarily by thick blocky bar sands (often referred to in the literature as the massive sandstones), with occasional thin sand and shale sequences between the blocky sands, which may represent abandoned

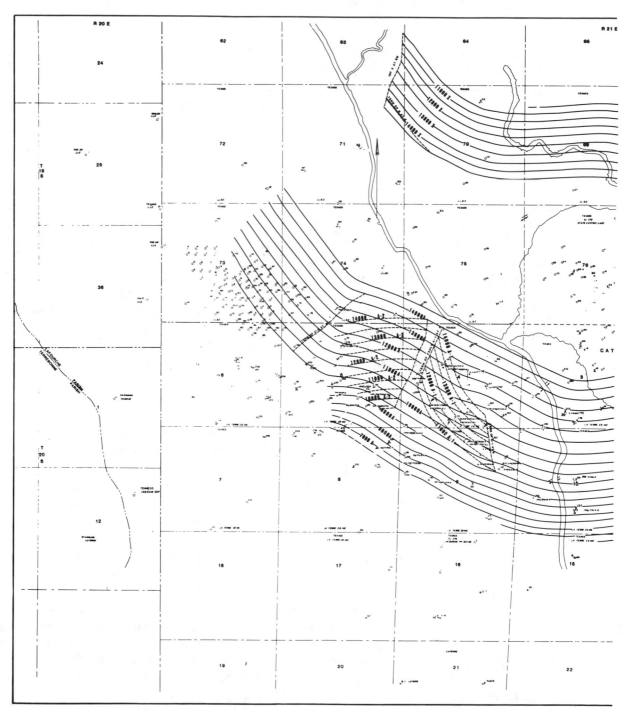

Figure 11-13 Fault surface map for Faults A, S, and X. (Published by permission of Texaco USA.)

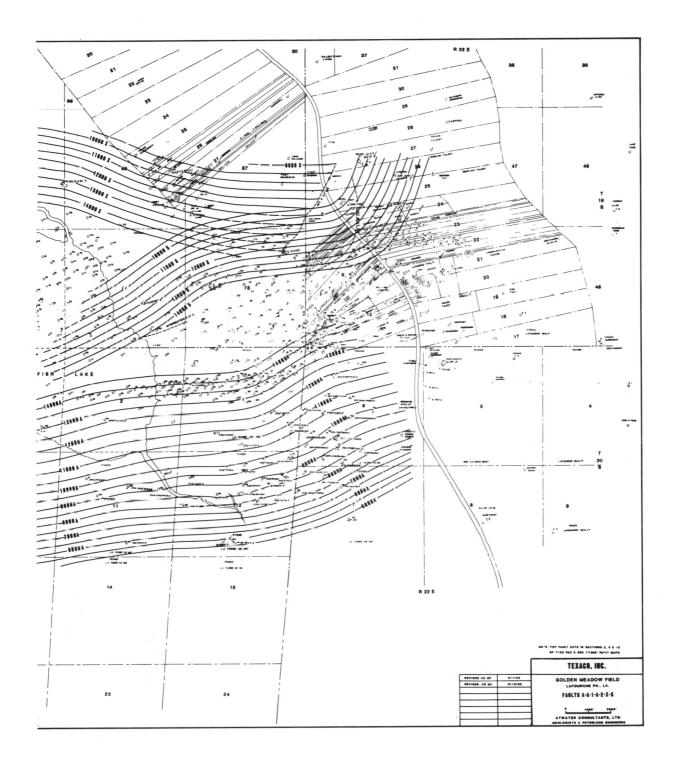

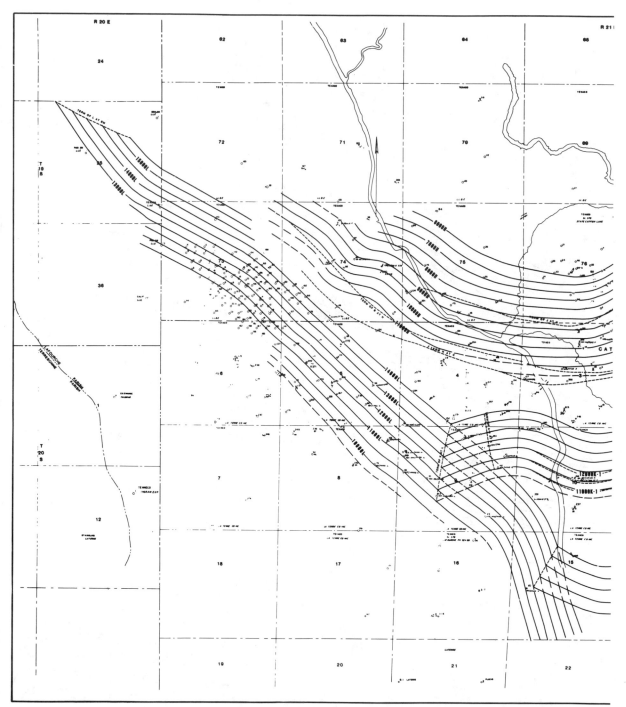

Figure 11-14   Fault surface map for Faults H, K, L, and Q. (Published by permission of Texaco USA.)

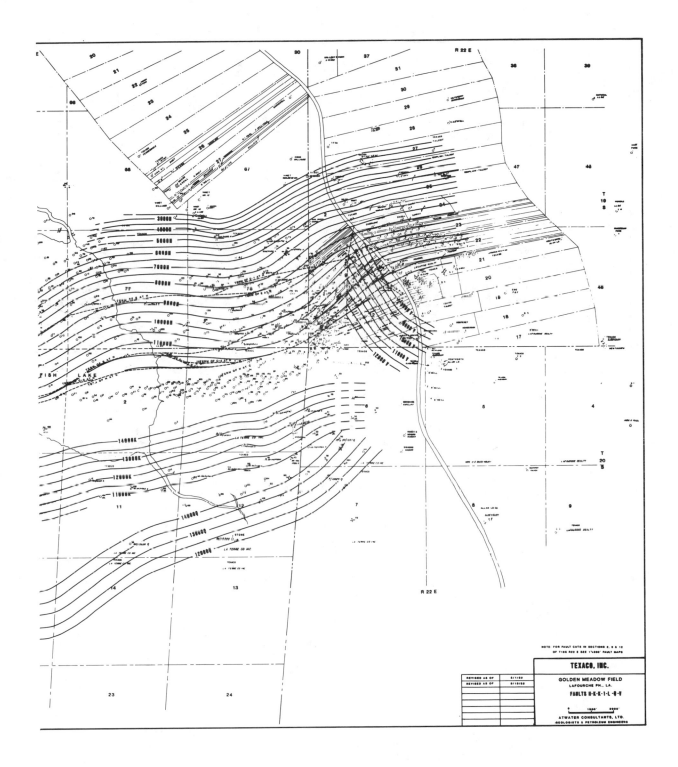

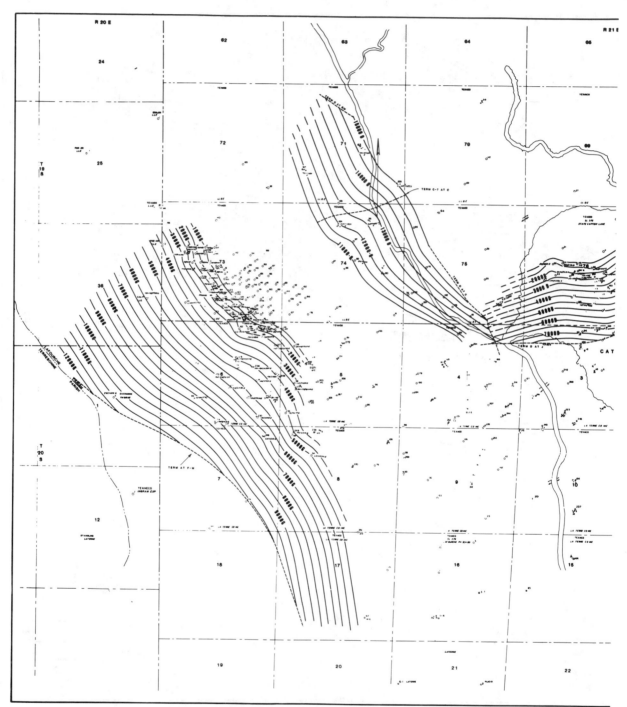

Figure 11-15   Fault surface map shows trend of secondary faults which do not strike east-west. (Published by permission of Texaco USA.)

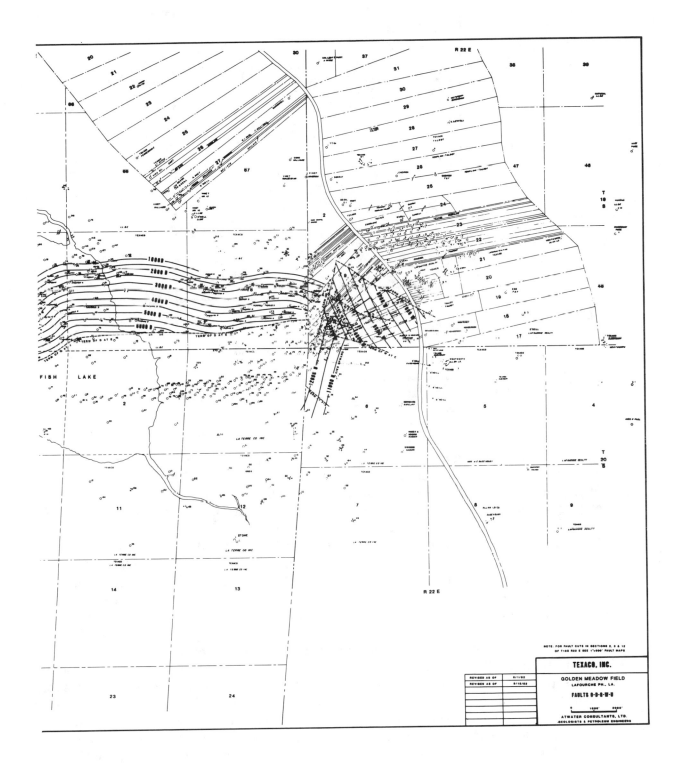

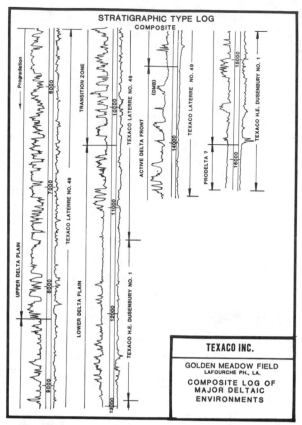

Figure 11-16 Composite stratigraphic type log for the Golden Meadow Field. (Published by permission of Texaco USA.)

channel fill and small crevasse splay deposits. The splay systems are rather small with thin discontinuous sand members, often uncorrelatable from well to well. Without extremely detailed sand analyses, the identification of the specific depositional environment relating to these localized sand bodies is interpretive because the log patterns for several subenvironments are often similar.

The most prolific sand in this section is the 2700-ft Sand, which has produced in excess of 20 million barrels of oil (Fig. 11-5). Numerous localized stratigraphically trapped hydrocarbons have produced from this zone, with the most significant area in Sections 77 and 78, T19S-R21E near Catfish Lake. The crestal graben system has been the most prolific producing area of the field in this zone.

Stratigraphic Zone from 5000 Feet to 8000 Feet.   Depositionally, this zone appears to be in the lower portion of the upper delta plain, characterized by thick blocky sands separated by thin shale sequences. In the lower portion of the zone, the shale layers become thicker and more consistent between the sand bodies. Occasionally, the massive sands are separated by shales and thin discontinuous sand bodies possibly associated with abandoned channel fill, backswamp deposits, overbank, and small crevasse splays.

At these depths, the field is still dominated by the crestal ridge system striking east-west across the field. This area is again the most prolific hydrocarbon-producing area in the field. Additional reservoirs are trapped by flanking faults, and a limited number result from small, shallow anticlines in the western portion of the field.

The sands in this zone show relatively consistent development. Though stringer sands are common in the upper portion of the zone, the majority of the production has come from well-developed sands which can be correlated fieldwide. Faulting is the most common trapping mechanism, although a limited number of reservoirs are partially defined by shale-outs or permeability barriers. Eight sands have produced more than one million barrels of oil each from this zone, with the 7600-ft Sand producing in excess of 12,500,000 barrels of oil.

Stratigraphic Zone from 8000 Feet to 13,000 Feet. Depositionally, the upper portion of this zone is a transition zone between the upper delta plain and the lower delta plain. The lower portion is representative of a lower delta plain environment. Well logs show thicker shale sections with the sandstones becoming thinner and more isolated, possibly indicating distributary channels that were fairly straight during deposition, producing fewer sequences of point bars. Point bars, distributary channels, and large bay fills are the primary depositional sand types in this stratigraphic zone. Most of the sand bodies appear continuous over laterally larger areas, and therefore fieldwide correlations are usually good.

Production from this zone has been substantial, with twelve individual sands having produced over one million barrels of oil each. As might be expected, reservoirs in this zone are generally larger than those in the shallower depths, and accumulations are most often defined by structural rather than stratigraphic features. The 11,800-ft Sand is the most prolific oil producer, with production in excess of eight million barrels of oil. The 13,000-ft Sand (McGowan Sand) is the most prolific gas producer, with production in excess of 81,000 MMcf (Fig. 11-8).

Below 13,000 Feet. Below 13,000 feet there is very limited well control across the field. Based on the available data, however, most of the zone appears to be representative of the active delta front environment progressing into deeper water sediments with depth. The primary depositional sand bodies are distributary mouth bars and bay fills.

Sand correlations in the upper portion of this zone are relatively good. The deep sands are difficult, if not impossible, however, to correlate fieldwide, and in some cases even in localized areas. The deepest sands encountered appear representative of deep water sediments. To date, the production from these deep sands has not been significant.

RESERVES

Reserves identified and included in this study are subdivided into two major categories: (1) proved developed primary reserves, and (2) hydrocarbon accumulations.

Proved developed primary reserves are further subdivided into 1-A proved producing reserves that can be expected to be produced from current perforations, and 1-B reserves behind casing in existing wells for which reserves have been reasonably proved and are expected to be produced in the predictable future as wells are worked up the hole (plugbacks) after lower completions are abandoned.

A hydrocarbon accumulation is the classification for proved undeveloped primary reserves which require significant investment for recovery. For the most part, these reserves have been identified as the result of analysis of produced reservoirs. They are subdivided into categories 2-A, accumulations which will require the drilling of a new well usually within offset distance of existing wells; and category 2-B, which differs from 2-A only in that a plugdown or deepening is required in a well not usually more than one location removed from an existing well.

TABLE 11-4

Cumulative Production, Estimated Reserves and Ultimate Recovery for Some of the Reservoirs in Golden Meadow Field

| Sand | Reservoir | Status | Cumulative production | | Proved developed reserves | | | | Accumulations | | | | Ultimate recovery | |
| | | | | | Producing 1-A | | Behind-pipe 1-B | | Attic | | Other | | | |
			Oil/cond. MBbls	Gas MMcf	Oil/cond. MBbls	Gas MMcf	Oil/cond. MBbls	Gas MMcf	Oil/cond. MBbls	Gas MMcf	Oil/cond. MBbls	Gas MMcf	Oil/cond. MBbls	Gas MMcf
5500L	D	P	108.8	105.6	2.2	1.1	—	—	—	—	75.0	38.3	186.0	145.0
5500	E	NP/D	61.3	20.8	—	—	—	—	—	—	—	—	61.3	20.8
5500	R	NP	41.5	9.6	—	—	56.0	34.3	—	—	—	—	97.5	43.9
5550	D	NP	16.7	17.7	—	—	—	—	16.0	16.3	—	—	32.7	34.0
5600	E	NP/D	73.5	35.8	—	—	—	—	—	—	—	—	73.5	35.8
5600	R	NP	1,118.0	1,098.9	—	—	—	—	—	—	90.0	55.0	1,208.0	1,153.9
5600S	R	NP/D	61.7	91.2	—	—	—	—	—	—	—	—	61.7	91.2
5908	E	NP	30.8	11.1	—	—	—	—	—	—	17.2	5.3	48.0	16.4
5950	E	NP/D	8.8	8.9	—	—	—	—	—	—	—	—	8.8	8.9
6000	E	NP	—	—	—	—	—	—	—	—	1.1	187.1	1.1	187.1
6000	E-1	NP/D	144.8	55.9	—	—	—	—	—	—	—	—	144.8	55.9
6000L	E	NP/D	13.1	8.9	—	—	—	—	—	—	—	—	13.1	8.9
6100	E	NP	52.6	77.4	—	—	—	—	—	—	—	—	52.6	77.4
6150	E	NP	792.1	624.6	—	—	—	—	—	—	—	—	792.1	624.6
6200	E-1	P	442.0	153.7	119.5	42.6	73.0	26.1	—	—	—	—	634.5	222.4
6220	E	NP/D	395.2	201.6	—	—	—	—	—	—	—	—	395.2	201.6
6250	E	NP	5.4	337.4	—	—	—	—	—	—	2.7	166.0	8.1	503.4
6250	E-1	P	353.7	183.6	35.0	17.9	—	—	—	—	—	—	388.7	201.5
6250	E-2	NP	—	—	—	—	38.0	23.3	—	—	—	—	38.0	23.3
6300	E	NP/D	13.1	8.1	—	—	—	—	—	—	—	—	13.1	8.1
6450	E-1	NP	—	—	—	—	27.4	19.6	—	—	—	—	27.4	19.6
6600	Strays	NP	—	—	—	—	—	—	—	—	—	—	—	—
6600	B-1	NP	3.8	672.4	—	—	—	—	25.0	16.3	—	—	28.8	688.7
6600	B-2	NP	—	—	—	—	24.7	65.9	—	—	—	—	24.7	65.9
6600	B-3	NP/D	29.3	23.4	—	—	—	—	—	—	—	—	29.3	23.4
6600	D-1	NP	—	—	—	—	—	—	—	—	216.6	154.7	216.6	154.7
6600	D/E	NP	71.0	63.8	13.6	8.2	—	—	—	—	—	—	84.6	72.0
6600	E-1	P	1,353.5	1,546.8	14.7	9.7	—	—	—	—	—	—	1,368.2	1,556.5
6600B	G-1	NP	2.8	373.6	—	—	—	—	—	—	9.0	1,302.0	11.8	1,675.6
6600D	G-1	NP	225.4	113.2	—	—	75.8	38.1	—	—	—	—	301.2	151.3
6600C	G-2	NP	53.0	44.8	—	—	—	—	—	—	—	—	53.0	44.8
6600A	G/B	P	55.5	24.7	38.7	17.3	—	—	—	—	—	—	94.2	42.0
6600	H-1	NP	—	—	—	—	—	—	—	—	257.2	170.5	257.2	170.5
6600	J/H	NP	—	—	—	—	—	—	—	—	10.0	1,018.0	10.0	1,018.0
6650	A	P	8.6	3.7	17.9	9.8	—	—	—	—	—	—	26.5	13.5
6900	E	NP	103.1	62.6	—	—	70.0	42.8	—	—	—	—	173.1	105.4
7000	B/G	P	2,506.8	1,307.6	329.0	208.0	—	—	—	—	—	—	2,835.8	1,515.6
7000	E-1	P	256.2	229.6	2.7	1.6	27.0	19.3	—	—	—	—	285.9	250.5
7000	J	P	94.4	57.3	32.0	19.7	—	—	77.7	47.5	—	—	204.1	124.5
7100	E	NP	53.7	3.1	—	—	—	—	—	—	5.0	2.6	58.7	5.7
7100	E-1	NP/D	111.6	144.0	—	—	—	—	—	—	—	—	111.6	144.0
7100	G-1	P	1,468.6	732.7	79.2	39.5	—	—	—	—	—	—	1,547.8	772.2
7200	E	P	1,862.4	1,704.9	16.0	14.7	—	—	—	—	—	—	1,878.4	1,719.6

Reservoir	Zone		(1)	(2)	(3)	(4)	(5)	(6)	(7)	(8)	(9)	(10)	(11)	(12)
7220	E	NP/D	.5	5.4	—	—	—	—	—	—	—	—	.5	5.4
7250	G-1	NP/D	85.1	72.3	—	—	—	—	—	—	—	—	85.1	72.3
7300S	E	P	46.1	23.7	20.6	10.0	—	—	—	—	—	—	66.7	33.7
7500	E-1	NP	5.6	537.0	—	—	14.0	475.9	—	—	—	—	19.6	1,012.9
7500	G-1	NP	—	—	—	—	—	—	—	—	24.1	203.0	24.1	203.0
7500L	E	P	55.4	65.6	11.9	10.9	—	—	—	—	67.5	56.5	134.8	133.0
7550	E	NP/D	18.3	14.5	—	—	—	—	—	—	—	—	18.3	14.5
7600	E	NP	763.2	77.5	—	—	—	—	—	—	—	—	763.2	77.5
7900	B/E	P	11,640.6	9,681.5	214.1	174.7	18.2	14.9	—	—	—	—	11,872.9	9,871.1
7900	B/H	NP	7.5	183.0	—	—	—	—	—	—	9.0	220.0	16.5	403.0
8100	E	P	617.5	607.4	74.3	62.8	—	—	—	—	100.0	91.8	791.8	762.0
8300	B	NP	—	—	—	—	—	—	—	—	16.5	67.3	16.5	67.3
8300	B/H	NP	60.0	31.9	—	—	14.6	7.4	—	—	43.7	22.2	118.3	61.5
8300	C	NP	—	—	—	—	—	—	—	—	0.8	78.3	0.8	78.3
8500	H/B	NP/D	1,330.4	1,145.7	—	—	—	—	—	—	—	—	1,330.4	1,145.7
8550	B	NP	152.3	367.7	—	—	—	—	—	—	2.5	421.0	154.8	788.7
8550	B-1	NP	—	—	—	—	40.3	1,227.8	—	—	—	—	40.3	1,227.8
8550	B/H	NP	178.3	170.4	—	—	—	—	—	—	—	—	178.3	170.4
8550	B/H-1	NP	91.3	40.5	—	—	—	—	—	—	46.7	52.4	138.0	92.9
8550	B/H-2 & B/H-3	NP	—	—	—	—	15.5	12.6	—	—	—	—	15.5	12.6
8600	B	P	84.3	45.6	5.0	0.3	—	—	—	—	—	—	89.3	45.9
8600	B/C-3	NP	2.9	3.1	—	—	—	—	—	—	—	—	2.9	3.1
8600	B/H	NP	6.2	55.5	—	—	—	—	—	—	—	—	6.2	55.5
8650	B	NP/D	5.9	3.7	—	—	—	—	—	—	—	—	5.9	3.7
8650	B-1	NP	—	—	—	—	31.0	25.3	—	—	—	—	31.0	25.3
8650	B-2	NP	4.9	3.8	—	—	—	—	—	—	—	—	4.9	3.8
8650	B-3	NP	—	—	—	—	—	—	—	—	—	—	—	—
8650	B/C-3	NP	—	—	—	—	—	—	—	—	27.6	22.5	27.6	22.5
8650	C	NP	1,178.3	861.4	—	—	—	—	—	—	—	—	1,178.3	861.4
8700	A	NP	—	—	—	—	—	—	—	—	—	—	—	—
8700	B	NP/D	—	—	—	—	—	—	—	—	47.0	43.1	47.0	43.1
8700	B-1	NP	401.5	590.6	—	—	—	—	—	—	60.0	57.6	461.5	648.2
8700	B/C-3	NP	143.5	174.9	—	—	—	—	—	—	—	—	143.5	174.9
8700	C	NP	98.9	269.8	—	—	—	—	—	—	49.5	118.2	148.4	388.0
8750	B/H	NP	111.3	1,177.1	—	—	—	—	—	—	—	—	111.3	1,177.1
8800	B	NP	.2	15.1	—	—	—	—	—	—	—	—	.2	15.1
13800	K/L	NP/D	12.0	798.4	1.6	167.7	—	—	—	—	—	—	13.6	966.1
13800	L	NP/D	—	—	—	—	—	—	—	—	—	—	—	—
13800	O/C-7	P	79.5	3,349.3	—	—	—	—	—	—	—	—	79.5	3,349.3
14000	K/L	NP	—	—	—	—	—	—	—	—	—	—	—	—
14000	O/C-7	NP/D	—	—	—	—	295.9	251.9	—	—	178.9	309.7	474.8	561.6
23 miscellaneous unmapped reservoirs		NP												
Totals, studied reservoirs			**86,951.5**	**219,012.8**	**3,146.3**	**10,872.5**	**1,774.3**	**7,146.4**	**2,581.9**	**554.4**	**2,623.6**	**11,693.1**	**95,050.1**	**251,306.7**
Totals, reservoir studies incomplete (See Appendix)		NP	9,320.1	29,646.9										
Total Texaco Working Interests			96,271.6	248,659.7										

P–Producing
NP–Nonproducing
D–Depleted
Accumulation—proved undeveloped primary reserve categories 2-A, which requires a new well, and 2-B, which requires a plugdown or deepening of an existing well.
S–Stray or Stringer

*–Less than 50 barrels or 50 Mcf
Gas volumes are expressed at 60° Farenheit and a pressure base of 14.73 pounds per square inch.
Source: Published by permission of Texaco, USA.
Totals reflect all reservoirs shown and not shown.

Reserves Determination

Proved developed reserves and accumulations were estimated by analysis of performance histories and the volumetric method using standard geologic and engineering procedures. The method or combination of methods used in the reserve estimates for each reservoir was determined by availability and quality of the basic data, and the stage of development of the reservoir.

Reservoir Packets

Upon completion of the analysis and determination of reserves, the supporting data were organized into individual self-contained packets for each reservoir. A total of *268 packets* were prepared. Each reservoir packet contained the following data:

1. geologic and engineering write-up
2. volumetric analysis
3. performance analyses
4. all data sheets including sand, fault, and correction data
5. all pertinent maps, including structure and isopach maps
6. detailed, analyzed 5-in. logs for each well penetrating the reservoir
7. remaining reserves
8. recommendations for the continued production and ultimate depletion of the reservoir.

Cumulative Production, Estimated Reserves, and Ultimate Recovery

The schedules shown in Table 11-4 show a selected number of reservoirs studied, the status (producing, nonproducing, depleted), cumulative production and estimated proved developed reserves, accumulations, and ultimate recovery. The totals for reserves, accumulations, and the ultimate recovery are for studied reservoirs only. The cumulative production to January 1, 1982, was 86,951,500 barrels of oil and condensate and 219,013 MMcf of gas from 268 studied reservoirs. In addition, 9,321,100 barrels of oil and condensate and 29,647 MMcf of gas had been produced from approximately 58 other reservoirs for which reserve studies were incomplete. Total Texaco working interest production from the field as of January 1, 1982, was 96,271,600 barrels of oil and condensate and 248,660 MMcf of gas (Table 11-4).

The estimated proved developed reserves total 4,920,600 barrels of oil and condensate and 18,019 MMcf of gas from 100 reservoirs. The estimated volumes in the accumulation classification total 3,178,000 barrels of oil and condensate and 14,275 MMcf of gas in 63 reservoirs. The estimated total ultimate recovery for the studied reservoirs is 95,050,100 barrels of oil and condensate and 251,307 MMcf of gas.

DEVELOPMENT AND EXPLORATORY PROSPECTS

A total of **28** development and exploratory prospects resulted from the Golden Meadow Field study. The well location map in Fig. 11-17 shows the proposed well locations recommended for each prospect. Each prospect packet contained a technical discussion and

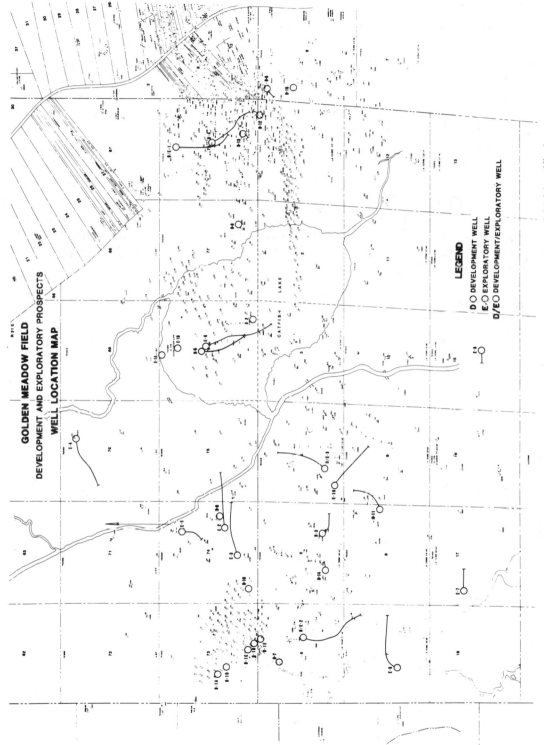

Figure 11-17 Well location map shows the proposed locations of 25 identified development and exploratory prospects. (Published by permission of Texaco USA.)

605

appropriate support material as required, such as geologic maps, reserve estimates, cross sections, electric logs, and other pertinent data. A summary of the prospects and reserves was shown in Table 11-2.

The development prospects were primarily directed toward the recovery of proved hydrocarbons that could not be recovered by existing wells. In some cases, development wells were proposed as alternate wells or to accelerate production. Fifteen development prospects were identified with a potential of 2,045,800 barrels of oil and condensate and 11,374 MMcf of gas.

The exploratory prospects were directed toward untested fault blocks and structural highs. Ten exploratory prospects were identified with unrisked potential of 38,127,000 barrels of oil and condensate and 249,461 MMcf of gas. These reserves were risked in order to arrive at a reasonable estimate of recoverable reserves. The risked reserves for the ten prospects were 3,823,800 barrels of oil and condensate and 30,116 MMcf of gas.

Three prospects were identified that had shallow development potential with deeper exploratory objectives. The estimated unrisked potential for these three prospects was 7,271,000 barrels of oil and condensate and 54,989 MMcf of gas. The risked reserves were estimated to be 1,269,800 barrels of oil and condensate and 6,868 MMcf of gas. The development segment of these prospects has a potential of 421,800 barrels of oil and condensate and 1,064 MMcf of gas.

Prospect Presentation

In this section, we review two of the prospects that were generated during the study. The presentation summarizes the important aspects related to each prospect.

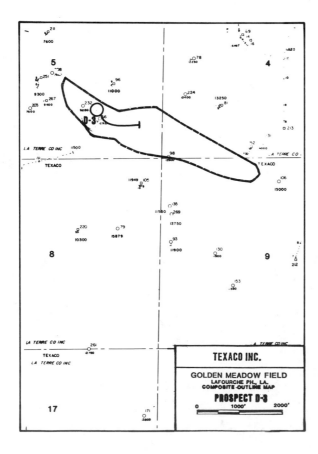

Figure 11-18 Composite outline map for the three prospective reservoirs to be encountered by well D-3. (Published by permission of Texaco USA.)

TABLE 11-5

Development Prospect No. 3 Oil and Solution Gas Reservoir Data and Reserves

Sand	*6600*	*8900*
Reservoir	B-1	A-1
Datum, feet	6,433	9,045
Original reservoir pressure, psia	2,988	4,103
Reservoir temperature, °R	614	647
Porosity, percent	31.5	31.3
Connate water, percent	30.0	18.1
Gravity, °API	36.0	36.0
Solution ratio, Mcf per Bbls.	650	1,100
Formation volume factor, RB/STB	1.34	1.50
Original oil in place, STB per ac.ft.	1,276	1,327
Recovery factor, percent	49.3	34.5
Recoverable oil, STB per ac.ft.	629	457.4
Productive acres	10	105
Net effective sand thickness	4	7
Productive acre ft.	40	735
Ultimate recoverable oil, MBbls.	25	336.2
Production to 1/1/82, MBbls.	—	196.5
Remaining oil reserves as of 1/1/82, MBbls.	25	100[a]
Remaining gas reserves as of 1/1/82, MMcf	16	110

Source: Published by permission of Texaco, USA.

[a] (2-A) Reserves: 336.2 MBO − (196.5 MBO cum. prod. oil zone) × discount factor = 100.0 MBO.

Summary of the oil and solution gas reservoir data and reserves for the 6600 ft sand reservoir B-1 and the 8900 ft sand, reservoir A-1.

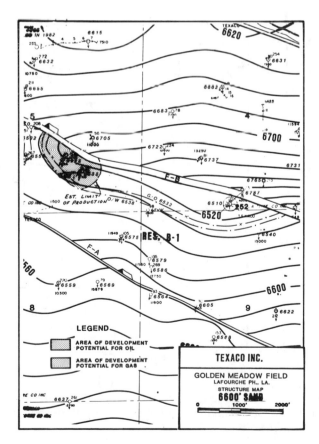

Figure 11-19 Structure map for the 6600-ft Sand, Reservoir B-1. Well No. 252 has produced in the eastern portion of the reservoir. (Published by permission of Texaco USA.)

Development Prospect No. 3

Objective: Development Prospect No. 3 was intended to recover hydrocarbons that could not be recovered from existing wells in three reservoirs with closure against Faults A, A-2, or B in the southwest portion of the field in Section 5, T20S-R21E. The prospect is highlighted on the field location map in Fig. 11-17.

The prospect well was designed to encounter the 6600-ft Sand about 300 ft east-southeast of LaTerre No. 66; thereafter, the well would be directionally drilled to the east-southeast to penetrate the 8900-Ft and 10,000-Ft Sands at optimum structural positions upthrown to Fault A. A reservoir composite outline map is shown in Fig. 11-18.

Development Potential

The 6600-Ft Sand, Reservoir B-1: The 6600-Ft Sand, Reservoir B-1, is an east-west trending elongate reservoir with a thin oil rim and gas cap, limited by Fault B to the north, with the remainder of the reservoir limited by an average original oil/water contact at 6538 ft as shown on the structure map in Fig. 11-19. LaTerre No. 252 is the only well that was completed in this reservoir. This completion was made in the eastern portion of the reservoir and apparently did not drain the western portion around LaTerre Nos. 66 and 232. Since these two wells were not available for plugback into this reservoir, the potential hydrocarbon accumulation of 25,000 barrels of oil and 16 MMcf of gas was assigned to Development Well No. 3.

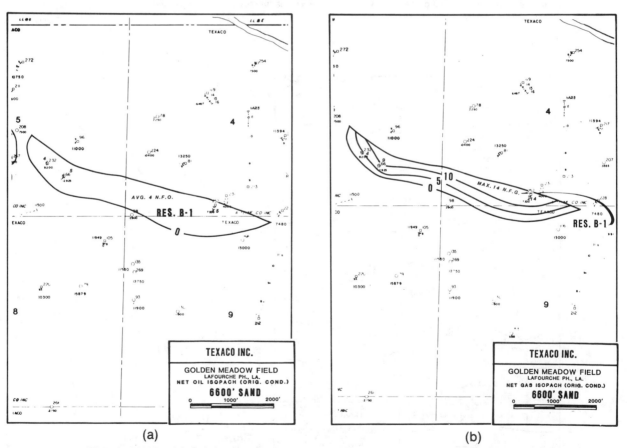

(a)   (b)

Figure 11-20 (a) Original conditions, net oil isopach map for the 6600-ft Sand, Reservoir B-1. (b) Original condition, net gas isopach map for the 6600-ft Sand, Reservoir B-1. (Published by permission of Texaco USA.)

The reserves are based on a completion in the oil zone. Since the oil zone is thin, the well could be gas productive with an estimated recovery of 1500 Mcf/Ac.Ft. A summary of the reservoir data and reserves is shown in Table 11-5. A new well at an optimum location could improve the recovery potential from this reservoir. The original net oil and gas isopach maps are shown in Figs. 11-20a and 11-20b. A sand identification log is shown in Fig. 11-21.

The 8900-Ft Sand, Reservoir A-1:   The 8900-Ft Sand, Reservoir A-1, is also a reservoir with a thin oil rim and gas cap. The reservoir is limited by Fault A to the north and northeast, by a permeability barrier to the west, with the remainder of the reservoir limited by the original oil/water contact at 9052 ft as shown on the structure map in Fig. 11-22.

The reservoir was produced by LaTerre Wells No. 98 and 232 which were off production at the time of the study. LaTerre No. 98 was completed as a gas well in August 1961, and produced 2.2 BCF of gas from two sets of perforations (9039-43 and 9024-36) before going off production with high water production in March 1971. In August 1975, the well was recompleted (perforations 9015-22) and flowed 200 barrels of oil per day with a gas oil ratio of 1734 cu ft per barrel. The well had a tubing pressure of 1900 psig and the production was water free.

In September 1975, LaTerre No. 232 was completed as an oil well and flowed 315 barrels of 35.6 API gravity oil per day with no water production. Both wells were shut in for a storm in July 1979, and could not be restored to production.

Based on the reservoir study, it appeared that the original gas cap was blown down by the initial gas production from LaTerre No. 98 from August 1961, to March 1971. The strong water drive in the reservoir apparently displaced the oil column into the volume originally occupied by the gas cap. In view of the reservoir performance and the location

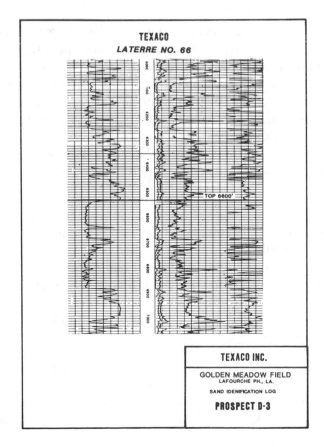

Figure 11-21 Log from Well No. 66 shows the 6600-ft Sand. (Published by permission of Texaco USA.)

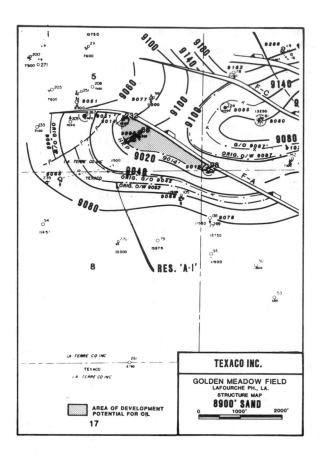

Figure 11-22 Structure map for the 8900-ft Sand, Reservoir A-1. (Published by permission of Texaco USA.)

of Wells No. 98 and 232, it appeared that additional oil could be recovered from this reservoir. Considering the available data and assuming that a residual oil saturation of 30% would be lost in the pore volume originally occupied by gas and applying an approximate 30% risk factor to the reserves, it was estimated that as much as 100,000 barrels of recoverable oil remained in the reservoir. The oil and solution gas reservoir data and reserves are shown in Table 11-5. The original net oil and gas isopach maps are shown in Figs. 11-23a and 11-23b. A sand identification log is shown in Fig. 11-24.

LaTerre No. 66 located near the apex of the structure was not available for recompletion in this sand because of junk in the hole. Therefore, a new well would be required to recover the reserves.

The 10,000-Ft Sand, Reservoir A/A-2: The 10,000-Ft Sand, Reservoir A/A-2, contains an oil reservoir with an associated gas cap. The reservoir limits are defined by Faults A and A-2 to the north and by an original oil/water contact at 9978 feet as shown on the structure map in Fig. 11-25. Wells No. 66, 98, and 106 have penetrated the reservoir. LaTerre Nos. 66 and 106 have produced with a cumulative production of 348,934 barrels of oil and 488,346 MMcf of gas, some of which was gas-cap gas. LaTerre No. 106 was producing at the time of the study with additional recoverable oil estimated at 43,000 barrels of oil based on performance data.

LaTerre No. 98 tested gas in August 1961, and was squeezed out of this zone. Based on the volumetric calculations, the attic gas cap had 2000 MMcf of remaining recoverable gas (see attic gas-cap isopach, Fig. 11-26c). The gas and condensate reservoir data and reserves are shown in Table 11-6. A 25% risk factor was applied to the 2000 MMcf resulting in an estimated reserve of 1500 MMcf of gas. Original net oil and gas isopach

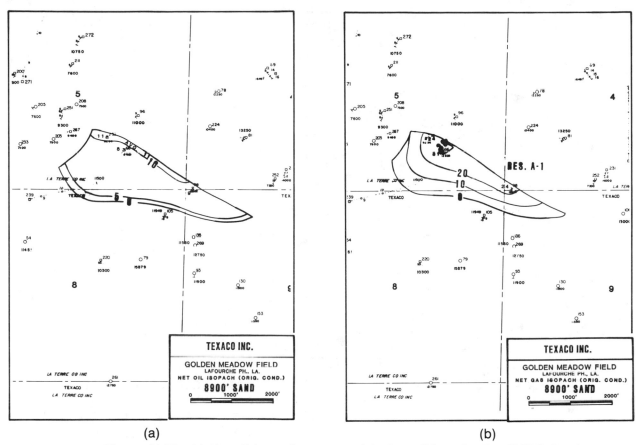

(a)   (b)

Figure 11-23 (a) Net oil isopach map at original conditions for the 8900-ft Sand, Reservoir A-1. (b) Original net gas isopach map for the 8900-ft Sand, Reservoir A-1. (Published by permission of Texaco USA.)

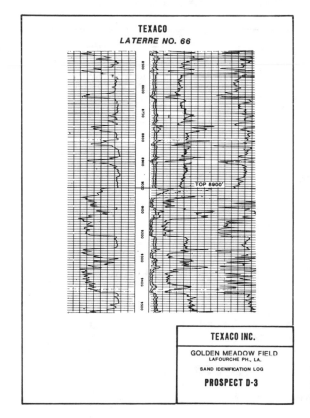

Figure 11-24 Portion of electric log from Well No. 66 shows the productive interval for the 8900-ft Sand. (Published by permission of Texaco USA.)

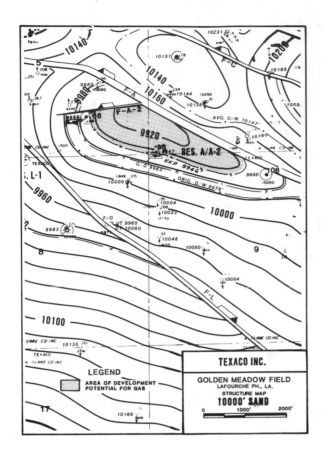

Figure 11-25 Structure map on the top of the 10,000-ft Sand, delineating Reservoir A/A-2. (Published by permission of Texaco USA.)

TABLE 11-6

Development Prospect No. 3
Summary of the Gas and Condensate Reservoir Data and Reserves for the 10,000 Ft Sand, Reservoir A/A-2

Sand	*10,000*
Reservoir	A/A-2
Datum, feet	9,965
Original reservoir pressure, psia	4,499
Reservoir temperature, °R	660
Porosity, percent	29.7
Connate water, percent	17.7
Separator gas gravity	62.0
Original Z factor	.9585
Condensate yield, Bbls. per MMcf	21.4
Original gas in place, Mcf per ac.ft.	2,621
Shrinkage factor	.985
Recovery factor, percent	69.6
Recoverable gas, Mcf per ac.ft.	1,797
Productive acres	81.8
Net effective sand thickness, ft.	12.3
Productive acre ft.	1,006
Ultimate recoverable gas, MMcf	2,000[a]
Production to 1/1/82, MMcf	—
Remaining gas reserves as of 1/1/82, MMcf	1,500[b]
Average anticipated condensate yield, Bbls. per MMcf	10
Remaining cond. reserves as of 1/1/82, MBbls.	15

Source: Published by permission of Texaco, USA.

[a] The 2,000 MMcf includes 1,808 MMcf of recoverable gas based on the current attic isopach map plus 192 MMcf of recoverable gas from the partially displaced original cap volume.

[b] A 25 percent risk factor has been applied to the reserves.

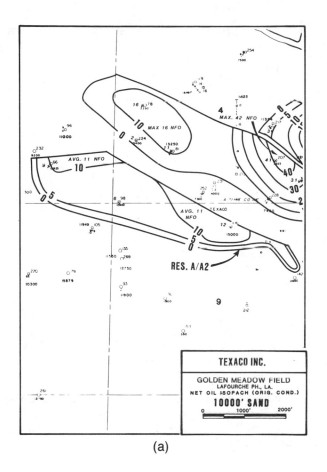

(a)

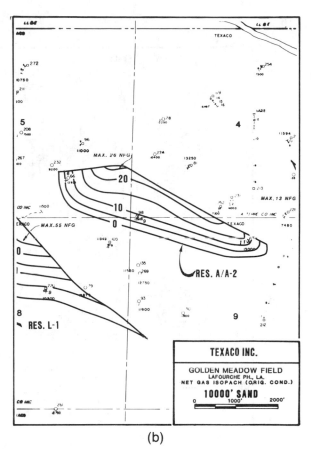

(b)

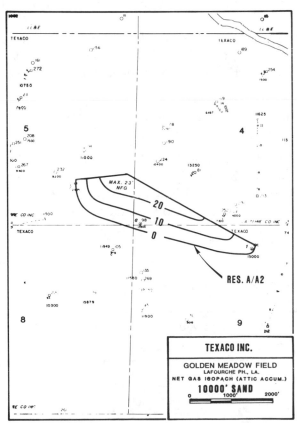

(c)

Figure 11-26 (a) Net oil isopach map based on original conditions in the 10,000-ft Sand, Reservoir A/A-2. (b) Net gas isopach map at original condition for the 10,000-ft Sand, Reservoir A/A-2. (c) Net gas isopach map delineating the remaining attic gas in the A/A-2 Reservoir. (Published by permission of Texaco USA.)

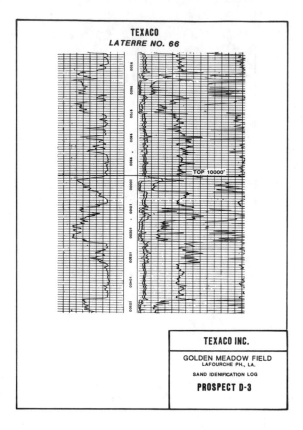

Figure 11-27 Portion of the electric log from Well No. 66 shows the productive interval for the 10,000-ft Sand, Reservoir A/A-2. (Published by permission of Texaco USA.)

maps are shown in Figs. 11-26a and 11-26b. A sand identification log is shown in Fig. 11-27.

Although no oil reserves were assigned to this reservoir, the chances that a new well would encunter a producible oil rim were good in view of the volumetric analysis, the strong water drive, and the fact that the two previously produced oil wells were at opposite ends of the reservoir.

Reserves: The reserve estimate for this prospect is based on the three reservoirs discussed. The risked development reserves are summarized in Table 11-7.

LaTerre Well No. 112: As a result of the study, LaTerre Well No. 112 was drilled and completed in the 10,000-Ft Sand, Reservoir A/A-2, on July 13, 1983. As of October 1, 1984, the well had produced 49,000 barrels of oil, and 138 MMCF of gas. A test on the well (9/1/84) showed the current production at 118 barrels of oil per day, 278 MCF of gas per day, and 620 barrels of water.

TABLE 11-7

Summary of Development Potential for Development Prospect No. 3

| Sand | Reservoir | Reserves | |
		Oil/cond.-MBbls.	Gas-MMcf
6,600	B-1	25	16
8,900	A-1	100	110
10,000	A/A-2	15	1,500
Total Reserves		**140**	**1,626**

Source: Published by permission of Texaco, USA.

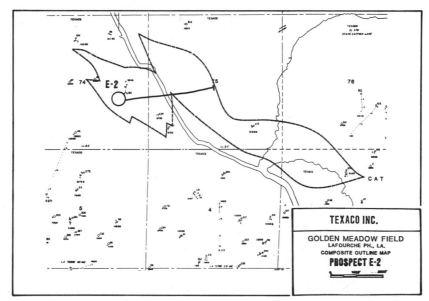

Figure 11-28 Reservoir composite outline map for Exploratory Prospect No. 2. (Published by permission of Texaco USA.)

The 10,000-Ft Sand, Reservoir A/A-2, produced as an oil rather than gas reservoir. As indicated in the previous section, although no oil reserves were assigned to this reservoir, the chances of encountering a producible oil rim were good considering all the data on the reservoir. No data were available on the ultimate recovery from this reservoir.

Exploratory Prospect No. 2

Objective: Exploratory Prospect No. 2 was recommended to test a series of sands with closures against Fault O in the northwest portion of the field in Sections 74 and 75, T19S-R21E. The location of this exploration prospect is highlighted on the well location map in Fig. 11-17. The well was recommended to be drilled to encounter the 10,500-Ft Sand at the optimum structural position downthrown to the trapping Fault O. Thereafter, the well would be drilled as a directional well to continue downthrown and parallel to Fault O in order to test the series of potentially productive sands through the 13,900-Ft Sand, which is the deepest objective. The reservoir composite outline map is shown in Fig. 11-28.

TABLE 11-8

Summary of Previous Production from the Objective Sands. All Four Sands Have Been Good Hydrocarbon Producers.

Sand	Reservoir	Cumulative production		Average production per reservoir	
		Oil/cond.-MBbls.	Gas-MMcf	Oil/cond.-MBbls.	Gas-MMcf
10,500	7	1,408.9	5,408.7	201.3	772.7
11,000	7	1,101.1	16,945.5	157.3	2,420.8
11,800	6	7,947.0	19,676.6	1,323.5	3,279.4
13,000	10	4,050.2	78,799.3	405.0	7,879.9

Source: Published by permission of Texaco, USA.

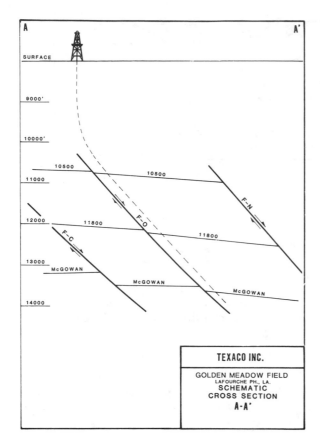

Figure 11-29 Schematic cross section A-A' shows the path of the proposed exploratory well. (Published by permission of Texaco USA.)

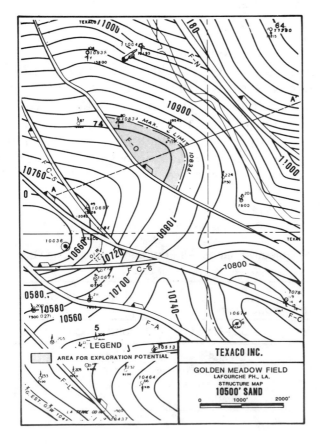

Figure 11-30 Structure map—Top of 10,500-ft Sand. (Published by permission of Texaco USA.)

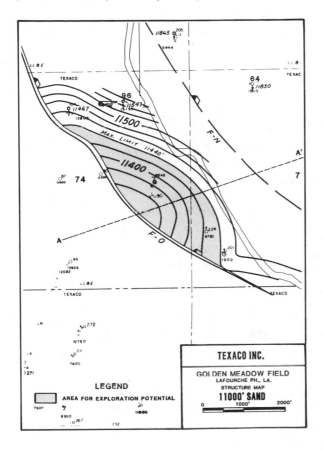

Figure 11-31 Structure map—Top of 11,000-ft Sand. (Published by permission of Texaco USA.)

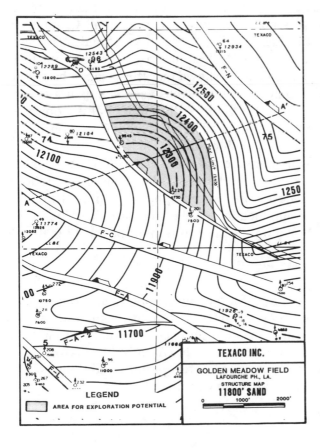

Figure 11-32 Structure map—Top of 11,800-ft Sand. (Published by permission of Texaco USA.)

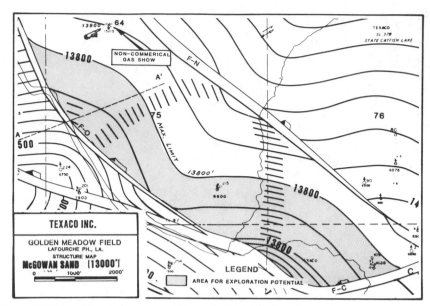

Figure 11-33 Structure map—Top of 13,000-ft (McGowan) Sand. Notice non-commercial gas shown in Well No. 64. (Published by permission of Texaco USA.)

Exploratory Potential: The four objective sands mapped for this prospect are the 10,500-Ft, 11,000-Ft, 11,800-Ft, and 13,000-Ft (McGowan) Sands. A schematic cross section and structure maps of the objective sands are shown in Figs. 11-29 to 11-33. The structure maps indicate the probability of structural closures downthrown to Fault O of sufficient size to justify an exploratory well. It is possible, however, that Fault O may not be big enough to separate thick sand sections (such as in the McGowan Sand). This factor was considered in the risking of the reserves. Table 11-8 summarizes the previous production history of the four mapped sands. The tabulation shows that these sands have been good producers and the potential for additional reserves in these sands appears favorable.

In addition to the four mapped sands, there are 13 other potentially productive sands in the objective section that have not been mapped. The objective section is shown on the sand identification logs in Figs. 11-34a, 11-34b, and 11-34c.

Three objective sands have downdip condensate shows based on core data. These sands are the Lower Authement, 13,000-Ft, and the 13,900-Ft Sands. Log sections showing the core data are in Fig. 11-35. These cores suggest the migration of hydrocarbons and possible accumulation at an optimum location on the structure. The location of Well No. 64, which has a gas show, is shown on the structure map for the 13,000-Ft Sand in Fig. 11-33.

Reserves: Potential reserves were estimated for the four mapped sands only. The reserve potential for the 13 unmapped sands may be greater than that estimated for the four mapped sands. The volumetric reservoir data and reserves for the mapped sands is shown in Table 11-9. The summary of potential reserves is shown in Table 11-10.

The drilling of this exploratory well has been deferred until additional seismic processing has been complete and further evaluation conducted.

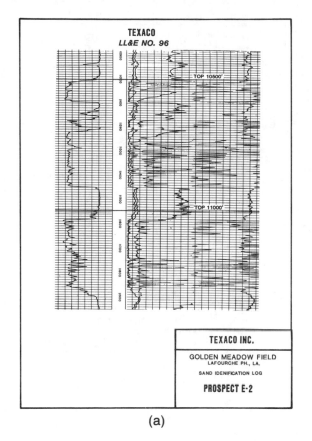

(a)

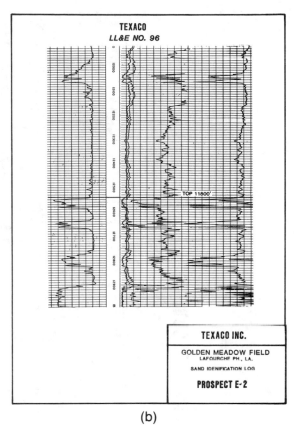

(b)

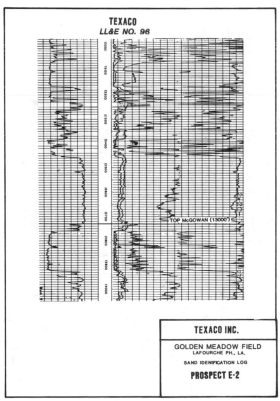

(c)

Figure 11-34 (a) Sand identification log for the 10,500-ft and 11,000-ft Sands. (b) Sand identification log for the 11,800-ft Sand. (c) Sand identification log for the 13,000-ft Sand. (Published by permission of Texaco USA.)

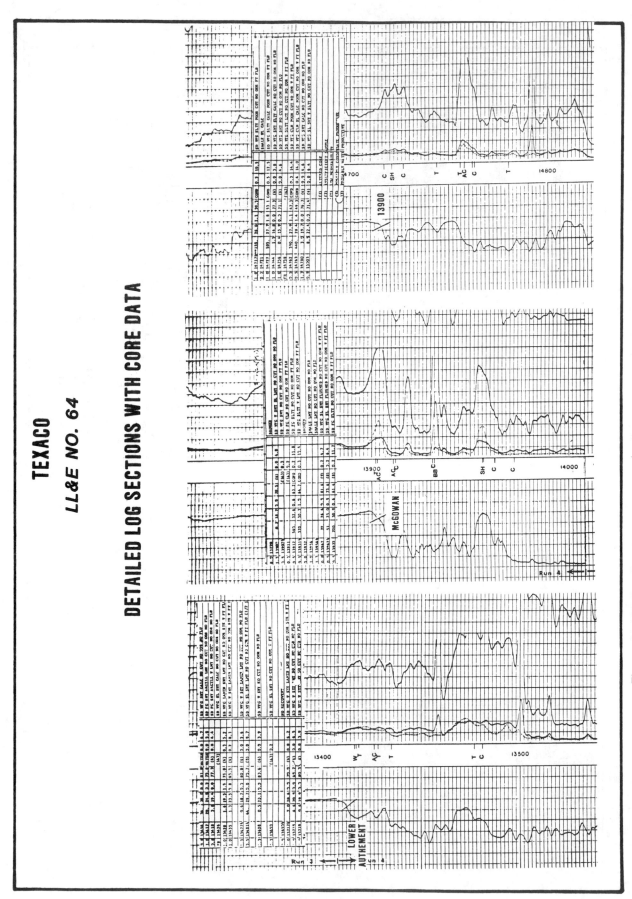

TEXACO
LL&E NO. 64

DETAILED LOG SECTIONS WITH CORE DATA

Figure 11-35 Detailed log section with core data for Well No. 64. (Published by permission of Texaco USA.)

TABLE 11-9

Volumetric Reservoir Data and Reserves for the Four Mapped Sands

Sand	10,500	11,000	11,800	13,000
Reservoir	0	0	0	0
Original oil (gas) per ac.ft.	952	(2,647)	1,030	(3,136)
Recovery factor	43	65	49	55
Recoverable oil (gas) per ac.ft.	409	(1,721)	505	(1,725)
GOR (yield)	1,000	(20.3)	1,600	(50)
Productive acres	44	106	72	349
Net effective sand thickness, ft.	18	26	20	70
Productive acre ft	792	2,756	1,440	24,430
Unrisked reserves in MBbls. (MMcf)	324	(4,743)	727	(42,142)
Confidence factor, percent	10	10	10	15
Risked reserves in MBbls. (MMcf)	32	(474)	73	(6,321)

Source: Published by permission of Texaco, USA.

TABLE 11-10

Tabulation of the Exploratory Potential by Sand, for Exploratory Well No. 2

Sand	Unrisked reserves		Risked reserves	
	Oil/cond.-MBbls.	Gas MMcf	Oil/cond.-MBbls.	Gas MMcf
10,500	324	324	32	32
11,000	96	4,743	10	474
11,800	727	1,163	73	117
McGowan	2,107	42,142	316	6,321
Total Reserves	**3,254**	**48,372**	**431**	**6,944**

Source: Published by permission of Texaco, USA.

CONCLUSION

The purpose of this final chapter is to show how the methods and techniques presented in this textbook can be designed and modified to conduct a variety of different geologic and engineering studies. In addition, we presented guidelines to the *Synergistic Approach* to conducting integrated geologic, geophysical, and engineering studies.

The Golden Meadow Field Case History shows, in a general way, how geologic and engineering methodology is tailored to meet the specific objectives of a field study. It also shows that additional hydrocarbon potential might result from such a detailed and integrated study. Each and every geologic and engineering study requires a unique approach because of such factors as the stage of field development, geologic complexities, number of wells drilled, and available data. Certain procedures and methods applied to one study may not necessarily be applied to another.

APPENDIX

GENERAL MAP SYMBOLS

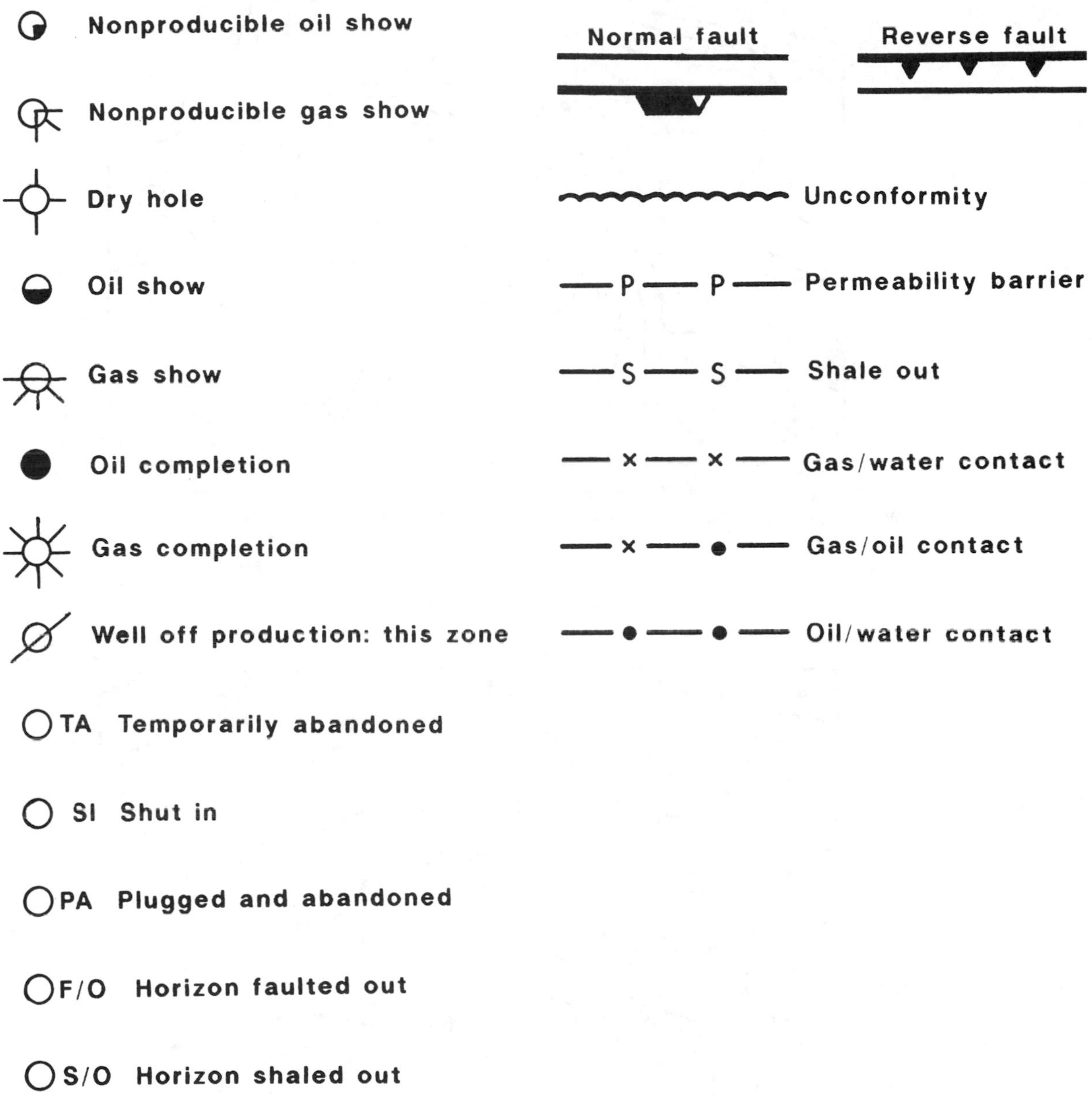

Symbol	Description		Symbol	Description
	Nonproducible oil show		**Normal fault** / **Reverse fault**	
	Nonproducible gas show			
	Dry hole		~~~~~~~	Unconformity
	Oil show		— P — P —	Permeability barrier
	Gas show		— S — S —	Shale out
	Oil completion		— × — × —	Gas/water contact
	Gas completion		— × — ● —	Gas/oil contact
	Well off production: this zone		— ● — ● —	Oil/water contact
○ TA	Temporarily abandoned			
○ SI	Shut in			
○ PA	Plugged and abandoned			
○ F/O	Horizon faulted out			
○ S/O	Horizon shaled out			

MAP SYMBOLS USED THROUGHOUT THE TEXT ARE NOT ALWAYS CONSISTANT WITH THOSE SHOWN HERE. THERE IS A WIDE VARIETY OF SYMBOLS USED BY DIFFERENT COMPANIES, AND THEREFORE NO CONSISTANT USE OF SYMBOLS THROUGHOUT THE INDUSTRY.

Figure A-1 Map symbols.

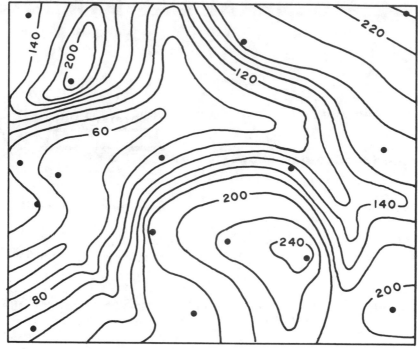

Figure A-2 Map constructed using the interpretive contouring technique. (Reproduced from *Analysis of Geologic Structures* by John M. Dennison, by permission of W. W. Norton & Company, Inc. Copyright © 1968 by W. W. Norton & Company, Inc.)

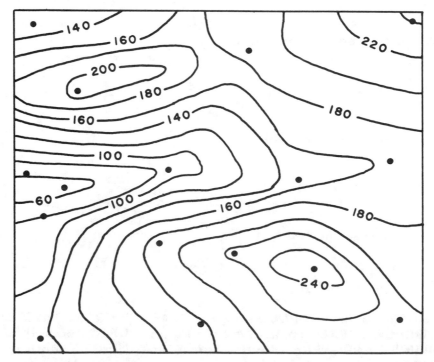

Figure A-3 Map constructed using the mechanical contouring technique. (Reproduced from *Analysis of Geologic Structures* by John M. Dennison, by permission of W. W. Norton & Company, Inc. Copyright © 1968 by W. W. Norton & Company, Inc.)

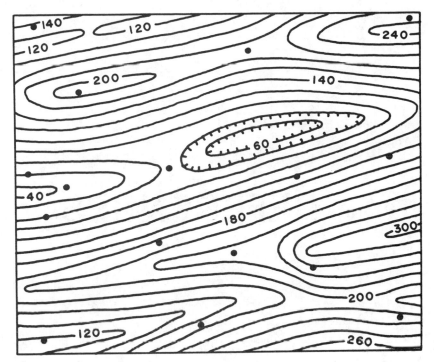

Figure A-4 Map constructed using the equal-spaced contouring technique. (Reproduced from *Analysis of Geologic Structures* by John M. Dennison, by permission of W. W. Norton & Company, Inc. Copyright © 1968 by W. W. Norton & Company, Inc.)

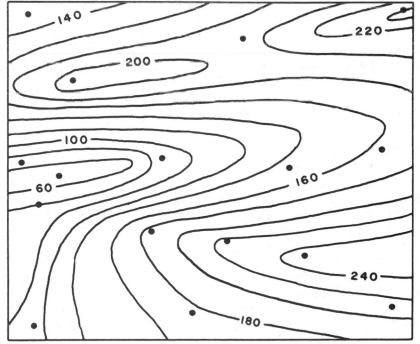

Figure A-5 Map constructed using the parallel contouring technique. (Reproduced from *Analysis of Geologic Structures* by John M. Dennison, by permission of W. W. Norton & Company, Inc. Copyright © 1968 by W. W. Norton & Company, Inc.)

REFERENCES

Acevedo, J. S., 1980, *Giant fields of the Southern Zone—Mexico:* Am. Assoc. of Petroleum Geol., Memoir 30, p. 339–385.

Adams, G. F., 1957, *Block diagrams from persective grids:* Journal of Geological Education, v. 5, No. 2, p. 10–19.

Allaud, L., 1976, *Vertical Net Sandstone Determination for Isopach Mapping of Hydrocarbon Reservoirs:* Am. Assoc. Petroleum Geol. Bull., v. 60, p. 2150–2153.

Anspach, D. H., S. E. Tripp, R. E. Berlitz, and J. A. Gilreath, 1987, *Postdevelopment analysis of producing shelf-slope environments of deposition, High Island area:* Gulf Coast Assoc. of Geol. Soc. Transactions, v. 37, p. 1–10.

Anstey, N. A., 1977, *Seismic Interpretation, The Physical Aspects.* IHRDC Press.

———, 1982, *Simple Seismic,* IHRDC Press.

Archie, G. E., 1942, *The electrical resistivity log as an aid in determining some reservoir characteristics,* Trans. AIME, Paper 155.

———, 1947, *Electrical resistivity an aid in core-analysis interpretation:* Am. Assoc. of Petroleum Geol. Bull., v. 31, No. 2, p. 350–366.

Arps, J. J., 1956, *Estimation of primary oil reserves,* Trans. AIME, Paper 207, p. 183–186.

Atwater, G. I., and M. J. Foreman, 1959, *Nature of growth of southern Louisiana salt domes and its effect on petroleum accumulation:* Am. Assoc. Petroleum Geol. Bull., v. 43, p. 2592–2622.

Atwater, G. I., and E. E. Miller, 1965, *The effect of decrease in porosity with depth on future development of oil and gas reserves in South Louisiana:* Am. Assoc. of Petroleum Geol. Bull., v. 49, p. 334.

Badgley, P. C., 1959, *Structural Methods for the Exploration Geologist and a Series of Problems for Structural Geology Students,* Harper & Bros., New York, NY.

Badley, M. E., 1985, *Practical Seismic Interpretation,* IHRDC Press.

Baldwin, B., and C. O. Butler, 1985, *Compaction curves:* Am. Assoc. of Petroleum Geol. Bull., v. 69, p. 622–626.

BALL, S. M., 1982, *Exploration application of temperatures recorded on log-headings—An up-the-odds method of hydrocarbon-charged porosity prediction:* Am. Assoc. of Petroleum Geol. Bull., v. 66, No. 8, p. 1108–1123.

BANKS, C. J., AND J. WHARBURTON, 1986, *"Passive-roof" duplex geometry in the frontal structures of the Kirthar and Sulaiman Mountain Belts:* Pakastan Jour. Structural Geol., v. 8, p. 229–237.

BEAL, C., 1946, *The viscosity of air, water, natural gas, crude oil and its associated gases at oil field temperatures and pressures,* Trans. AIME, Paper 165.

BECKWITH, R. H., 1941, *Trace-slip faults:* Am. Assoc. of Petroleum Geol. Bull., v. 25, No. 12, p. 2181–2193.

———, 1947, *Fault problems in fault planes:* Geol. Soc. of Am. Bull., v. 58, p. 79–108.

BELL, W. G., 1956, *Tectonic Setting of Happy Springs and Nearby Structures in the Sweetwater Uplift Area,* Central Wyoming: in Geological Record of the American Association of Petroleum Geologists, Rocky Mountain Section.

BENGSTON, C. A., 1980, *Statistical curvature analysis methods for interpretation of dipmeter:* Oil and Gas Jour. (June 23).

BENNISON, G. M., 1975, *An Introduction to Geological Structures and Maps,* third ed., Edward Arnold LTD, London, England.

BERG, R. R., 1967, *Point bar origin of Fall River sandstone reservoirs, northeastern Wyoming:* Soc. of Petroleum Engineers of Aime, Paper No. 1953.

BERG, O. R., AND D. G. WOOLVERTON, 1985, *Seismic stratigraphy II—an integrated approach to hydrocarbon exploration:* Am. Assoc. of Petroleum Geol., Mem. 39.

BILLINGS, M. P., 1954, *Structural Geology,* second ed., Prentice-Hall, New York, NY.

———, 1972, *Structural Geology,* third ed., Prentice-Hall, Englewood Cliffs, NJ.

BIOT, M. A., 1961, *Theory of folding of stratified viscoelastic media and its implication in tectonics and orogenesis:* Geol. Soc. of Am. Bull., v. 72, p. 1595–1620.

BISCHKE, R. E., 1990, *Applied structural balancing:* Gulf Coast Assoc. Geol. Soc.—Short Course (GCAGS 40th Annual Convention in Lafayette, LA).

BISCHKE, R. E., AND J. SUPPE, 1990a, *Calculating sand/shale ratios from growth normal fault dips on seismic profiles:* Trans. Gulf Coast Assoc. Geol. Soc., v. 40.

BISCHKE, R. E., AND J. SUPPE, 1990b, *Geometry of rollover: origin of complex arrays of antithetic and synthetic crestal faults:* Am. Assoc. of Petroleum Geol., Abstract with programs.

BISCHKE, R. E., J. SUPPE, AND R. DEL PILAR, 1990, *A new branch of the Philippine Fault system, as observed from aeromagnetic and seismic data:* Tectonophysics—Special Issue, Geodynamic Evolution of the Eastern Eurasian Margin.

BISHOP, M. S., 1960, *Subsurface Mapping,* John Wiley & Sons, New York, NY.

BLOOMER, R. R., 1977, *Depositional environments of a reservoir sandstone in west-central Texas:* Am. Assoc. of Petroleum Geol. Bull., v. 61, No. 3, p. 344–359.

BOECKELMAN, W., unpublished notes.

BOOS, C. M., AND M. F. BOOS, 1957, *Tectonics of eastern flank and foothills of front range, Colorado:* Am. Assoc. of Petroleum Geol. Bull., v. 41, No. 12, p. 2603–2676.

BOTT, W. F., JR., AND T. T. TIEH, 1987, *Diagenesis and high fluid pressures in the Frio and Vicksburg shales, Brooks County, Texas:* Gulf Coast Assoc. of Geol. Soc. Transactions, v. 37, p. 323–334.

BOWER, T. H., 1947, *Log map, new type of subsurface map:* Am. Assoc. of Petroleum Geol. Bull., v. 127, No. 7, p. 340–349.

BOYER, S. E., 1986, *Styles of folding within thrust sheets: examples from Appalachian and Rocky Mountains of the U.S.A. and Canada:* Jour. of Struct. Geol., v. 8, p. 325–339.

BOYER, S. E., AND D. ELLIOTT, 1982, *Thrust systems:* Am. Assoc. of Petroleum Geol. Bull., v. 66, No. 9, p. 1196–1230.

BRISTOL, H. M., 1975, *Structural geology and oil production of northern Gallatin County and southernmost White County, Illinois:* Ill. State Geol. Surv., Ill. Petrol., No. 195.

BRISTOL, H. M., AND J. D. TREWORGY, 1979, *The Wabash Valley fault system in southeastern Illinois:* Ill. State Geol. Surv., Circular 609.

BROWN, G. G., D. L. KATZ, ET AL, 1948, *Natural Gasoline and Volitale Hydrocarbons:* National Gasoline Association of America, Tulsa, OK.

BROWN, L. F., AND W. L. FISHER, 1980, *Stratigraphic interpretation and petroleum exploration:* Am. Assoc. of Petroleum Geol., Continuing Education Course Note Series 16.

BROWN, R. L., 1981, *Thickness Corrections for Deviated Wells—A Simplified Technique Utilizing the Apparent Dip along a Well Path:* Contribution to the 1981 Geological/Petrophysical Conference, Houston.

BROWN, R. N., 1980, *History of exploration and discovery of Morgan, Ramadan and July Oil Fields, Gulf of Suez, Egypt, in* A. D. Miall, ed., *Facts and Principles of World Petroleum Occurrence:* Canadian Soc. Petroleum Geol., Mem. 6, p. 733–764.

BROWN, W., 1984a, *Working with folds:* Am. Assoc. of Petroleum Geol., Structural Geology School Course Notes.

———, 1984b, *Basement involved tectonics: "Foreland area":* Am. Assoc. of Petroleum Geol., Continuing Education Course Series No. 26.

BROWN, W. G., 1982, *New Tricks for Old Dogs—A Short Course in Structural Geology.* American Assoc. of Petroleum Geol., Southwest Sections Meeting, Witchita Falls, Texas, 79 pgs.

BRUCE, C. H., 1973, *Pressured shale and related sediment deformation: mechanism for development of regional contemporaneous faults:* Am. Assoc. Petroleum Geol. Bull., v. 57, p. 878–886.

BUCHER, W. H., 1933, *The Deformation of the Earth's Crust,* Princeton University Press, Princeton, NJ.

BUSCH, D. A., 1959, *Prospecting for stratigraphic traps:* Am. Assoc. of Petroleum Geol. Bull., v. 43, No. 12, p. 2829–2843.

———, 1974, *Stratigraphic traps in sandstones—exploration techniques:* Am. Assoc. of Petroleum Geol., Mem. 21.

BUSCH, D. A., AND D. A. LINK, 1985, *Exploration Methods for Sandstone Reservoirs:* Oil and Gas Consultants, Inc., Tulsa, OK.

BUSK, H. G., 1929, *Earth Flexures, Their Geometry and Their Representation and Analysis in Geological Section, with Special Reference to the Problem of Oil Finding,* Cambridge University Press, Cambridge, England.

———, 1957, *Earth Flexures, Their Geometry and Their Representation and Analysis in Geological Section, with Special Reference to the Problem of Oil Finding,* William Trussell, England.

CAPE, D. C., S. McGEARY, AND G. A. THOMPSON, 1983, *Cenozoic normal faulting and the shallow structure of the Rio Grande Rift near Socorro, New Mexico:* Geol. Soc. of Am. Bull., v. 94, p. 3–14.

CHALLINOR, J., 1933, *The "throw" of a fault:* Geological Magazine, v. LXX, No. 9, p. 384–393.

CHAMBERLIN, T. C., 1897, *The method of multiple working hypothesis:* Jour. of Geol., v. 5, p. 837–848.

CHAMBERLIN, R. T., 1910, *The Appalachian folds of central Pennsylvania:* Jour. of Geol., v. 18, p. 228–251.

CHENOWETH, P. A., 1972, *Unconformity traps:* Am. Assoc. of Petroleum Geol., Mem. 16, p. 42–46.

CHEW, J., AND C. A. CONNALLY, JR., 1958, *A Viscosity Correlation for Gas-Saturated Crude Oils:* presented before the Society of Petroleum Engineers of AIME at the annual meeting in Houston, Texas.

CHRISTENSEN, A. F., 1983, *An example of a major syndepositional listric fault, in* A. W. Bally, ed., *Seismic expression of structural styles:* Am. Assoc. Petroleum Geol., Studies in Geol., No. 15, v. 2, p. 2.3.1–36 to 2.3.1–40.

CLARK, S. K., 1943, *Classification of faults:* Am. Assoc. of Petroleum Geol. Bull., v. 27, No. 9, p. 1245–1265.

CLARK, J. C., 1959, *Problems of fault nomenclature:* Am. Assoc. of Petroleum Geol. Bull., v. 43, No. 11, p. 2653–2674.

CLOOS, E., 1942, *Distortion of stratigraphic thickness due to folding:* Proceed. Nat. Acad. of Sci., v. 28, No. 10, p. 401–407.

———, 1968, *Experimental analysis of Gulf Coast fracture patterns:* Am. Assoc. Petroleum Geol. Bull., v. 52, No. 3, p. 420–444.

COATES, J., 1945, *The construction of geological sections:* The Quarterly Jour. of the Geological Mining and Metallurgical Soc. of India, v. 17, No. 1, p. 1–11.

COLE, W., 1969, *Reservoir Engineering Manual,* second ed., Gulf Publishing Co., Houston, TX.

COLEMAN, J. M., 1982, *Deltas: Processes of Deposition and Models for Exploration,* IHRDC, Boston, MA.

———, 1988, *Dynamic changes and processes in the Mississippi River Delta:* Geol. Soc. of Am. Bull., v. 100, p. 999–1015.

CRAFT, B. C., AND M. F. HAWKINS, 1959, *Applied Petroleum Reservoir Engineering,* Prentice Hall, Englewood Cliffs, NJ.

CRANS, W., G. MANDL, AND J. HAREMBOURE, 1980, *On the theory of growth faulting: a geomechanical delta model based on gravity sliding:* Jour. of Petroleum Geol., v. 2 & 3, p. 265–307.

CROWELL, J. C., 1948, *Template for spacing structure contours:* Am. Assoc. of Petroleum Geol. Bull., v. 32, No. 12, p. 2290–2294.

———, 1959, *Problems in fault nomenclature:* Am. Assoc. of Petroleum Geol. Bull., v. 43, No. 11, p. 2653–2674.

CURRIE, J. B., 1952, *Three-dimensional method for solution of oil field structures:* Am. Assoc. of Petroleum Geol. Bull., v. 36, No. 5, p. 889–890.

———, 1956, *Role of concurrent deposition and deformation of sediments in development of salt-dome graben structures:* Am. Assoc. of Petroleum Geol. Bull., v. 40, p. 1–16.

DAHLBERG, E. C., *Exploration Contouring, Mapping and Data Interpretation Methods,* ECD Geol. Specialists LTD.

DAHLEN, F. A., AND J. SUPPE, 1988, *Mechanics, growth, and erosion of mountain belts:* Geol. Soc. of Am., Special Paper 218, p. 161–178.

DAHLEN, F. A., J. SUPPE, AND D. DAVIS, 1984, *Mechanics of fold-and-thrust belts and accretionary wedges: Cohesive Coulomb Theory,* Jour. of Geophysics Res., v. 89, p. 10087–10101.

DALSTROM, C. D. A., 1969, *Balanced cross sections:* Can. Jour. Earth Sci., v. 6, p. 743–757.

———, 1970, *Structural geology in the eastern margin of the Canadian Rocky Mountains:* Bull. Canadian Petroleum Geol., v. 18, p. 332–406.

DAVIS, D., AND T. ENGELDER, 1985, *The role of salt in fold-and-thrust belts:* Tectonophysics, v. 119, p. 67–88.

DAVIS, D., J. SUPPE, AND F. A. DAHLEN, 1983, *Mechanics of fold-and-thrust belts and accretionary wedges:* Jour. of Geophysics Res., v. 88, p. 1153–1172.

DAVIS, T. L., 1987, *Seismic facies analysis: pitfalls and applications in cratonic basins:* Geophysics: The Leading Edge, v. 6, No. 7, p. 18–23, 37.

DAVISON, I., 1986, *Listric normal fault profiles: calculation using bed-length balance and fault displacement:* Jour. Structural Geol., v. 8, p. 209–210, p. 1560–1580.

———, 1987, *Normal fault geometry related to sediment compaction and burial:* Jour. of Structural Geol., v. 9, No. 4, p. 393–401.

DE LANGE, J. I., G. D. NIJEN TWILHAAR, AND J. J. PELGROM, 1987, *Accurate surveying—an operator's point of view,* SPE, Publication 822.

DE PAOR, D. G., AND G. EISENSTADT, *Stratigraphy and structural consequences of fault reversal: an example from the Franklinian Basin, Ellesmere Island:* Geology, v. 15, p. 948–949.

DENNIS, J. G., 1972, *Structural Geology,* Ronald Press, New York, NY.

DENNISON, J. M., 1968, *Analysis of Geologic Structures,* W. W. Norton & Co., San Francisco, CA.

DICKEY, P. A., 1981, *Petroleum Development Geology,* second ed., PennWell Publishing Co., Tulsa, OK.

———, 1986, *Petroleum Development Geology,* third ed., PennWell Publishing Co., Tulsa, OK.

DICKINSON, G., 1954, *Subsurface interpretation of intersecting faults and their effects upon stratigraphic horizons:* Am. Assoc. of Petroleum Geol., v. 38, No. 5, p. 854–877.

DICKINSON, W. R., AND D. R. SEELY, 1979, *Structure and stratigraphy of Forearc Region:* Am. Assoc. of Petroleum Geol. Bull., v. 63, No. 1, p. 2–31.

DIXON, J. M., 1975, *Finite strain and progressive deformation in models of diapiric structures:* Tectonophysics, v. 28, p. 89–124.

DOBROVOLNY, J. S., 1951, *Descriptive geometry for geologists:* Am. Assoc. of Petroleum Geol. Bull., v. 35, No. 7, p. 1674–1686.

DOEBL, F., AND R. TEICHMULLER, 1979, *Zur Geologie und heutigen Geothermik mittleren Oberrhein-Graben:* Fortscher, Geol. Rheinol. u. Westf., v. 27, p. 1–17.

DOLLY, E. D., *Geological techniques utilized in Trap Spring Field discovery, Railroad Valley, Nye County, Nevada, in* G. W. Newman and H. D. Goode, eds., Basin and Range Symposium and Great Basin Field Conference: Rocky Mountain Assoc. of Geologists and Utah Geological Assoc., p. 455–467.

DONATH, F. A., AND R. B. PARKER, 1964, *Folds and folding:* Geol. Soc. of Am. Bull., v. 75, p. 45–62.

DONN, W. L., AND J. A. SHIMER, 1958, *Graphic Methods in Structural Geology,* Appleton-Century-Crofts, New York, NY.

DOVETON, J. H., 1986, *Log Analysis of Subsurface Geology,* John Wiley & Sons, New York, NY.

DURAN, L. G., 1951, *Trigonometric and graphical solution of problems in structural mapping*, World Oil, v. 133, No. 6, p. 94–98.

DURY, G. H., 1952, *Map Interpretation*, Pittman, London, England.

EARDLEY, A. J., 1938, *Graphic treatment of folds in three dimensions:* Am. Assoc. of Petroleum Geol. Bull., v. 22, No. 4, p. 483–489.

EBANKS, W. J., JR., AND J. F. WEBER, 1982, *Development of a shallow heavy-oil deposit in Missouri:* Oil and Gas Jour. (Sept 27), p. 222–234.

ELLIOTT, D., 1976, *The energy balance and deformation mechanisms of thrust sheets:* Phil. Trans. Roy. Soc. London, Part A, v. 283, p. 289–312.

EMERY, K. O., 1980, *Continental margins—classification of petroleum prospects:* Am. Assoc. of Petroleum Geol. Bull., v. 64, no. 3, p. 297–315.

ENGELDER, T., AND R. ENGELDER, 1977, *Fossil distortion and decollement tectonics of the Appalachian Plateau:* Geology, v. 5, p. 457–460.

EWING, T. E., 1983, *Growth faults and salt tectonics in the Houston Diapir Province—relative timing and exploration significance:* Trans. Gulf Coast Assoc. Geol. Soc., v. 33, p. 83–90.

FAILL, R. T., 1969, *Kink band structure in the Valley and Ridge Province, Central Pennsylvania:* Geol. Soc. of Am. Bull., v. 80, p. 2539–2550.

———, 1973, *Kink band folding, Valley and Ridge Province, Pennsylvania:* Geol. Soc. of Am. Bull., v. 84, p. 1289–1314.

FALLAW, W. C., 1973, *Grabens on anticlines in Gulf Coastal Plain, and thinning of sedimentary section in downthrown fault block:* Am. Assn. of Petroleum Geol. Bull., v. 57, No. 1, p. 198–203.

FORGOTSON, J. M., JR., 1960, *Review and classification of quantitative mapping techniques:* Am. Assoc. of Petroleum Geol. Bull., v. 44, No. 1, p. 83–100.

FREY, M. G., AND W. H. GRIMES, 1970, *Bay Marchard-Timbalier Gay-Callou Island Salt Complex, Louisiana:* Am. Assoc. of Petroleum Geol., Mem. 14, p. 277–291.

GARB, F. A., 1985, *Oil and gas reserves, classification, estimation, and evaluation:* Jour. of Petroleum Technology, v. 37, No. 3, p. 373–390.

GATEWOOD, L. E., 1970, *Oklahoma City Field—anatomy of a giant:* Am. Assoc. Petroleum Geol., Mem. 14, p. 223–254.

GEIKIE, A., 1903, *Text-Book of Geology*, fourth ed., Macmillan and Company, New York, NY, v. 1, p. 689–694.

GEISER, P. A., 1988, *The role of kinematics in the construction and analysis of geological cross sections in deformed terrains:* Geol. Soc. of Am., Special Paper 222, p. 47–76.

GHIGNONE, J. I., AND G. DeANDRADE, 1970, *General geology and major oil fields of Reconcavo Basin, Brazil:* Am. Assoc. of Petroleum Geol., Mem. 14, p. 337–358.

GIBBS, A. D., 1983, *Balanced cross section construction from seismic sections in areas of extensional tectonics:* Jour. Structural Geol., v. 5, No. 2, p. 153–160.

———, 1984, *Structural evolution of extensional basin margins:* Jour. Geol. Soc. London, v. 141, p. 609–620.

GILES, A. B., AND D. H. WOOD, 1983, *Oakwood salt dome, east Texas: geologic framework, growth history, and hydrocarbon production:* Texas Bureau of Econ. Geol., Geol. Circular 83-1.

GILL, W. D., 1953, *Construction of geological sections of folds with steep-limb attenuation:* Am. Assoc. of Petroleum Geol. Bull., v. 37, No. 10, p. 2389–2406.

GOGUEL, J., 1962, *Tectonics*, second ed., W. H. Freeman and Co., San Francisco, CA.

GORDY, P. L., AND F. R. FREY, 1975, *Geological-Cross Sections Through the Foothills: Foothills Fieldtrip Guide Book*, Canadian Soc. of Petroleum Geol./Canadian Soc. of Exploration Geophysicists, Calgary, Canada.

GORDY, P. L., F. R. FREY, AND D. K. NOSSIN, 1977, *Geological guide for C.S.P.G. and 1977 Notes For Glacier Park Field Conference:* Canadian Soc. Petroleum Geol., Calgary, Canada.

GROSHONG, R. H., JR., 1975, *"Slip" cleavage caused by pressure solution in a buckle fold:* Geology, v. 3, p. 411–413.

———, 1989, *Half graben structures: balanced models of extensional fault-bend folds:* Geol. Soc. of Am. Bull., v. 101, p. 96–105.

GWINN, V. E., 1964, *Thin-skinned tectonics in the Plateau and Northwestern Valley and Ridge Province of the Central Appalachians:* Geol. Soc. of Am. Bull., v. 75, p. 863–900.

HALBOUTY, M. T., 1967, *Salt domes, Gulf Region, United States and Mexico,* Gulf Publishing Co., Houston, TX.

————, 1979, *Salt Domes Gulf Region, United States and Mexico,* second ed., Gulf Publishing Co., Houston, TX.

————, 1982, *The deliberate search for the subtle trap:* Am. Assoc. of Petroleum Geol., Mem. 32.

HAMBLIN, W. K., 1965, *Origin of "reverse drag" on the downthrown side of normal faults:* Geol. Soc. of Am. Bull., v. 76, p. 1145–1164.

HANDLEY, E. J., 1954, *Contouring is important:* World Oil, v. 138, No. 4, p. 106–107.

HARDING, T. P., 1973, *Newport-Inglewood Trend, California—an example of wrenching style of deformation:* Am. Assoc. Petroleum Geol. Bull., v. 57, p. 97–116.

————, 1974, *Petroleum traps associated with wrench faults:* Am. Assoc. Petroleum Geol. Bull., v. 58, p. 1290–1304.

————, 1976, *Predicting productive trends related to wrench faults:* World Oil, v. 182, No. 7, p. 64–69.

————, 1976, *Tectonic significance and hydrocarbon trapping consequences of sequential folding synchronous with San Andreas faulting, San Joaquin Valley, California:* Am. Assoc. of Petroleum Geol. Bull., v. 60, No. 3, p. 356–378.

————, 1984, *Graben hydrocarbon occurrences and structural style:* Am. Assoc. of Petroleum Geol. Bull., v. 68, No. 3, p. 333–362.

HARDING, T. P., AND J. D. LOWELL, 1979, *Structural styles, their plate-tectonic habitats, and hydrocarbon traps in petroleum provinces:* Am. Assoc. of Petroleum Geol. Bull., v. 63, No. 7, p. 1016–1058.

HARRINGTON, J. W., 1951, *The elementary theory of subsurface structural contouring:* Trans Am. Geophys. Union, v. 32, No. 1, p. 77–80.

HAY, J. T. C., 1978, *Structural development in the northern North Sea:* Jour. of Petroleum Geol., v. 1, p. 65–77.

HEROLD, S. C., 1933, *Projection of dip angle on profile section:* Am. Assoc. of Petroleum Geol. Bull., v. 17, p. 740–742.

HILL, M. L., 1942, *Graphic method for some geologic calculations:* Am. Assoc. of Petroleum Geol. Bull., v. 26, No. 6, p. 1155–1159.

————, 1947, *Classification of faults:* Am. Assoc. of Petroleum Geol. Bull., v. 31, No. 2, p. 1669–1673.

————, 1959, *Dual classification of faults:* Am. Assoc. of Petroleum Geol. Bull., v. 43, No. 1, p. 217–237.

HILLS, E. S., 1963, *Elements of Structural Geology,* Methuen & Co., London, England.

HORSFIELD, W. T., 1977, *An experimental approach to basement-controlled faulting:* Geologie en Mijnbouw, v. 56, p. 363–370.

————, 1980, *Contemporaneous movement along crossing conjugate normal faults:* Jour. of Structural Geol., v. 2, p. 305–310.

HOSSACK, J. R., 1983, *A cross-section through the Scandinavian Caledonides constructed with the aid of branch-line maps:* Jour. of Structural Geol., v. 5, No. 2, p. 103–111.

HUBBERT, M. K., AND W. W. RUBEY, 1959, *Role of fluid pressure in mechanics of overthrust faulting:* Geol. Soc. of Am. Bull., v. 70, p. 115–166.

HULL, C. E., AND H. R. WARMAN, 1968, *Asmari Oil Fields of Iran,* British Petroleum Co., London, England, p. 428–437.

ILLIES, J. H., 1981, *Mechanism of graben formation:* Tectonophysics, v. 73, p. 249–266.

Introduction to Directional Drilling, Eastman Whipstock Inc., Houston, TX.

IVANHOE, L. F., 1956, *Integration of geological data on seismic sections:* Am. Assoc. of Petroleum Geol. Bull., v. 40, No. 5, p. 1016–1023.

JACKSON, M. P. A., and W. E. GALLOWAY, 1984, *Structural and depositional styles of Gulf Coast Tertiary continental margins: application to hydrocarbon exploration:* Am. Assoc. Petroleum Geol., Continuing Ed. Course Note Series, No. 25.

JACKSON, M. P. A., and S. J. SENI, 1983, *Geometry and evolution of salt structures in a marginal rift basin of the Gulf of Mexico, east Texas:* Geology, v. 11, p. 131–135.

JAGELER, A. H., AND D. R. MATUSZAK, 1972, *Use of well logs and dipmeters in stratigraphic-trap exploration:* Am. Assoc. of Petroleum Geol., Mem. 16, p. 107–135.

JENYON, M. K., 1988, *Fault-salt wall relationships, southern North Sea,* Oil & Gas Jour. (Sept. 5), p. 76–81.

JOHANSON, D. B., 1987, *Structural evolution of Grand Lake Field, Cameron Parish, Louisiana:* Gulf Coast Assoc. of Geol. Soc. Trans., v. 37, p. 113–122.

JONES, P. B., 1971, *Folded faults and sequence of thrusting in Alberta Foothills:* Am. Assoc. of Petroleum Geol. Bull., v. 55, p. 292–306.

———, 1982, *Oil and gas beneath east-dipping underthrust faults in the Alberta Foothills,* Rocky Mtn. Assoc. of Geol., v. 1, p. 61–74.

———, 1988, *Balanced cross-sections—an aid to structural interpretation:* Geophysics: The Leading Edge, v. 7, No. 8, p. 29–31.

KATZ, D. L., M. W. LEGATSKI et al., 1966, *How Water Displaces Gas from Porous Media:* Oil and Gas Journal (January 10).

KAY, M., 1945, *Paleogeographic and palinspastic maps:* Am. Assoc. of Petroleum Geol. Bull., v. 29, No. 4, p. 426–450.

———, 1954, *Isolith, isopach, and palinspastic maps (Geological Note):* Am. Assoc. of Petroleum Geol. Bull., v. 38, No. 5, p. 916–917.

KLEIN, G., *Sandstone Depositional Models for Exploration for Fossil Fuels,* third ed., IHRDC, Boston, MA.

KRUMBEIN, W. C., 1942, *Criteria for subsurface recognition of unconformities:* Am. Assoc. of Petroleum Geol. Bull., v. 26, No. 1, p. 36–62.

———, 1948, *Lithofacies maps and regional sedimentary stratigraphic analysis:* Am. Assoc. of Petroleum Geol. Bull., v. 32, No. 10, p. 1909–1923.

———, 1952, *Principles of facies map interpretation:* Jour. Sed. Pet., v. 22, No. 4, p. 200–211.

KUHME, A. K., 1987, *Seismic interpretation of reefs:* Geophysics: The Leading Edge, v. 6, No. 8, p. 60–65.

LAFAYETTE GEOLOGICAL SOCIETY, 1964, *Typical oil and gas fields of Southwestern Louisiana:* Lafayette Geological Society, Lafayette, LA, v. 1.

———, 1970, *Typical oil and gas fields of Southwestern Louisiana:* Lafayette Geological Society, Lafayette, LA, v. 2.

———, 1973, *Offshore Louisiana Oil & Gas Fields:* Lafayette Geological Society, Lafayette, LA.

LAMB, C. F., 1980, *Painter Reservoir Field—giant in Wyoming Thrust Belt:* Am. Assoc. of Petroleum Geol. Bull., v. 64, p. 638–373.

LAMERSON, P. R., 1982, *The fossil basin and its relationship to the Absaroka thrust system, Wyoming and Utah, in* R. B. Powers, ed., *Geologic Studies of the Cordilleran Thrust Belt:* Denver, Rocky Mountain Assoc. of Geologists, p. 279–340.

LANGSTAFF, C. S., AND D. MORRILL, 1981, *Geologic Cross Sections,* International Human Resources Development Corp., Boston, MA.

LAUBSCHER, H. P., 1961, *Die Fernschubhypothese der Jurafaltung:* Ecolog. Geol. Helvetiae, v. 54, p. 222–282.

———, 1977, *Fold Development in the Jura:* Tectonophy, v. 37, p. 337–362.

LEBLANC, R. J., 1972, *Geometry of sandstone reservoir bodies:* Am. Assoc. of Petroleum Geol., Mem. 18, p. 133–190.

LELEK, J. J., 1982, *Geologic factors affecting reservoir analysis, Anschutz Ranch East Field, Utah-Wyoming:* SPE 57th Ann. Offshore Tech. Conf. Paper, SPE 10992.

LEROY, L. W., 1950, *Subsurface Geologic Methods—A Symposium,* second ed., Colorado School of Mines, Golden, CO.

LEROY, L. W., AND J. W. LOW, 1954, *Graphic Problems in Petroleum Geology,* Harper and Bros., New York, NY.

LEROY, L. W., AND D. O. LEROY, 1977, *Subsurface Geology,* fourth ed., Colorado School of Mines, Golden, CO.

LEROY, L. W., AND D. O. LEROY, 1987, *Subsurface Geology,* fifth ed., Colorado School of Mines, Golden, CO.

LEVORSEN, A. I., 1943, *Discovery thinking:* Am. Assoc. of Petroleum Geol. Bull., v. 27, No. 7, p. 887–928.

———, 1954, *Geology of Petroleum,* W. H. Freeman and Co., San Francisco, CA.

LEY, H. H., 1930, *Structure contouring:* Am. Assoc. of Petroleum Geol. Bull., v. 14, No. 1, p. 103–105.

LINK, P. K., 1982, *Basic Petroleum Geology:* Oil and Gas Consultants International, Inc., Tulsa, OK.

LINK, T. A., 1949, *Interpretations of foothills structures, Alberta, Canada:* Am. Assoc. of Petroleum Geol. Bull., v. 33, p. 1475–1501.

LISTER, G. S., M. A. ETHERIDGE, AND P. A. SYMONDS, 1986, *Detachment faulting and the evolution of passive continental margins:* Geology, v. 14, p. 246–250.

LOCK, B. E., 1989, *Subsurface Geological Investigations,* unpublished course notes.

LOCK, B. E., and S. L. VOORHIES, 1988, *Sequence stratigraphy as a tool for interpretation of the Cock Field/Yegua in Southwest Louisiana:* Gulf Coast Assoc. of Geol. Soc., v. 39, p. 123–131.

LOW, J. W., 1951, *Subsurface Maps and Illustrations,* Subsurface Geologic Methods Symposium, compiled by L. W. LeRoy, Colorado School of Mines, Dept. of Publications, Golden, CO, p. 894–969.

LOWELL, J. D., 1985, *Structural Styles in Petroleum Exploration:* Oil & Gas Consultants International, Tulsa, OK.

LYLE, H. N., 1951, *Southwest Texas faults:* Oil & Gas Journal, v. 49, No. 41, p. 108–112.

MacKENZIE, D. B., 1972, *Primary stratigraphic traps in sandstone:* Am. Assoc. of Petroleum Geol., Mem. 16, p. 47–66.

MACKIN, J. H., 1950, *The down-structure method of viewing geologic maps:* Jour. Geology, v. 58, p. 55–77.

MACURDA, D. B., JR., 1987, *Seismic interpretation of transgressive and progradational sequence:* Geophysics: The Leading Edge, v. 6, No. 4, p. 18–21.

Magnetic Single-Shot Operations Manual, Eastman Whipstock, Houston, TX.

MALHASE, J., 1927, *Constructing geologic sections with unequal scales:* Am. Assoc. of Petroleum Geol. Bull., v. 11, p. 755–757.

MANNHARD, G. W., AND D. A. BUSCH, 1974, *Stratigraphic trap accumulation in southwestern Kansas and northwestern Oklahoma:* Am. Assoc. of Petroleum Geol. Bull., v. 58, No. 3, p. 447–463.

MARSHAK, S., AND G. MITRA, 1988, *Basic Methods of Structural Geology,* Prentice Hall, Englewood Cliffs, NJ.

MARTIN, R. G., 1980, *Distribution of salt structures in the Gulf of Mexico: map and descriptive text:* U.S. Geol. Survey Map MF-1213, #1284.

MAYUGA, M. N., 1970, *Geology development of California's giant-Wilmington Oil Field:* Am. Assoc. of Petroleum Geol., Mem. 14, p. 158–184.

McKAY, B. A., 1974, *Laboratory Studies of Gas Displacement from Sandstone Reservoirs Having Strong Water Drive:* APEA Journal.

MEDWEDEFF, D. A., 1988, *Structural Analysis and Tectonic Significance of Late-Tertiary and Quaternary Compressive-Growth Folding, San Joaquin Valley, California,* PhD Thesis, Princeton University.

———, 1989, *Growth fault-bend folding at southeast Lost Hills, San Joaquin Valley, California:* Am. Assoc. of Petroleum Geol. Bull., v. 73, p. 54–67.

MEDWEDEFF, D. A., AND J. SUPPE, 1986, *Growth-fault bend folding-precise deformation of kinematics, timing and rates of folding and faulting from syntectonic sediments:* Geol. Soc. of Am. Abstract with Programs, v. 18, p. 692.

MERRITT, J. W., 1946, *Geotechniques of oil exploration:* Oil Weekly, v. 121, No. 5, p. 17–26.

MERTIE, J. B., JR., 1947, *Calculation of thickness in parallel folds:* Geol. Soc. of Am. Bull., v. 58, p. 779–802.

———, 1947, *Delineation of parallel folds and measurement of stratigraphic dimensions:* Geol. Soc. of Am. Bull., v. 58, p. 779–802.

———, 1948, *Application of Brianchon's theorem to construction of geologic profiles:* Geol. Soc. of Am. Bull., v. 59, p. 767–786.

MERTOSONO, S., 1975, *Geology of Pungut and Tandun Oil Fields Central Sumatra:* Indonesian Petroleum Assoc. (June), p. 165–179.

MEYER, H. J., AND H. W. McGEE, 1985, *Oil and gas fields accompanied by geothermal anomalies in Rocky Mountain region:* Am. Assoc. of Petroleum Geol. Bull., v. 69, No. 6, p. 933–945.

MITRA, S., 1986, *Duplex structures and imbricate thrust systems: geometry, structural position, and hydrocarbon potential:* Am. Assoc. of Petroleum Geol. Bull., v. 70, p. 1087–1112.

———, 1988, *Three-dimensional geometry and kinematic evolution of the Pine Mountain thrust system southern Appalachians:* Geol. Soc. of Am. Bull., v. 100, p. 72–95.

MITRA, S., AND J. NAMSON, 1989, *Equal-area balancing:* Am. Jour. Sci., v. 289, p. 563–599.

MOODY, J. D., 1973, *Petroleum exploration aspects of wrench-fault tectonics:* Am. Assoc. of Petroleum Geol. Bull., v. 57, No. 3, p. 449–476.

MOSAR J., AND J. SUPPE, 1988, *Fault propagation folds: models and examples from the Pre-Alps and the Jura:* 6ieme Reunion du Groupe Tectonique Suisse, Neuchatel, 8/9.

MOUNT, V. S., J. SUPPE, AND S. C. HOOK, 1990, *A Forward Modeling Strategy for Balancing Cross Sections,* Am. Assoc. of Petroleum Geol. Bull., (in review).

MURRAY, G. E., 1968, *Salt structures of Gulf of Mexico basin—a review:* Am. Assoc. Petroleum Geol., Mem. 8, p. 99–121.

NAMSON, J., 1981, *Structure of the western Foothills Belt, Miaoli-Hsinchu area Taiwan: (1) Southern Part:* Petroleum Geol. Taiwan, No. 18, p. 31–51.

NELSON, P. H. H., 1980, *Role of reflection seismic in development of Nembe Creek Field, Nigeria:* Am. Assoc. of Petroleum Geol., Mem. 30, p. 565–576.

NEW ORLEANS GEOLOGICAL SOCIETY, 1965, *Oil & Gas Fields of Southeast Louisiana,* v. 1, New Orleans, LA.

———, 1967, *Oil & Gas Fields of Southeast Louisiana,* v. 2, New Orleans, LA.

———, 1983, *Oil & Gas Fields of Southeast Louisiana,* v. 3, New Orleans, LA.

———, 1988, *Offshore Louisiana Oil & Gas Fields,* v. 2, New Orleans, LA.

O'BRIEN, C., 1988, *Pragmatic migration: a method for interpreting a grid of 2D migrated seismic data:* Geophysics: The Leading Edge, v. 7, No. 2, p. 24–29.

OCAMB, R. D., 1961, *Growth faults of south Louisiana:* Trans. Gulf Coast Assoc. of Geol. Soc., v. 11, p. 139–175.

PAGE D., 1859, *Handbook of Geological Terms and Geology,* William Blackwood & Sons, Edinburgh.

PAYTON, C. E., ed., 1977, *Seismic stratigraphy—applications to hydrocarbon exploration:* Am. Assoc. of Petroleum Geol., Mem. 26.

PELTO, C. R., 1954, *Mapping of multicomponent systems:* Jour. of Geol., v. 62, No. 5, p. 501–511.

PENNEBAKER, PAUL E., 1972, *Vertical Net Sandstone Determination for Isopach Mapping of Hydrocarbon Reservoirs:* Am. Assoc. Petroleum Geol. Bull., v. 56, p. 1520–1529.

PERRIER, R., AND J. QUIBLIER, 1974, *Thickness changes in sedimentary layers during compaction history; Methods for quantitative evaluation:* Am. Assoc. of Petroleum Geol. Bull., v. 58, No. 3, p. 507–520.

PERRY, W. J., D. H. ROEDER, AND D. R. LAGESON, *North American thrust-faulted terraines:* Am. Assoc. of Petroleum Geol., Reprint Series No. 27.

PETERS, W. C., 1987, *Exploration and Mining Geology,* fourth ed., Colorado School of Mines, Golden, CO.

PETTIJOHN, F. J., P. E. POTTER, AND R. SIEVER, 1972, *Sand and Sandstone,* Springer-Verlag, Heidilberg.

———, 1987, *Sand and Sandstone,* second ed., Springer-Verlag, New York, NY.

POTTER, P. E., 1963, *Late Paleozoic sandstone of the Illinois Basin, Report of Investigation 217, Urbana, IL:* Illinois State Geol. Surveys.

PRICE, R. C., 1986, *The southeastern Cordillera: thrust faulting, tectonic wedging and delamination of the lithosphere:* Jour. of Structural Geol., v. 8, p. 239–254.

RAGAN, D. M., 1985, *Structural Geology, An Introduction to Geometrical Techniques,* John Wiley & Sons, New York, NY.

REES, F. B., 1972, *Methods of mapping and illustrating stratigraphic traps:* Am. Assoc. of Petroleum Geol., Mem. 16, p. 168–221.

REID, H. F., 1909, *Geometry of faults:* Geol. Soc. of Am. Bull., v. 20, p. 171–196.

REID, H. F., W. M. DAVIS, A. C. LAWSON, F. L. RANSOME, AND COMMITTEE, 1913, *Report of the committee on the nomenclature of faults:* Geol. Soc. of Am. Bull., v. 24, p. 163–186.

REITER, W. A., 1947, *Contouring fault planes:* World Oil, v. 126, No. 7, p. 34–35.

RETTGER, R. E., 1929, *On specifying the type of structural contouring:* Am. Assoc. of Petroleum Geol. Bull., v. 13, No. 11, p. 1559–1560.

RICH, J. L., 1932, *Simple graphical method for determining true dip from two components and for constructing contoured maps from dip observations:* Am. Assoc. of Petroleum Geol. Bull., v. 16, No. 1, p. 92–94.

———, 1934, *Mechanics of low-angle overthrust faulting as illustrated by Cumberland thrust block, Virginia, Kentucky, and Tennessee:* Am. Assoc. of Petroleum Geol. Bull., v. 18, p. 1584–1596.

————, 1935, *Graphical method for eliminating regional dip:* Am. Assoc. of Petroleum Geol. Bull., v. 19, No. 10, p. 1538–1540.

RIDER, M. H., 1978, *Growth faults in Carboniferous of Western Ireland:* Am. Assoc. of Petroleum Geol. Bull., v. 62, No. 11, p. 2191–2213.

ROBINSON, J. P., 1982, *Petroleum exploration in southeastern Arizona: anatomy of an overthrust play, in* Powers, R. B., ed., *Geologic studies of the Cordilleran thrust belt:* Rocky Mtn. Assoc. Geol., p. 665–674.

ROCKY MTN. ASSOC. OF GEOLOGISTS, 1982, *Geologic Studies of the Cordilleran Thrust Belt,* v. 6, Denver, CO.

————, 1982, *Geologic Studies of the Cordilleran Thrust Belt,* v. 7, Denver, CO, p. 475–976.

ROEDER, D. H., 1973, *Subduction and orogeny:* Jour. Geophysicl Research, v. 78, p. 5005–5024.

————, 1983, *Hydrocarbons and geodynamics of folding belts:* Rocky Mtn. Assoc. of Geol., Continuing Education Short Course Notes.

ROUX, W. F., JR., 1978, *The development of growth fault structures:* Am. Assoc. Petroleum Geol., Structural Geology School Notes.

ROWAN, M. G., AND R. KLIGFIELD, 1989, *Cross section restoration and balancing as aid to seismic interpretation in extensional terrains:* Am. Assoc. of Petroleum Geol. Bull., v. 73, p. 955–966.

ROYSE, F., JR., M. A. WARNER, AND D. C. REESE, 1975, *Thrust belt structural geometry and related stratigraphic problems, Wyoming-Idaho-northern Utah, in* Bolyard, D. W., ed., *Symposium on deep drilling frontiers in the central Rocky Mountains:* Rocky Mtn. Assoc. of Geol., p. 41–54.

SANFORD, A. R., 1959, *Analytical and experimental study of simple geologic structures,* Geol. Soc. of Am. Bull., v. 70, p. 19–52.

SCHLUMBERGER, 1987, *Log Interpretation Principles/Applications,* Schlumberger Educational Services.

SCHMID, C. F., AND E. H. MACCANNELL, 1955, *Basic problems, techniques and theory of isopleth mapping:* Jour. Am. Statistical Assoc., v. 50, No. 269, p. 220–239.

SCHOLLE, P. A., AND D. SPEARING, 1982, *Sandstone depositional environments, Tulsa:* Am. Assoc. of Petroleum Geol. Bull., v. 2.

SCLATER, J. G., AND P. A. F. CHRISTIE, 1980, *Continental stretching: an explanation of the post-mid-cretaceous subsidence of the central North Sea Basin:* Jour. Geophysics Res., v. 85, p. 3711–3739.

SEBRING, L., JR., 1958, *Chief tool of the petroleum exploration geologist: the subsurface structural map:* Am. Assoc. of Petroleum Geol. Bull., v. 42, No. 3, p. 561–587.

————, 1958, *Subsurface map: underground guide for oil men:* Oil and Gas Jour., v. 56, No. 27, p. 186–189.

SENGBUSH, R. L., *Petroleum Exploration: A Quantitative Introduction,* IHRDC, Boston, MA.

SENI, S. J., AND M. P. A. JACKSON, 1983, *Evolution of salt structures, east Texas Diapir Province, part 1: Sedimentary record of halokinesis:* Am. Assoc. Petroleum Geol. Bull., v. 67, p. 1219–1244.

SERRA, O., 1985, *Sedimentary Environments From Wireline Logs,* Schlumberger.

SETCHELL, J. 1958, *A Nomogram for Determining True Stratum Thickness: Shell Trinidad.* EP 28884, Abstract in PA Bulletin, No. 127/128, May/June 1958. N. V. DeBataafache Petroleum Maatschappij, The Hague, Production Department. p. 8.)

SHELTON, J. W., 1984, *Listric normal faults: an illustrated summary:* Am. Assoc. Petroleum Geol. Bull., v. 68, p. 801–815.

SHERIFF, R. E., 1973, *Encyclopedic Dictionary of Exploration Geophysics:* Soc. of Exploration Geophysics.

————, 1980, *Seismic Stratigraphy,* IHRDC, Boston, MA.

————, 1989, *Geophysical Methods,* Prentice-Hall, Englewood Cliffs, NJ.

SILVER, A., 1982, *Techniques of Using Geologic Data,* IED Exploration, Tulsa, OK.

SITUMORANG, B., SISWOYO, E. THAJIB, and F. PALTRINIERI, 1976, *Wrench fault tectonics and aspects of hydrocarbon accumulation in Java:* Indonesian Petroleum Assoc. (June), p. 53–67.

SMITH, D. A., 1966, *Theoretical consideration of sealing and non-sealing fault:* Am. Assoc. of Petroleum Geol. Bull., v. 50, No. 2, p. 363–374.

————, 1980, *Sealing and nonsealing faults in Louisiana Gulf Coast Salt Basin:* Am. Assoc. Petroleum Geol. Bull., v. 64, p. 145–172.

SMITH, D. A., and F. A. E. REEVE, 1970, *Salt piercement in shallow Gulf Coast salt structures:* Am. Assoc. Petroleum Geol. Bull., v. 54, p. 1271–1289.

SMOLUCHOWSKI, M., 1909, *Folding of the earth's surface in formation of mountain chains:* Acad. Sci. Craiovie Bull., v. 6, p. 3–20.

SNEIDER, R. M., F. H. RICHARDSON, D. D. PAYTNER, R. E. EDDY, AND I. A. WYANT, 1977, *Predicting reservoir rock geometry and continuity in Pennsylvanian reservoirs, Elk City Field, Oklahoma:* Jour. of Petroleum Technology, p. 851–866.

SPENCER, E. W., 1977, *Introduction to the Structure of the Earth,* second ed., McGraw-Hill, New York, NY.

STANDING, M. B., 1952, *Volumetric and Phase Behavior of Oil Field Hydrocarbon Systems,* Reinhold Publishing Company, New York, NY.

———, 1977, *Volumetric and phase behavior of oil field hydrocarbon system,* SPE of AIME, Dallas, TX.

STANDING, M. B., AND D. L. KATZ, 1942, *Density of Natural Gases:* Trans. AIME, p. 144, 146.

STEPHENSON, E. A., AND D. D. HAINES, 1946, *Preparation of contour maps:* Oil and Gas Jour., v. 45, No. 15, p. 115.

———, 1946, *Use of contour maps:* Oil and Gas Jour., v. 45, No. 16, p. 131.

STEPHENSON, M., 1984, *Program challenges directional survey accuracy claims:* Oil & Gas Jour., PennWell Publishing Company.

STEWART, W. A., 1950, *Unconformities, A Subsurface Geologic Methods Symposium,* second ed., compiled by L. W. LeRoy, Colorado School of Mines, Dept. of Publication, Golden, CO.

STRAHLER, A. N., 1948, *Geomorphology and structure of the West Kaibab fault zone and Kaibab Plateau, Arizona:* Geol. Soc. of Am. Bull., v. 59, p. 513–540.

STRALEY, W. H., III, 1932, *Some notes on the nomenclature of faults:* Studies for Students, University of Chicago, Chicago, IL, p. 756–763.

STOCKWELL, C. H., 1947, *The use of plunge in the construction of cross-sections of folds:* Geol. Assoc. of Canada, v. 3, p. 97–121.

STONE, D. S., *Wrench faulting and Rocky Mountain tectonics:* The Mountain Geologist, v. 6, No. 2, p. 67–79.

STUDE, G. R., 1978, *Depositional environments of Gulf of Mexico South Timbalier Block 54 salt dome and salt dome growth models:* Trans. Gulf Coast Assoc. Geol. Soc., v. 28, p. 627–646.

SUPPE, J., 1980, *Imbricate structure of western Foothills Belts, south-central Taiwan:* Petroleum Geol. Taiwan, No. 17, p. 1–16.

———, 1983, *Geometry and kinematics of fault-bend folding:* Am. Jour. Sci., v. 283, p. 684–721.

———, 1985, *Principles of Structural Geology,* Prentice-Hall, Englewood Cliffs, NJ.

———, 1988, *Short course on cross-section balancing in petroleum structural geology, Pacific section:* Am. Assoc. of Petroleum Geol.

SUPPE, J., AND Y. L. CHANG, 1983, *Kink method applied to structural interpretation of seismic sections, western Taiwan:* Petrol. Geol. Taiwan, No. 19, p. 29–47.

SUPPE, J., AND D. A. MEDWEDEFF, 1984, *Fault-propagation folding, Abstracts with Programs:* Geol. Soc. of Am. Bull., v. 16, p. 670.

SUPPE, J., AND J. NAMSON, 1979, *Fault-bend origin of frontal folds in the western Taiwan fold-and-thrust belt:* Petroleum Geol. Taiwan, No. 6, p. 1–18.

TANNER, J. H., III, 1967, *Wrench fault movements along Washita Valley Fault, Arbuckle Mountain area, Oklahoma:* Am. Assoc. of Petroleum Geol. Bull., v. 51, No. 1, p. 126–141.

TEARPOCK, D. J., 1990, *Quantitative Mapping Techniques:* Houston Geological Society, Continuing Education Short Course Notes.

TEARPOCK, D. J., AND R. E. BISCHKE, 1980, *Structural analysis of the Wissahickon Schist near Philadelphia, Pennsylvania:* Geol. Soc. of Am. Bull., v. 94 (November 1980).

TEARPOCK, D. J., AND R. E. BISCHKE (1990), *Mapping throw in place of vertical separation: a costly subsurface mapping misconception,* Oil and Gas Journal, July 16, V. 88, No. 29, p. 74–78.

TEARPOCK, D. J., AND J. HARRIS, 1987, *Subsurface Geological Mapping Techniques—A Training Manual,* Tenneco Oil Co., Houston, TX.

TEARPOCK, D. J., AND J. HARRIS, 1990, *Isopach maps and their application in subsurface mapping:* Lafayette Geological Soc., Continuing Education Short Course Notes.

TEARPOCK, D. J., AND J. HARRIS, 1990, *Applied subsurface mapping techniques:* Gulf Coast Assoc. Geol. Soc.—Short Course (GCAGS 40th Annual Convention, Lafayette, LA).

TEARPOCK, D. J., AND H. POUSSON, 1990, *A three-dimensional correction factor equation for directionally drilled wells:* Trans. Gulf Coast Assoc. Geol. Soc., v. 40.

THOMAS, W. A., 1968, *Contemporaneous normal faults on flanks of Birmingham anticlinorium, central Alabama:* Am. Assoc. Petroleum Geol. Bull., v. 52, p. 2123–2136.

THOROGOOD, J. L., 1986, *Well surveying: past progress, current status and future needs,* World Oil (Jan.), p. 87–91.

———, 1988, *Instrument performance models and their application to directional surveying operations,* SPE, Paper 18051, 63rd Annual Technical Conference, Houston, TX.

THORSEN, C. E., 1963, *Age of growth faulting in southeast Louisiana:* Trans. Gulf Coast Assoc. Geol. Soc., v. 13, p. 103–110.

TODD, R. G., AND R. M. MITCHUM, 1977, *Seismic stratigraphy and global changes of sea level, Part 8: identification of Upper Triassic, Jurassic, and Lower Cretaceous seismic sequences in Gulf of Mexico and offshore West Africa, in* C. E. Payton, ed., *Seismic Stratigraphy—Applications to Hydrocarbon Exploration:* Am. Assoc. Petroleum Geol., Mem. 26, p. 145–163.

TRAVIS, RUSSELL B., 1978, *Graphic Determination of Stratigraphic and Vertical Thicknesses in Deviated Wells:* Am. Assoc. Petroleum Geol. Bull., v. 63, p. 845–866.

TRUSHEIM, F., 1960, *Mechanism of salt migration in northern Germany:* Am. Assoc. Petroleum Geol. Bull., v. 44, p. 1519–1541.

TUCKER, P. M., 1982, *Pitfalls revisited:* Soc. of Exploration Geophysicists.

———, 1988, *Seismic contouring: A unique skill:* Soc. of Exploration Geophysicists, v. 53, No. 6, p. 741–749.

TUCKER, P. M., AND H. J. YARSTON, 1973, *Pitfalls in seismic interpretation:* Soc. of Exploration Geophysicists, Monograph No. 2.

TURK, L. B., 1950, *Significance and use of lap-out maps in prospecting for oil and gas* (Abstract): Am. Assoc. of Petroleum Geol. Bull., v. 34, No. 3, p. 625.

USDANSKY, S. I., AND R. H. GROSHONG, JR., 1984, *Analytical extrapolation of cross sections of vertical drape folds by digital computer:* Geol. Soc. of Am., Abstract with Programs, Part A, v. 16, p. 258.

USDANSKY, S. I., AND R. H. GROSHONG, JR., 1984, *Comparison of analytical models for dip-domain and fault-bend folding:* Geol. Soc. of Am., Abstract with Programs, Part B, v. 16, p. 680.

VANCE, H., 1950, *Petroleum Subsurface Engineering,* Educational Publishers, St. Louis, MO.

VOGLER, H. A., AND B. A. ROBISON, 1987, *Exploration for deep geopressure gas: Corsair Trend, offshore Texas:* Am. Assoc. of Petroleum Geol. Bull., v. 71, p. 777–787.

WADSWORTH, A. H., JR., 1953a, *Percentage of thinning chart—new techniques in subsurface geology:* Am. Assoc. of Petroleum Geol. Bull., v. 37, No. 1, p. 158–162.

———, 1953b, *The percentage of thinning chart:* Oil and Gas Jour., v. 51, No. 43, p. 72–73.

WEBBER, K. J., AND E. DAUKORA, 1976, *Petroleum geology of the Niger Delta:* Ninth World Petroleum Congress, v. 2, p. 209–221.

WEBER, K. J., G. MANDL, W. F. PILAAR, F. LEHNER, AND R. G. PRECIOUS, 1978, *The role of faults in hydrocarbon migration and trapping in Nigerian growth fault structures:* 10th Annual SPE of AIME Offshore Technol. Conf. Preprint No. OTC-3356, p. 2643–2653.

WEISS, L. E., 1972, *The Minor Structures of Deformed Rocks, A Photographic Atlas,* Springer-Verlag, New York, NY.

WERNICKE, B., 1985, *Uniform-sense normal simple shear of the continental lithosphere:* Candaian Jour. Earth Sci., v. 22, p. 108–125.

WERNICKE, B., P. L. GUTH, AND G. L. AXEN, 1984, *Tertiary extensional tectonics in the Sevier thrust belt of southern Nevada, in* J. Lintz, Jr., ed., *Western Geological Excursions,* Ann. Mtg. Geol. Society America, v. 4, Mackay School of Mines, Reno, Nevada, p. 473–510.

WEST, J., AND H. LEWIS, 1982, *Structure and palinspastic reconstruction of the Absaroka Thrust, Anschutz Ranch area, Utah and Wyoming:* Rocky Mt. Assoc. Geol. 1982 Ann. Symposium, p. 633–639.

WHARTON, J. B., JR., 1948, *Isopachous maps of sand reservoirs:* Am. Assoc. of Petroleum Geol. Bull., v. 32, No. 7, p. 1331–1339.

WHEELER, R. L., 1980, *Cross-strike structural discontinuities: possible exploration tool for natural gas in Appalachian Overthrust Belt:* Am. Assoc. of Petroleum Geol. Bull., v. 64, p. 2166–2178.

WHITE, N. J., J. A. JACKSON, AND D. P. McKENZIE, 1986, *The relationship between the geometry of normal faults and that of the sedimentary layers in their hanging walls:* Jour. Structural Geol., v. 8, p. 897–909.

WIGGINS, G. B., AND D. J. TEARPOCK, 1985, *Methods in oil and gas reserves estimation: prospects, newly discovered, and developed properties:* Am. Assoc. of Petroleum Geol., Short Course (1985 Am. Assoc. of Petroleum Geol. Annual Convention, New Orleans, LA).

WILCOX, R. E., T. P. HARDING, and D. R. SEELY, 1973, *Basic wrench tectonics:* Am. Assoc. of Petroleum Geol. Bull., v. 57, No. 1, p. 74–96.

WILSON, C. W., AND R. G. STEARNS, 1958, *Structure of the Cumberland Plateau, Tennessee:* Geol. Soc. of Am. Bull., v. 69, p. 1283–1296.

WINKER, C. D., AND M. B. EDWARDS, 1983, *Unstable progradational clastic shelf margins:* SEPM Special Publ., No. 33, p. 139–157.

WINKER, C. D., R. A. MORTON, T. E. EWING, AND D. D. GARCIA, 1983, *Depositional setting, structural style, and sandstone distribution in three geopressured geothermal areas, Texas Gulf Coast:* Texas Bureau of Econ. Geol., Rept. of Inv. 134.

WOLFF, C. J. M., and J. P. DE WARDT, 1981, *Borehole position uncertainty—analysis of measuring methods and deviation of systematic error model,* Jour. of Petroleum Technology (Dec.), p. 2339–2350.

WOODBURY, H. O., I. B. MURRAY, JR., AND R. E. OSBOURNE, 1980, *Diapirs and their relation to hydrocarbon accumulation, in* A. D. Miall, ed., *Facts and Principles of World Petroleum Occurrence:* Canadian Soc. Petrol. Geol., Mem. 6, p. 119–142.

WOODWARD, N. B., 1987, *Stratigraphic separation diagrams and trust belt structural analysis:* Thirty-Eighth Field Conference—1987 Wyoming Geol. Assoc. Guidebook, p. 69–77.

WOODWARD, N. B., S. E. BOYER, AND J. SUPPE, 1985, *An outline of balanced cross sections:* University of Tennessee Dept. of Geol. Sci. Studies in Geology 11, second ed.

WYOMING GEOL. ASSOC., 1987, *The Thrust Belt Revisited,* Casper, WY.

WYOMING GEOL. ASSOC., MONTANA GEOL. SOC., UTAH GEOL. SOC., 1977, Rocky Mountain Thrust Belt Geology and Resources, Casper, WY.

XIAO, H., F. A. DAHLEN, AND J. SUPPE, 1988, *Mechanics of Extensional Wedges,* EOS, v. 69, p. 470.

XIAO, H., AND J. SUPPE, 1988, *Origin of rollover:* Geol. Soc. of Am., Abstracts with Programs, p. A109.

————, 1989, *Role of compaction in the listric shape of growth normal faults:* Am. Assoc. of Petroleum Geol. Bull., v. 73, No. 6, p. 777–786.

————, 1990, *Origin of rollover:* Am. Assoc. of Petroleum Geol. (in review).

YOGUEL, J., 1962, *Tectonics,* W. H. Freeman and Co., San Francisco, CA.

INDEX

DANIEL J. TEARPOCK
(*Petroleum Consultant*)

Daniel J. Tearpock is currently the owner of SUBSURFACE CONSULTANTS & ASSOCIATES, a geological, geophysical, and engineering consulting firm in Lafayette, Louisiana. He received his Bachelors Degree in geology from Bloomsburg University and his Masters Degree from Temple University.

He has held part-time teaching positions at several small colleges and universities. He was an adjunct Associate Professor at Tulane University in New Orleans in the Petroleum Engineering Department.

Mr. Tearpock served as a Geothermal Energy Development Specialist with Vickers in Jackson, Mississippi. He conducted geothermal studies in California, Nevada, Utah, and geopressured studies in the Gulf of Mexico. With his main interest in the area of oil and gas exploration and development, he became a petroleum consultant in New Orleans, Louisiana. As a consultant, he conducted and supervised a broad range of geological, engineering, and economic studies both onshore and offshore. In addition, he helped organize and co-taught an industry course in subsurface mapping. His primary area of interest is in the evaluation, remapping and redevelopment of older oil and gas fields (*to find the hydrocarbons that others have left behind*).

As a Project Geological Engineer for Tenneco Oil Company, he conducted detailed field studies in the Central Gulf Divisions older offshore fields, which resulted in the generation of both new development and exploratory drilling locations. He was also responsible for the organization and development of the company's subsurface mapping training program which included the development of a subsurface mapping manual and the teaching of subsurface mapping courses to Tenneco's worldwide staff of geologists, geophysicists, and petroleum engineers.

Mr. Tearpock has authored and co-authored publications in the areas of geothermal energy, structural geology, and petroleum subsurface mapping. As a consultant, he has worked for major and independent oil and gas companies, governmental agencies, and landowners. In addition to offering a wide range of consulting services, Mr. Tearpock's consulting firm offers industry and in-house training courses in Applied Subsurface Geological Mapping, Applied Structural Balancing, and several one and two day seminars on various subsurface mapping topics. He is a certified petroleum geologist and a member of the AAPG, GCAGS, SPE, GSA, LGS, NOGS, and HGS.

RICHARD E. BISCHKE
(*Research Staff—Princeton University*)

Richard (Dick) Bischke, who is known as "Dr. Dick" in Asia, graduated from the University of Wisconsin-Milwaukee with a B.S. in Geology and M.S. in Structural Geology. He received a Ph.D. at Columbia University in Tectonophysics.

Dr. Bischke became an Associate Professor of Geology and Geophysics at Temple University and an adjunct Professor of Engineering at Drexel University in Philadelphia. During this time he became associated with the consulting firm of International Exploration (INTEX) where he worked mainly with major oil companies and large independents on frontier basins. He eventually joined the firm to become Chief Geophysicist.

After serving several years in the Philippines on a basin evaluation project sponsored by the Philippine Bureau of Energy Development, the Philippine National Oil Company, and the World Bank, Dr. Bischke worked with John Suppe on the prospect generation of several Philippine fold belts.

Dr. Bischke's interest in complexly deformed structures resulted in a position at Princeton University, where he joined John Suppe as a member of the Research Faculty. His most current work focuses on balancing compressional and extensional structures, as applied to the petroleum industry, which was sponsored by Texaco USA and other oil companies.

Dr. Bischke has conducted studies in many parts of the world including Offshore China, the Italian Alps, Eastern and Western Overthrusts Belts (USA), Indonesia, Philippines, North Greenland Sea, the North Slope of Alaska, Offshore California, Offshore Baltimore Canyon, and the U.S. Gulf of Mexico.

Through SUBSURFACE CONSULTANTS & ASSOCIATES, he teaches a course on applied structural balancing which includes techniques for compressional and extensional areas. He has authored a number of papers on tectonics, earthquakes, and subsurface mapping.